Structures Technology for Future Aerospace Systems

Structures Technology for Future Aerospace Systems

Edited by
Ahmed K. Noor
Center for Advanced Computational Technology
University of Virginia

Volume 188
PROGRESS IN
ASTRONAUTICS AND AERONAUTICS

Paul Zarchan, Editor-in-Chief
Charles Stark Draper Laboratory, Inc.
Cambridge, Massachusetts

Published by the
American Institute of Aeronautics and Astronautics, Inc.
1801 Alexander Bell Drive, Reston, Virginia 20191-4344

ISBN 1-56347-384-4

Progress in Astronautics and Aeronautics

Table of Contents

Chapter 4 Extreme Environment Structures

Donald B. Paul, Christopher L. Clay, Brett Harber, and Harold Croop, *Air Force Research Laboratory, Dayton, Ohio*; and David Glass and Steven Scotti, *NASA Langley Research Center, Hampton, Virginia*

Chapter 5 Gossamer Structures: Space Membranes, Inflatables, and Other Expandables

A. B. Chmielewski, *Jet Propulsion Laboratory, Pasadena, California*; and C. H. Jenkins, *South Dakota School of Mines and Technology, Rapid City, South Dakota*

Chapter 6 Smart Air and Space Structures

Janet M. Sater, *Institute for Defense Analyses, Alexandria, Virginia*; C. Robert Crowe, *Virginia Polytechnic Institute and State University, Blacksburg, Virginia*; Richard Antcliff, *NASA Langley Research Center, Hampton, Virginia*; and Alok Das, *Air Force Research Laboratory, Kirtland AFB, New Mexico*

Chapter 7 Computational Structures Technology

Ahmed K. Noor, *Center for Advanced Computational Technology University of Virginia, NASA Langley Research Center, Hampton, Virginia*

Preface

Structures technology encompasses a wide range of component technologies—from materials development to analysis, design, testing, production, and maintenance. Materials and structures have been largely responsible for major performance improvements in many aerospace systems. The maturation of computational structures technology and the development of advanced composite materials, witnessed during the past 40 years have improved structural performance, reduced operational risk, and shortened development time. In addition to being enabling technologies for future aeronautical and space systems, materials and structures continue to be key elements in determining the reliability, performance, testability, and cost effectiveness of these systems. Although a large number of journal articles and conference proceedings have been devoted to certain aspects of structures technology for future aerospace systems, no recent volumes have been published on the subject. The present volume is an attempt to fill this void. The volume is divided into seven chapters written by experts in the field. The first chapter identifies the major characteristics of future aerospace systems, provides an overview of the structures technology, and describes some of the opportunities provided by the synergistic coupling of structures technology with other technologies.

Chapter 2 provides an overview of the emerging structural concepts and technologies available to the designer of composite aircraft in the future. Although the discussion is focused on fighter fixed-wing implementation, the concepts described can be applied to other military, civil, and transport aircraft, as well as to space vehicles and to civil infrastructure.

Chapter 3 focuses on aging systems. It describes currently used sustainment technologies developed by the Federal Aviation Administration, NASA, and United States Air Force to maintain safe and economical operation of the aging fleets. Technology development needed for future operation of the aging systems is also identified.

Chapter 4 deals with extreme environment structures technology, focusing on current and recent efforts in the development of fully reusable high-speed vehicle applications where the load environment includes extremes of temperatures (both hot and cold) or extreme temperature combined with significant acoustic loading. Exposure to extreme radiation environments and radiation resistant materials and structures are not discussed in this chapter.

Gossamer structures are discussed in Chapter 5. These include space membranes, inflatables, and other expandables. The various applications, advanced concepts, material systems, analysis techniques, and testing methods of these structures are described, and future requirements are identified.

Chapter 6 deals with smart air and space structures. It reviews the current status, results to date, and issues associated with the integration of sensors, actuators, and controllers with structures into smart aerospace systems.

Chapter 7 summarizes the history, recent developments, and trends in computational structures technology (CST). Development of CST software and goals of CST activities are described, along with research areas that have a high potential for meeting future technological needs.

The editor expresses his sincere thanks to the many individuals and organizations who contributed to this volume: in particular, to the lead authors—William G. Baron, John W. Lincoln, Donald Paul, Artur B. Chmielewski, Christopher H. M. Jenkins, Janet M. Sater, and C. Robert Crowe; to the authors of the different sections for the generous contribution of their overtaxed free time and for their cooperation; to the reviewers for their useful comments and suggestions, and to the members of the AIAA Structures Technical Committee for their suggestions and encouragement. Special thanks go to Rodger Williams, Jodi Bergeman Glasscock, and the staff of the AIAA books program for their cooperation in publishing this volume; and to Mary Fox for her secretarial support. It is my earnest hope that this effort will broaden awareness and place in perspective the structures technology for future aerospace systems, and will serve the profession well.

Ahmed K. Noor
August 2000

Chapter 1

Perspectives on Structures Technology

Ahmed K. Noor
Center for Advanced Computational Technology
University of Virginia
NASA Langley Research Center
Hampton, Virginia 23681

I. Introduction

THE twenty-first century will witness a new era of aerospace missions and vehicles. Aerospace vehicles will get both bigger and smaller. They will be faster and go farther than we ever thought possible.[1–5] In aeronautics, future systems will include very large subsonic aircraft with innovative designs such as blended wing[6] (Fig. 1.1), fully integrated aircraft equipped with simple pilot interface, artificial vision and collision avoidance facilities, high-speed aircraft for transoceanic travel, hypersonic vehicles with sharp leading edges (Fig. 1.2), and micro-flying machines with a variety of sensors. Design drivers will include safety, affordability (initial and lifecycle costs), and environmental compatibility. For military aircraft, emphasis is shifting from best performance to low cost at acceptable performance. The military metrics include affordability, responsiveness, flexibility, lethality, survivability, logistic support, and standardization/interoperability.[7] A number of advanced configuration concepts and vehicles are being considered such as box wing (Fig. 1.3), tailless aircraft (Fig. 1.4), vehicles with multistealth capabilities (optical, acoustic, thermal, radar), and hypersonic vehicles operating in, and transiting, both air and space.

In space, large spacecraft will be supplemented or replaced by large numbers of robotic small and nano-spacecraft (Fig. 1.5) for global telecommunications, sustained earth observations, and in-depth scientific exploration of the near reaches of the universe. Sustained in-depth scientific studies will be performed in increasingly remote environments and will have themes as compelling as the search for life in the universe. The goals of these activities include vigilant intelligent presence in the solar system, exploration of interstellar space (Fig. 1.6), and discovery of earth-like planets around nearby stars[8] (Figs. 1.7 and 1.8). The latter will use a telescope powerful enough to determine signs of life 1,000 trillion miles away.

An integrated human-robotic exploration strategy, far more advanced than that associated with the space station, will be developed for the solar system and beyond.

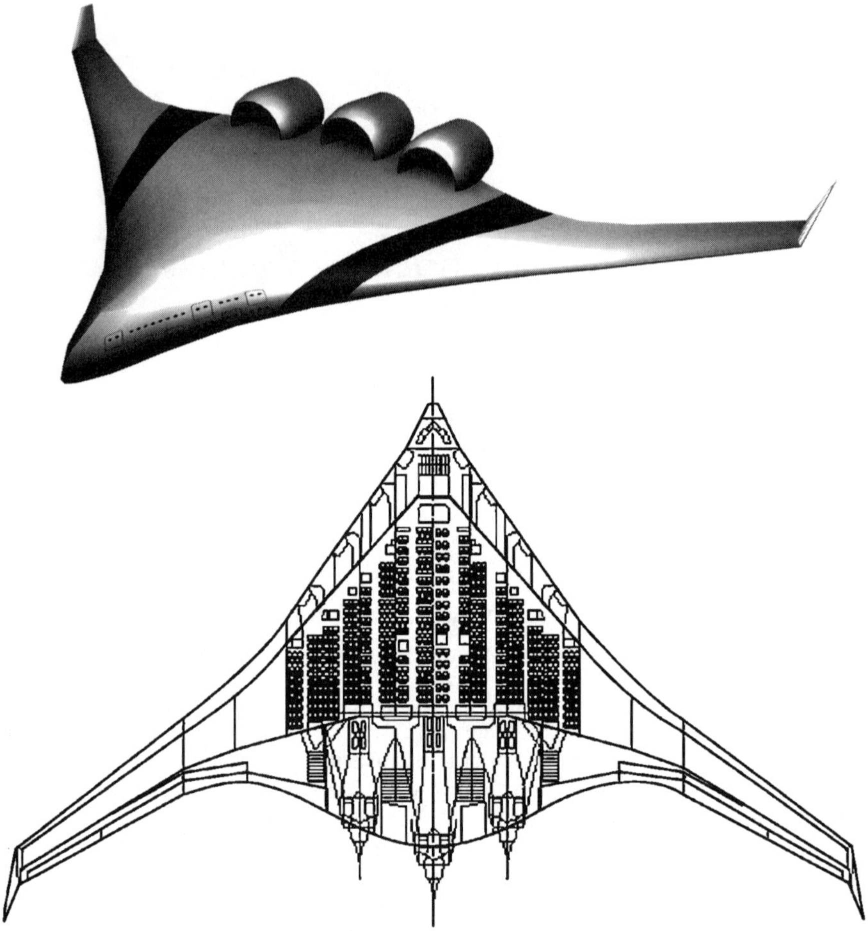

Fig. 1.1 Blended-wing-body subsonic aircraft concept combines a number of structures, aerodynamic and system advantages over classical configurations. Blending the wing and fuselage structures into a lifting body results in efficient deep sections and favorable span loading. Also, moving the engine to the rear of the aircraft, above the wing, provides noise shielding.

This strategy focuses on enhancing safety, improving performance, expanding objectives, and minimizing life-cycle costs.

Future space transportation systems will include fully reusable space liners for safe, routine, earth-to-orbit transportation (Fig. 1.9). An order of magnitude reduction in cost and two orders of magnitude increase in reliability will be achieved for earth-to-orbit transportation. Challenges include reusability, design certification, certification of successive missions, and minimum operational infrastructure. Advanced concepts and systems will be developed for rapid human and robotic transportation to the planets and nearby celestial bodies and interstellar missions (Fig. 1.10). Routine passenger space travel is expected before the middle of the century.

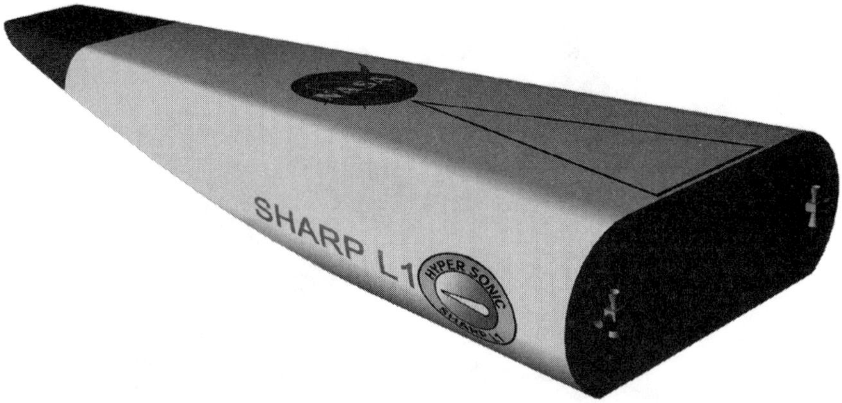

Fig. 1.2 Hypersonic vehicle with sharp leading edge using ultra-high temperature ceramic material. The vehicle is more aerodynamically efficient and requires smaller engines than blunt body vehicles (courtesy of NASA's Advanced Space Transportation Program). (See color section.)

Fig. 1.3 Box and joined wing concepts. a) box wing concept for military cargo aircraft (courtesy of Lockheed Martin Aeronautics Company), b) joined wind airborne surveillance aircraft with radar apertures integrated into the four joined wings (courtesy of the Boeing Company). (See color section.)

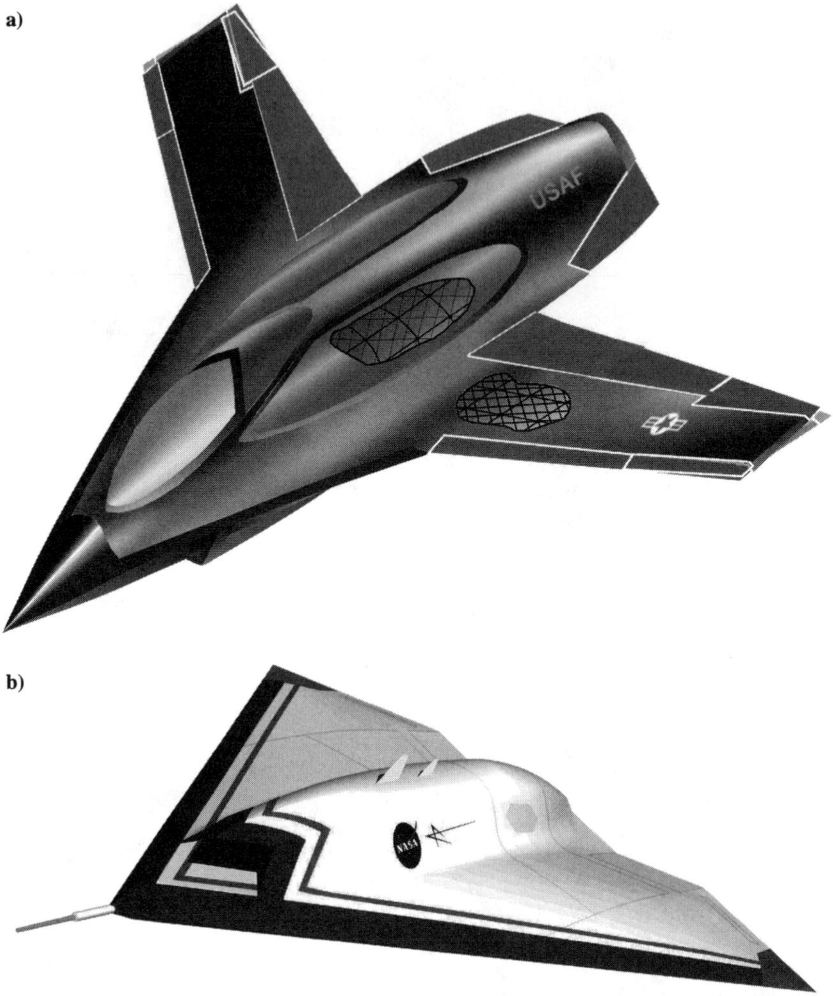

Fig. 1.4 a) Future tailless military aircraft will use smart materials (checkerboard area), twisting wing, expandable fuel cell (pink area), and compliant trailing edge (red area) (courtesy of Air Force Research Laboratory, Wright Patterson Air Force Base, Ohio), b) uninhabited aircraft with innovative hingeless aerodynamic effectors that will increase the maneuverability and performance of the vehicle with reduced signature (courtesy of Lockheed Martin Aeronautics Company). (See color section.)

Materials and structures have been largely responsible for major performance improvements in many aerospace systems.[9,10] The maturation of computational structures technology and the development of advanced composite materials, witnessed during the past 40 years, have improved structural performance, reduced operational risk, and shortened development times in numerous aerospace systems. For future aeronautical and space systems, materials and structures, in addition to being enabling technologies, continue to be key elements in determining the reliability, performance, testability, and cost effectiveness of these systems. For some of the future air vehicles, the development and deployment of new structures

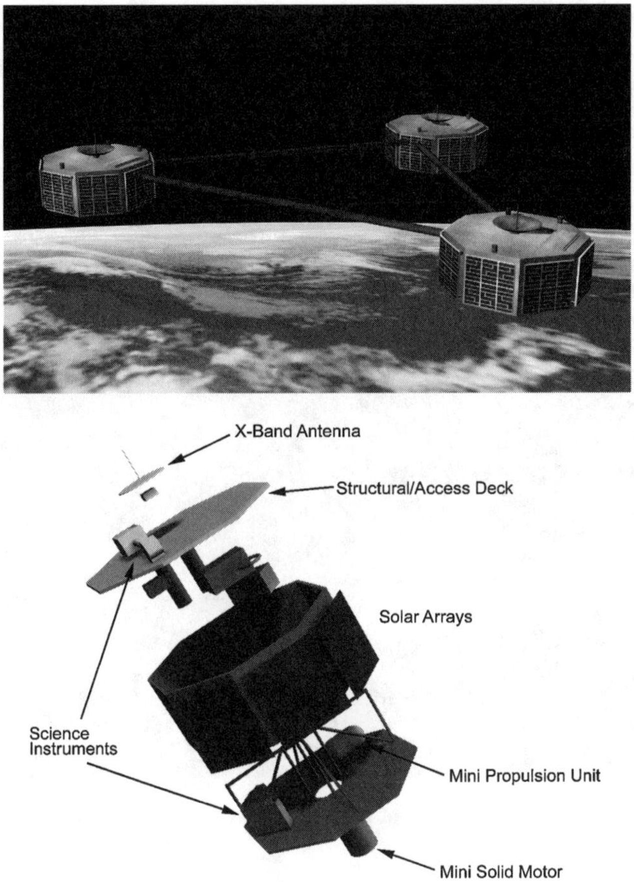

Fig. 1.5 Nanosat constellation concept of the NASA New Millennium Program's Space Technology Project ST5 (each spacecraft is highly autonomous, and is in the shape of hexagons, 40 cm across and 20 cm high and weighs only 44 pounds). (See color section.)

technologies can have more impact on reducing the operating cost and the gross weight than any other technology area.[11] This is depicted in Figs. 1.11 and 1.12. A United Kingdom Foresight report has identified seven technology priority categories for future civil and military aircraft, one of which is materials and structures.[12]

Structures technology encompasses a wide range of component technologies from materials development to analysis, design, testing, production, and maintenance. The treatment of future directions in structures technology in a single book must necessarily be selective. The following six categories of component technologies were selected for this book because they can improve the vehicle performance, advance the technology exploitation process, and reduce system lifecycle costs:

- Affordable composite structures,
- Aging systems and sustainment technology,
- Extreme environment structures,
- Flexible load-bearing structures,

Fig. 1.6 Artist's rendition of an interstellar probe. (See color section.)

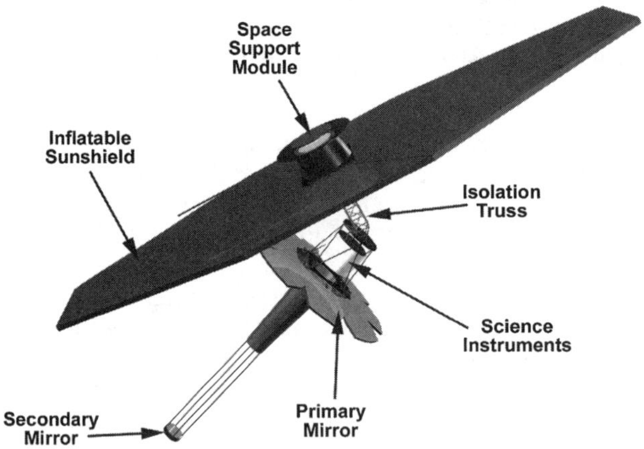

Fig. 1.7 NASA Yardstick Model of the Next Generation Space Telescope (NGST). (See color section.)

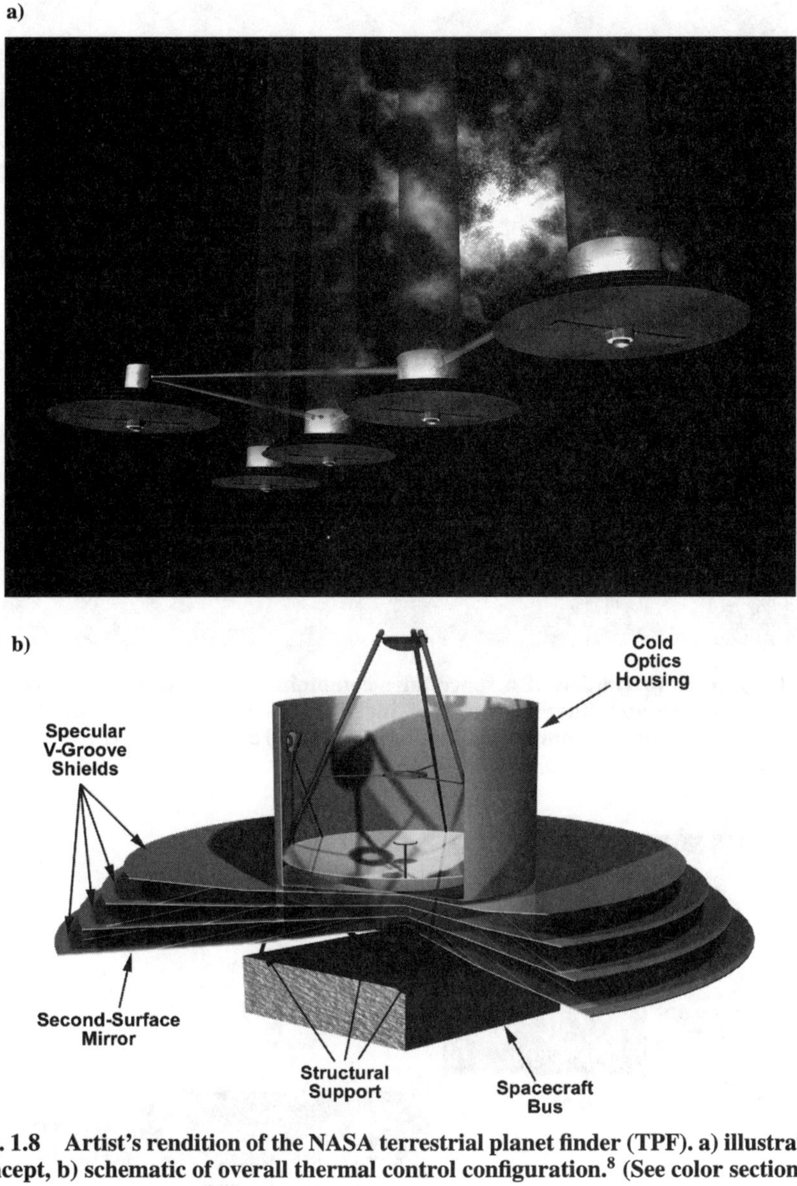

Fig. 1.8 Artist's rendition of the NASA terrestrial planet finder (TPF). a) illustrative concept, b) schematic of overall thermal control configuration.[8] (See color section.)

- Smart air and space structures,
- Computational structures technology.

The development of each of the component technologies is a multidisciplinary activity that involves tasks in other disciplines. The status and trends of each of these component technologies are described in the succeeding chapters, along with their applicability to future vehicles. Some topics are not addressed in this volume and others are not fully covered. These include structural acoustics;[13] structural testing,

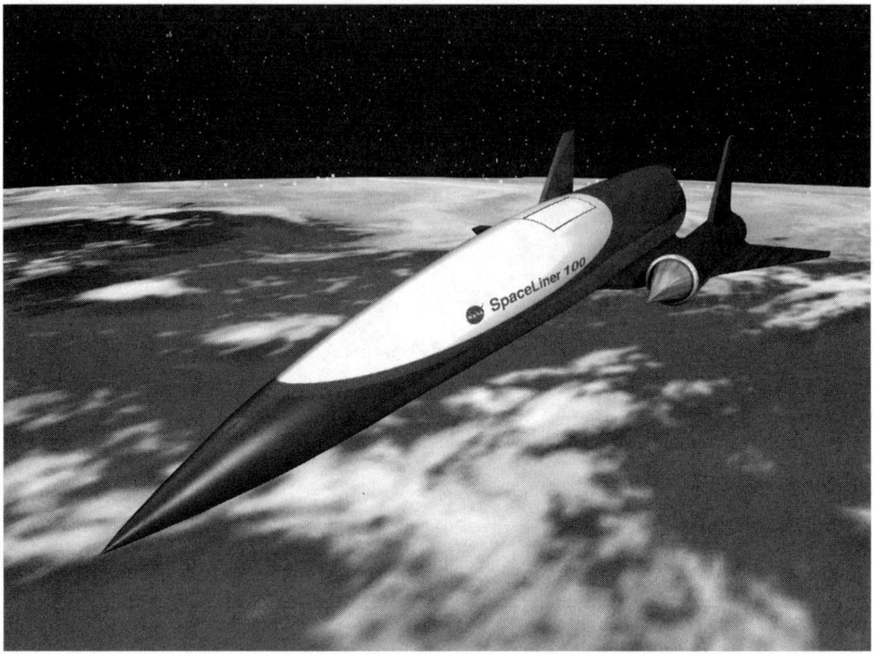

Fig. 1.9 Artist's rendition of a space liner combining a number of advanced materials, structures and other technologies, including ultra-high temperature leading edges, advanced stitched composites, intelligent TPS and autonomous NDE. (See color section.)

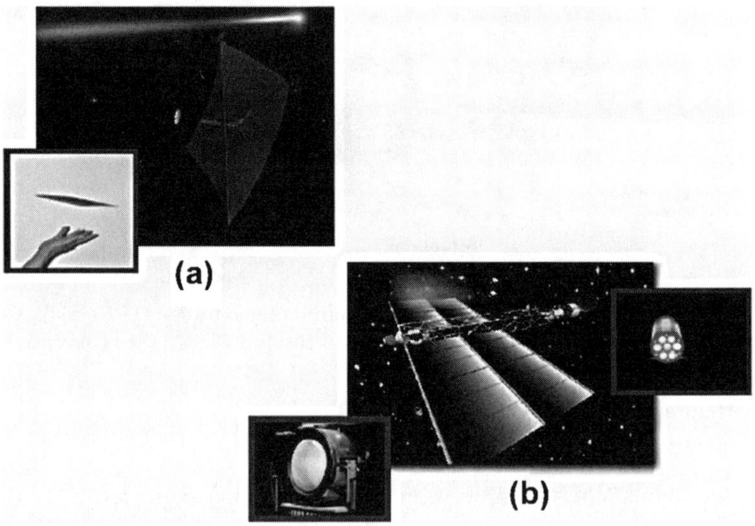

Fig. 1.10 Advanced space transportation concepts. a) solar sails, b) nuclear electric propulsion (courtesy of NASA's Advanced Space Transportation Program and Sandia National Laboratories). (See color section.)

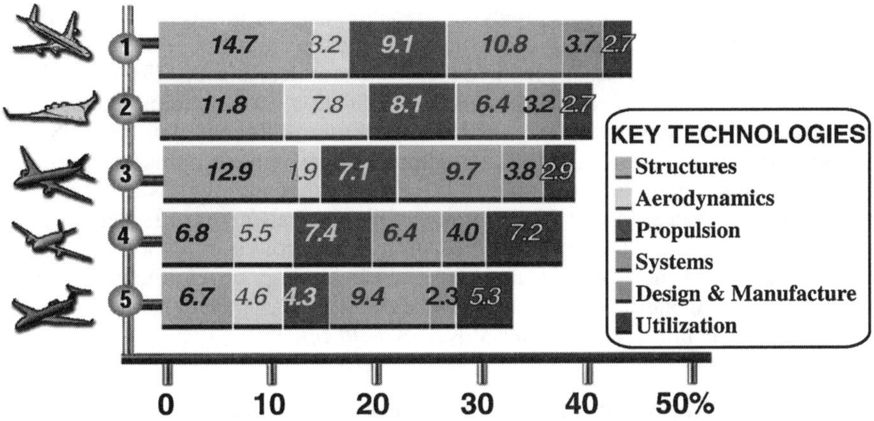

Fig. 1.11 Projected percentage reduction in subsonic transport operating cost in 2020 resulting from deploying new technologies. Long-haul high capacity: (1) conventional, (2) blended-wing body, (3) long-haul capacity conventional, (4) medium range inter-continental, (5) regional jet (provided by NASA Intercenter Systems Analysis Team[11]). (See color section.)

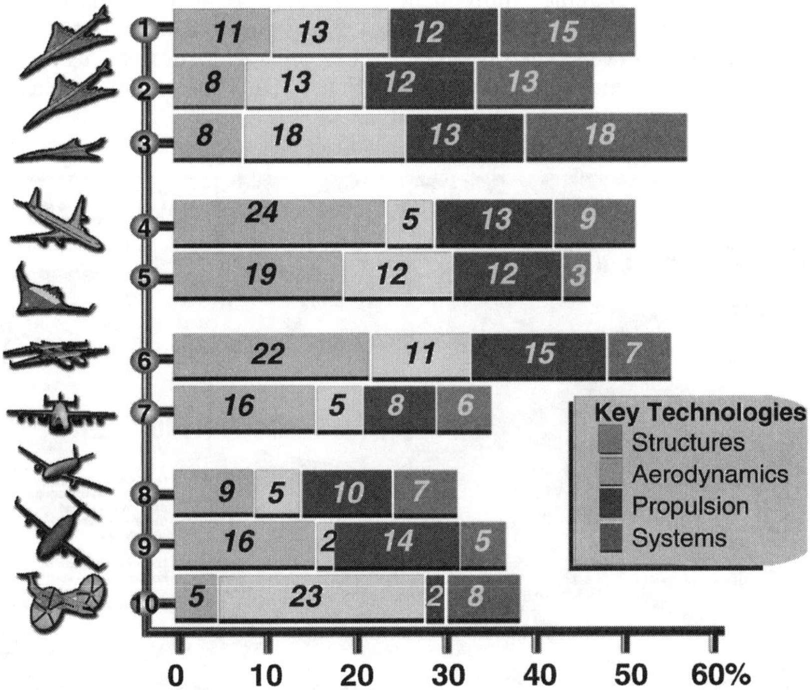

Fig. 1.12 Projected vehicle total gross weight reduction percent. Supersonic: (1) long-haul, (2) premium service, (3) business jet; long-haul, high capacity subsonic, (4) conventional, (5) blended-wing body; global air cargo, (6) long haul, (7) short haul; STOL, (8) medium range inter-continental, (9) short-haul high capacity; short-haul vehicle lift, (10) tiltrotor (provided by NASA Intercenter Systems Analysis Team[11]). (See color section.)

including full-scale tests;[14,15] lifecycle management (including manufacturing, inspections and applications of NDE methods, repairs and maintenance—see, for example, Refs. 16–19); lessons learned (see, for example, Refs. 20–27); and effects of new methods for risk assessment and certification on structures technology. Materials technologies for future aerospace systems are discussed in Refs. 28 and 29. An overview of some of the U.S. government-sponsored programs on structures technology is given in Ref. 30. Some of the recent European activities are described in conference proceedings.[31] NASA has established the Center of Excellence for Structures and Materials at Langley Research Center for coordination, planning, and assessment of the structures and materials technology development activities throughout the agency.[32] The cited reference includes an inventory of structures and materials technical capabilities at NASA field centers.

II. Major Characteristics of Future Aerospace Systems

The realization of the ambitious aeronautical and space goals, with the current budget constraints, will require new kinds of aerospace systems that use novel materials and structures concepts as well as other technologies, and manage risks in new ways. Future aerospace systems must be autonomous, evolvable, resilient, and highly distributed. They must also be able to perform their missions with no ground support (or extremely infrequent ground support); exploit and use local resources; and routinely close decision loops in real time to handle contingencies and replan mission tasks when necessary. These characteristics are highly coupled and require integrated technologies[33] (Fig. 1.13). These characteristics are described subsequently.

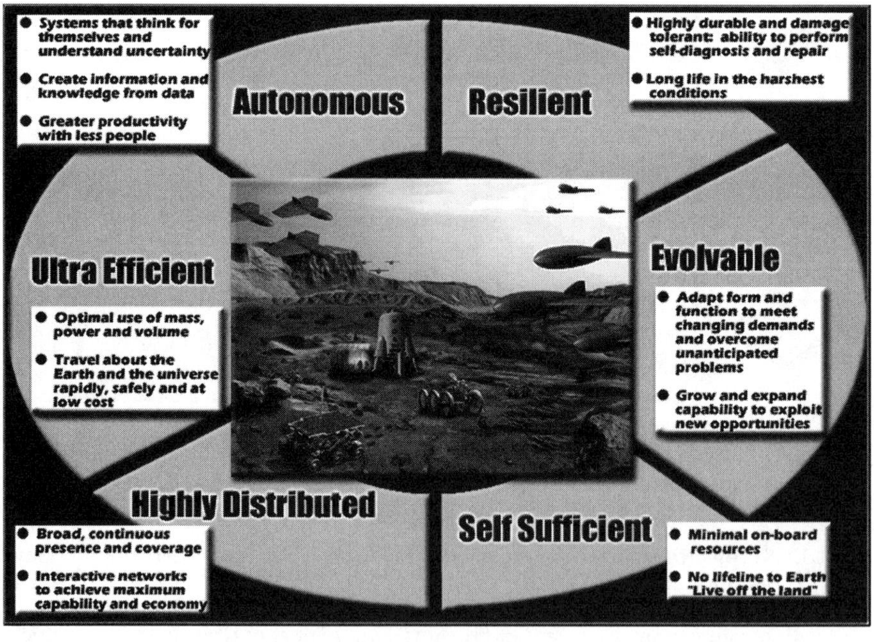

Fig. 1.13 Major characteristics of future space systems. (See color section.)

A. Autonomy

Autonomy refers to a system's ability to learn and adapt its performance to changing environments. Applications in aeronautics include onboard autonomous fault isolation and repair, and intelligent air traffic management systems, with automated planning and scheduling for airport ground operations, and automated landing procedures. These can be combined with autonomous support systems to assist pilots and air traffic controllers in enhancing the safety, capacity, and performance of the aviation system, and in reducing the cost of air travel.

In deep-space missions, the uncertainty about hazardous terrain, the great distances from Earth, and the attendant communication delays will require that the system (probe or in-situ platform such as a rover) be capable of autonomously navigating, selecting, and tracking targets, and be capable of maneuvering over a wide variety of surfaces. It would also have to independently survey and evaluate potential sites and science opportunities, perform data evaluation and observation planning, and gather samples.

In human–robotic exploration missions, the goal is to find the right way to combine human and machine intelligence into a single, autonomous system that can perform several useful support functions for scientists, astronauts, and ground operations personnel. Among these functions are automated planning, scheduling and resource management, condition-based maintenance and diagnosis, and some forms of intermediate science data analysis.

Space system autonomy will provide three interrelated benefits: reduced mission lifecycle cost; improved mission operations; and bolder, unprecedented mission concepts.[34,35] Reduced lifecycle cost will result from moving some traditionally ground-based functions to the space system. This will enable a paradigm shift from large ground teams for each mission to smaller ground teams shared among several missions.

Improved mission operations will be achieved through near real-time analysis of science data onboard the spacecraft. This onboard analysis offers two advantages. First, it will make more efficient and flexible use of the limited communication links between the ground and space platform, through down link prioritization and, in some cases, the onboard construction of more compact and useful science information from the raw data. Second, it will capture transient opportunities that require the quick, reliable recognition of scientifically important events. The goal is to provide a direct link between principal investigators on Earth and the onboard software directing mission activities.

Bolder mission concepts result when new space exploration is carried out on missions that naturally entail autonomy. We are already seeing the first of these new mission classes—the in-situ missions, in which space platforms experience ongoing interactions with a planetary environment; and the constellation missions, where several space platforms must coordinate to accomplish mission objectives.

Autonomy does imply a trade between predictability and robustness in execution. Current assessments are examining its technological maturity, risk, feasibility (from a systems-engineering viewpoint), and actual benefits.

B. Evolvability

Space systems can manifest evolvability in a variety of ways. They can, for example, dynamically change their architecture, structure, or function to improve the performance of certain tasks. They can exhibit adaptive reconfiguration for long life and purposeful survival (against degradation, for example). Or, they can create new functionality for changes in requirements, and/or environment (for example, provide stealth on demand for military vehicles). The adaptive reconfiguration and dynamic changes of such systems apply to both the software and hardware.

Evolvability will likely have a major impact on deployable systems for missions that must provide optimal performance during long periods in unknown harsh and/or changing environments, such as outer solar system exploration, missions to comets and planets with severe environments, and applications needing adaptive information processing. Such applications range from adaptive hardware for the Internet to self-diagnosing/self-repairing engines.

Revolutionary advances in component system robustness could include onboard learning capability and adaptable materials and structures. These advances, combined with evolvability, can lead to space platform lifetimes of unprecedented length—from a few decades to a century.

To be able to continue a mission despite unexpected external events and failure, a system must be fault tolerant and capable of self-repair. This requires fully automated onboard self-assessment and self-reconfiguration as defense against faults. In addition, onboard reasoning can be used as an extension of hardware redundancy and other forms of fault tolerance. The contribution of immune systems to the survivability of biological systems can provide inspiration in this regard.

C. Highly Distributed Systems

To achieve maximum capability for Earth observation, space exploration, and communication, mission planners are considering the use of large fleets of small low-cost intelligent vehicles (such as micro-air vehicles or nano-spacecraft). The vehicles cooperate with one another, adapt to mission goals, and evolve over time to extend and expand the capability. An example of highly distributed systems is the concept of robotic outposts—a remote, highly distributed scientific research station. It is similar to insect or human society, but operates autonomously using multiple cooperating robotic platforms. The highest level of autonomy would entail the use of intelligent robots that are given only the top-level goals. They would, in turn, determine their own immediate objectives in pursuit of those goals, using only occasional consultation with remote human directors.

The notion of such outposts arose from the extensive work done on multiagent systems and cooperative robots. These areas, which are a branch of distributed artificial intelligence, include distributed diagnosis, resource management and execution, and coordination of heterogeneous reasoning methods.

A key challenge in the development of robotic outposts is the creation of systems that exhibit fault tolerance, reliability, and adaptivity. A fault-tolerant system should be able to detect and gracefully compensate for partial system failures, thereby minimizing its vulnerability to individual robot outages. A reliable cooperative system should guarantee that its mission will be accomplished, within certain operating constraints, each time it is used. An adaptable robot team should be able to dynamically modify its actions as the environment or individual robot characteristics change over time.

Robotic outposts would extend human senses into the solar system. With occasional resupply, they would be permanent and self-sustaining. They could be

Fig. 1.14 Artist's rendition of an evolvable rover on Mars' surface. (See color section.)

deployed as expandable intelligent stations in space, on near-Earth objects, such as asteroids and comets, and on the moon, Mars, or elsewhere. They could conduct in-situ planetary studies or remote astrophysical observations, and they could set the stage for later human participation.

Examples of the mission concepts that would fully utilize space systems that have autonomous, evolvable, highly distributed characteristics are: science stations on the surface of Mars with a network of orbiting communication satellites to enable coordination of an array of heterogeneous, autonomous evolvable rovers (Fig. 1.14), and a Europa cyrobot/hydrobot mission. The latter would involve melting through the ice surface, then releasing an underwater submersible to reach and explore the ocean floor looking for signs of life (Fig. 1.15). To perform its mission, the submersible would require a high degree of autonomy, including onboard algorithms embodying knowledge of bio-signatures.

III. Paradigm Change in Structures Technology

In the past, the structures and other technology programs were derived from the missions and vehicles selected to meet the enterprise/organization objectives. In the future, the enterprise/organization objectives will drive the structures technology, which in turn will expand the mission horizons. The missions and vehicles will evolve from the convergence of objectives and technology (Fig. 1.16). The paradigm change is driven/aided by three categories of forces (Fig. 1.17):

Fig. 1.15 Artist's rendition of a hydrobot for exploring Europa's surface. The spacecraft will be exposed to deep space cold and radiation for an extensive period of time. (See color section.)

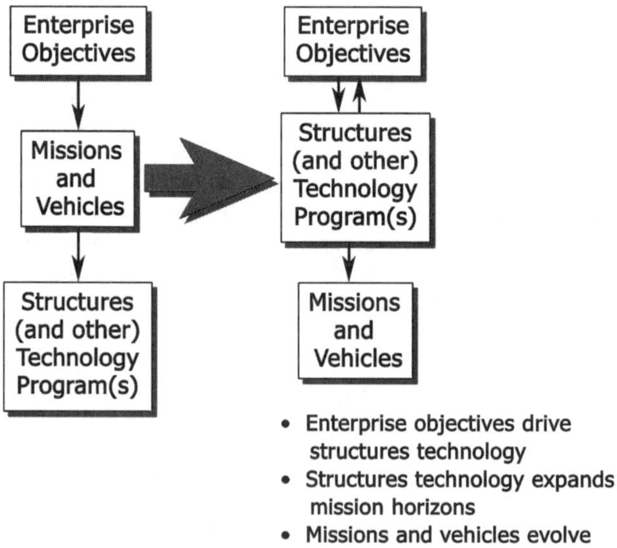

Fig. 1.16 Paradigm change in structures technology.

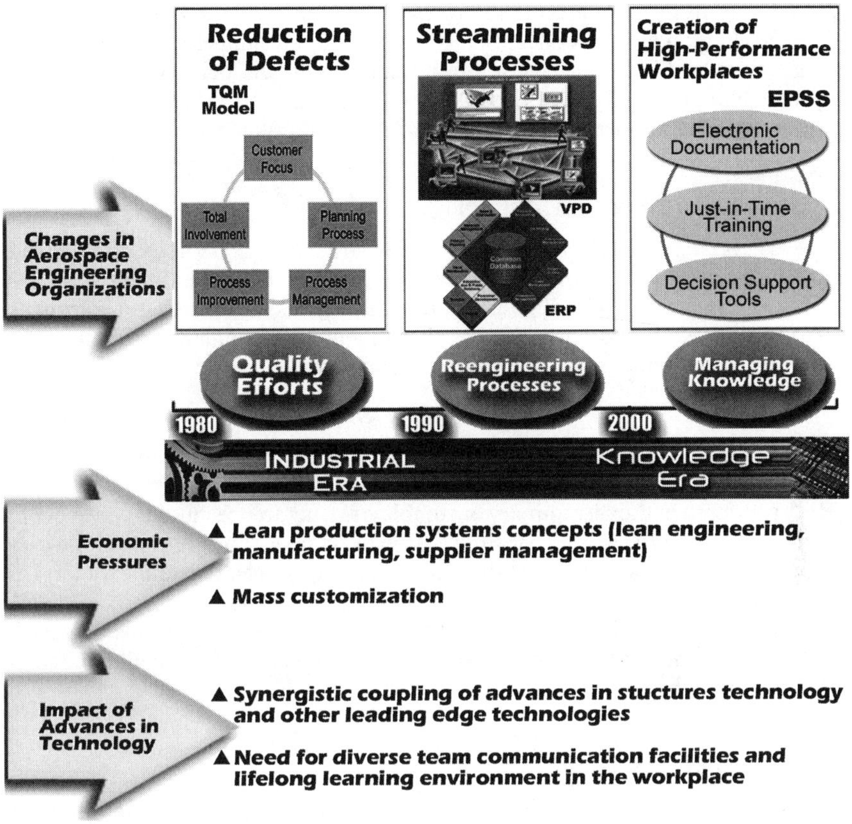

Fig. 1.17 Forces driving a paradigm change in engineering practice. (See color section.)

A. Changes in Aerospace Engineering Organizations

In the 1980s, the focus of aerospace and other engineering organizations was on quality through reduction of defects and use of total quality management models. In the 1990s, the focus shifted to reengineering and streamlining the processes through the use of virtual product development and enterprise resource planning. As we move from the industrial to the knowledge era, the engineering organizations are getting bigger through mergers, and are making radical changes to create high-performance workplaces. The focus is shifting to managing knowledge. The workplace is being transformed from stationary offices centered around desktop computers and workstations into an intelligent, networked, distributed environment that enables the structures experts to collaborate in real time in immersive multisensory environments with other diverse, geographically dispersed teams.

B. Economic Pressures

The declining military procurement budgets, along with customer requests for cheaper, better, faster products, have driven aerospace and other engineering organizations from mass production to mass customization, and to the adoption of lean production system concepts—getting the right things to the right place at the right time, the first time, while minimizing waste and being open to change. They have integrated structures simulation and design tools with other tools and facilities for lean engineering, manufacturing, and supplier management.

C. Impact of Advances in Technology

The synergistic coupling of advances in the structures technology with those in other leading-edge technologies can have a significant impact on future aerospace systems and missions. To realize the potential of this synergism, engineering enterprises will have to provide effective diverse team communication facilities and a conducive lifelong learning environment in the workplace.

IV. Opportunities Provided by Synergistic Coupling of Structures Technology with Other Technologies

Not every area in the disciplinary ensemble of structures technology is expected to yield breakthroughs in the near future. However, steady progress in all the component technologies will improve the efficiency of aerospace systems. Synergistic coupling of structures technology with some of the other leading-edge technologies can significantly enhance the performance of future aerospace systems. Examples are provided in the succeeding subsections, as well as in the succeeding chapters of this monograph.

A. Novel Materials Technologies

Several materials technologies, when coupled with novel structural concepts, offer extraordinary opportunities for future aerospace systems.[36,37] Among these are functionally graded, bio-mimetic and nano-phase materials.

1. Functionally Graded Materials

For FGMs, novel processing techniques are used to produce engineered, gradual transitions in microstructure, composition, and properties to satisfy spatially varying functional performance requirements within a single component. Use of FGMs can alleviate the high gradients of internal stresses and strains resulting from the different local deformation fields induced when dissimilar materials are joined to form a component. This technology could allow researchers to develop complicated structures without using conventional methods of joining. Besides being less costly, jointless gradient materials are much more robust as stresses are distributed throughout the volume of the material rather than being concentrated at the joint. Current FGM applications include piezoelectric devices, thermoelectric/thermoionic energy conversion systems, metal to ceramic joining, and magnetic devices.[38–40] Two types of processes have been used to produce functionally graded metals and ceramic composites: constructive, in which the gradient is produced by stacking material in a programmed fashion, and those that use transport of mass, heat, and fluid to create gradients within materials.

The ultimate goal of FGM is to couple materials design with materials processing. To realize FGM's potential, high-fidelity modeling is needed for the processing, response, and failure of these materials, as well as low-cost processing approaches.

2. *Function-integrated Materials*

In these materials (an extension of smart materials), sensing functions that use photons, mechanical forces, and magnetic or electric fields are built into the molecular structure.[41,42] Among their potential applications are sprayable and adhesive batteries or solar cells for aircraft wings, to convert solar energy to electrical power and store it, and sprayable structural composites that have a switchable antenna function to receive and process information or to provide low observability on demand. Nano-phase and bio-mimetic materials, described subsequently, hold promise for function integration.

3. *Nano-phase Materials*

These are produced by consolidating ultra-fine particles of the same atoms as their common forms. The atoms are arranged in nanometer-size clusters, which become the constituent grains of the new materials. The grains in conventional materials range from microns to 1 mm in diameter and contain several billion atoms. By contrast, the grains in nano-phase materials are less than 100 nm in diameter and contain fewer than tens of thousands of atoms. Mechanical, optical, chemical, magnetic, and electrical properties of nano-phase materials can be tailored to meet specific needs through controlling the size of their constituent grains.[43–45] In addition, reduced grain size reduces the processing temperature for ceramic materials.

These materials represent one of the important frontiers of material technology. Nano-structured ceramics are being considered as thermal barriers in airframe and propulsion systems. Ceramic–metal nano-phased materials may also enable the technology of highly efficient, miniaturized antenna arrays on aircraft.

4. *Bio-mimetic Materials*

Bio-mimetic technology aims at producing new materials by mimicking the synthesis, processing, and properties of materials found in biological systems. Multifunctionality, hierarchical organization, self-repair, adaptability, and durability are among the unique and useful characteristics of such systems. Moreover, biological structural systems do not distinguish between materials and structures.[46,47] The design and development of natural organisms is an integrated process in which component functions are multiple and result in a cost-effective, durable structure whose performance matches the demands brought upon the living system. Recent examples include bio-mineralization and spider silk.

Biological composites, such as bones and teeth, consist of a polymeric matrix reinforced by an inorganic phase. Synthetic factors comprising bio-mineralization include strong binding by the organic matrix of the inorganic reagents, good solvation reagents by the polymer, and an ordered polymer environment to induce nucleation. Inorganic crystal phases normally requiring high temperatures and pressures can be prepared at room temperature and atmospheric pressure by this method.

Understanding the mechanisms involved in growth of biological materials, along with experience gained in developing advanced materials, may lead to material/structural concepts far superior to those currently used in aerospace systems.

B. Multifunctional Structures

Multifunctional structures (MFS), in addition to supporting loads, use sensors to detect and evaluate loads or failure, and to interact with the surrounding electromagnetic environment. MFS represents a new manufacturing and integration technology by which communications and electronics equipment are integrated into conformal load-bearing structures. Advances in large-scale integrated electronics packaging, lightweight composite structures and high conductivity materials enable the technology. In MFS, electronic assemblies (multichip modules), miniature sensors, and actuators are embedded into load-carrying structures, along with associated cabling for power and data transmission. This level of integration effectively eliminates traditional boards and boxes, large connectors, bulky cables, and thermal base plates, yielding major weight, volume, and cost savings.

Current MFS research for aircraft addresses the antenna/airframe proliferation, integration, and maintenance. Present flight vehicle designs have almost 100 antennas occupying roughly 60 apertures (Fig. 1.18). To accommodate electromagnetic windows, these antennas require local reinforcement of the airframe structure, increasing structural weight and cost. Externally mounted antennas degrade the aerodynamic performance and require extensive maintenance. Integrating antennas and other electronics into load-bearing skin structures will be a common concept in future air vehicles. It will yield lower-cost, lighter-weight airframes, increased antenna performance, and lower manufacturing and maintenance costs. These multifunctional structures may also enable greater data transfer rates and increased aircraft surface area for additional sensors and transmitters.

MFS technology offers significant savings in the mass and volume of spacecraft by eliminating electrical chassis and cabling, and placing most sensors and batteries on the bus structure, which also provides structural and thermal control (see, for example, Refs. 48 and 49 and Fig. 1.19). The technology offers several benefits to future spacecraft: its inherent modularity supports low-cost mass production and assembly, significantly reducing life-cycle costs; it eliminates cables and

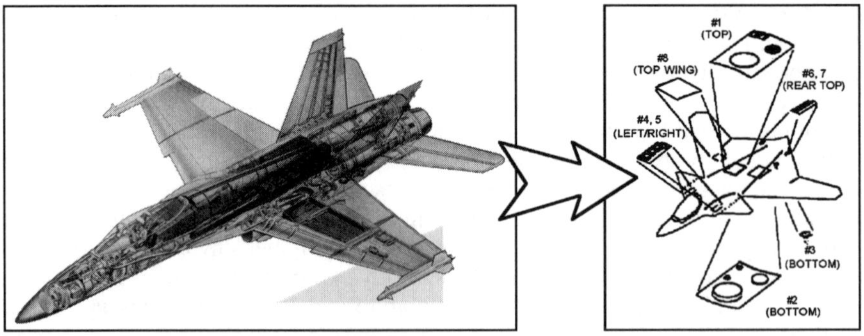

Fig. 1.18 RF multifunctional structural panel (courtesy of Air Force Research Laboratory, Wright Patterson Air Force Base, Ohio).

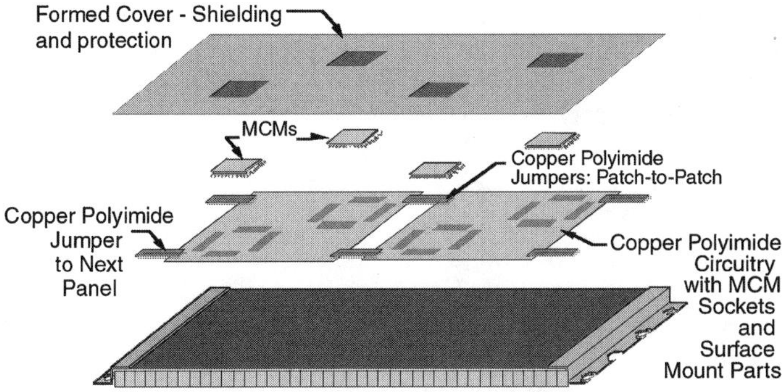

Fig. 1.19 Multifunctional structural panel with integral electronic, structural and thermal control (courtesy of Air Force Research Laboratory). (See color section.)

connectors, minimizing "touch labor" needed during final spacecraft integration, whose robustness and reliability it increases; it enables reworkable "plug-n-play" spacecraft; and it provides integrated thermal control, flex interconnect architecture for power distribution, and data storage and electronics into load-bearing structures.

Structural integration of complex aero-fluidic and thermo-mechanical subsystems into future aerospace vehicle design is being studied. The system-level payoffs of integrating these technologies will be realized only through simultaneous development across all disciplines involved in designing future aircraft. Structural integration must begin early in the preliminary design phase of the system development for these payoffs to materialize. Advanced subsystem devices, novel structural concepts, design analysis methods, and manufacturing approaches will be assimilated to quantify reductions in weight and assembly costs.

Technologies of interest for structural integration are flexible skin panels, integral cooling for thermal management, fluidic jets for aerodynamic control, integral electric and hydraulic lines, and actuators whose attachment hardware may be built directly into the airframe substructure. Composite materials will be used extensively in these integral designs because of their unique thermal and mechanical tailorability. These properties may eventually allow for incorporation of highly conductive elements, embedding of electrical sensors and shape memory materials, and integration in the structural design of microactuators and subsystems.

Payoffs to be realized by future aerospace vehicles through the use of multifunctional structures include reduced operational costs and weight and improvements in mission effectiveness. For future air vehicles, advanced structural concepts will be combined with integral flight systems to provide significant reduction in part count and increased volumetric efficiency. The savings realized in operational and sustainment costs for systems with increased robustness and durability may offset the probable increase in cost per pound of the airframe, attributed to inclusion of electronic or mechanical subsystems. The maximum payoffs to be measured in efficiencies at the system level can only be attained by considering the airframe to be a multifunctional structure early in the development of the conceptual design and simultaneously optimizing the design across all disciplines. Previously unimagined design efficiencies and synergies will be achieved. Ultimately, the transition of multifunctional aircraft structural concepts to next generation aircraft will require resources focused on development of technologies to ensure their affordability, durability, and supportability in the operational environment. These quantities will be essential to continued development of future systems including uninhabited aircraft and hypersonic vehicles.

C. Micro Air Vehicles

Micro Air Vehicles (MAVs) are a new category of aircraft that are significantly smaller than the currently available Unmanned Aerial Vehicles (UAVs). In the U.S., the Defense Advanced Research Projects Agency (DARPA) has funded several research laboratories and universites to develop MAVs. The characteristics of MAVs as envisioned by DARPA are: a maximum size of approximately 6 in., weight less than 4 oz., speed up to 35 mph, 20–60 min of endurance, and range of 20–40 miles. MAVs can serve as intelligence-gathering platforms seeking real-time information "over the hill" (i.e., beyond visual range). MAVs may have to be operated in diverse environments such as open terrain, dense forests, and urban areas. In the traditional role, MAVs would overfly the area of interest. However, other MAVs may have to land and hide for a time or drop sensors off at selected locations. Some versions may even have to enter the interior of buildings. Instrumentation on-board MAVs may also vary depending on their role or mission. Video cameras might be used in open terrain, infrared sensors in forests, acoustic sensors inside buildings, and biochemical sensors may be used to identify biochemical warheads. In another role, MAVs might be used as overhead laser relay stations providing secure and stealthy communications. They might be used to control laser-guided weapons while the troops are a safe distance away from the enemy. Consequently, a variety of different missions could be envisioned for MAVs. In addition to military applications, MAVs can have nonmilitary uses, such as civilian rescue, traffic surveillance, and finding sources of dangerous toxins or chemicals. They can also be used for planetary explorations.

A successful MAV will require the synergy of several different subsystems into an integrated whole. The major subsystems in MAVs might be loosely classified as: 1) aerodynamics and controls; 2) propulsion and power; and 3) electronics and instrumentation. Potential technology developments for each of these subsystems must first be identified and then investigated. After this is accomplished, the different technologies can be integrated into a MAV. However, the optimum configuration or even the technical approach to be taken may differ depending on the particular mission or role for the MAV.

The purpose of the aero-controls subsystem is to sense, generate, and control the lift and directional stability required for controlled flight. Because of the stringent size, weight, and power requirements, a single component has to fulfill all or many of the tasks for the entire aero-control subsystem. Some components may even have to cross subsystem boundaries. A number of concepts have been proposed for the aero-control subsystem. The first is to use flapping wings (mimicking bird and insect flight), thereby combining aerodynamics and propulsion into a single entity. The second concept involves the use of rotary wing for hovering flight. A third concept involves the use of adaptive fixed-wing technology—smart materials and structures whose shape and/or size change with the application of electric current. Other concepts include disc-wing design with microflight control system

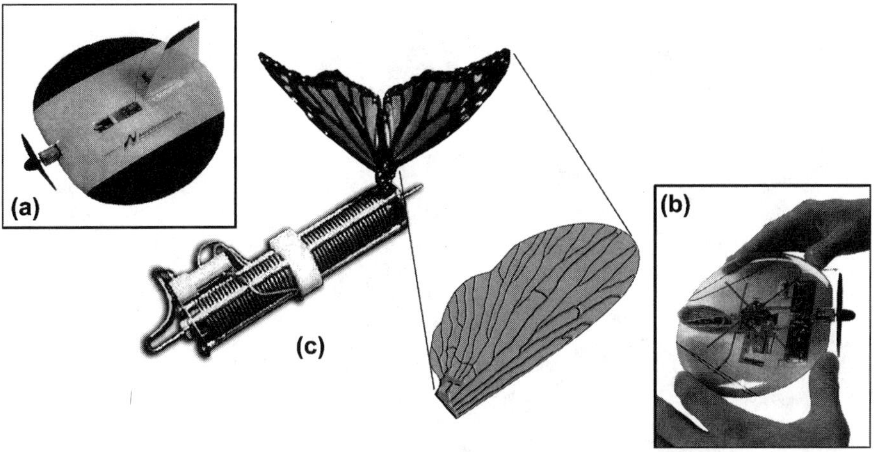

Fig. 1.20 Micro air vehicles (MAVs). a) A variant of the Black Widow disc-shaped MAV featuring rudder and elevator control (courtesy of AeroVironment, Inc.); b) view of the mockup of the flex circuit installed in the Black Widow MAV. Note all electrical connections and antennas are on the flex circuit. Only the batteries, motor, propeller and structure need to be added to the integrated flex avionics package to complete the operational MAV (courtesy of AeroVironment, Inc.); c) entomopter MAV using meso-scale flapping wing with circulation controlled airfoil (courtesy of R. Michelson, Georgia Institute of Technology). (See color section.)

for remote operation of the front-mounted propeller and two control flaps (see Fig. 1.20).

D. Biologically Inspired Structures

During the past few years, a growing number of engineers have been seeking inspiration from biological systems in an effort to augment the capabilities of their engineered systems. The goal is to make these systems more adaptive; that is, capable of undergoing modifications according to changing environmental circumstances, thereby ensuring their continued functionality. Such systems evolve, develop, and learn—they adapt on their own with no human intervention.

Biomorphic explorers are among the biologically inspired aerospace systems that the NASA Jet Propulsion Laboratory is studying. These small, dedicated, low-cost devices capture some of the key features of biological explorers, including reconfigurable units with versatile mobility (such as aerial, surface and sub-surface explorers). They might also feature control by adaptive, fault-tolerant, bio-inspired algorithms to match changing ambient/terrain conditions.

These features enable comprehensive exploration at lower cost with broader coverage through cooperative organization of landers, rovers, and a variety of inexpensive low-mass biomorphic explorers (Fig. 1.21), such as biomorphic gliders, balloons, and powered aircraft. Specific science objectives targeted for biomorphic explorer missions include atmospheric information gathering by distributed multiple site measurements, close-up imaging for geological site selection, deployment

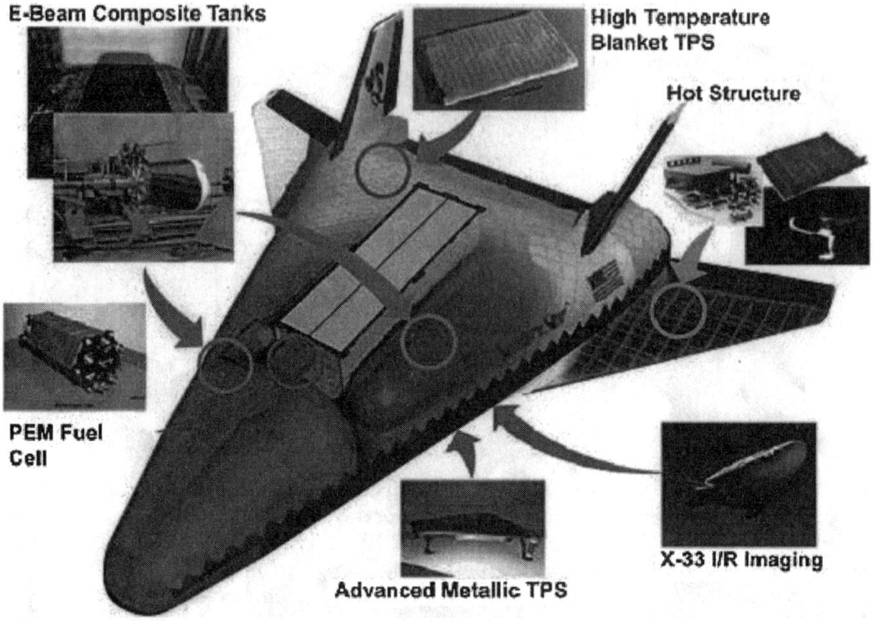

Fig. 1.21 Integration of airframe system technologies for a reusable launch vehicle (courtesy of NASA's Advanced Space Transportation Program).

of surface payloads such as instruments, surface experiments and sample return reconnaissance. Other examples of biological structures that utilize composite materials and adhesives are given in Ref. 50.

V. Concluding Remarks

Future research issues in structures technology involve integrating existing and new materials and other leading edge technologies into functional aerospace systems with high quality and low-cost features. Examples of this integration are shown in Figs. 1.4, 1.9, and 1.21 for a tailless military aircraft, a space liner, and a reusable launch vehicle. Future efforts will address advanced load path management, innovative materials processing, low-cost fabrication and other technology challenges to enable more affordable, lighter, stronger and stiffer, safer and more durable vehicles for different flight regimes, and for planetary atmospheric entry and flights throughout the solar system. These activities will dispel the notion that high performance can be achieved only at high cost. Today, structures technology is derived from mission and vehicle requirements. A cultural shift is needed to effect a change to missions and vehicles that are enabled by innovative structures technology. This century will see aerospace vehicles made from computationally designed materials at the atomic/molecular level to accomplish a variety of missions. Programmable multifunctional materials and structures will be able to adjust their shape and their mechanical, electromagnetic, optical and acoustic properties on demand. Distributed networks of space observatories and scientific research stations will be developed, consisting of multiple cooperating robotic platforms and small low-cost intelligent vehicles with self-healing, reconfigurable, multifunctional structures, and miniaturized instrumentation.

Acknowledgment

The author appreciates useful and valuable suggestions by many of his colleagues, including Michael Mohaghegh, Boeing Commercial Airplanes, Gary Lyles, George C. Marshall Space Flight Center, Donald B. Paul, Air Force Research Laboratory, WPAFB, Douglas S. Stetson, Jet Propulsion Laboratory, and Robert Michelson, Georgia Institute of Technology. The author also appreciates the diligent work of Alan McCoy and Derek Wilson of the University of Virginia in enhancing the quality of the artwork.

References

[1]"New World Vistas—Air and Space Power for the 21st Century," USAF Scientific Advisory Board, Dept. of the Air Force, Washington, D.C., Vols. 1–13, 1995.

[2]Noor, A. K., and Venneri, S. L. (eds.), *Future Aeronautical and Space Systems*, Progress in Astronautics and Aeronautics, Vol. 172, AIAA, Reston, VA, 1997.

[3]Engstrom, F., "Space Transport Systems for the 21st Century," *ESA Bulletin*, No. 95, Aug. 1998, pp. 54–61.

[4]Kuczera, H., and Johnson, C., "The Major Results of the FESTIP System Study (Future European Space Transportation Investigation Program)," *Proceedings of the Ninth AIAA Int. Space Planes and Hypersonic Systems and Technologies Conference*, Norfolk, VA, Nov. 1–5, 1999, Collection of Technical Papers, American Institute of Aeronautics and Astronautics, Reston, VA, 1999.

[5]Bell, T. E., and Dooling, D. (writers), Fouke, J. (ed.), *Engineering Tomorrow: Today's Technology Experts Envision the Next Century*, IEEE Press, Piscataway, NJ, 2000.

[6]Liebeck, R. H., Page, M. A., and Rawdon, B. K., "Blended-Wing-Body Subsonic Commercial Transport," 36th Aerospace Sciences Meeting and Exhibit, Reno, NV, Jan. 12–15, 1998, Report No. 98-0438, American Institute of Aeronautics and Astronautics, Reston, VA, 1998.

[7]*Future Aerospace Technology in the Service of the Alliance*, Proceedings of the AGARD Symposium held April 14–17, 1997, Palaiseau, France, Advisory Group for Aerospace Research and Development, North Atlantic Treaty Organization, AGARD Conference Proceedings No. 600, 1997.

[8]Beichman, C. A., Woolf, N. J., and Lindensmith, C. A. (eds.), "The Terrestrial Planet Finder (TPF): A NASA Origins Program to Search for Habitable Planets," JPL Publication 99-3, May 1999 (also available on the web, URL address: http://tpf.jpl.nasa.gov/).

[9]Davies, G. A. O., "Aircraft Structures," *The Aeronautical Journal*, Vol. 100, No. 1000, Dec. 1996, pp. 523–529.

[10]Noor, A. K. (ed.), *Structures Technology: Historical Perspective and Evolution*, AIAA, Reston, VA, 1998.

[11]"Maintaining U.S. Leadership in Aeronautics: Scenario-Based Strategic Planning for NASA's Aeronautics Enterprise," National Research Council, National Academy Press, Washington, D.C., 1997.

[12]"Technology Foresight—Progress Through Partnerships," Defence and Aerospace, Office of Science and Technology for HMSO, Vol. 12, April 1995.

[13]Ohayon, R., and Soize, C., *Structural Acoustics and Vibration: Mechanical Models, Variational Formulations and Discretization*, Academic Press, San Diego, CA, 1998.

[14]Mohov, V. F., "Methodological Textbook on Preparing and Conducting Static Tests on Full-Scale Airplane Structures," Central Aerodydrodynamic Institute (TsAGI), CWA 22 Corporation, Zhukovsky, Russia, 1998.

[15]Harris, H. G., and Sabnis, G. M., *Structural Modeling and Experimental Techniques*, CRC Press, Boca Raton, FL, 1999.

[16]*Non-destructive Evaluation of Aging Materials and Composites III*, Proceedings of the Meeting, March 3–5, 1999, Newport Beach, CA, Society of Photo-Optical Instrumentation Engineers, Bellingham, WA, SPIE Proc., Vol. 3585, 1999.

[17]*CEAS Forum on Life Extension—Aerospace Technology Opportunities*, Proceedings, March 23–25, 1999, Cambridge, U.K., Royal Aeronautical Society, London, 1999.

[18]*Proceedings of the Third Joint FAA/DoD/NASA Conference on Aging Aircraft*, Albuquerque, NM, Sept. 20–23, 1999, FAA William J. Hughes Technical Center, Atlantic City Airport, NJ, 1999.

[19]Hagemaier, D., Jappe, B., and Wood, N., "Advances in Nondestructive Inspection Methods," *Aerospace Engineering*, Vol. 19, No. 10, Oct. 1999, pp. 23–25.

[20]Ryan, R., Blair, J., Townsend, J., and Verderaime, V., "Working on the Boundaries: Philosophies and Practices of the Design Process," NASA Technical Paper 3642, July 1996.

[21]Ryan, R. S., "A History of Aerospace Problems, Their Solutions, Their Lessons," NASA Technical Paper 3653, Sept. 1996.

[22]Aljabri, A. S., Bernard, D. E., Dvorak, D. L., Man, G. K., Pell, S., and Starbird, T. W., "Infusion of Autonomy Technology Into Space Missions—DS1 Lessons Learned," *Proceedings of the 1998 IEEE Aerospace Conference*, Aspen, CO, Mar. 21–28, 1998, Institute of Electrical and Electronics Engineers, Inc., Piscataway, NJ, 1998, Vol. 2, pp. 315–329.

[23]van der Ha, J. C., "Lessons Learned From Small Satellite System Design and Operations," Proceedings of the 21st International Symposium on Space Technology and Science, Omiya, Japan, May 24–31, 1998, 21st ISTS Publications Committee, Tokyo, Japan, 1998, Vol. 1, pp. 986–991.

[24]Wilcox, B., and Nguyen, T., "Sojourner on Mars and Lessons Learned for Future Planetary Rovers," 28th International Conference on Environmental Systems, Danvers, MA, July 13–16, 1998, SAE Paper 981695.

[25]Dunford, P. J., MacDonald, T. L., and Rudzis, J., "V-22 'Osprey' Engineering-Manufacturing-Development (EMD) Flight Test—Challenges and Successes," *1998 Report to the Aerospace Profession—Society of Experimental Test Pilots*, Proceedings of the 42nd Symposium, Beverly Hills, CA, Sept. 1998, Society of Experimental Test Pilots, Lancaster, CA, 1998, pp. 105–117.

[26]Long, J., and Koyama, W., "F/A-18A/B/C/D 9g Flight Test Program," *Proceedings of the 1999 IEEE Aerospace Conference*, Aspen, CO, Mar. 6–13, 1999, Institute of Electrical and Electronics Engineers, Inc., Piscataway, NJ, 1999, pp. 111–125.

[27]Kawaguchi, J., "Lessons Learned—Lunar, Earth and Solar Gravity Assist in 'Nozomi' Mars Spacecraft," 50th International Astronautical Congress, Amsterdam, Netherlands, Oct. 4–8, 1999, IAF/IAA Paper 99-A708.

[28]Peel, C. J., "Advances in Materials for Aerospace," *The Aeronautical Journal*, Vol. 100, No. 1000, Dec. 1996, pp. 487–503.

[29]Noor, A. K., Spearing, S. M., Adams, W. W., and Venneri, S. L., "Frontiers of the Material World," *Aerospace America*, Vol. 36, No. 4, 1998, pp. 24–31.

[30]Noor, A. K., and Malone, J. B. (compilers), "Government-Sponsored Programs on Structures Technology," NASA CP-97-206241, 1997.

[31]*Proceedings of the European Conference on Spacecraft Structures, Materials and Mechanical Testing*, Braunschweig, Germany, Nov. 4–6, 1998, ESA, Noordwijk, Netherlands, 1999.

[32]Harris, C. E. (ed.), "Implementation Plan for the NASA Center of Excellence for Structures and Materials," NASA TM-1998-208714, Aug. 1998.

[33]Noor, A. K., Doyle, R. J., and Venneri, S. L., "Autonomous, Biologically Inspired Systems for Future Space Missions," *Advances in Engineering Software*, Vol. 31, No. 7, June 2000, pp. 475–482.

[34]Doyle, R. J., "Spacecraft Autonomy and the Missions of Exploration," *IEEE Intelligent Systems*, Sept./Oct., 1998, pp. 36–44.

[35]Glass, B., and Zornetzer, S., "Intelligent, Autonomous Technologies for Space Systems Development," 50th IAF International Astronautical Congress, Amsterdam, The Netherlands, Oct. 4–8, 1999, IAF Paper 99-U201, 1999.

[36]Jacobs, J. A., and Kilduff, T. F., *Engineering Materials Technology: Structures, Processing, Properties and Selection*, Prentice Hall, Upper Saddle River, NJ, 1997.

[37]Good, M. L., "Materials in the 21st Century's Global Economy: Innovations," *Advanced Materials and Processes*, Vol. 157, No. 1, Jan. 2000, pp. 32–36.

[38]Neubrand, A., and Rodel, J., "Gradient Materials: An Overview of a Novel Concept," *Zeitschrift fur Metallkunde*, Vol. 88, No. 5, May 1997, pp. 358–371.

[39]Miyamoto, Y., "Worldwide Trends of Functionally Graded Material Development," *Metals and Technology*, Vol. 67, No. 2, Feb. 1997, pp. 5–16.

[40]Kaysser, W. A., "Functionally Graded Materials 1998," *Materials Science Forum*, 1999, pp. 308–311.

[41]Newnham, R. E., and Ruschau, G. R., "Smart Electroceramics," *The American Ceramic Society Bulletin*, Vol. 75, No. 10, Oct. 1996, pp. 51–61.

[42]Newnham, R. E., "Molecular Mechanisms in Smart Materials," *Materials Research Society (MRS) Bulletin*, Vol. 22, No. 5, May 1997, pp. 20–34.

[43]Goldstein, A. N. (ed.), *Handbook of Nanophase Materials*, Marcel Dekker, NY, 1997.

[44]Komarneni, S., Parker, J. C., and Wollenberger, H. J. (eds.), *Nanophase and Nanocomposite Materials II*, Proceedings of the symposium held Dec. 2–5, 1996, Boston, MA, Materials Research Society, Pittsburgh, PA, Vol. 457, 1997.

[45]Wang, Z. L. (ed.), *Characterization of Nanophase Materials*, Wiley-VCH, Weinheim, Germany, 2000.

[46]Sarikaya, M., Aksay, I. A., and Sarikaya, M. (eds.), *Biomimetics: Design and Processing of Materials*, AIP Press, Woodbury, NY, 1995.

[47]Simmons, W. C., "Biomimetics and Smart Materials," *Smart Electronics and MEMS*, Proceedings of the 1997 Far East and Pacific Rim Symposium, Adelaide, Australia, Dec. 11–13, 1997, Society of Photo-Optical Instrumentation Engineers, Bellingham, WA, SPIE Proc., Vol. 3242, 1997, pp. 2–9.

[48]Rawal, S. P., Barnett, D. M., and Kustas, F. M., "Multifunctional Composite Structures Designs," *Advanced Materials and Processes – Preparing for the New Millennium*, Proceedings of the 31st International SAMPE Technical Conference, Chicago, IL, Oct. 26–30, 1999, Society for the Advancement of Material and Process Engineering, Covina, CA, Int. SAMPE Technical Conference Series, Vol. 31, 1999, pp. 307–318.

[49]Boller, C., "Smart Structures and Materials in Aerospace Applications of Next Generation," Advisory Group for Aerospace Research and Development, Structures and Materials Panel, Neuilly-Sur-Seine, France, NASA Report No. 19990014383, 1997.

[50]Renton, W. J., "Aerospace and Structures: Where Are We Headed?," *International Journal of Solids and Structures* (special issue honoring the 70th birthdays of Charles Bert and Jack Vinson), in press.

Chapter 2

Affordable Composite Structures

William Baron
Air Force Research Laboratory, Dayton, Ohio

Tia Benson Tolle
Air Force Research Laboratory, Dayton, Ohio

Dan Arnold
Boeing, Seattle, Washington

Gary Renieri
Boeing, Saint Louis, Missouri

and
Larry Bersuch
Lockheed Martin, Fort Worth, Texas

I. Introduction

A. Purpose of the Chapter

THIS chapter provides an overview of emerging structural concepts and technologies that will be available to the designer of composite aircraft in the future. Structural concepts and associated material processes that enable the design approaches discussed represent various levels of technical maturity and are not necessarily ready for transition. The authors have attempted to provide insight to engineering confidence that exists with these emerging technologies but recognize that, in some instances, significant development still needs to be undertaken before the designer can confidently include such structural systems in the manufacturing of aerospace systems. This chapter is not intended to be a design guide, but merely directed at identifying future trends in affordable composite structures that will offer reduced cost and increased performance. Concepts presented need to be considered tools for the designer to enhance producibility and improve the overall capability of future aerospace vehicle airframes. The structural concepts have limited quantification of their low-cost benefits due to their early state of development. It was felt by the authors that these concepts will enable affordable designs as they mature. Technologies discussed are primarily directed at fighter fixed-wing aircraft implementation; however, these emerging concepts will have wide applicability to military rotor, civil, and transport aircraft; space vehicle applications; and civil infrastructure.

Specific focus is given to emerging concepts that can be used to solve many of the challenges that have prevented increased use of composite materials. The benefits and present applications of composite materials will be reviewed, with a focus on how composites are integrated to the design process. This treatment will discuss the successes of integration but will also depict current limitations. Key considerations such as the building block design process, load path management, subsystem integration requirements, and manufacturing quality will be addressed to define the environment that the designer must acknowledge while developing an advanced airframe. A discussion of key material processes will also be reviewed to show their enabling influence on novel concepts. These processes will include review of resin infusion practices, low-temperature cure materials, and emerging fiber preforms. Major progress has been accomplished in the development of concepts to improve the out-of-plane properties of composites. This chapter provides a thorough review of advancements in discrete reinforcement techniques and results from the use of carbon textile preforms to enhance joint performance. As joints become more robust, the ability to unitize the assembly and increase spar spacing has become an increasingly desirable way to reduce cost. Therefore, the need to improve buckling resistance through low-cost skin stiffening techniques has become an area of focus in the development of future airframes. Improvements in sandwich technology are discussed as alternatives to conventional approaches of stiffening with hats, blades, and I-beams. It will be shown how many of these concepts have been considered in the integration of advanced airframes to enable reduced weight and cost. This will include identifying key requirements that must be met in the development of advanced airframes but will also discuss how the intrinsic failure attributes of these concepts can be used to further enhance performance. Novel airframe configurations and associated concepts considered to enable these airframes are discussed. This will include assessing the benefits of airframes with extensive use of primary sandwich structure, concepts that exploit the ability to tailor composite properties in large cocured structure, and configurations that eliminate many of the hard points that typically inhibit the incorporation of composite materials. The authors have attempted to discuss the primary drivers of structural concept development. Many of the detailed benefits of composites such as aeroelastic tailoring, stiffness coupled design practices, and crashworthiness have not been addressed since these concepts are out of the scope of the chapter. Additionally, the need for repair technology development cannot be understated. This chapter presents many emerging concepts; however, specific repair practices have yet to be developed for most, which offers a significant opportunity for further research.

B. Structural Affordability: Low Cost Plus Performance

Affordability is only achieved through the process of developing a design that is low in cost but meets its intended functional requirements. While this may seem simple, it is quite complex since the tradeoff between high performance and cost involves many variables. This is especially true in the case of military systems that require that specific war-fighting capabilities be achieved. For example, suppose a fighter aircraft was developed to perform in environments that presented ballistic threats and was extremely low cost to produce but it possessed no structural

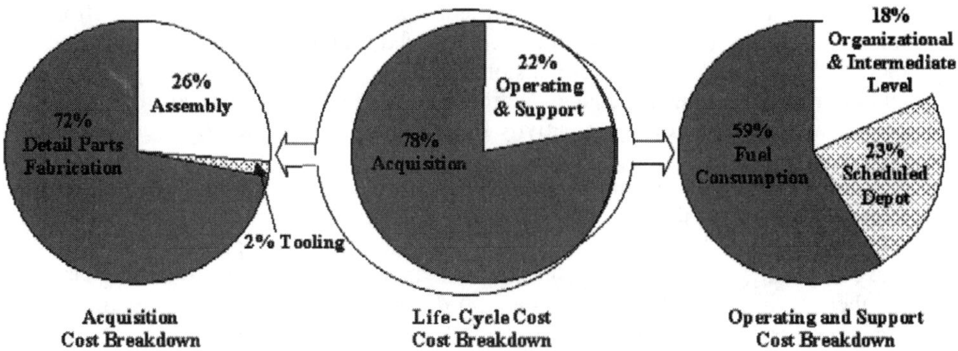

Fig. 2.1 Life-cycle vehicle costs.

survivability or low observable capabilities. While this system may be very low in cost, it most likely would not be very affordable to the military user since its ability to deliver ordnance would be limited. Therefore, the design of affordable structural systems must involve a process of synthesizing requirements into a structural design that offers a low-cost approach to development, production, and operational support. Generally, these costs can be minimized with systems that meet intended requirements and capabilities in the smallest physical size vehicle possible. The major overall cost of a system is driven primarily from the cost associated with operational support and acquisition, as shown for a typical fighter vehicle in Fig. 2.1. Approximately 59% of operational costs are derived from fuel consumption alone, and 72% of acquisition cost is in detail parts fabrication. Therefore, major cost savings can be achieved by reducing weight and part count. Airframe weight reduction can be achieved through two primary methods: 1) the use of lighter materials and 2) more efficient structural designs. Composites offer significant reductions in material density and the strength-to-weight ratio is higher, providing the basis to develop composite structure technology for increased integration to the airframe.

Advanced structural composites provide the aerospace community with an unprecedented opportunity to achieve fractional costs and improved weapon system performance over existing airframes. Significant reductions in airframe cost will make larger fleets economically viable and the associated improvements in weight will offer dramatic improvements in range, payload, and maneuverability. Advances in affordable composite structure will impact all air-vehicle designs, including fighters, transports, and new classes of unmanned vehicles. The design and manufacturing capabilities that are offered by composite structures must be exploited to achieve these benefits.

II. Background

A. Benefits of Composites

Composite materials have demonstrated significant performance benefits over conventional materials when used in airframe, missile, launch vehicle, or spacecraft structures and are currently a critical part of many such systems. Composites have inherent characteristics that lead to system level payoffs, such as very high specific stiffness and strength; are fatigue and corrosion resistant, can be conductive or nonconductive; have tailorable properties; and can be formed to complex shapes.

These attributes result in payoffs such as decreased structural weight fractions for higher speed, better fuel economy, improved maneuverability and response, increased structural stiffness, dimensional stability, and decreased life-cycle costs and offer opportunities to reduce electromagnetic signature. These attributes also provide for increased design creativity by enabling the designer to tailor material properties for specific applications.

While composite design opportunities continue to be explored, the cost to manufacture composite structure has proved to be a significant obstacle to their widespread use.[2] This is due to design and manufacturing approaches that utilize composite materials in the conventional "metals fashion" of assembling large numbers of mechanically fastened parts. A design approach involving larger, integrated components to maximize producibility, quality, and design efficiency will enable affordable composite airframe structures.

The properties of composite materials are directionally dependent and offer the ability for aeroelastic tailoring the strength and stiffness of a structure in the directions to allow the most efficient management of airframe loads. Advancements in manufacturing using fiber placement, adhesive bonding, textile structures, and low-cost tooling enable the designer to fully exploit the benefits of composite materials. Naturally occurring composite structural members have evolved with extremely complex directionally dependent load paths for system performance. This is prevalent in the skeletal formation in bird wings, the damage tolerant structure of beetle shells, and the directionality of fibers in tree limb to trunk attachments. The ability to tailor material properties, exploit simplified manufacturing tooling, and incorporate advanced materials allows emerging structural designs to achieve similar levels of complexity as observed naturally.

B. Current Use of Composites

1. Current Air-Vehicle Application

Current use of composites is extensive and has provided significant benefits to military aircraft. Composite applications have been primarily limited to flat and slightly curved or doubly curved surfaces. This is primarily because these parts can be fabricated with consistent tolerance, low porosity with uniform resin content, and with limited risks associated with tooling. In addition, it has been highly desirable to fabricate smaller parts as opposed to large cocured components due to the cost of discarding structure that encounters problems during the curing operation, such as loss of vacuum pressure from bag tears. These "piece-parts" have required mechanical fasteners for assembly, resulting in limitations of design efficiency. Historical requirements of military vehicles have also dictated many access panels for subsystem maintenance, detachable wings, and ease of repairability given in-service damage. These requirements directly inhibit the designer from taking full advantage of unitized structure. Current vehicles have evolved so that many of the composite component applications are developed so that they are replaceable to prevent the elimination of the aircraft from service due to damage. In addition, historical composite concepts available have been limited to laminated structure preventing use of composites in applications that are subjected to high out-of-plane loads. At the current time, composite applications in highly loaded structure are limited primarily to aircraft wing and fuselage skins. Therefore, composite material integration has been limited to approximately 24% by weight for fighter

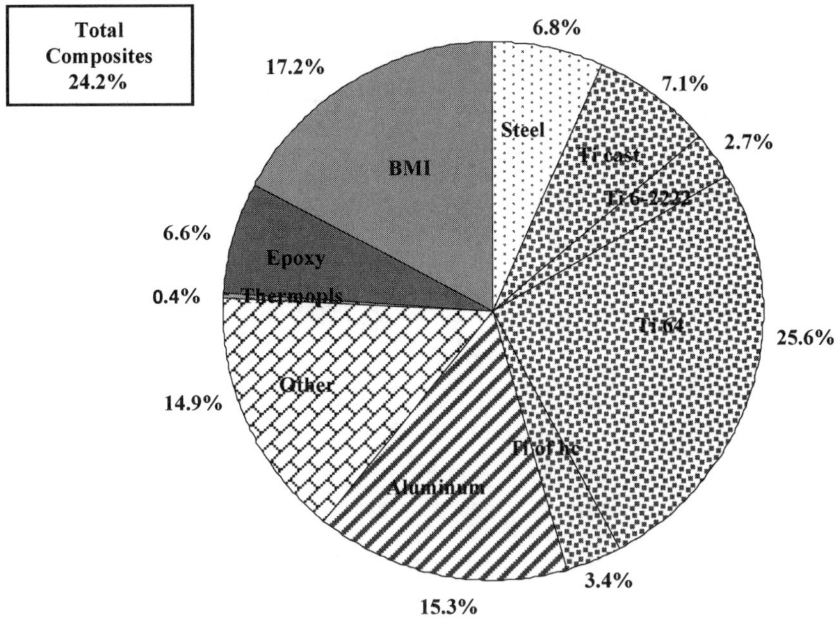

Fig. 2.2 Percentage composite by weight in F-22.

aircraft as shown in Fig. 2.2 for the Air Force F-22 fighter aircraft. However, since composites are much lighter than metals, the percentage of composite is understated in assessments by weight. Evaluating this percentage based on composite part count or in terms of density would likely show an increase in usage.

Because of loading conditions and requirements of fighter aircraft, the percent of composite has been less than that achieved in other aircraft. This trend is shown for various aircraft in Fig. 2.3.[3] Initial structural applications on the fighters shown are low due to incorporation of composites on secondary structure or structures not subject to loading environments seen in primary structural applications.

2. *Design Development Process*

To exploit the weight and cost benefits of composites, the requirements, design, and manufacturing approach must be considered in parallel. The best solution to exploit composite structures can only be achieved by modifying the design space, not through the substitution of composite for metal structure. This needs to be directed at changing load paths, modifying joining/assembly techniques, and looking at unique approaches for subsystem integration. These alternatives must be considered without adverse impact to system functionality in terms of operations and support. For this to occur, a fully coordinated design approach needs to be undertaken that is baselined and defined from the system level. This requires composites to be considered as early in the conceptual/preliminary design processes as possible so that load paths are defined that offer manufacturability and do not penalize the efficiency of the composite structure.

Before a structural concept can be considered in the development of an aircraft, it must be developed to a sufficient level of structural confidence so that

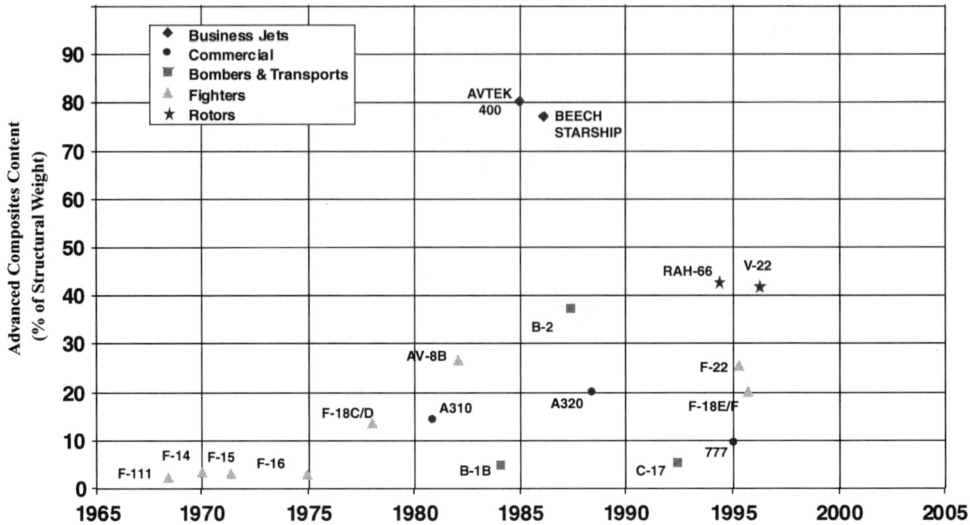

Fig. 2.3 Trends in composite usage on aircraft. (See color section.)

the risk of incorporation can be minimized. This process involves a number of "building block" steps that allow for the progressive understanding of a new structural concept. The generalized development process is depicted in Fig. 2.4. Before a structural concept can be considered for implementation in an aircraft, it must be shown to satisfactorily meet requirements defined in the building block approach. A major limitation in the development of a new structural concept is the ability to analyze its performance. Through the use of empirical testing and subsequent

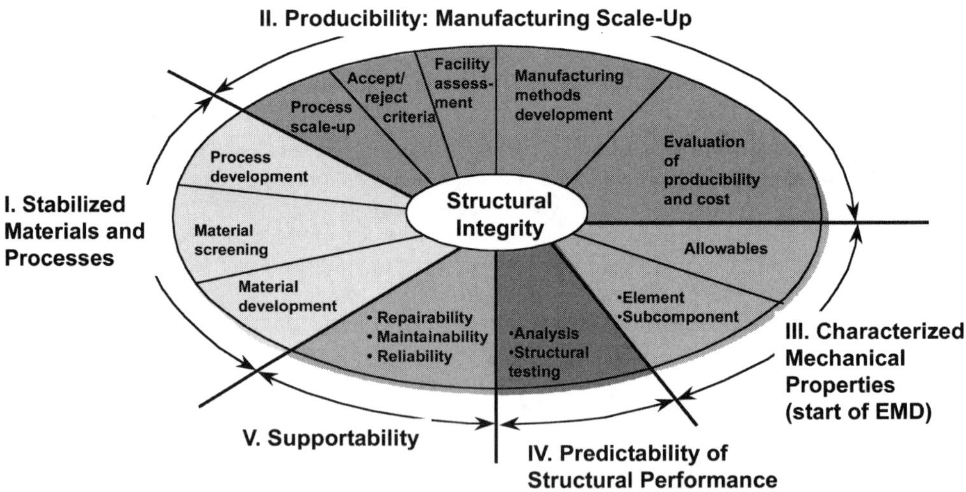

Fig. 2.4 Building block development process.

development of analyses methodology, the performance can be predicted. If a concept must rely solely on empirical efforts to develop a refined design, the cost can quickly outweigh the benefit of integration. Therefore, this building block methodology must be closely adhered to during the development of novel concepts.

The design process typically used involves three distinct stages of development, including 1) conceptual design, 2) preliminary design, and 3) detail design. During the conceptual design phase of development, the use of composites has very little impact on configuration considerations. During the initial activities within conceptual design, composites are not directly considered. Instead issues such as requirements, vehicle size, planform and layout, aerodynamics, weights, propulsion, and performance are the primary factors considered. During this process, optimization is conducted in an iterative manner to develop a refined configuration[4] that can perform the stated mission. The conceptual design is achieved by using weight factors to represent application of composite material. Therefore, this stage of the design process largely ignores the design limitations of composites such as out-of-plane strength and manufacturing limitations of new concepts. During the preliminary design phase, the structural configuration is further refined and results in a "freeze" of the structural configuration. This is provided to the detail design effort, which designs actual pieces to be fabricated and tested. During detail design, unanticipated issues with producibility, quality control, and out-of-plane loads sometimes occur, but at this time the design space has been largely fixed, preventing redesign of the composite component. This has generally resulted in the elimination of some composite applications with metallic structure used as a backup alternative. This trend has been observed to result in airframes with reduced amounts of composites between preliminary design and final detail design configurations.

3. *Benefit of Computer-Aided Design on Composite Structure*

The increased capability of computer-aided design (CAD) is assisting in further integration of composites to airframe structures. The use of CAD is allowing the ability to address the impact of all systems integrated to the airframe much earlier in the design cycle. Upon definition of the basic structural layout, the integration of weapon stores, subsystems, internal access, and the associated effects on load paths can be directly assessed. This enables the elimination of cutouts and subsystem routing that would result in high load concentrations or introduce out-of-plane load challenges. As the use of CAD practices become capable of addressing multidisciplinary issues of design, the three stages of the design process will become less distinct, with the increased ability to address composite structure shortfalls early in the design process. This will result in increased use of composites throughout the airframe.

C. Challenges of an "All-Composite" Air Vehicle

The challenges of implementation are primarily dictated by the requirements of the weapon system. If the weapon system is only intended for limited life service, quality control, required maintenance, and definition of material failure can be modified/reduced to achieve major reductions in cost. For manned fighter aircraft, a balance between requirements and cost must be achieved if a low-cost airframe

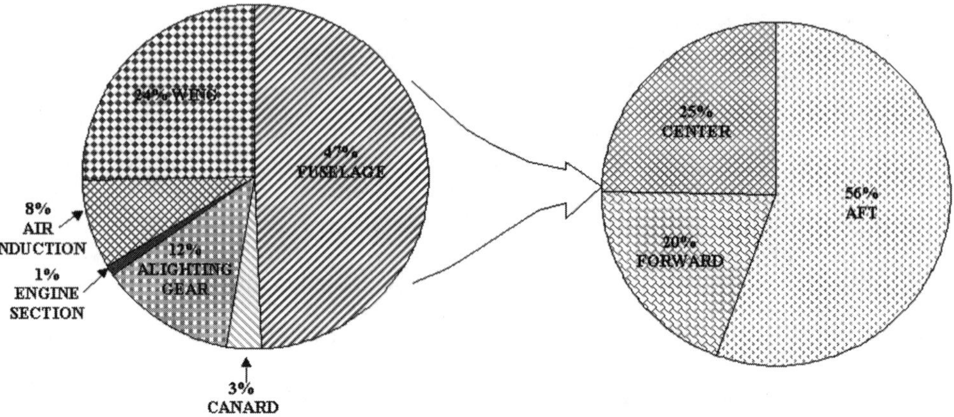

Fig. 2.5 Typical fighter aircraft weight breakdown. (See color section.)

is to be achieved. One of the largest payoffs in terms of weight from composite usage will be in the incorporation of composite to the wing–fuselage carry-through structure. The carry-through generally represents the highest loaded structure in an airframe, is the most weight intensive, and is the source of the highest cost. The fuselage of a fighter typically represents approximately 47% of the vehicle weight and the wings make up approximately 24%, as shown in Fig. 2.5 for a generic fighter configuration.

This figure shows that incorporation of lighter weight material that is structurally more efficient will yield large reductions in overall vehicle weight.[5] Fuselage and wing substructure application of composite material requires the acceptance by the user community and the associated development of key composite structure technologies to enable concepts that can be fabricated at low cost and meet rigorous loading challenges.

The airframe of a fighter is highly loaded, weight critical, durability driven, and is highly packed with subsystems requiring accessibility. Therefore, many of the materials and structures practices that exist in the civil aviation community are not directly applicable because they do not offer the performance required of modern fighter aircraft. Many functional requirements exist for "all-composite" vehicles such as ballistic survivability (which will be discussed later) that represent design "show-stoppers" if not adequately addressed. However, as discussed by Taggart,[2] key design/manufacturing challenges exist that require solutions to enable composites for use in "all-composite" fighter vehicle application.

1) Out-of-plane loading,
2) Low-cost assembly,
3) Intersecting structure,
4) Turning loads, and
5) Panel stability.

The challenge of *out-of-plane loading* is evident in composite applications where loads introduce stresses transverse to the laminate. This is one of the key reasons for composites not achieving their theoretical benefits. The planar load-carrying capability of composites (as in a skin or shear web) is primarily a fiber property dominated condition, which is well suited for composites. Unfortunately, planar loads must ultimately be transferred through a three-dimensional joint and into adjacent structure (as in a spar or bulkhead). These three-dimensional joints are subject to high shear, out-of-plane tension, and out-of-plane bending loads, all of which are matrix property dominated conditions, resulting in significant knockdowns in load-carrying capability. This has required the use of metallic substructure, which is typically attached to composite panels or skins with fasteners. The composite part must then be sized to handle bearing loads from fasteners, thereby reducing any weight savings gained by the use of composite materials. These loads tend to be the primary design driver in establishing the configuration of a wing layout. If the wing is to be constructed with composite skin and substructure, the fuel pressure design requirement introduces significant out-of-plane loading to the laminate in the joint. Therefore, pull-off loads that can be achieved in the composite joint are a primary design driver dictating the maximum spar spacing that can be used, which has a direct effect on the cost and number of spars and bulkheads required to react the loads. Later in this chapter, three-dimensional composites will be discussed to show how they are being used to increase durability and the out-of-plane strength.

Coprocessing is a critical technology enabler because it will directly eliminate fasteners, which represent a significant cost center. *Low cost assembly* through methods of coprocessing such as cobonding and cocuring make unitized structure possible. The drawback of cocuring is that it typically requires very costly complex tooling. This tooling provides tolerance control and compaction pressures that establish the quality of the final part. In cobonding, much of the cost of tooling can be eliminated because it typically involves bringing many pieceparts together in a low-cost bonding operation. However, other variables such as tolerance control need to be addressed through the use of temporary holding fixtures.

Intersecting structure occurs at many locations on an airframe. Each of these intersections introduce concentrated loads and represent sources of structural inefficiency. The conventional approach of using mechanical fasteners and cutting "mouse holes" for structure to pass through the intersection result in loads being "routed" around the discontinuity.

Designing larger, integrated components allows the designer the freedom to establish "cleaner" load paths without stress concentrations. These load paths, otherwise known as *turning loads*, can be easily established using modern manufacturing equipment such as fiber placement. However, significant work needs to be accomplished to ensure manufacturing quality and repeatability.

The need for *panel stability* stems from the desire to increase bulkhead spacing, which directly complements technology development in robust joining and increased out-of-plane strength. As bulkhead spacing increases, the upper surface compression panels become more stability critical. Many concepts for stiffened skins have been developed and several will be discussed in this chapter. The key challenge is to develop a skin-stiffening concept that offers a high level of structural efficiency, is low in cost, and has sufficient durability in the field environment. Sandwich structure concepts offer the buckling stability required; however, low-cost concepts and environmental tolerant cores need to be further developed.

III. Design Freedom Enabled with Composite Structures

Because composite material mechanical properties are defined at the same time the structure is designed, the designer has a unique freedom to tailor the performance of a structural system. This allows the designer to tailor material properties to manage loads in the most efficient manner possible, lay-down material where it is required, and establish complex geometry that would be extremely difficult with metal structures. For these reasons, composite structures have become widely adopted in the home-built aviation community. While the military research and development community has heralded these design freedom benefits, production, manufacturing, design criteria, analyses, and the lack of suitable inspection techniques have inhibited full exploitation of composites in modern military air vehicles.

A. Design Philosophy

The fundamental design philosophy of advanced vehicle concepts needs to be directed at exploiting the constitutive mechanical characteristics of composite structures.[6] This is aimed at avoiding design and manufacturing approaches that utilize composite materials in the conventional fashion of assembling large numbers of mechanically fastened parts. A design approach involving larger, integrated components to maximize producibility, quality, and design efficiency is expected to enable affordable composite airframe structures. Integrated composite structural designs eliminate machining, reduce part interfaces (eliminating drilling for bolts), reduce waste material, and offer a significant reduction of manufacturing labor hours. The conventional and unitized approach for composite design are illustrated in Fig. 2.6. The conventional approach is an assembly of mechanically fastened components, where the unitized system incorporates the use of complex skins/bulkheads with integrated features and low fastener count.

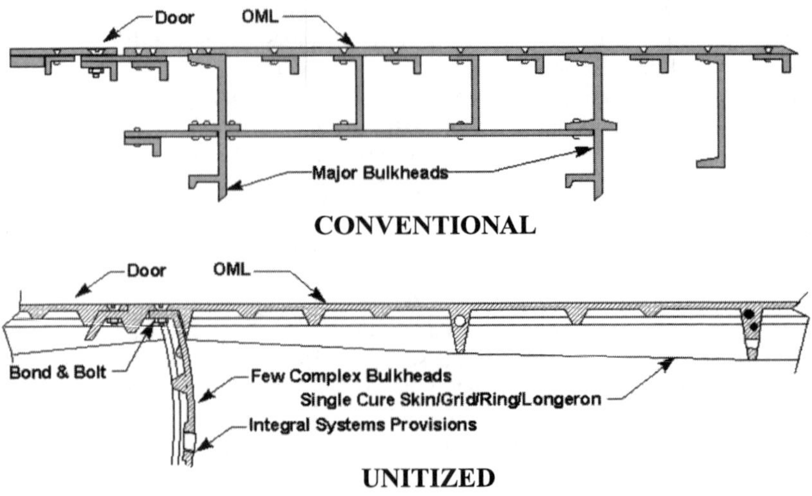

Fig. 2.6 Structural concept philosophy.

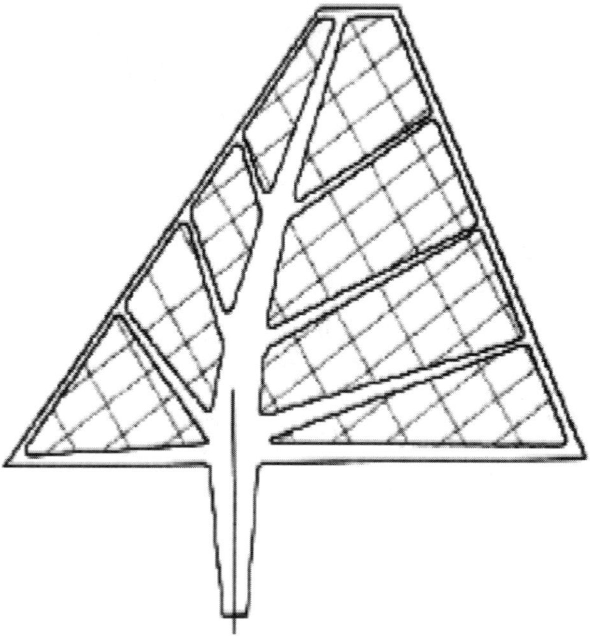

Fig. 2.7 Advanced load path management.

The designer needs to specifically seek approaches to exploit optimized load paths by using the multidirection capabilities of fibers through blending skin into the substructure in skin details, minimizing fasteners/interfaces, and providing safe, redundant load paths that offer robustness with respect to modest manufacturing flaws. One load path management strategy investigated used a "skeletal" approach to route the loads from the substructure to a central pivot shaft, as shown in Fig. 2.7, of a conceptual canard structure. The skins in this concept are further stabilized by integrating a local "geodesic grid" stiffening arrangement, enabled by placing intersecting tows of material into the cure of the skin. The viability of this concept to meet the specific requirements of the vehicle would need to be determined in a preliminary/detail design effort. However, this concept demonstrates the versatility of designing with composites.

To further demonstrate the flexibility that the designer has with composites, recent research has been conducted using nonuniform grid stiffening to exploit the directional tailoring of composite material. Fiber placement allows the ability to establish cocured stiffening arrangements to direct loads around access holes, provide skin stiffening, and incorporate integrated hard points. An example of this approach is shown in Fig. 2.8. In this concept, the skin uses a combination of small and large stiffening elements to direct primary loads and provide buckling stability, respectively. The metallic fitting shown is introduced to the composite while it is being built up, allowing fiber paths to be wrapped around the fitting for improved load-transfer capability.

The use of integrated hard points, inclusion of subsystem routing, and conduits to the airframe, is an area of design that could change the way structure is

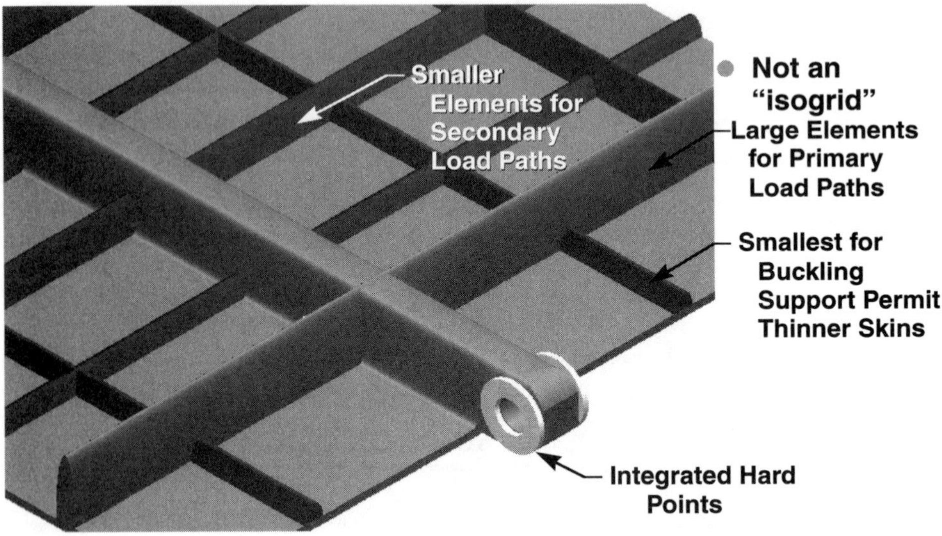

Fig. 2.8 Nonuniform grid stiffening.

viewed. Incorporating system functions into the layup and cure of a composite allows the structure to do more than carry loads. Other concepts involve designing the structure to manifold fuel to limit the number of required fuel pumps and sensors. The structure would become a "living" system, in the sense that it would provide multifunctional capability. Through the integration of subsystems to the airframe, many of the clips, brackets, and fastening requirements will not be needed, which will further enhance the efficiency of the structure. By incorporating the structural design effort into the subsystem design activity, major improvements in affordability could be achieved. These systems would no longer be independent from one another and would provide mutual improvements in efficiency.

IV. Materials and Processes: Enabling for Novel Concepts

A. Materials and Processes Overview

When designing with composites, both challenges and opportunities lie in the fact that the very material can be designed and built with the structure. Therefore, this section deals specifically with emerging materials and processes about which the designer of the future must be cognizant. Past experience has shown that successful composite structure development requires direct interaction among the designer and materials/process engineers, making concurrent engineering approaches critical for effective structure development.[7] Design freedom starts at the basic constituent (resin, fiber, interface) level, as the very determination of composite constituents should be dictated by the application and requirements of the structure. As an example, this includes characteristics such as the mechanical properties, shape, size, cost, use, temperature, and conductivity requirements but continues through the processing selection. Both the material and the process can be enabling to composite designs. The purpose of this section is merely to introduce the variety of emerging materials and processing concepts at a cursory level that may enable advanced structural concepts and not to distract the reader from the primary objective of emerging structural concept development. The reader is encouraged to explore available texts and conference proceedings for in-depth treatment of these materials and processes.

Because of the historical motivation of performance in the application of composites, the dominant material form has relied on unidirectional prepreg material—long, continuous, highly aligned, fiber-reinforced matrices, typically in tape, tow, or fabric forms. Various commingled tow product forms, discontinuous fiber forms, pultruded rods, sandwich cores, and other innovative material forms have been investigated. Each type of product form offers its own attributes or limitations and may lend itself more to one type of structural concept. As composites become increasingly accepted and better understood, material forms are being explored for the design freedoms and innovation that they offer. The key with any composite material is to process it to achieve high-quality structures (low void contents) and the final desired shape with high tolerance control. It is believed (by these authors) that as the science of design for composites continues to evolve, significant reductions in cost can be achieved through designing a structure with "sufficient quality" and "tolerance friendly" structural concepts.

B. Curing Processes

For aerospace applications, where quality and performance have been paramount, autoclave processing has been the dominant process and is often still used as the baseline for most high-performance structures. This process provides good quality parts, is reliable, and has facilitated the introduction and use of advanced composites in many applications. However, autoclave processing may have inherent limitations for certain structural concepts. Maximum part size, processing cost, low production rates, and inherent thermal stresses associated with autoclave processing pose barriers for certain classes of composite applications.

Size limitations of autoclaves preclude very large integrated structures. Autoclave processing is a harsh environment, and the temperature and pressure requirements are very demanding on tooling and ancillary materials. Autoclave tools can be costly as well because they must maintain vacuum integrity, dimensions at high temperatures, and withstand repeated thermal cycling. The nonrecurring costs associated with such tooling can be especially significant for low-volume productions where tool cost cannot be easily amortized. Analyses indicate that 60 to 70% of the manufacturing cost for a two-ship prototype program can be required for tooling development.[8] Autoclave processing is also rate limited due to the time associated with processing cycles. Vacuum bagging and leak checking can be time consuming and typical autoclave cycles last many hours due to heat-up and cool-down rates required to prevent material exotherm or tool thermal mass considerations. Given tool turnaround times, production rates of one part per day or less are typical. Small differences in heat-up rates between different sections of a part or tool due to differences in thermal mass or airflow can result in thermal gradients, which may lead to built-in thermal stresses and strains. This can be unacceptable for precision structures such as antennae and optical benches.

The benefits associated with processing out of the autoclave have been realized in many industries but have only recently been aggressively explored within the aircraft structures community due to an increased emphasis on cost. By using lower-temperature/lower-pressure curing composites, the costs of low rate production can be alleviated because tooling can be fabricated quickly out of relatively inexpensive materials. An additional benefit associated with such tools is that tooling lead times, and cycle times can be reduced, increasing flexibility in the design phase and providing tolerance for design changes. Processes that do not rely on autoclaves may provide opportunities to expand concepts into very large integrated structures beyond current autoclave size limitations as well. Other benefits and attributes that may result in more affordable structural concepts are associated with individual nonautoclave approaches as discussed below.

C. Nonautoclave Low-Temperature/Low-Pressure Curing

Nonautoclave processes of promise include low-temperature processing and radiation processing. Again, the material and processes are intimately tied because the material selected must be processed by the nonautoclave technique selected. The material must also deliver the necessary quality and structural performance, typically equivalent to autoclave/aerospace standards but are application specific. To date, most nonautoclave materials provide slightly lower properties than autoclave-processed materials, but continued development is focused on reaching the goal of mechanical property equivalence. However, available materials may provide sufficient properties for many applications at their current levels or may enable specific designs that are not possible within the autoclave.

Low-temperature processes still rely on thermal energy for the initiation of polymerization. Temperatures employed are typically low enough such that local heating or ovens suffice. Resins for these processes are typically formulated to cure at temperatures in the realm of 150°F and vacuum pressure, therefore enabling the use of relatively inexpensive tooling materials such as wood, foam, or fiberglass. Innovative approaches that are enabled with such processes and tooling materials can be incorporated to the design, as illustrated by the modular design in Fig. 2.9. This configuration features cellular structures for each of the bays that are brought together to a unitized structural configuration. The reduced part count translates directly to reduced assembly costs. Tooling required to fabricate this structure is trapped by the upper flanges subsequent to the cure; therefore, a complex tool design to allow removal would normally be required. Conventional autoclave processing would likely require a multipiece numerically controlled machined set of aluminum blocks in each of the bays, which would be very costly and have significant lead time to fabricate. The use of low-temperature/low-pressure curing composites enabled the bays to be tooled with castable expanding foam, which can be carved out after the structure is cured.

Low-temperature/low-pressure curing material systems were developed and demonstrated during the late 1990s. The mechanical properties of these systems are below those of autoclave-cured composites; however, material improvements are continuing to be developed. Data generated under a 1992 Air Force sponsored development program entitled the Low Cost Composite Processing (LCCP) program are shown in Figs. 2.10 and 2.11. Tension, compression, compression strength after impact (CSAI), interlaminar shear (ILS), and open hole tension

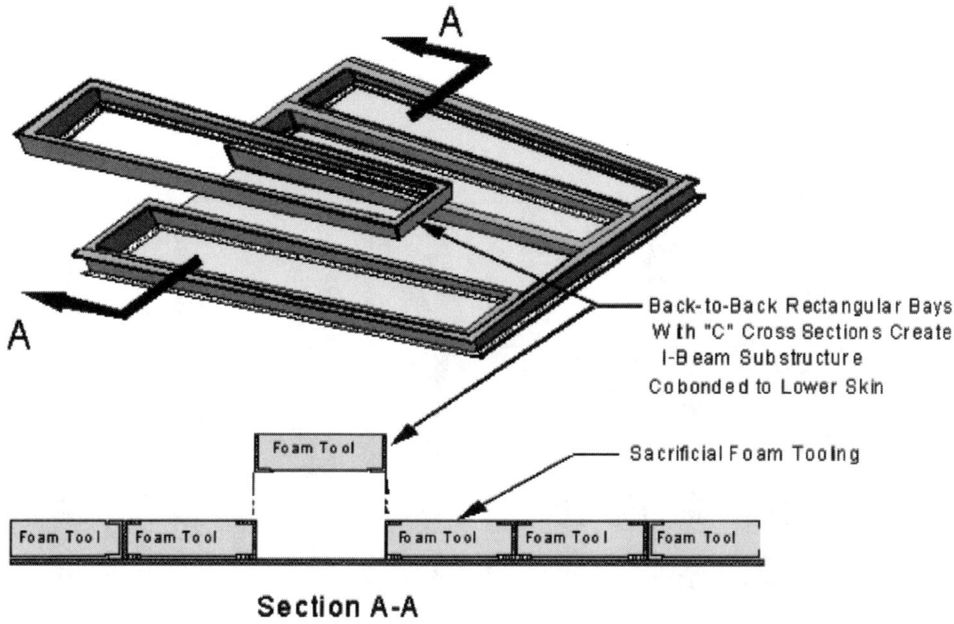

Fig. 2.9 Modular wing design enabled through nonautoclave processing.

(OHT) properties have been normalized against 3501-6 composite material in these figures.

D. Electron Beam Processing

Radiation processes have been utilized in many industrial applications to initiate polymerization. Electron beam, or e-beam, processing is a radiation process that offers unique benefits to aerospace structural designs. E-beam processing has been used since 1931 for drying oils, the fabrication of paper coatings, pressure-sensitive adhesives, flocked products, and sterilization procedures.[9] Initial adaptation of the process for curing aerospace structural composites was demonstrated by the French company, Aerospatiale, for full-scale production of rocket motor cases.[10] E-beam processes offer designers many of the same potential benefits as other nonautoclave processes, such as enabling very large structures, integration of parts, and low-cost tooling materials due to lower processing temperatures.

In the e-beam process, high-energy electrons (typically 10 MeV) are used to initiate polymerization in the matrix as opposed to thermal energy. The equipment used in electron beam curing includes an electron source, a delivery system, composite handling system, shielding, and control system. The electron source is a heated filament. Electrons are evolved in an evacuated chamber and are accelerated through a positively charged aperture. The electrons are focused into a beam that can be directed onto or scanned across the uncured composite part via manipulation with magnetic fields. The electrons deposit energy that initiate polymerization of the uncured composite almost instantaneously, making this an

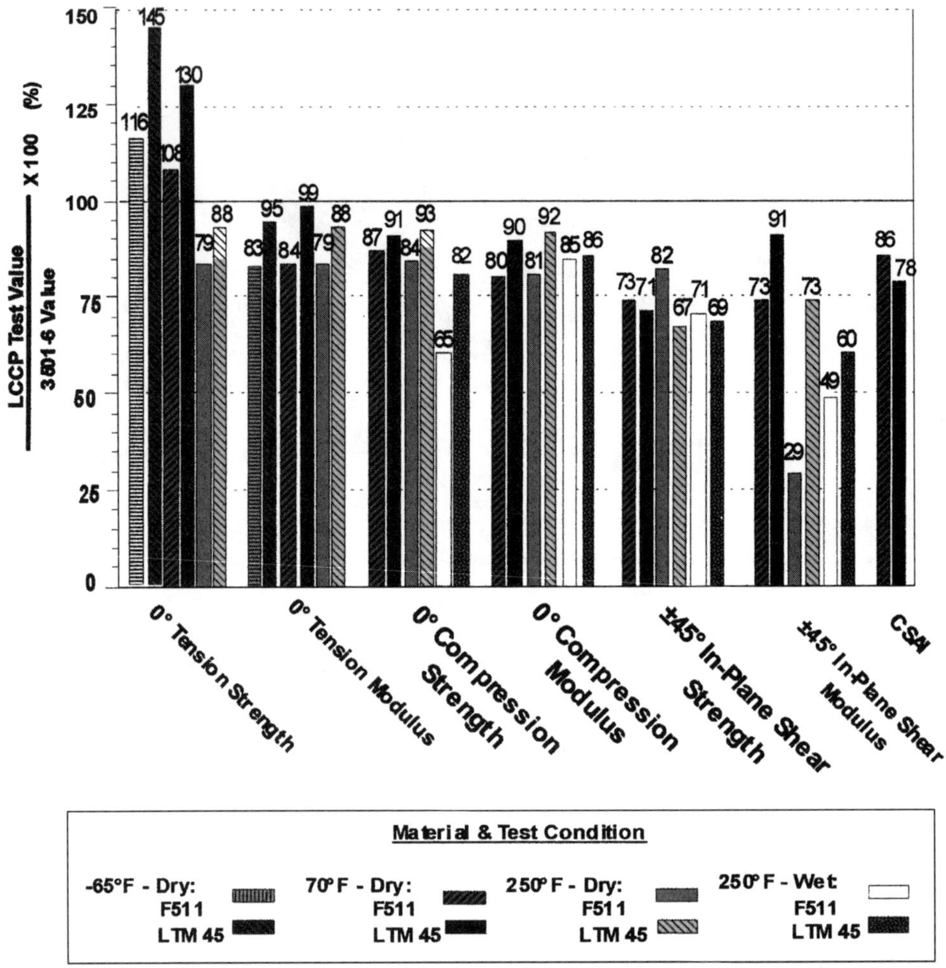

Fig. 2.10 Low-temperature low-pressure curing composites: woven material.

extremely fast process. Low-temperature increases in the composite part can occur during the e-beam cure due to scattering of electrons and exotherms. The maximum temperature reached during the process is primarily a function of the thermal conductivity of the part, part shape, energy of the electrons, dose, and dose rate (which can be varied to tailor the maximum temperature). Typical aerospace-grade composites experience maximum temperatures low enough to enable the use of simplified and low-cost tooling materials and concepts. The penetration of the electrons depends on thickness and density of the material. Typical carbon-fiber-reinforced composite materials (fiber volume fraction in the 60% range) about 1 in. thick can be cured with 10-MeV electrons. However, parts can be cured from both sides if thicker. The e-beam can also be used to generate x rays to penetrate much thicker composites. The efficiency of x-ray processing is much lower than that of e-beam and so the cure can be much slower, but thickness up to about 1 ft can be cured

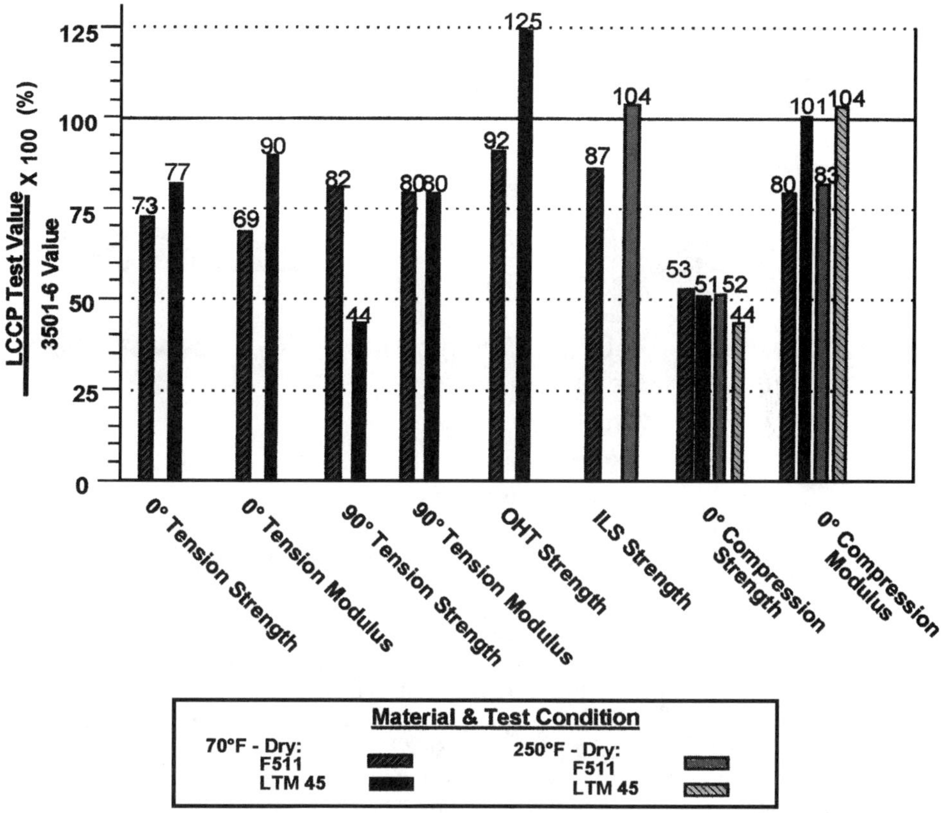

Fig. 2.11 Low-temperature low-pressure curing composites: unidirectional. (See color section.)

in this manner. The process is suited for manufacturing environments where the structure to be cured can be placed on a conveyer system (rotated or otherwise passed in front of a beam) while being irradiated with the beam scan horn as shown in Fig. 2.12.

E-beam processing also offers some unique advantages over low-temperature thermal processes: it is an extremely fast process, on the order of minutes as opposed to hours, and it is a "line-of-sight" process, which enables selected area processing. This can provide for innovative assembly techniques that could enable various design approaches using dissimilar materials processed together, exploited in large integrated structural designs. The processing parameters, such as e-beam dosage, can be tailored to achieve different processing temperatures, which make it suitable to a large variety of materials. Examples of research and development structure cured with e-beam are illustrated in Figs. 2.13 and 2.14.

E-beam processing can be coupled with many material placement techniques, such as hand layup, filament winding, resin infusion methods, and tape placement.[11,12] Tight control of dimensional accuracy and gaps can be achieved with this process enabled by bonding and assembly approaches using selective area

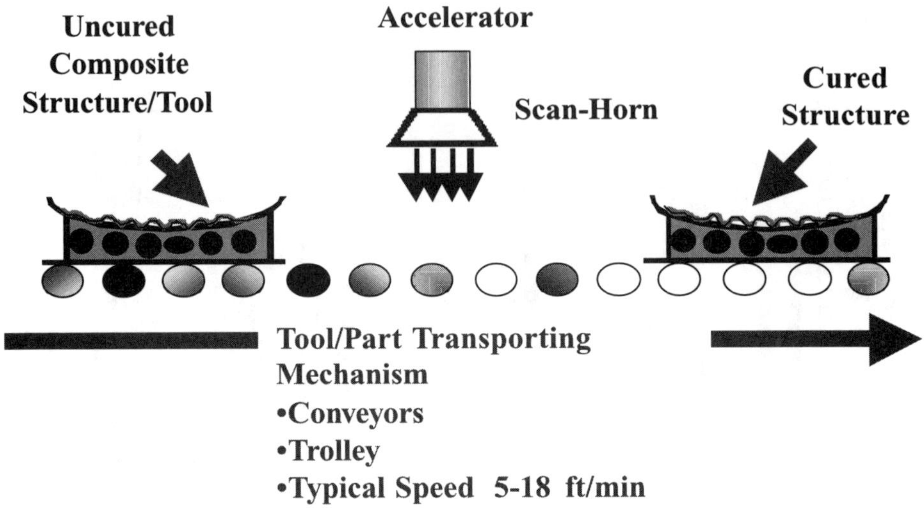

Fig. 2.12 Electron beam process.

curing, cocuring, or cobonding.[13] Selective area curing can be accomplished by masking or shielding parts of the structure from the e-beam for future mating and assembly and is illustrated in Fig. 2.15. E-beam materials can also be B-staged, which allows "piece-parts" to be partially cured to aid in assembly of complex structures as well as provide for processing flexibility.

Innovative shapes offered by composites in general can be coupled with low-cost tooling and assembly methods enabled by e-beam. An innovative application of foam tooling that uses conformable prepreg fabrics and e-beam processing is

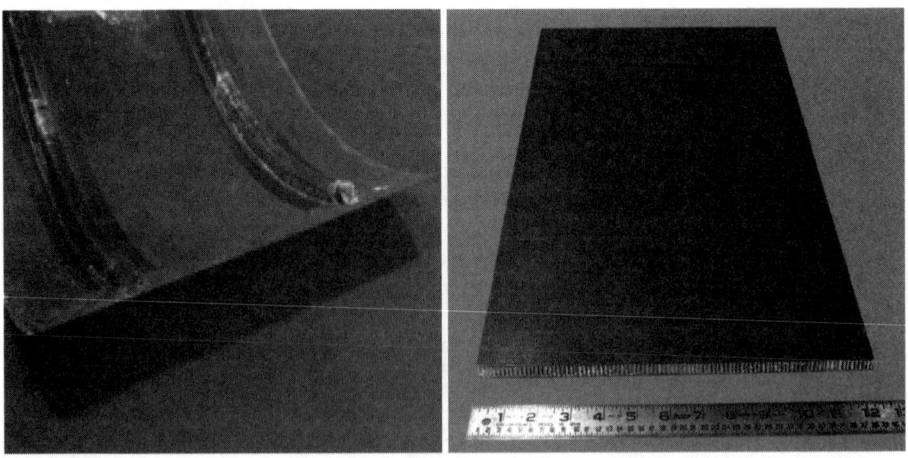

a) b)

Fig. 2.13 Coprocessing of dissimilar materials with e-beam: a) different resins co-processed and b) aluminum core coprocessed with facesheets.

Fig. 2.14 Fiber placed skin and vacuum resin infused Keelson.

shown in Fig. 2.16. Demonstration of the integration of this complex contoured structure into a subscale wing substructure is shown in Fig. 2.17. Each section was stage-cured (80% degree of cure) then assembled with e-beam processable adhesive.[12] The e-beam bonding process completed the cure of the structure and adhesive simultaneously (a clear plastic skin was used for one of the wing skins in this figure to allow for viewing of the substructure).

Current properties of e-beam curable resins are not at the level of autoclave-cured materials. However, several efforts have been undertaken to mature this technology to a level to be transitioned to the aerospace community.

E. Induction Processing

Induction processing provides a means for the curing and joining of composites at high-temperatures and pressures without an autoclave. Induction processes share some of the benefits associated with nonautoclave processes, such as the ability to fabricate large components. The process specifically avoids some of the issues associated with autoclave processing of high-temperature capable composites. Benefits

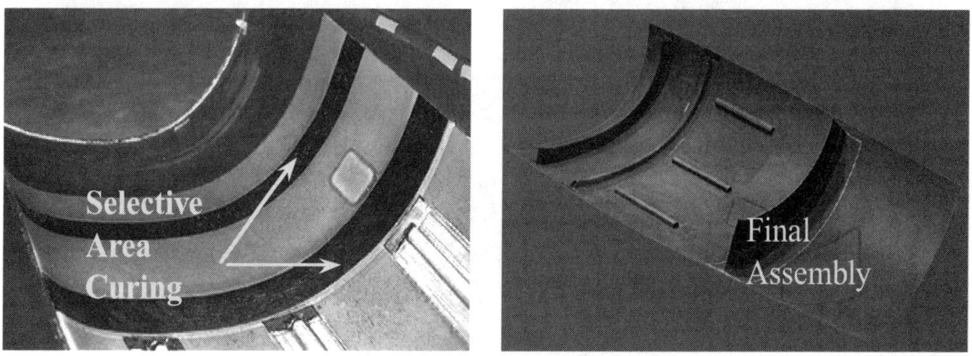

Fig. 2.15 E-beam selective area curing.

Fig. 2.16 Foam tooling for e-beam structure.

associated with this process include the processing of multiple materials or multiple parts, localized heating, rapid cycling, large part sizes, and high-temperature resin curing.

The induction process relies on resistance heating. In general, an oscillating electric current is passed through a coil, creating an oscillating magnetic field. The composite to be cured incorporates an embedded susceptor. The oscillating magnetic field induces a current in the susceptor generating local heating, providing temperatures required for curing. The susceptor remains in the composite

Fig. 2.17 E-beam enabled design and assembly concept.

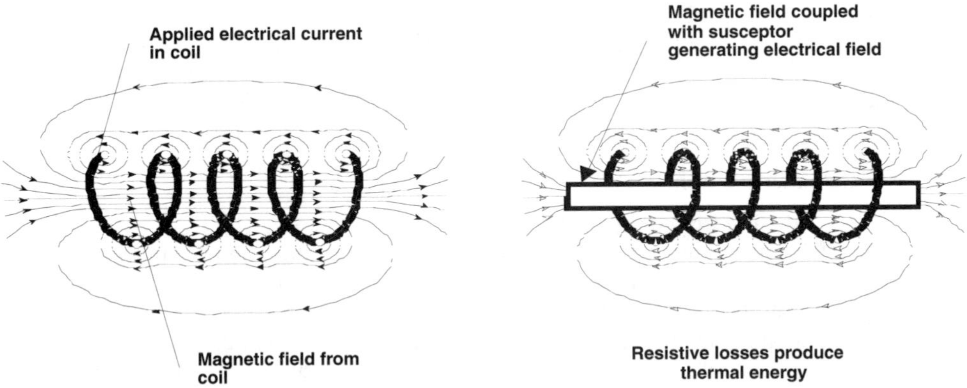

Fig. 2.18 Induction processing.

subsequent to the cure, so the designer must understand any mechanical performance implications resulting from this element. Other approaches have used a susceptor under the entire composite part to provide for large area or global heating. This process is illustrated in Fig. 2.18.

Achieving sufficient pressure for the process relies on the tooling approach. In some systems, large induction coils have been integrally cast to rod-reinforced, ceramic tools.[14] Tooling costs for such approaches that have been demonstrated are comparatively low compared to standard autoclave tooling.

Induction processing is applicable to most thermally cured composites, including high-temperature materials. High-temperature polymeric composite materials typically require high processing pressures and temperatures and long process cycles. Induction heating provides the capability to process such materials at high temperatures and pressure without many of the difficulties associated with autoclave processing. In some cases, the actual processing time can be reduced over autoclave cycles.[14] Typical autoclave and induction processing cycles for thermoplastic and bismaleimide (BMI) materials are shown in Fig. 2.19 and indicate the potential for process time reduction. This attribute may enable more cost-effective fabrication and application of high-temperature composites within structural concepts.

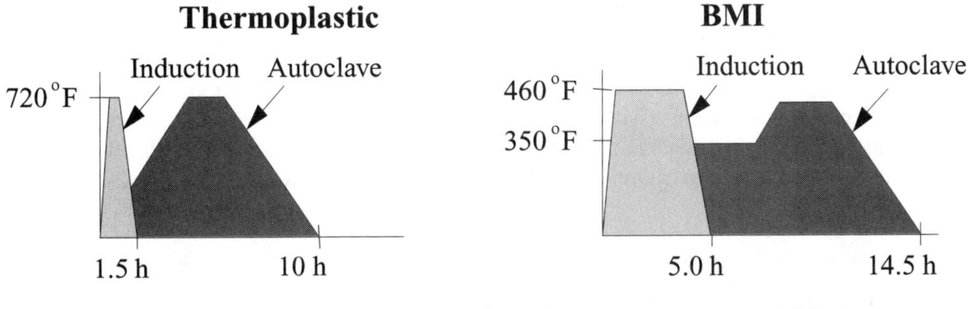

Fig. 2.19 Thermal cycle comparisons (temperatures not to scale).

The process flexibility afforded by induction processing can be exploited for fastenerless structural designs, reducing both weight and cost of composite structures. A large portion of the acquisition cost of composite structures is due to assembly costs, which can account for as much as 19 to 42%.[15]

F. Resin Infusion Methods

Resin infusion technology has been utilized in commercial applications for over 20 years for its low-cost benefits associated with high-rate production and the net shape/low scrap aspect of the process. It has recently been considered a process of great potential and benefit for the aerospace industry. Benefits associated with this process include net shape and close tolerance parts, which can result in lower assembly costs, potential for low material costs and processing, reduction in need for ancillary materials, and the ability to fabricate structures of high complexity. In most performance-driven processes, a dry fiber or reinforcement preform is designed, it is placed into a tool or mold, the tool is closed, the resin is injected, and then it is cured. Variations exist on the temperatures, pressures, and types of tooling that can be used.

1. *Resin Transfer Molding*

With resin transfer molding (RTM), dry or tackified fiber preforms are placed into closed molds and resin is injected under heat and pressure. Although it is a process that has been used for many years in industrial and sporting goods markets, the application to aerospace parts has resulted in significant changes to the process. RTM has been shown to produce parts with fiber volumes between 55 and 65% with properties commensurate with autoclave-processed materials. RTM enables complex part shapes and geometries that can incorporate ply drop-offs and controlled fiber orientation, resulting in structures that maximize load-carrying capability.[16] RTM can utilize a variety of reinforcement preforms such as two-dimensional fabrics or three-dimensional braids and weaves. Developments in preform technology could enable highly tailored structures and the potential for improved out-of-plane loading capabilities. Foams, cores, structural members such as stiffeners, and other inserts can be integrated within the preform to enable integrated designs, with the flexibility of thick sections. The process offers the opportunity for consolidation of many small parts into larger integrated structures for reduced assembly cost. Various models and monitoring processes have become available that can sense the degree of flow and cure[17] and allow for real-time monitoring of cure progression. Many secondary processes such as shimming and machining can be eliminated through excellent dimensional control and tolerances that are achieved. Tooling costs can be high if low volumes are required because the process requires matched metal tools that can withstand the temperature, tool pressures, and high-pressure, resin injection. RTM structures can also be size limited by tooling mass and resin flow limits. Infusion approaches that are not limited by these constraints are currently being explored and will be discussed below.

2. *Vacuum-Assisted Resin Transfer Molding*

Vacuum-assisted resin transfer molding (VARTM) enables resin infusion under vacuum pressure only and is often coupled with resins that can be injected and cured at low temperatures. This coupling of low temperature and low pressure alleviates many of the costs and limitations associated with conventional RTM, especially those associated with tooling. The process has its roots in nonaerospace industries, and one example is SCRIMP™. In the VARTM process, a vacuum bag is used for one tool surface as opposed to a closed mold. Resins are then infused to dry fiber preforms using vacuum pressure. This process enables complex shapes and geometries and integrated buildups and allows inserts and very thick sections similar to RTM. Because only vacuum is required to infiltrate the preform with resin, low-cost tools can generally be used. Fiber volume fractions can be low but recent developments suggest that fiber volume fractions will continue to improve, offering the potential for high properties in the future.

3. *Resin Film Infusion*

Resin film infusion (RFI) uses plaques of resin that are applied to a plane of fiber preforms, then heat and pressure are applied so that the resin infuses through the thickness of the preform. Low-temperature curing resins are typically employed. RFI enables moderately complex shapes, primarily limited by what type of preform can be laid up. RFI can also enable unitized structures such as integral spar caps, therefore low assembly costs due to reduced part counts. Integrated cores and elements are possible as with the other infusion processes. The utilization of grouped ply-stacks to obtain high material lay-down rates has been demonstrated and provides for low handling costs due to the ply packs. Stitched preforms can be used to obtain improved damage tolerance and through-thickness properties.[18]

G. Innovative Reinforcements—Discontinuous Fiber Product Forms

The majority of composites currently used rely on highly aligned, continuous, fiber-reinforced prepreg materials that are layed up. However, there are many innovative textile reinforcement materials that can be exploited that provide both cost and weight savings in airframe structures. Braided fuselage frames, woven/stitched fuselage panels, stitched multiaxial warp wing skins, and braided stiffeners have all been demonstrated.[19,20] Many of these emerging structural material concepts are discussed in this chapter. Examples of some other methods of reinforcement are also provided to show the breadth of possibilities beyond the long, continuous, aligned fiber reinforcement.

1. *Discontinuous Fiber Product Forms*

Discontinuous fibers were investigated in the 1980s and demonstrated potential for cost-effective manufacturing of complex structural parts. Discontinuous fiber prepreg materials such as long discontinuous fiber thermoplastics from DuPont demonstrated improved formability of complex beaded structures and the potential for lower fabrication costs. To obtain the theoretical maximum load-carrying capability, the fiber length must be greater than a critical value that is dependent on the fiber diameter, tensile strength, and the fiber–matrix interface strength. Variables such as the process used to create discontinuous fibers and the quality of the processed laminate will dictate if the theoretical strength is achieved.

Processing advantages associated with discontinuous materials make them attractive for near-term application on secondary and lightly loaded structures. As performance improves with continued development, primary structure application may be viable. Recent developments have investigated the use of oriented discontinuous fibers in preform manufacturing for resin infusion processes.[21,22] Benefits of such preforms include the potential for dramatically lower costs when automated processes are utilized, applicability to corrugated and beaded structures, very large structures, and embedment or integration of various reinforcing elements such as metal inserts and fiber preforms.

An emerging discontinuous fiber product form is the programmable powdered preform process (P4A). This process produces discontinuous fiber preforms for subsequent resin infusion and cure. P4A is an automated process that chops and sprays fibers onto the tooling surface. This process allows for real-time changing of fiber alignment and length at the preform level. This offers the designer the ability to provide local reinforcement in a structure without the cost and labor-intensive activity of manually orienting individual plies to achieve desired structural properties. After a preform is "built-up" from the spraying (material lay-down) operation, proven techniques of resin infusion can be used to produce complete structural articles. Because this process has the potential of being fully automated using numerically programmable systems, the risk of human error involved in constructing a preform can be greatly reduced. A panel with bead stiffeners is shown in Fig. 2.20 that was successfully fabricated using the P4A process coupled with the VARTM resin infusion process.

H. Material Lay-Down Methods

Material placement techniques can provide design flexibilities or limitations in themselves. For prepreg materials, hand layup has been the historical benchmark

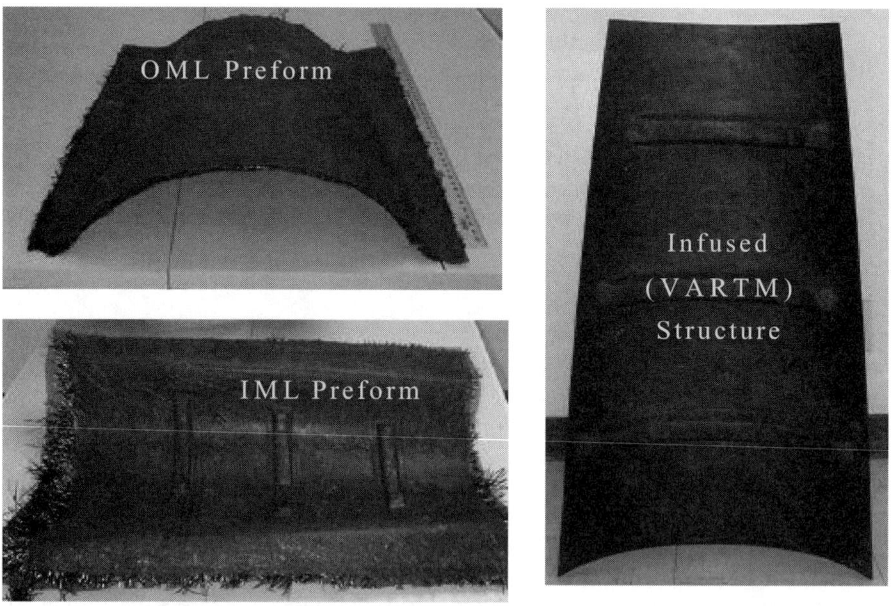

Fig. 2.20 Discontinuous fiber preform (P4A). (See color section.)

and has associated high labor costs and limitations. Hand layup can still be an appropriate and cost-effective method for certain structures, such as mildly countoured structures of low or limited production numbers. Laser ply alignment can assist this process and improve quality, but the process still requires significant labor. There are a variety of automated techniques for the placement of prepreg material products. Any of these techniques could conceivably be coupled with a secondary cure operation or an in situ process for potentially lower cost fabrication.[23]

Filament winding is a process that has been used over the years primarily for cylindrical sections such as rocket motors. The process consists of winding the reinforcing tow or other prepreg product form onto a mandrel for subsequent curing. Wet winding can also be performed, in which dry fibers are fed through a resin bath before winding, which can be a very rapid and automated process. High control over orientation and uniformity is possible. Winding can be performed over various structural elements such as bulkheads or ribs to create integrated cocured assemblies. Isogrid and other geodesic winding patterns can be very easily formed using this process. However, shapes are typically limited to bodies of revolution, and very low or very high-fiber angles cannot be wound efficiently.

Automated tape placement is a process in which tape prepreg, typically about 6 in. wide, is placed and compacted onto a tool by the tape placement machine head. It offers low cost through automation and low scrap and is applicable for mildly contoured surfaces. Plies can be dropped off or added to achieve precise ply thickness control where required. Automated tow or fiber placement differs from tape placement primarily in the prepreg used, which is tow–preg as opposed to tape. This smaller product form enables increased part contours than what can be achieved with tape.

The majority of the above has discussed materials and processes based on thermoset polymers. This is due to the fact that current aerospace structures rely mostly on thermoset polymer systems. However, thermoplastic composites offer many unique attributes such as high toughness and impact resistance as well as unique processing methods that also enable affordable structural concepts. Impulse and resistant welding, consolidation on the fly, stamping, thermoforming, and diaphragm forming are a few processes that are enabled by the unique attributes of thermoplastics. Some of these processes have been demonstrated to provide innovative and affordable design concepts. Diaphragm forming of thermoplastics enables the ability to form structurally efficient shapes such as beaded stiffened concepts that can provide weight and cost savings over other designs. This was effectively demonstrated on the F-22.[24] Robotic placement of thermoplastic tape or tow coupled with in situ consolidation offers the potential for considerable cost savings through high rate capability, low touch labor, and moderate structural complexity. Studies have shown the ability to process at rates of 0.0126 kg/s (100 lb/h[25]), are possible.

These processes were provided as examples to express the fact that the selection of the material lay-down process as well as the fabrication process in general for composites may enable certain structural concepts. The selection of the particular process depends greatly on the structure's complexity, size, and quantity (as well as equipment availability). Bringing the material and process early into the design phase can provide for affordable high-performance structural concepts.

V. Emerging Structural Concepts

A. Three-Dimensional Composites in Aircraft Joints

The use of three-dimensional composite structural concepts will enable effective management of out-of-plane loads. This key feature opens the door to novel composite airframe design with both breakthrough performance improvement and cost reduction. Composite integration in state-of-the-art fighter air vehicles has been generally limited to under 25% by vehicle weight, as previously discussed. The primary barriers to increased composite utilization are fabrication and assembly cost, overly conservative design criteria, lack of high-fidelity design tools, poor ballistic survivability, and poor interlaminar/out-of-plane mechanical properties. The use of three-dimensional composite structural concepts will help to eliminate these barriers.[26,27] However, the propulsion community has successfully incorporated the use of highly tailored structure for helicopter blades, turbine fan blades, and propellers for civil aircraft to minimize these barriers. These structures wrap/integrate material around fittings at the root of the components, optimizing the material capability by improving load transfer.

A variety of structural concepts based on complex, yet low-cost, woven and braided fiber textile reinforcements are currently under development.[28] Intersecting structure with continuous fiber reinforcement through the intersection and woven three-dimensional fiber preforms for highly loaded composite wing carry-through bulkheads are being investigated. The development of z-direction or through-the-thickness (Z-Fiber™) reinforcement to greatly increase the interlaminar strength and toughness of laminated composite structures is enabling the local tailoring of out-of-plane strength. Emphasis is being placed on coupling complex textile based fiber architecture with low-temperature RTM, resin film infusion, and e-beam curing materials.

1. Three-Dimensional Woven and Braided Composite Technology

Three-dimensional woven and braided preforms can be fabricated into many versatile configurations, offering the designer significant freedom in developing a structure. To date, preforms have offered the best performance in joints where load concentrations and the need for toughness are the greatest. Previous attempts to build entire assemblies from woven/braided structures have resulted in inefficient designs because of the loss of in-plane properties. Therefore, by using laminated materials in skin structures for superior properties and integrating textiles to the joints, a highly efficient, durable, and cost-effective design can be developed. An example of a woven preform for a joint application is shown in Figs. 2.21 and 2.22.

Joint design concepts have been developed for application in center and aft fuselage components. Configurations in development vary, depending on part loading and assembly requirements. Bulkhead-to-skin joints may incorporate bulkheads with integral attachment flanges or bonded-on, double-flanged pi-clips (consolidated woven preforms) for bolting to the skins. Lightly loaded substructure (e.g., longitudinal shear webs and intermediate spar webs) attached to the skins may be handled through single-flanged, preform clips that are either precured and embedded in the skin covers or cocured with the skins. Special cruciform preforms can be fabricated to transfer load through intersections such as in shear webs to bulkheads. These preforms can be woven to make the fibers continuous through the load path intersection and provide a location for attaching parts during assembly.

Fig. 2.21 Woven three-dimensional "PI" preform for joints.

Because joint designs are driven by the assembly sequence to be used, the overall manufacturing assembly plan must be formulated early to aid selection of the joint concepts.

Woven three-dimensional preforms can be produced by a number of techniques, including the following:

1) *Orthogonal pattern*: z-direction reinforcements are pulled through the warp (0° direction) and fill (90° direction) fiber, intersecting the layers at a 90° angle.

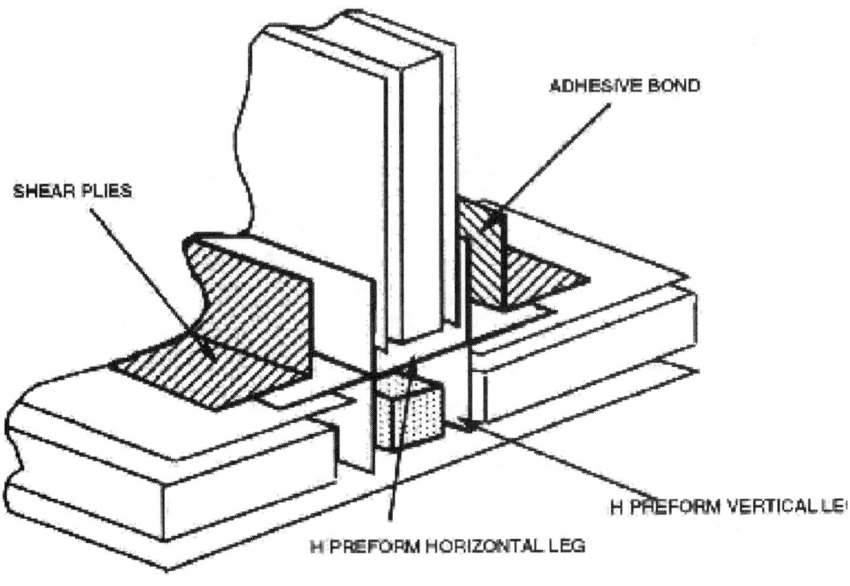

Fig. 2.22 Woven preform joint integration.[29]

2) *Through-the-thickness angular interlock*: z-direction fibers are woven through the layers of the fabric, intersecting the layers at an angle. These z-direction fibers may be the same type and filament count as the warp and fill fibers or tailored for the application. In some cases, a finer fiber is preferred for through-thickness reinforcement because it allows tight packing of the warp and fill fibers.

3) *Layer-to-layer interlock*: warp fibers are woven into adjacent layers within the weave at designated intervals. This process improves interlaminar shear strength but does not provide true through-thickness reinforcement.

Braiding technology for three-dimensional preforms[30] offers extensive capability to develop highly complex geometries. Several types of braiders may be employed for preform production, including round, flat, and cartesian. Braiding allows fiber architectures with 0° and +/− degree angle fibers. Fiber angles are generally controlled by the takeup speed of the preform at the formation point, and preform thickness can be varied with the fiber tow size. Mandrels are used for braiding a wide variety of random net shapes. Flat braiding allows incorporation of continuous fiber intersections for grid networks. Flat and cartesian braiders allow integrally braided cruciforms and round braider preforms can be stitched and shaped into a variety of sections such as Ts and Is. A wider array of fiber orientation and distribution is achievable with braiding compared to weaving; however, the properties for each process and application must be understood before selecting a method.

To make three-dimensional woven and braided preforms more affordable will require standardization of shapes (T, cruciform, special shapes) in stepped sizes for various loads. These could be standardized in a manner similar to aluminum extrusions. To reduce storage costs, preforms can be impregnated with resin just before cocure into laminated composite structure. Reducing the cost of three-dimensional preform weaving will require automation of more complex shape preforms and the addition of bias (+/− 45° plies) for efficient shear of loads into web structure.

2. *Z-Fiber™ Reinforcement Technology*

The "Z-Fiber™ reinforcement" process incorporates through-thickness reinforcement to preimpregnated and/or staged composite details. The Z-Fiber™ reinforcement process is illustrated in Fig. 2.23. A foam block containing Z-Fibers™ is placed on the component before curing at specific locations where damage tolerance and/or through-thickness reinforcement is desired. Z-Fibers™ are small composite pins made by pultruding unidirectional tows through a die, and can be made from fiber reinforcement and matrix as required for the application. Pressure is applied to the foam block, allowing compaction, resulting in the Z-fibers™ to be pressed into the component. The required pressure to allow foam compaction has been demonstrated using autoclave pressure, and handheld portable ultrasonic devices.

Alternative methods of Z-Fiber™ insertion are also under development, which will enable improved incorporation to the manufacturing environment. Figure 2.24 shows an approach that uses an ultrasonic insertion method with a numerically controlled gantry system. This system allows for repeatable placement of Z-Fiber™ reinforcements, in a "hands-off" operation, to reduce cost of insertion and improve quality of the three-dimensional laminate.

Z-fiber™ reinforcement can be used in damage-prone areas such as blade and hat-stiffened laminates, in monolithic/sandwich joints where three-dimensional

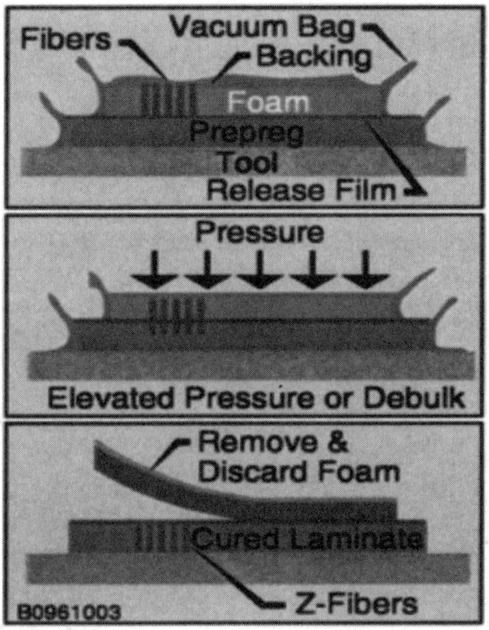

Fig. 2.23 Z-fiber™ insertion.

Fig. 2.24 Automated Z-fiber™ insertion.

woven preforms and two-dimensional materials intersect, and as edge reinforcements to prevent delamination. Z-Fiber™ reinforcement provides a number of advantages in composite structures. Empirical testing has shown only minimal decreases in in-plane properties compared to two-dimensional components with no through-thickness reinforcement. Testing has shown that, in typical applications, Z-Fiber™ reinforcement provides a 30- to 50-fold improvement in interlaminar fracture toughness. In quasi-isotropic laminates using Z-Fiber™, tensile strength degradation using 20-mil-diam pins is 3.5% when the insertion area is 4%. In the case of compression, strength degradation is about 5%, with 4% insertion area coverage. Z- Fiber™ reinforcement is especially important for hat- and blade-stiffened structure where Z-Fibers™ improve initial and ultimate pull-off failure loads. In addition, they provide significant improvements in damage tolerance and ballistic survivability improvement. One of the key attributes of Z-Fiber™ technology is that transverse pins inhibit catastrophic failure that is typical in composite cocured joints after initial failure occurs. Generally, as the interface between the upstanding leg and the skin laminate of a joint starts to fail, it "unzips" in a rapid fashion, completely losing all load-carrying capability. The performance of a typical z-pinned subelement is shown in Fig. 2.25. The fail-safe characteristic shown after the onset of failure offers a new opportunity to improve ballistic survivability[31] and eliminate redundant fasteners included in bonded joints that are intended to provide a secondary load path if initial failure occurs.

Z-Fiber™ integrated to a typical blade-stiffened structure that would be fabricated using a low-cost approach employing precured webs and "green" or uncured skin and preform elements is shown in Fig. 2.26. In addition, Z-Fibers™ will eliminate the need for radius blocks and fasteners at the stiffener ends where high peel loads are commonly encountered. These improvements will prevent catastrophic failure from propagation of a manufacturing flaw or in-service damage.

The use of three-dimensional structures will allow unitization, e.g., reduced fastener and part count. In addition, long-term support costs may be reduced since the majority of repair actions in fielded fighter air vehicles are conducted on

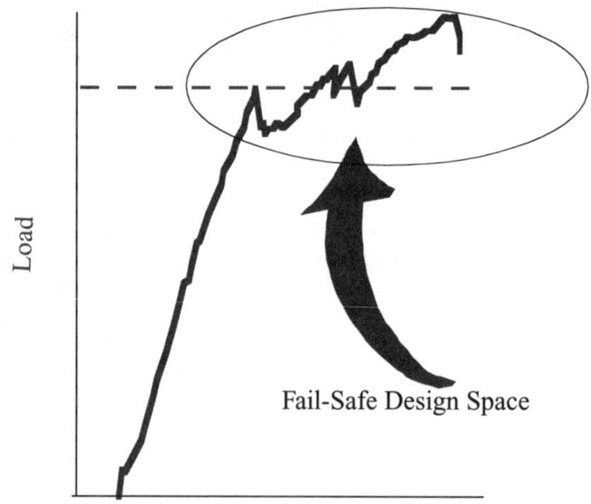

Fig. 2.25 Typical Z-fiber™ joint failure.

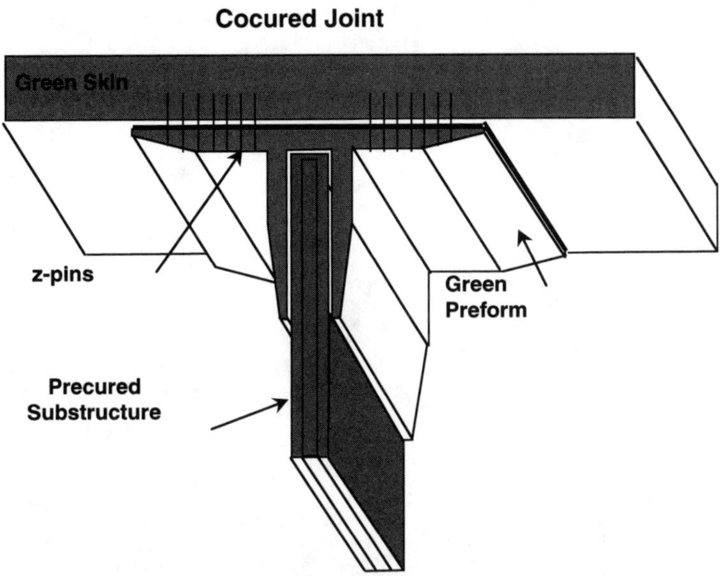

Fig. 2.26 Z-fiber™ insertion in a cocured blade stiffener.

mechanically fastened joints, which will be reduced through the use of unitized structure. However, for this to become possible, changes to failure criteria must be established that allow for fail-safe three-dimensional joints with initial failure to be considered safe for flight.

B. Three-Dimensional Composite Application for Bulkheads and Frames

Composite utilization in the heaviest parts of the airframe such as highly loaded fuselage bulkheads and moderately loaded frames has been virtually non-existent because of poor interlaminar properties of conventional laminated or two-dimensional reinforced composites. Because of these low interlaminar properties, composites have not been used in structures subjected to high out-of-plane or complex loads without costly integration of metallic fittings and fasteners. The development of composite structures that can react out-of-plane loads will greatly reduce the reliance on bolted joints and will offer improved structural efficiency. A moderately loaded bulkhead that reacts engine hammer shock and longitudinal body bending pass-through loads with three-dimensional preforms is shown in Fig. 2.27.

In this bulkhead, preforms are staged and then cocured with conventional laminated composite details during final cure. The staged details are net shape before the final cocure assembly. Three-dimensional fibers in the out-of-plane direction are capable of carrying the high engine hammer shock and fuselage bending pass-through loads without costly machining of fittings and fastener installation. Insert data indicate that three-dimensional inserts carry approximately twice the load of the baseline laminated configuration.

To achieve a structure with optimal performance, future design concepts should consider assemblies that use available three-dimensional concepts together to

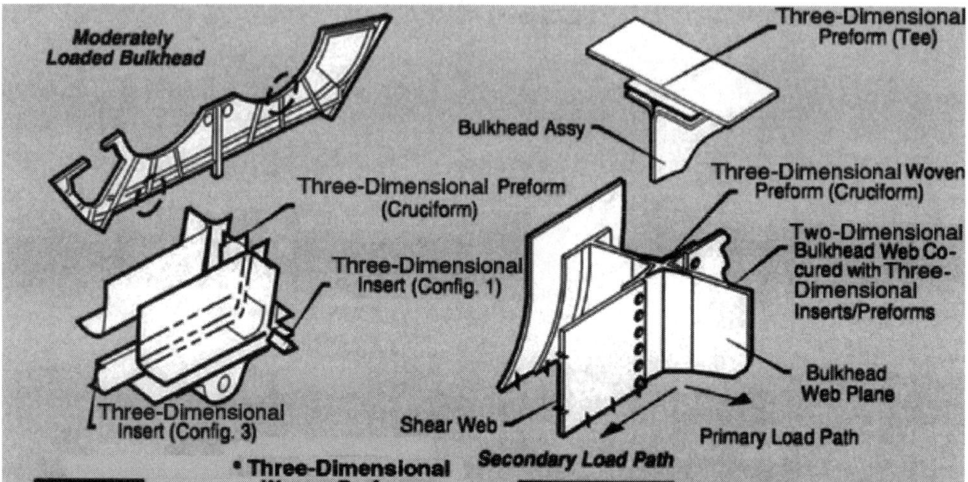

Inserts

Configuration	Material	Performance, lb/in.
1	IM7/977-3 (uni-tape)	540
1	IM7/PR500 (RTM)	825
1	IM7/977-3 (solvent)	989
3	IM7/PR500 (RTM)	1142

Cruciforms

Load Direction	Joint	Performance, lb/in.
Bonded Joint (Baseline)		1900
Primary	Flat	6100
Secondary	Flat	5100

Fig. 2.27 Laminated composite bulkhead with embedded preforms.

improve skin and joint durability. As shown in Fig. 2.28, Z-Fiber™ can be used in the skins to inhibit delamination due to impact events and low-cost textiles with woven intersections used to react the local concentrated loads in the stiffeners. By using continuous intersections made with textile, discontinuities are eliminated, resulting in improved overall structural efficiency and the hand work that is involved in cutting "mouse-holes" is eliminated.[32]

C. Bonding of Primary Structure

As previously discussed, one of the key challenges of affordable composite structure is low-cost assembly. Bonding has always been an alternative to cocuring that designers have considered, primarily because it largely eliminates the dependence on complex costly tooling. However, application of bonding in production military air vehicles has been extremely limited due to the many variables that are introduced to the assembly sequence. These variables include void content in the bond line, quality surface preparation to ensure good adhesion, control of bonding pressures, tolerance control of the bond lines, nondestructive inspection criteria, and fail-safe failure characteristics if bonds become damaged during service and go undetected. Bonding can be used in many different structural configurations to reduce the cost of assembly. Bonding has been extensively investigated using a single lap bond but these joints have only been considered applicable to very lightly loaded structures due to the presence of high peel loads that are introduced from the eccentric loading of the adherends. An advanced bonded joint with

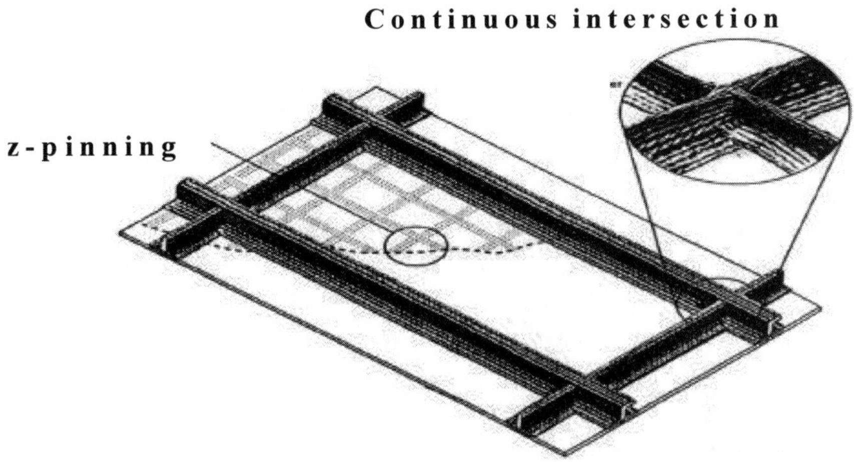

Fig. 2.28 Continuous intersection textile stiffeners with Z-fiber™.

precured skins and webs assembled in a bonding operation with a preimpregnated textile preform is shown in Fig. 2.29. This joint offers significant improvement in durability due the incorporation of the textile. The joint incorporates a double lap bond with beveled adherends to further reduce peel stresses in the bond line and improve structural performance. However, the skin is still held in place with the adhesive bond alone so, if damage goes undetected during manufacturing or occurs in service, it would be prone to failure.

Other approaches to fabricating joints of this nature have also been evaluated. This includes approaches to fabricate the skin and double lap clevis in a single cocured element and then assembling a precured web into a bonded clevis joint. This approach takes advantage of the ability to fabricate large cocured skins with complex details but requires that the tolerance/positioning of the clevis be "exact," otherwise large bond line thickness discontinuities can occur. If the clevis is misaligned relative to the web structure, the adhesive will be forced to one side of the joint. The web creates a "snowplowing" effect, resulting in the bond line on one side to be thin and the other side to be thick.

Advanced bonded joints have been demonstrated to offer structural performance similar to bolted joints. The response of a bolted and bonded sandwich joint test is shown in Fig. 2.30. These joints were loaded using simply supported end conditions on the horizontal legs and pulled in tension by clamping to the upstanding leg. The width of the specimens was designed to represent the expected flexure that would be present during joint loading in a wing structure from internal fuel pressure. Both of these sandwich joints include the use of textile "pi" preforms, and have been sized similarly for comparative purposes.

The bolted joint experienced a noticeable initial failure early in the loading profile at approximately 1950 lb. This was believed to be from initial bearing failure in the matrix material of the hole. Subsequent to this event, the two specimens carried load in a linearly elastic manner, with the bolted configuration showing some early decrease in stiffness due to the effects of bending. The "yield" point of the two specimens is approximately the same, which is based on the initial onset of failure, but the postinitial failure characteristics are very different. In the case

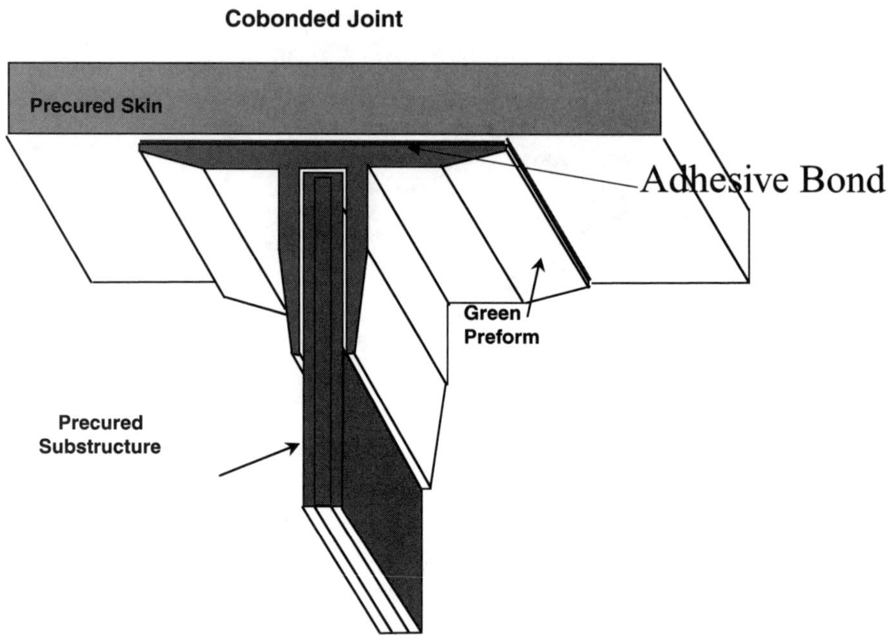

Fig. 2.29 Adhesively bonded clevis joint.

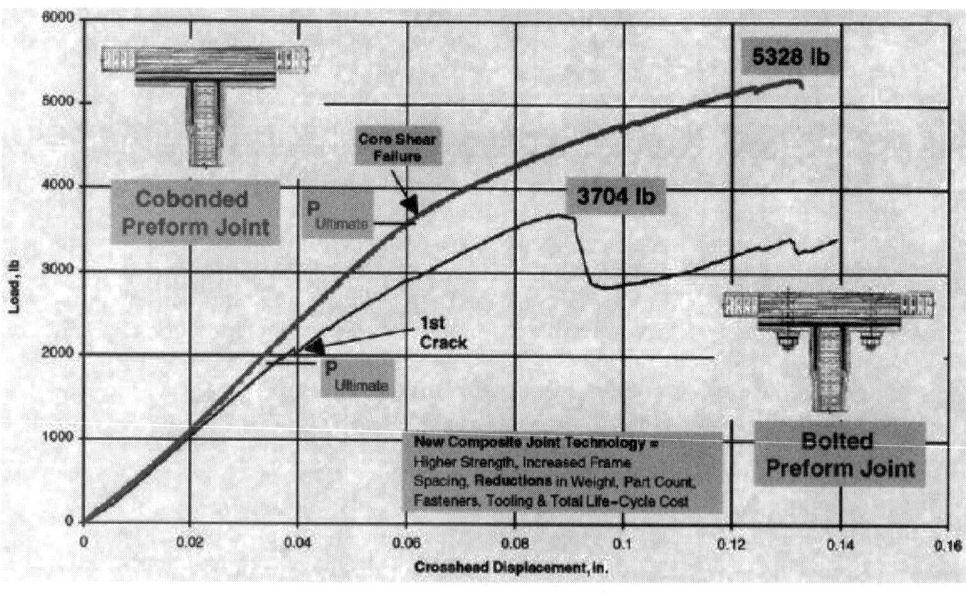

Fig. 2.30 Bolted and bonded joint pull-off comparison. (See color section.)

of the bonded joint, the sandwich core adjacent to the hard laminate under the foot of the upstanding leg experienced core shear failure. This resulted in a controlled progressive failure characteristic, indicated by the stiffness of the bonded joint gradually leveling off. It is interesting to note that the bonded joint achieved a significantly higher ultimate strength, and its mode of failure indicates that fail-safe performance, improved ballistic survivability, and crashworthiness may be possible. This testing has shown that bonded joints can be designed to offer a high level of performance compared to conventional bolted joints. Investigations of this nature have demonstrated that the design of a bonded joint must consider how the skin effects the failure mechanism. In the case of this sandwich test, the failure characteristics were largely governed by the core failure, enabling a progressive failure. It has also been shown that, when monolithic bonded joints are tested, the skin stiffness and support span (which governs bending effects) can alter the modes and mechanisms of failure. For primary bonded structure to be adopted into advanced design practice it will be necessary to develop guidelines to govern the selection of skin thickness and layup so that consistent performance can be predicted.

The versatility of curing complex shapes from composite and completing assembly with bonding gives the designer many options to consider when developing a structural configuration. A large-scale design development activity was undertaken to demonstrate affordable bonded primary structure.[33,34] The baseline structure for this development was the V-22 wing, which represents a deep section structure that is highly loaded. The goal in this program was to develop an all-composite bonded primary structure that met an aggressive requirements suite derived from the V-22 wing. The complex assembly for the wing developed in this activity is shown in Fig. 2.31. This design incorporates the use of precured parts, geometry that reduces

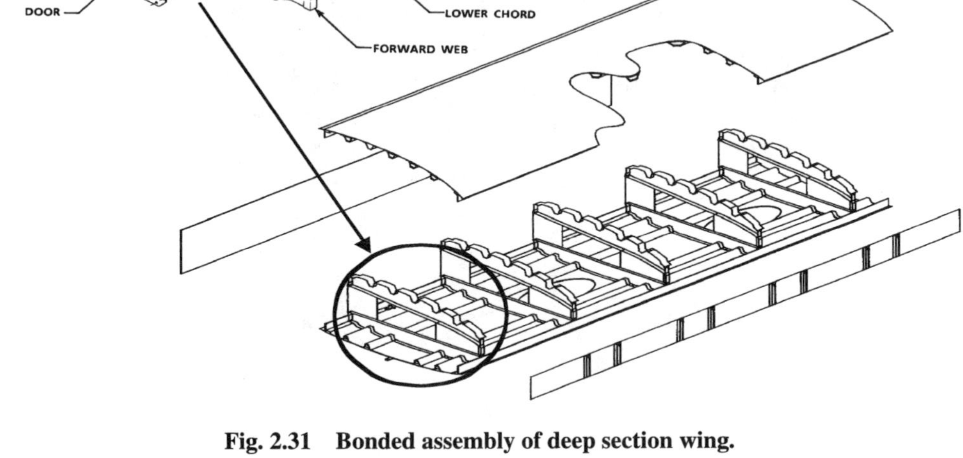

Fig. 2.31 Bonded assembly of deep section wing.

the peel loads generally present in bond lines, and uses continuous stiffeners from "tip to tip" for added structural efficiency.

The ribs in this component were designed with notches to provide an interlocking bonding surface for continuous stiffeners. These notches allow the pull-off loads to be reacted in shear in the bond as opposed to peel. An investigation was conducted to evaluate the effects of variances in bond line thickness to determine the mechanical sensitivity. This investigation involved measuring bond line variations that existed between the rib notch and continuous hat stiffeners and conducted an extensive subelement evaluation. These results showed that thick bond lines of 0.020 ± 0.010 in. did not have a significant effect on the structural performance of the design. This indicates that the current practice of requiring bond line control of 0.010 ± 0.005 in. could be overly conservative. Therefore, current approaches for shimming and the need for precision tooling could be a cost center that offers significant opportunity to be reduced. The notches were formed as an integral part of the rib structure by forming compliant fabric material around rib tools. Each of the ribs were fabricated as an assembly of several parts to assist in maintaining dimensional control. Each rib assembly consists of an upper and lower rib surface that is tooled to mate directly with the inner mold line of the continuous integral stiffeners on the cocured skins. These partial ribs are then assembled with a stiffened shear web to form a complete assembly.

The skin of this structure is shown in Fig. 2.32. This skin is made in a cocuring operation as an integral assembly. The web of the skin is made with a "soft" layup (predominately ±45° material) with the spanwise hat stiffeners constructed primarily

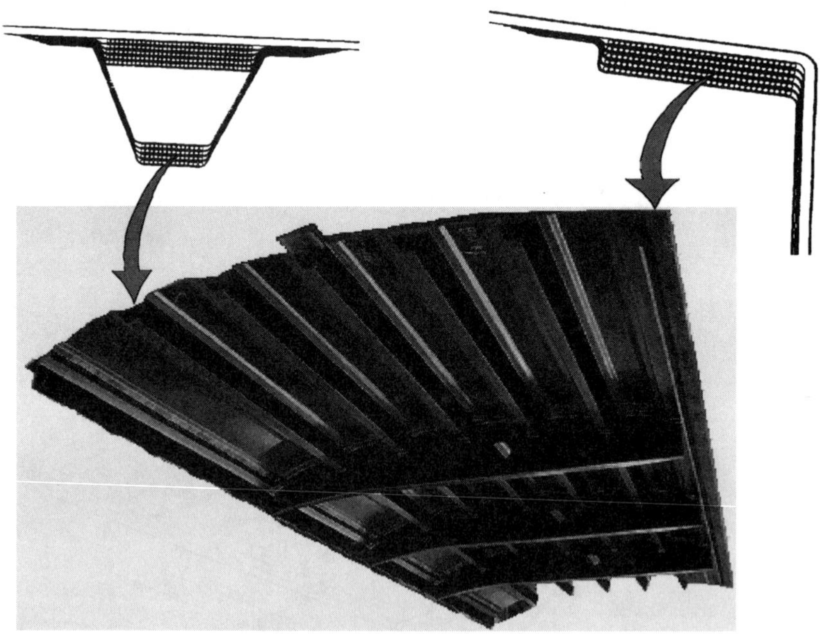

Fig. 2.32 Cocured stiffened skin.

Fig. 2.33 Pultruded carbon rods in hat stiffener.

of "hard" 0° ply packs. These stiffeners run the entire span of the wing in a wing–fuselage carry-through configuration, eliminating the weight and cost-intensive wing attach fittings. Because stiffeners are continuous from wing "tip to tip," the skin does not directly react peel loads typically encountered when stiffeners "run out" into the skin.

The ply packs were developed using pultruded carbon rods developed by Neptco™ to improve efficiency of the skin. Compressive strength is improved since fibers in the rods are drawn under tension during pultrusion, yielding much straighter fibers than in unidirectional tape. The hat-stiffener concept with pultruded carbon rods in a Syncore™ pack is shown in Fig. 2.33. By incorporating these rods to the Syncore™, a low-cost load path can be established that can be laid down in the manufacturing sequence quickly with repeatable placement. The Syncore™ material provides a medium for the rods to shear loads into the skin.

A detailed cost investigation was conducted on the design–manufacturing concept under the Design and Manufacturing of Low Cost Composites (DMLCC) program. The results have been compared in Fig. 2.34, where each of the major components and the cost of the assembly have been identified. It is shown that the bonded structure offers a significant reduction in costs for all of the composite components and assemblies when compared to the baseline V-22 configuration. In comparing the cost of the Phase III demonstration article to projected full-scale development (FSD) implementation of the concept, it is shown that cost savings are consistently lower with a slight additional reduction in the fabrication of the spars due to volume and improved automation.

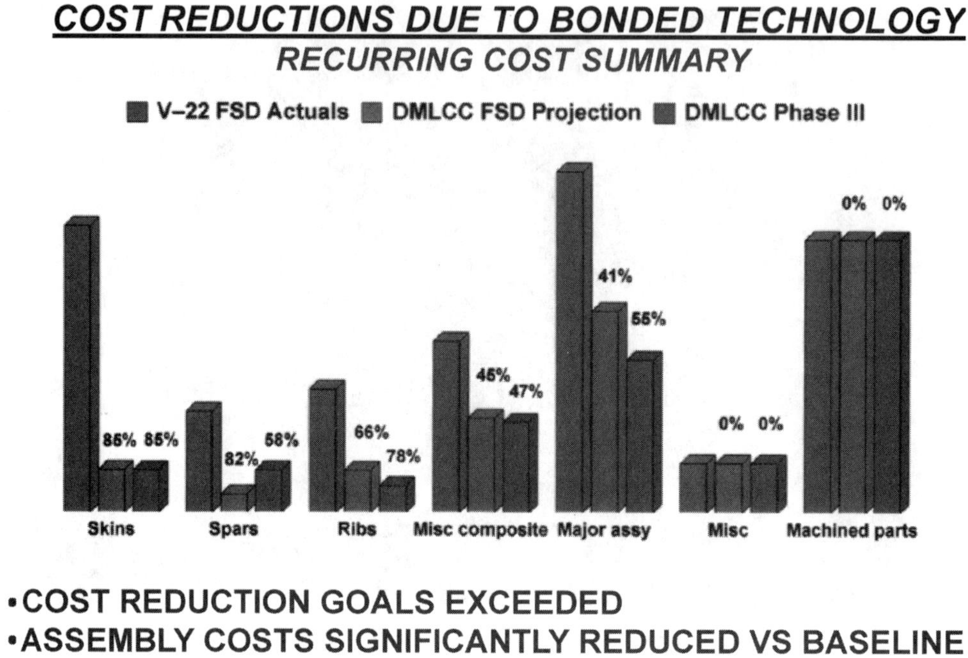

Fig. 2.34 Cost-assessment benefits of bonded wing. (See color section.)

Gridlock™[35] technology is a unique development in bonding because it enables assembly of large numbers of parts without extensive tooling fixtures. Composite elements that make up the assembly are fabricated on simple flat or slightly curved tooling surfaces with a high degree of quality and tolerance control. These piece-parts are then integrated to a bonded assembly by inserting the elements to mating grooves that have been irrigated with adhesive. In the ideal fabrication situation, Gridlock™ enables a "toolless" assembly approach, making a "snap" together structure possible. This general approach is shown in Fig. 2.35. Webs and skins are held together with the use of a clevis joint type feature, which provides self-alignment during assembly. The clevis functions as a double lap shear joint loading the adhesive primarily in shear, eliminating the peel loads.

This concept still requires a significant level of development but will offer improved damage tolerance due to the interlocking characteristics of the tongue-in-groove approach. The primary challenge to transition this technology will be sufficient quality control during the bonding operation to ensure that bonds are free from voids and appropriate tolerance on bond line thickness has been achieved. A specific challenge will be to develop procedures for certification without the dependence on ultrasonic inspection due to the limited accessibility of these design configurations.

These technology demonstrations have shown that major reductions in cost and weight can be achieved with bonded structure. Further use of bonded structure will continue as engineering confidence with design and manufacturing practice is fostered. Much of the reluctance in adopting increased use of bonded structure is due to apprehension associated with tolerance control of the bond line and reliable

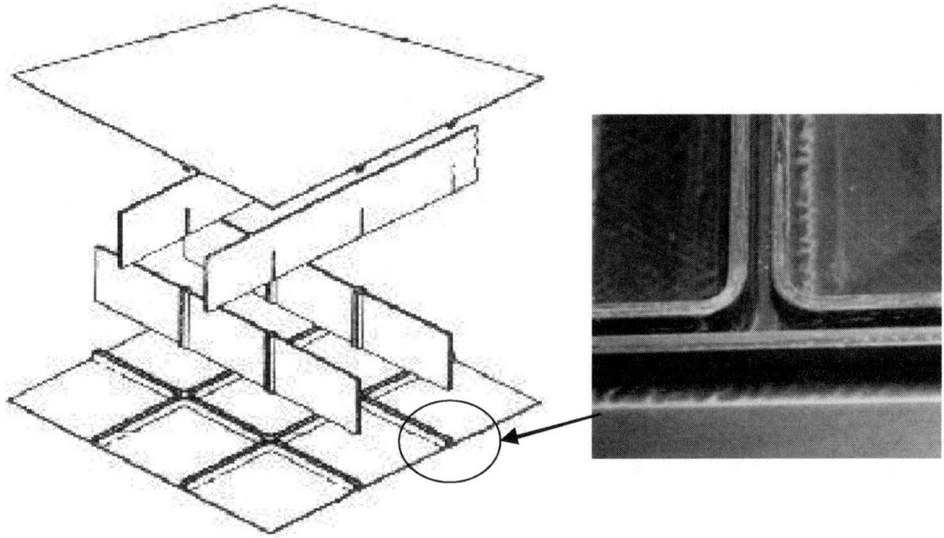

Fig. 2.35 Bonded Gridlock™ concept.

inspection to allow conventional certification methods. It is believed that this has stemmed from the inability to accurately analyze and predict the behavior of bond line variations. The majority of analyses activity to date has been directed towards analyzing uniform bond lines; therefore, the designer has gravitated towards design and manufacturing approaches that require high-fidelity control of the bond line.

D. Skin-Stiffening Concepts

1. Stiffened Panels

The skins of an aircraft make up a significant portion of the overall airframe weight; therefore, the selection of the most efficient skin-stiffening technique can have a direct effect on the vehicle structural weight fraction. In general, if the skin-stiffening approach offers a high level of stability, spar and bulkhead spacing can be increased significantly, allowing reduction of substructure weight. However, as substructure is eliminated, ballistic survivability becomes increasingly difficult to achieve because secondary load paths are eliminated. The designer must take into consideration many variables when selecting the appropriate skin design for a vehicle; therefore, the most efficient may not be the best solution from a system perspective. The skin-stiffened design must allow for the integration of structural access panels for access to subsystems without cost impacts and disruption of primary load paths. Many techniques have been investigated to develop low-cost stiffening concepts. Several of the widely used concepts are solid skin, J-stiffeners, I-stiffeners, hat stiffeners, geodesic, and sandwich. Current buckling criteria for fighter aircraft requires that there will be no failure from stability at the limit load. The designer must consider the trade between the number of stiffeners and the onset of buckling instability. As the stiffener spacing is decreased the panel becomes more stable, resulting in lower weight. The weight trend for nonstiffened

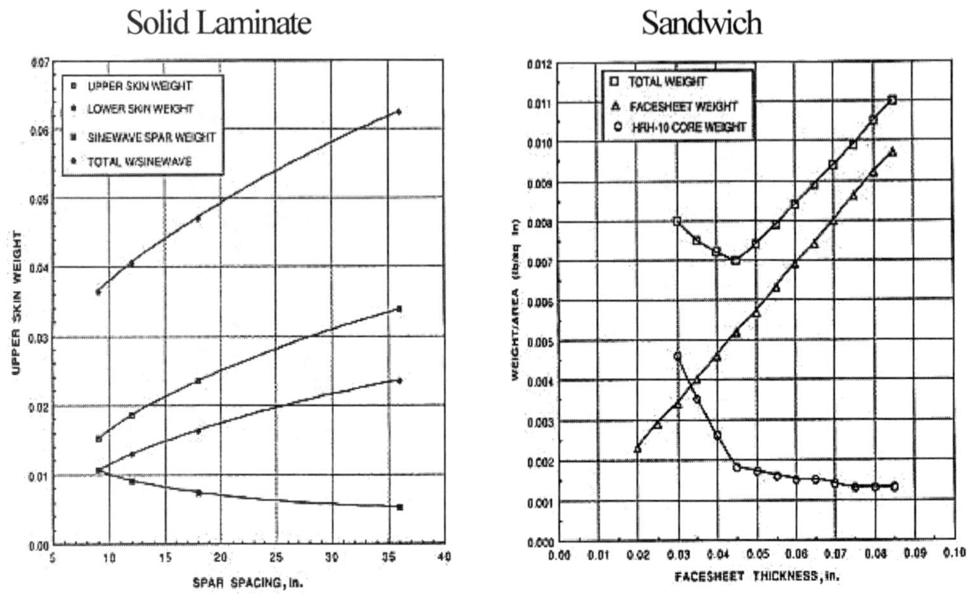

Fig. 2.36 Solid laminate and sandwich weight study.

solid skins and sandwich skins is shown in Fig. 2.36.[5] The J-/I-/hat-/geodesic-stiffened skins tend to fall in between the solid skin, and sandwich skin-stiffening approaches in terms of structural efficiency.

This study was completed for loads representative of a fighter wing structure. The unstiffened design results in an inefficient skin configuration due to added material to provide buckling stability. As spar spacing is decreased in an unstiffened design, the associated weight of the substructure increases slightly; however, decreases in skin weight yield a much lighter design. By incorporating stiffeners, the buckling stability is increased, resulting in increased structural efficiency.[36] In the case of the sandwich panel investigation, the thickness of the core provides the primary means of preventing panel buckling, but the density of the core needs to be considered to provide a foundation to the skin. If the core does not offer sufficient support, the skin will be subject to wrinkling and dimpling. The sandwich design provides the lowest weight-stiffening technique. The sandwich provides approximately a 35% weight savings over the stiffened design and 45% over the unstiffened. The weight savings of the three stiffening techniques applied to a wing structure are shown in Fig. 2.37,[5] and shows the weight savings potential of the honeycomb (HC) panel. This study evaluated the weight of the three stiffening techniques against spar spacing.

The sandwich configuration is clearly the lightest weight concept and offers significant opportunity to be the lowest cost since the detail parts manufacturing "may" be the least through the benefit of unitization. *However, since sandwich structure has had historical challenges associated with environmental durability, the designer must carefully assess the functionality of the structural system before adopting a primary sandwich structure design.*

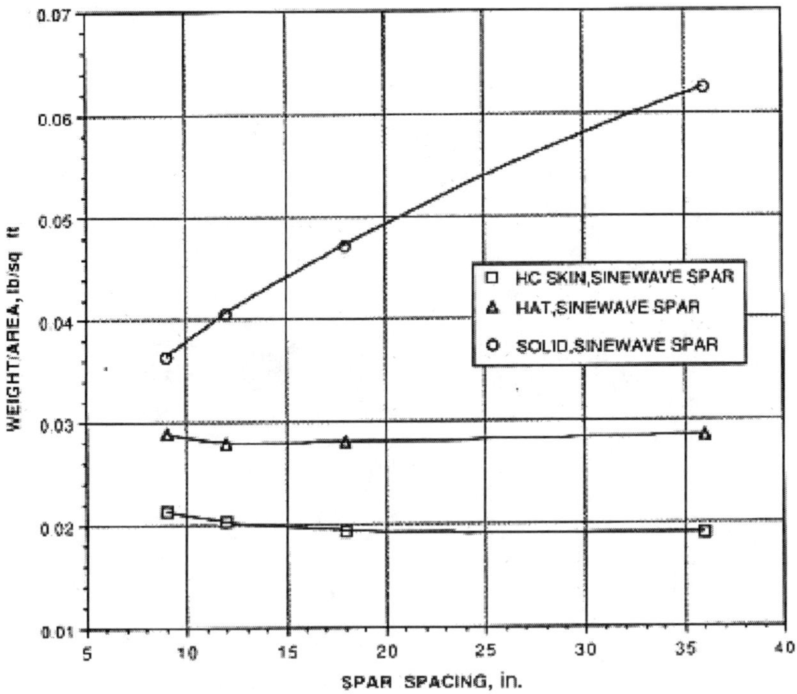

Fig. 2.37 Weight vs spar spacing.

2. *Sandwich Concepts*

Prior attempts at using extensive sandwich structure in aerospace applications have resulted in costly manufacturing and supportability problems. These problems have resulted due to the integration of materials and concepts that had a high affinity for moisture, resulting in corrosion and mechanical degradation of skin to core bond. In addition, challenges associated with cost of fabrication, joining, and load transfer through hard points further restricted their usage. Trade studies have traditionally shown that the sandwich configurations demonstrate large improvements in weight efficiency. Unfortunately, these large weight savings are not exploited due to the lack of engineering confidence in sandwich that has evolved from past experiences. This lack of confidence has established a sandwich structure design paradigm, which has prevented the full use of current-day structures and materials technology to be applied to sandwich designs. Through the development of composite materials, novel core concepts, tougher adhesives, textile preforms, and Z-Fiber^{TM} reinforcement, a potential opportunity exists to reduce weight and costs through unitized sandwich structure without the historical maintainability problems.

Structural configurations have recently been developed that feature the use of sandwich in highly loaded mid- and aft-fuselage structure. Integration of sandwich in the wing and fuselage will significantly reduce weight and cost, but the ability to accomplish this largely depends on robust joining concepts. The joints in a primary sandwich structure are required to react higher pull-off loads than in conventionally

Fig. 2.38 Complex sandwich assembly with preform joints.

stiffened structure. The internal wing pressure induces joint pull-off loads that must be reacted by fewer bulkheads; therefore, the joint running loads increase. The use of sandwich panel construction with low-cost joining to produce a one-piece component with complex joint intersections is demonstrated in Fig. 2.38. This manufacturing demonstration component includes the outer mold-line skin, rib, fuel-floor, and bulkhead intersection. The component was fabricated with precured sandwich panels cobonded with a preimpregnated resin textile preform.

To improve structural efficiency and eliminate structural discontinuities at intersections, longerons, spars, and bulkhead caps can be embedded in the skins using either unidirectional tape or pultruded carbon rods to provide unbroken primary load paths. The rods offer attractive features for sandwich construction—options that have just recently become available. Neptco™ rods can stiffen a sandwich skin for reinforcement over a bulkhead as depicted in Fig. 2.39. In this case, the bulkhead is of sandwich construction. Maintenance would likely be reduced for two reasons: 1) there are substantially fewer joints to fatigue and 2) most of the fasteners in the lower skins have been eliminated so fuel leak paths are greatly reduced.

A great deal of development has occurred to develop higher-strength, damage-tolerant sandwich joint concepts. Advancements in materials and preform technologies that have occurred over the last 20 years have enabled concepts that have demonstrated significant potential for application in primary aircraft structure. Through selection of textile inserts, Z-Fiber™, robust core concepts, and innovative methods of manufacturing, including bonding and cocure, joints have been demonstrated that can offer the performance required for the demanding requirements of primary structure application. Many of the variations of the joints

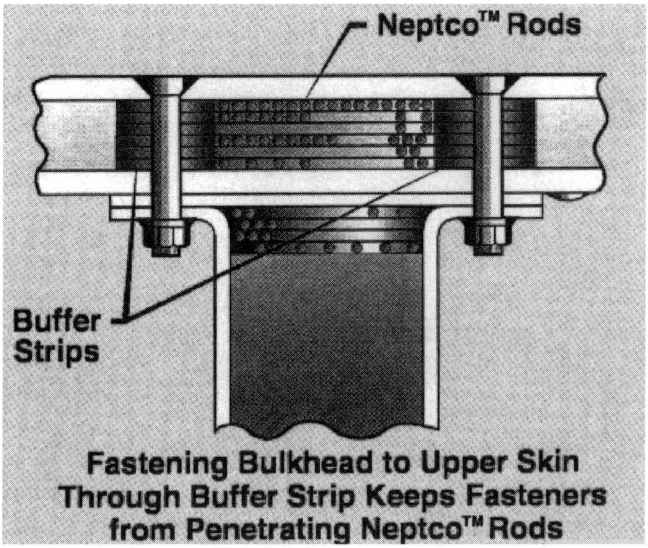

Fig. 2.39 Sandwich concept using embedded Neptco™ rods.

that are possible are shown in Fig. 2.40. It is important to appreciate the failure modes and mechanisms that each joint configuration offers since small changes in joint geometry and configuration can have major changes in the loading characteristics and overall structural capability. The joints shown cannot be applied randomly on the airframe; some of the configurations are better suited for secondary structure application. Mechanical fasteners are included in several of the joint configurations identified but are used only where drilling through solid laminate is possible. This eliminates the problems typically faced with drilling through sandwich core such as sealing the core, potting the core, or using reinforcement sleeves to react bolt preloads and prevent core crushing. The use of textile preforms is extensive in the sandwich concepts identified. The woven preforms are specifically directed at improving ultimate strength, reacting concentrated out-of-plane loads, and improving overall robustness. In concept four, the textile preform extends through the core to form the outer mold-line skin. While this approach offers the greatest level of ballistic survivability, it represents significant increases in manufacturing complexity due to the need to lay down skin material in multiple steps. Concept twelve is designed with a textile preform to react the local out-of-plane stresses at the corner of the core induced from the skin ramp. This concept provides improved bearing capability for the mechanical fastener and seals the core from potential moisture intrusion.

Highly damage-tolerant composite sandwich concepts are currently under development. The skin-to-core structural interface has long been the "weak link" of sandwich structures since the structural integrity is maintained by the bond line, the impact resistance of the skin, and the HC core itself. In-service experience with sandwich structure has shown that when moisture enters the core it degrades the quality of the bond line. While attempts have been made to inhibit moisture intrusion through sealing, success has been limited due to leak paths forming through fastener holes or microcracks. Under severe impact conditions, the bond

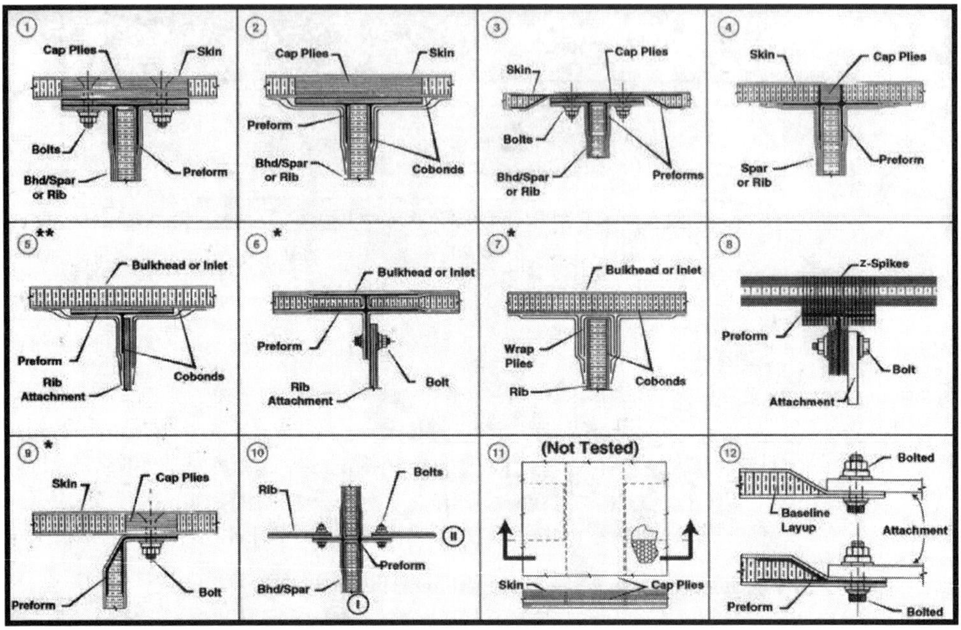

Fig. 2.40 Joint concepts in recent development.

has been prone to local failure, especially when its properties have been degraded by moisture. Under subsequent temperature, pressure, and flexure from loading, these local skin-to-core bond failures can grow to cause catastrophic failure of the structural system. Therefore, emerging concepts that eliminate moisture intrusion, corrosion, and face-sheet debond and improve overall damage tolerance will enable sandwich structure for primary structural applications. A novel concept has emerged in recent years referred to as "hollow core." This core is constructed

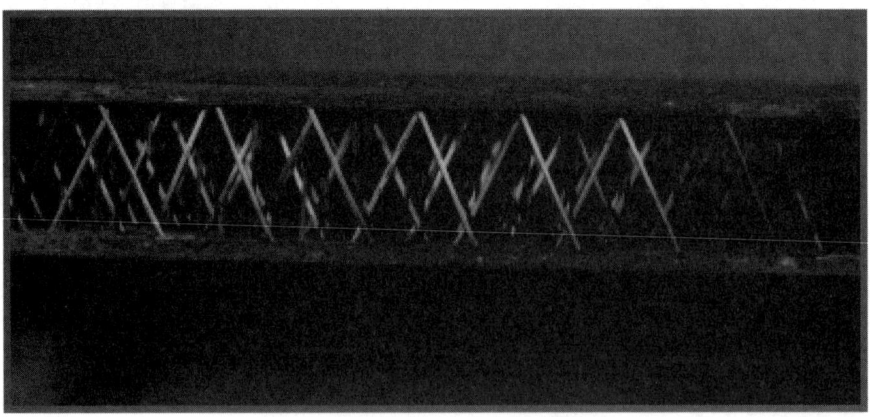

Fig. 2.41 Hollow x-core concept.

Fig. 2.42 Integrally woven sandwich core.

with carbon fiber pins embedded through the core and anchored in the sandwich facesheets. The hollow core concept is shown in Fig. 2.41.

These pins provide a mechanical linkage between the skin and core; hence, no debond can occur because there is no adhesive. The pins function as truss, elements, reacting tension and compression, allowing the structure to react shear loads. Because minimal shear load is directly reacted by individual pins, the primary modes of failure are from pin pull-out and buckling. This results in a progressive failure mode rather than catastrophic failure of the core from shear buckling that is typical in HC. The hollow core can also be fabricated with foam core to further enhance the structural capability by stabilizing pins from buckling and directly improving shear strength of the sandwich.

Another concept recently developed involves the use of textile three-dimensional woven preforms impregnated and cured into truss core. These sandwich concepts also eliminate moisture gain and core debond because the face sheets are integrally woven with the core in a single, unitized panel. This concept can be fabricated with or without foam, depending on the structural performance required. The concept with foam integrated to the truss structure is shown in Fig. 2.42.

VI. Three-Dimensional Structure Analysis and Certification

To develop the ability to predict structural failures, the structural certification of manned aircraft relies on the "building block" process. This process validates failure analysis through successive test and analysis of coupons, elements, subcomponents, and full-scale components. This approach accounts for impact, fatigue, and environmental conditions and thus ensures that the airframe will be structurally reliable throughout its designed service life.

The fundamental building block of the certification process is the laminated coupon. However, fabrication methods and structural forms are now being developed for which a simple representative coupon cannot be defined. For instance, in structural parts fabricated through fiber steering, fiber architecture varies continuously. Also braided and woven components contain multidimensional fiber architectures that cannot be simplistically characterized through any subdivision. For these structures, the mechanical response is based on the whole design. Methods need to be developed for characterizing these types of components and certifying structures employing them.

Certification issues and design methodologies for three-dimensional composites center around the measurable quality of as-manufactured structure, resulting allowables, and analysis prediction capability. The quality of a part is typically measured by nondestructive inspection (NDI) techniques such as ultrasonic and x ray and is assured through in-process controls. The NDI of three-dimensional composite joints will be less critical because of increased strength and enhanced damage tolerance. The ability to resolve flaws in complex three-dimensional joints (such as those using inserts, preforms, Z-Fiber™, and cocured assemblies) will, to a large extent, drive the manner in which allowables will be calculated for a particular joint technology. Analysis tool and methodology development, physical tests, and design and certification advisors will all reflect flaw-resolution limitations imposed by quality inspection capabilities. Though advanced inspection techniques such as laser ultrasonics can be readily applied to inspect three-dimensional composites with a high degree of accuracy, structural load, stress, and strain allowables development need to reflect manufacturing variability so that flaws produced by typical manufacturing processes are accounted for in allowables determination. The many advantages of three-dimensional composites have already been discussed above. However, there is much work to be done to provide a production structural certification capability to allow the structural designer to take advantage of all of these promises.

Analysis development is a key part of developing confidence in the certification of three-dimensional structures. The calculation of woven or braided properties, particularly at structural intersections, has proven to be very difficult. To develop a priori certification techniques, the properties and allowables for woven or braided inserts alone (not in a joint) must be developed, along with allowables for an insert to a joint. This is particularly important when calculating margins of safety for the insert and when specifying material response properties for detailed finite element analyses.

A. Testing of Three-Dimensional Woven Preforms

As previously noted, mechanical allowables will be dependent on processing methods, structural concept, and materials. Because of the lack of maturity in complex three-dimensional composite material response prediction, a comprehensive testing and analysis correlation will be needed in the near term. The response for IM7/977 three-dimensional woven preforms that have been cured is shown in Fig. 2.43. Understanding the mechanical characteristics of the preform by itself without being integrated to a joint is the first step in developing an understanding of the global performance of the joint. Data show the loads carried in each of the legs of the preforms. The results of these preforms cocured into sandwich subelement joints are shown in Fig. 2.44.[37]

The methodology and criteria must be updated to account for the effects of manufacturing defects and delaminations in Z-Fiber™/textile-reinforced structure if the advantages of the delamination arrestment capabilities are to be realized. Static test programs to date have shown that significant gains in delamination

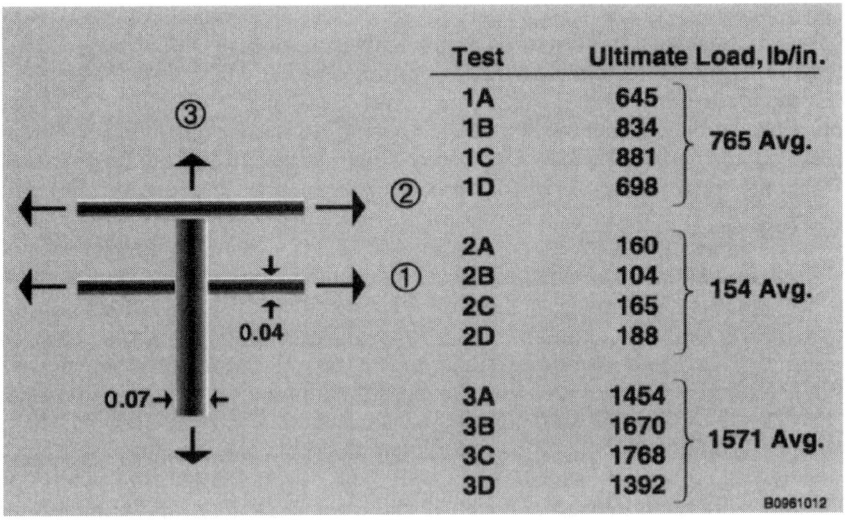

Fig. 2.43 Tension test results for three-dimensional woven preform.

arrestment are possible. However, a comprehensive test program to document the behavior of postdelaminated, three-dimensional, reinforced joints and components must include not only static strength but simulated service life testing for multiple lifetimes under realistic loading spectrums. Once the behavior of postdelaminated Z-Fiber™-reinforced structure has been completely characterized, the possibility of changing aircraft certification criteria to allow flight delaminated structures will become a possibility. This will make possible the potential weight and safety benefits demonstrated by Z-Fiber™-reinforced structural tests.

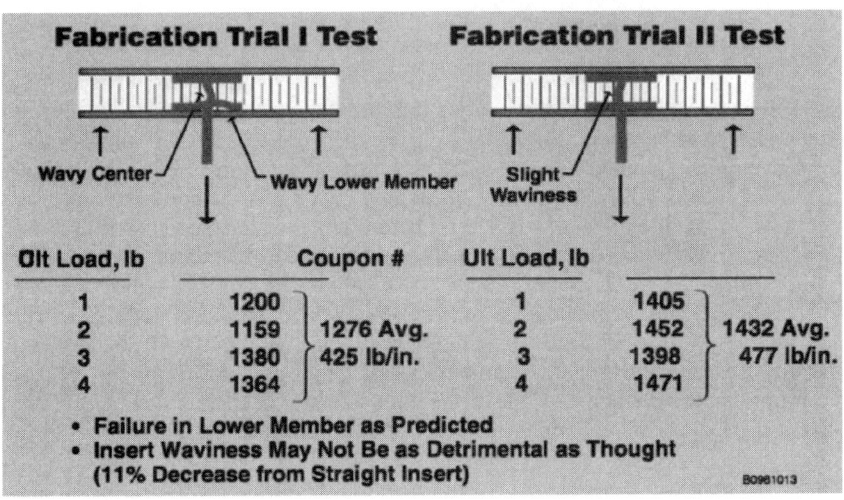

Fig. 2.44 Test results for subelement joint with embedded three-dimensional preform.

There are many gaps in the capabilities of three-dimensional composite joint analysis tools and methodologies. As previously mentioned, tools to reliably calculate material properties will provide the cornerstone that will enable detail design and structural certification. Both two-dimensional (plane stress/strain and shell elements) and three-dimensional (solid elements) FEA analyses techniques must be developed and verified by test comparisons. Two-dimensional modeling idealizations include microplane stress/strain modeling where individual or groups of plies are modeled. Macroplane stress/strain where entire materials or aspects of a structure (such as face sheet, insert, core) are modeled with one layer of elements, and shell element modeling where the laminate (or sandwich) is represented by matrices that represent the membrane, bending, and coupling responses of the structure. Three-dimensional modeling techniques similarly are micro- or macroapproaches. In all of the two- and three-dimensional modeling techniques fully characterized (and therefore accurate), three-dimensional properties are essential to the success of the analyses. Emerging methods of analysis, including variational models, may also offer the ability to further develop methodologies as these techniques become integrated to the design environment.

Testing will perform a crucial role in the development of three-dimensional structural certification tools and methodology. Each of the aspects of three-dimensional materials and joints must be characterized both within a structure and individually. Examples already discussed include woven or braided inserts and Z-Fiber™ reinforcement. One of the most difficult aspects of testing is the accurate determination of out-of-plane properties for three-dimensional laminates. This is only made more complicated by the extremely complex geometry of fiber placement in woven and braided intersections. A comprehensive program to test these properties combined with the development of mathematical material response property calculation tools will provide the optimum approach to the development of tools and methodologies capable of accurate predictions of structural strength and margins of safety. Once this has been completed, physical tests can be minimized and maximum advantage gained by virtual certification tools and methodologies.

B. Needs for Certification of Three-Dimensional Structure

It is clear from the above discussions that the design and certification of three-dimensional woven insert joints will require the application of a large body of knowledge gained from test and analysis as well as structural and life verification of three-dimensional woven joints. Durability and damage-tolerance approaches will have to be a mix of experience based on heuristics and analyses until much more sophisticated mathematical models are available. The complex body of knowledge for three-dimensional composite joint structural certification, design technology, and methodology is very suitable to automation in either a procedural or rule-based advisor system. The advisor will enable information and skill sharing in the integrated design–analysis environment in today's product development approach by minimizing the chance that one of the many aspects of three-dimensional composite joint design and certification are not considered.

VII. Affordable Considerations for Ballistic Survivability

Military airframes must be low in cost but are also required to survive against the threats that will be present in combat environments per the congressionally mandated live fire test law. Recent live fire test experiences with emerging flight vehicles have indicated that current structural design practice for "all-composite" designs do not achieve the desired level of survivability when subjected to high-explosive incendiary (HEI) threats. Therefore, "all-composite" structure has been perceived to be less survivable than more conventional structural configurations, without specific regard to design changes that should be considered to exploit composite materials. Given that this requirement has been recognized as a primary inhibitor in using composites, the authors chose to include recent developments in the design for survivability.

A. Description of Ballistic Event

The primary forces present during a ballistic event are from the hydrodynamic ram phenomena that occurs when an HEI, armor-piercing incendiary (API), or missile fragment enters a fuel-containing structure and imparts its kinetic and/or explosive energy to the fuel, causing extremely high fluid pressures. The fluid pressure directly couples with the structural fuel cell, resulting in extensive damage as shown in Fig. 2.45. The effects of hydrodynamic ram from the ballistic event have been characterized.[38]

The ballistic event is a complex series of loading effects that have extensive interaction, especially in the case of HEIs. Resulting structural failure characteristics are not predictable with significant fidelity; therefore, testing is required to be conducted under simulated combat conditions. To accurately evaluate the survivability of a design, it is necessary that the size of the structure be established with respect to threat lethality to allow redistribution of loads around the damage zone. This ensures that boundary conditions have been realistically represented and the interaction of individual threat effects is captured in the test. Historically, these tests have proven to be expensive and have prevented the opportunity to evaluate improved design configurations early in the design process of new airframes. This has resulted in structural designs becoming "locked in" during system

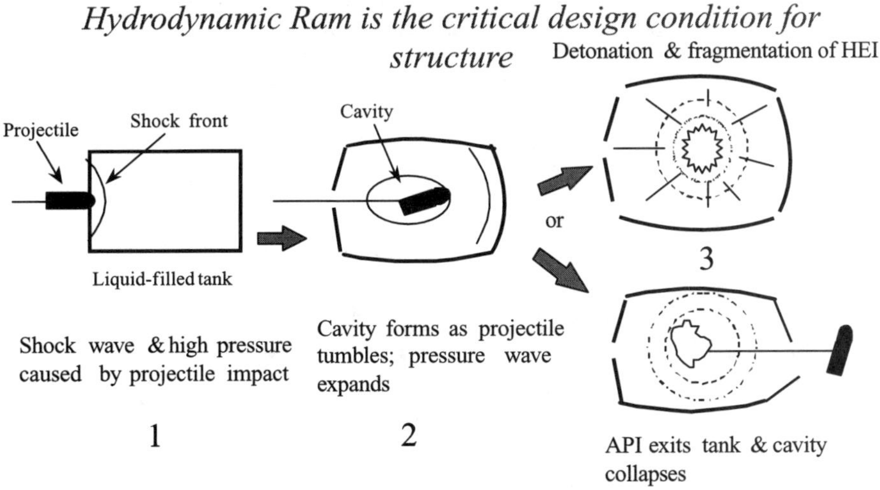

Fig. 2.45 Hydrodynamic ram event.

development, with the available design space needed to make improvements very limited. Attempting to redesign or retrofit primary structure after a design has been developed is not cost effective. Therefore, vulnerability reduction features must be integrated to the design early in the development cycle. The development of a survivable composite structure requires that it be tolerant to the combat environment, including the effects of airflow, structural load, hydrodynamic ram, projectile damage, blast pressure, and fragmentation from HEI threats. The effects of airflow have been shown to cause extensive interlaminar peel damage.[39] Subsequent to impact by the threat, lamina are susceptible to peeling induced by airflow over the surface.

B. Conventional Design Practice

Current design practice for composite structure involves the use of panel construction with mechanical fasteners for assembly and to provide containment from multibay ballistic damage. Multiple rows of fasteners have been used to provide a secondary load path between skin and substructure and maintain the integrity of the joint if the skin fails along the first row of fasteners. The primary drawback with fasteners is their inherent cost of installation in addition to inhibiting the designer from exploiting the properties of composite material by introducing concentrated forces. During the ballistic event of a traditionally designed structure, forces are reacted through the mechanically fastened interfaces instead of relying on the fiber architecture of the composite structure. Hydrodynamic ram pressure induced forces and the associated mode of failure on a conventional wing structure is similar to that shown in Fig. 2.46.[31]

These forces introduce bolt pull-through loads that directly impart out-of-plane stresses in the area of countersunk fasteners. To improve the capability of these joints, significant work has been conducted to improve the toughness of composite joints with the use of fiberglass or Kevlar™ material. This has resulted in joints with enhanced toughness but these hybrid systems typically involve additional manufacturing steps, leading to increases in cost. Without the use of these toughening elements, significant interlaminar failure can be expected. An ultrasonic inspection of a composite panel from a mechanically fastened wing box tested

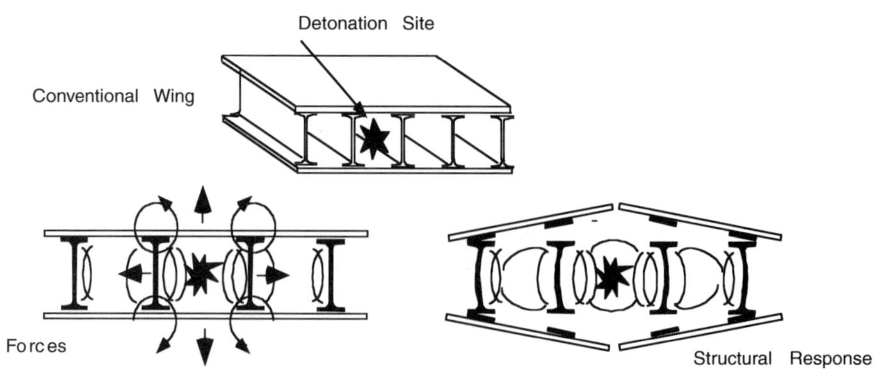

Fig. 2.46 Conventional failure effects.

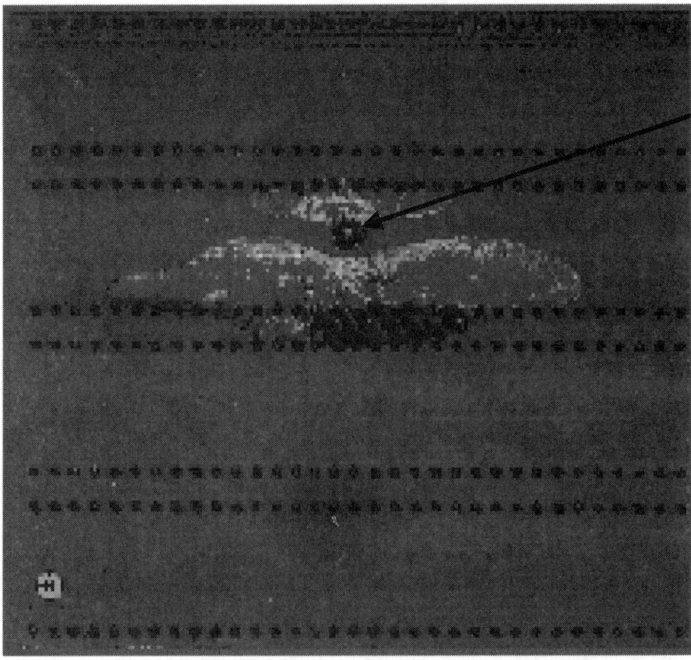

Fig. 2.47 Ultrasonic inspection of fastened panel (after API impact). (See color section.)

against an API threat is shown in Fig. 2.47.[40] This wing box was constructed in a conventional approach incorporating composite skins with metallic substructure. The zone of damage around the bolt pattern is typical of what can be expected from these types of design configurations. The bolt pull-through forces from the fasteners induced the nonvisible interlaminar damage shown in this ultrasonic scan. The delamination damage initiates at the bolted joint and travels towards the projectile penetration.

As the fasteners pull through the structural skins during a ballistic impact event, loads are directly translated to the next spars. The pressure that the adjacent bays react can be quite significant due to the minimal amount of energy absorbed by the spar webs that first see the pressure pulse. Because fluids transfer energy efficiently, pressures that are communicated throughout the wing structure can be significant, often times exceeding 1000 psi several bays from the point of detonation. Past research and development efforts have used parasitic compressible mediums to disrupt the chordwise coupling of hydrodynamic ram shock waves. These techniques have offered limited benefit because they only provide a short time delay in the reaction of pressure loads in adjacent bays and have little reduction in the actual magnitude of loads that occur. In addition, these concepts would also incur extensive supportability costs, reduced fuel volume, and increased weight. If the extent of damage is to be controlled, joints must be developed to prevent skin separation from the substructure. In the case of the HEI threat, significant skin and substructure failure can be expected. The failure of entry skin to an HEI threat is shown in Fig. 2.48. The damage exhibited involved extensive delamination to the

Fig. 2.48 HEI damage.

skin and bolt pull-through on the spar attachments adjacent to the detonation and substructure spar/web damage.

C. Design Considerations for "All-Composite" Structure

Specific guidelines for low-cost survivable composite structure have yet to be developed. Many factors have prevented the development of design standards and practices such as lack of analytical tools, robust joining concepts, low-cost processes, and design practices imposed by quality control requirements. The complexity of phenomena associated with structural failure from the ballistic event limits the benefit that the designer can gain from conducting analytical investigations; therefore, design guidelines that are currently evolving are primarily based on limited empirical results. Several design approaches/considerations are presented in the following discussion. These concepts are based on a sufficient number of tests to provide repeatable failure characteristics. However, these concepts could perform quite differently when applied to alternative structural configurations involving different spar spacing, skin/web thickness changes, or varying levels of fuel at the time of impact.

To achieve a low-cost survivable composite structure, the designer must consider the use of a unitized structure with limited reliance on the number of fasteners. There are three primary strategies that have been demonstrated with significant success to control failure in composite structure from these threats. These approaches include

1) Robust compliant joints,

2) Established, zones/lines of controlled failure through stiffness mismatch, and

3) Transfer of hydrodynamic ram pressure loads to the fiber architecture of the structure.

As described in the conventional design practice discussion, the joints of the structural system control the extent, mode, and mechanism of failure to the skin and substructure. Maintaining the integrity of the joint is critical if shear loads between the skin and substructure are to be reacted and loads redistributed around the damage zone. Therefore, the joints become the "Achilles' heel" of the structural system and must be addressed specifically during design if the failure is to be controlled. If the joint offers a significant amount of compliance and toughness, it will tend to rotate around the spanwise axis of the spar, changing the predominant failure mode of the structural system. Instead of the skin separating from bolt pull-through as in the conventional case, the joint will tend to rotate, allowing the skin to be placed in bending and membrane tension.

The designer must consider how the structure will fail during the impact event. Stiffness gradients should be created in the layup of a structure by introducing variations in ply orientation. A significant amount of work has been conducted using concepts that introduce preweakened failure paths considered "blow-out panels" in mechanically fastened skins. The results of this approach have not offered the level of improvement desired. The bolts tend to pull through the skins before the preweakened structure has time to fail. However, if the designer can successfully improve the survivable performance of the joints, the weak link of the structural system can be directed to controlled failure in the skin. It is not necessary that weak spots be introduced to the structure, only that a stiffness gradient exists to allow energy to quickly dissipate in the form of mechanical failure before joint separation. Several of the references have investigated the use of 0° plies or "hard laminates" over spar caps with the skins predominantly constructed of 45° plies or "soft" layups. Other techniques that have been used include the use of hybrid material systems involving the use of glass or Kevlar™ to form "buffer strips".[41] Stiffness gradients introduced by toughened materials assist in forcing the failure along designated zones to maintain secondary load paths. Novel low-cost approaches have been recently developed to establish "hard" load paths. Ultra-high-strength pultruded carbon fiber rods featuring extremely straight fiber alignment have been incorporated and demonstrated in advanced designs to enable highly controlled and tailored load path management and survivability improvements.

Historically, an extensive amount of research and development has been conducted to develop low-cost bonded/cocured composite structure by using the matrix material or adhesive to simplify assembly. While these design configurations have often proven to be low in cost and meet static performance requirements, they typically have not satisfied durability or survivability requirements. This is primarily because loads associated with these requirements are reacted in the matrix or bond line and not in the fiber architecture. Given the magnitude of stresses that occur during ballistic or durability related events, it is essential that low-cost designs be developed that allow these forces to be reacted in fibers of the composite structure. Recent exploratory development work has been conducted to investigate the effects of wrapping composite laminates around the fuel cell formed by the skin

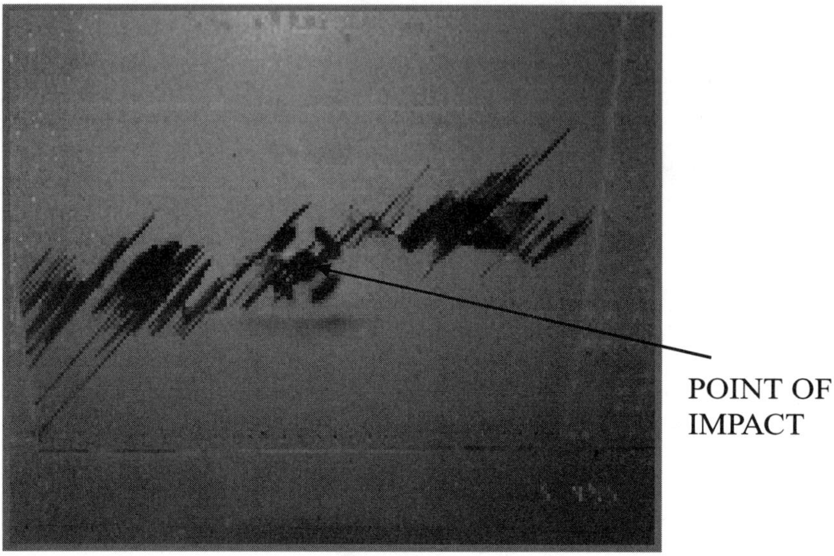

Fig. 2.49 Cocured entry panel.

and spar webs, allowing the web plies to form the inner mold-line of the skin. The performance of a wing box that was composite with a cocured entry panel and mechanically fastened exit surface is shown in Fig. 2.49. This box was designed to the same overall stiffness requirements as the wing box in Fig. 2.48 and featured a design where the inner lamina blended from the spar web substructure into the inner mold-line of the entry panel skin. This provided a high level of rotational compliance under hydrodynamic ram loading from the HEI threat.

During this test, joints experienced a significant amount of interlaminar tension as the cell attempted to pattern its geometry around the circular shape of the expanding pressure field. While local interlaminar failure was observed in joints of this test, joints offered a sufficient level of compliance, permitting rotation during

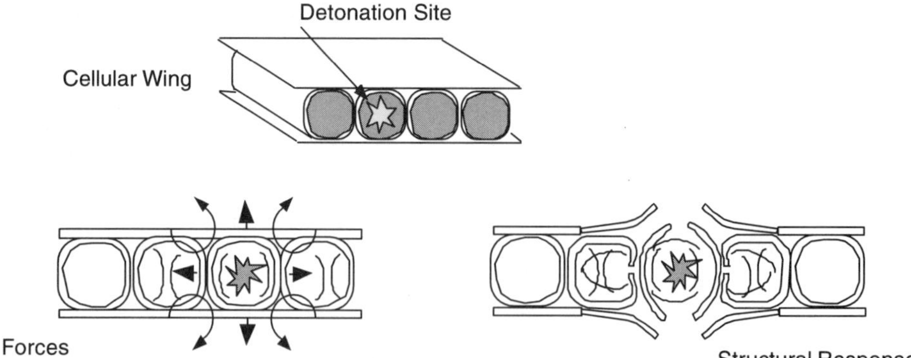

Fig. 2.50 Cellular wing with HEI damage.

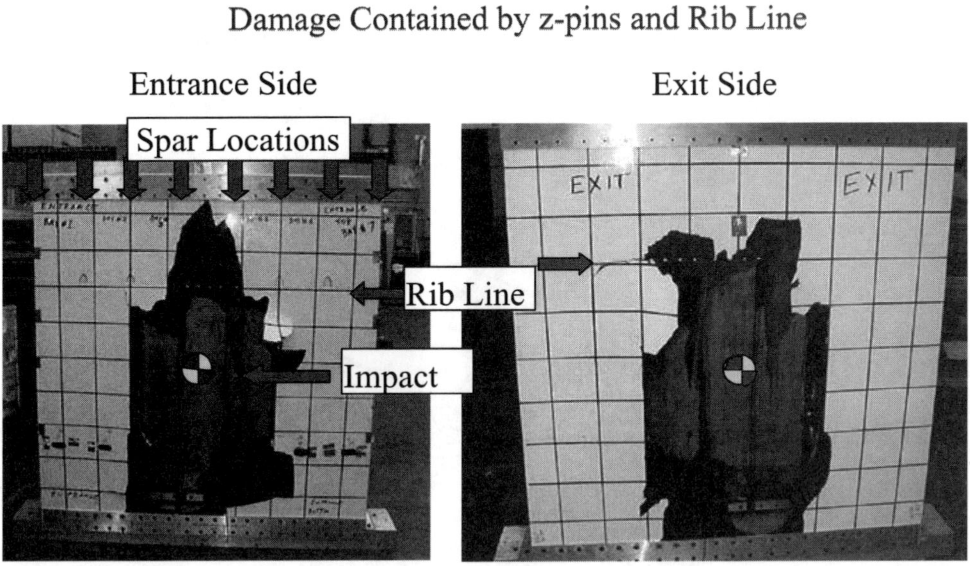

Fig. 2.51 Large-scale cellular wing test.

the event, and allowed loads to be transferred to membrane tension in the wing box skin. This resulted in the spanwise crack observed in the center of the impacted bay entry skin while the mechanically fastened exit surface failed in the conventional manner as shown in the earlier HEI damage discussion.

This design approach has been investigated and extended by Childress et al.[31] In this work, a tooling approach was developed to enable a full cocure of an upper and lower skin with the substructure, where the lamina are blended around the entire cell of the wing bays. The forces and associated response of a structure designed in this manner are shown in Fig. 2.50.

A large-scale cellular wing impacted with a HEI threat typical of today's combat environment is shown in Fig. 2.51. The overall damage observed in this test was determined to be within limits of what could be reacted without catastrophic loss of the structure.

By applying the aforementioned practices of compliant joints, toughened materials, and strength tailoring of the laminates, significant improvements in the ballistic performance of the composite structure can be attained.

VIII. Novel Airframe Configurations

A. Structural Definition

To develop an affordable structural airframe for composites, the design needs to be developed from the system level so that all key requirements are considered up front in the definition of primary load paths. To accomplish this, variables such as fuel management, subsystem integration and maintenance accessibility, engine integration, roll rate and refueling induced internal pressures, weapons stores integration, concentrated landing gear loads, etc., must all be considered

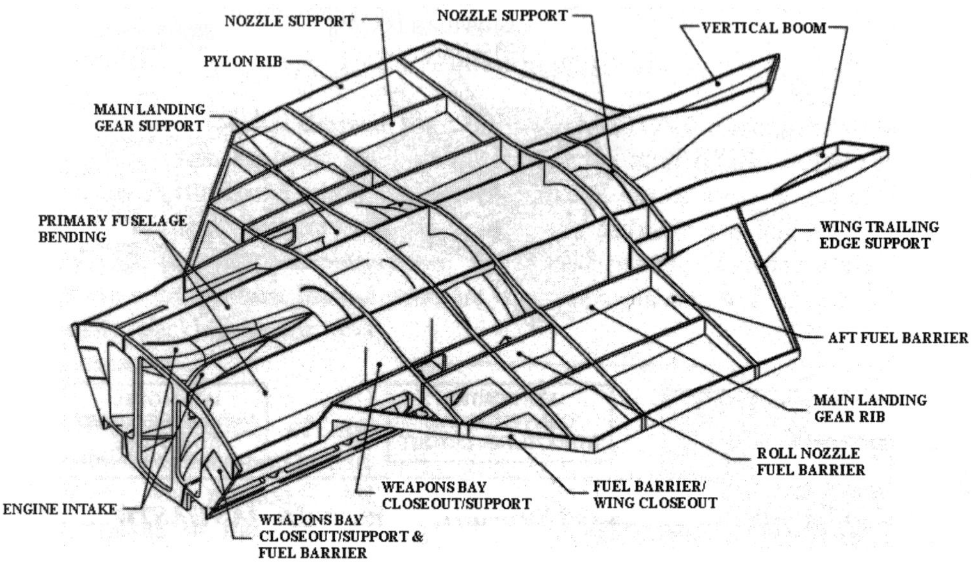

Fig. 2.52 Minimum structure required.

to ensure that the airframe can react all required loads but still offer low-cost benefits. Therefore, an approach must be taken to identify the minimum required structure to enable these performance requirements while introducing manageable loads that can be reacted by composite structures. This allows the designer to evaluate the functionality of each respective structural member and determine the best performing load paths. Design development was conducted in this manner,[2] resulting in the vehicle configuration shown above in Fig. 2.52.

B. Unitized Composite Structure

As previously discussed, the design of an airframe must take into consideration the function and required capabilities of the entire vehicle, which imposes limitations on the level of unitization that can be achieved. The basis for unitized structure is to reduce both fabrication and assembly costs. Design concepts and approaches that emphasize the development of larger and more complex parts can be employed, resulting in fewer total parts, which will directly reduce these costs. To accomplish this with conventional materials, the tooling approach must be carefully considered, because the benefits of reduced part cost could quickly be negated with large increases in tooling cost.

The advent of low-temperature curing materials and simplified tooling for low-temperature processing as well as improved methods of assembly have broadened the design space to enable complex unitized structure that offer system functionality. One of the key attributes of emerging low-temperature materials and processes such as RTM, VARTM-and e-beam resins is the elimination of complex tooling. These materials can be fabricated on low-cost foam tools to establish load paths in three dimensions, unlike conventional tooling, which has historically

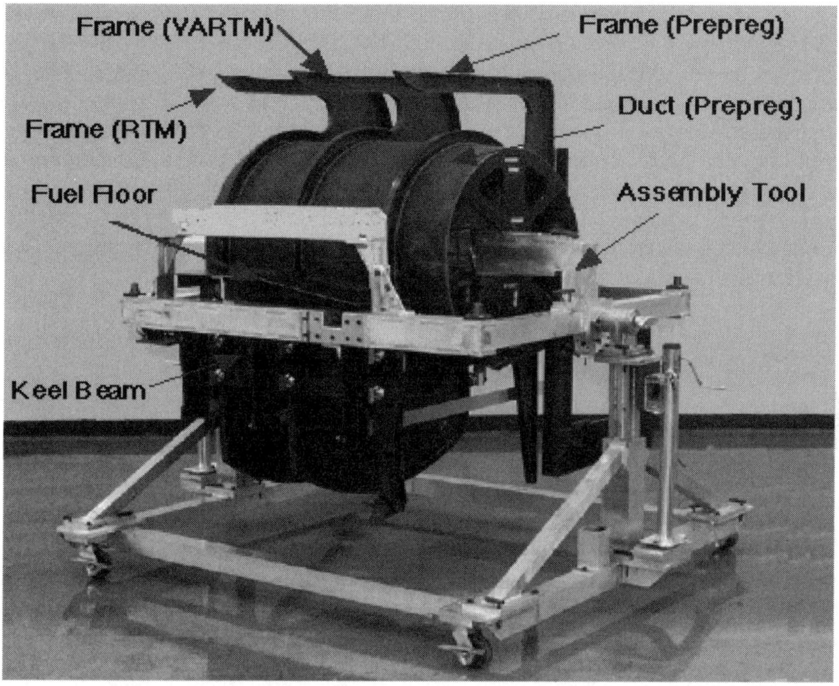

Fig. 2.53 Integrated duct structure.

been used "affordably and predictably" to fabricate panels with slight curvature. The reduction in temperature associated with cure of these systems allows the designer/manufacturer to have better control of warping and tolerances. Material can be placed specifically where it is most beneficial to react the loads of the airframe, permitting routing of loads around discontinuities such as critical access panels.

By using advanced materials and processes in combination, structural configurations have been developed that would not otherwise be possible. A one-piece duct structure that has been fabricated using RTM, VARTM, and prepreg materials is shown in Fig. 2.53.[29] The complexity of this one-piece structure is achieved in a single component by exploiting the inherent properties of each of these fabrication processes.

The benefits of unitized structure are clear; however, the structure must not be designed to result in a system not economically maintained and repaired when fielded. A number of key supportability considerations were directly addressed to ensure that the system could remain affordable when fielded, including:[42]

1) Reduction in aluminum parts in critical areas to reduce corrosion potential,

2) Reduction of fasteners and incorporation of redundant sealing to reduce leaks,

3) Fastener reduction to reduce fastener maintenance action,

4) Provisions for operational level repair of fuel leaks through removable fasteners and seal grooves, and

5) Insertion of a fiberglass barrier layer between all carbon epoxy and aluminum parts.

The corrosion resistance of composite materials, the elimination of fasteners, and the leak resistance of cocured/bonded structure directly improve the supportability of an "all-composite" unitized structure. However, before highly unitized structure is fielded, a significant amount of development work needs to be undertaken to define appropriate repair procedures and criteria. As previously discussed, three dimensionally reinforced composites may offer the potential to reduce the required level of support and frequency of repair, but the difficulty associated with implementing repairs will certainly increase. In highly integrated designs where blended load paths between the skin and substructure require repair, repair facilities must have procedures to reestablish functional capability.

C. Primary Sandwich Structure

Sandwich structure is the most weight-efficient design for a large percentage of aircraft structure because of buckling stability requirements. Unfortunately, the use of sandwich structure is limited to state-of-the-art aircraft due to high manufacturing and supportability costs experienced with metallic HC structure on earlier aircraft. The B-58, which used a high percentage of aluminum sandwich construction, is shown in Fig. 2.54. Mission requirements of the B-58 dictated a low structural weight fraction of 0.24. In addition, the vehicle required good thermal and sonic fatigue properties; the only way that these capabilities could be achieved was with extensive integration of sandwich construction.

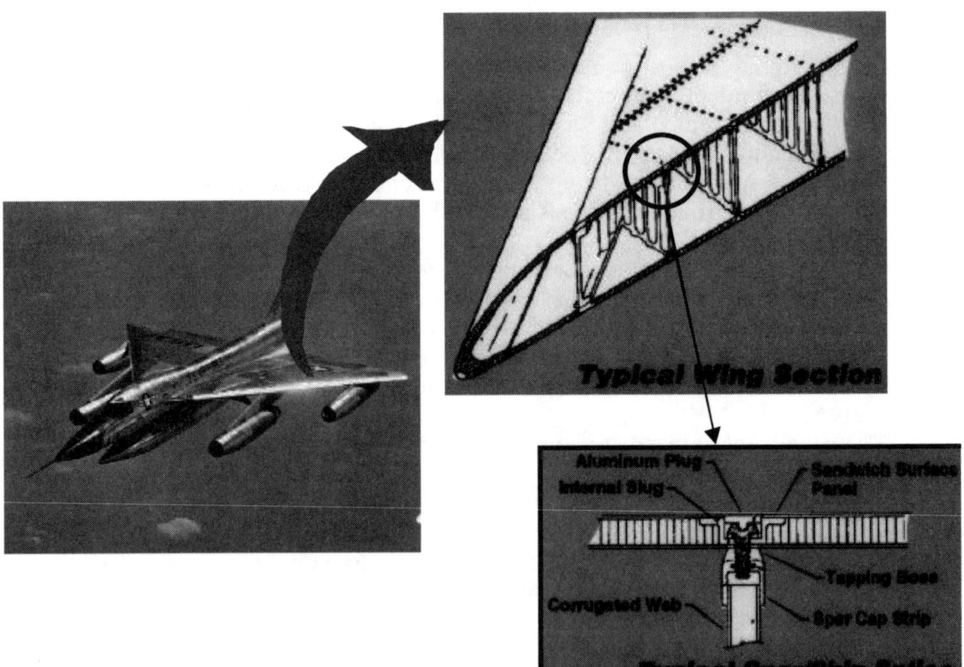

Fig. 2.54 B-58 construction approach.

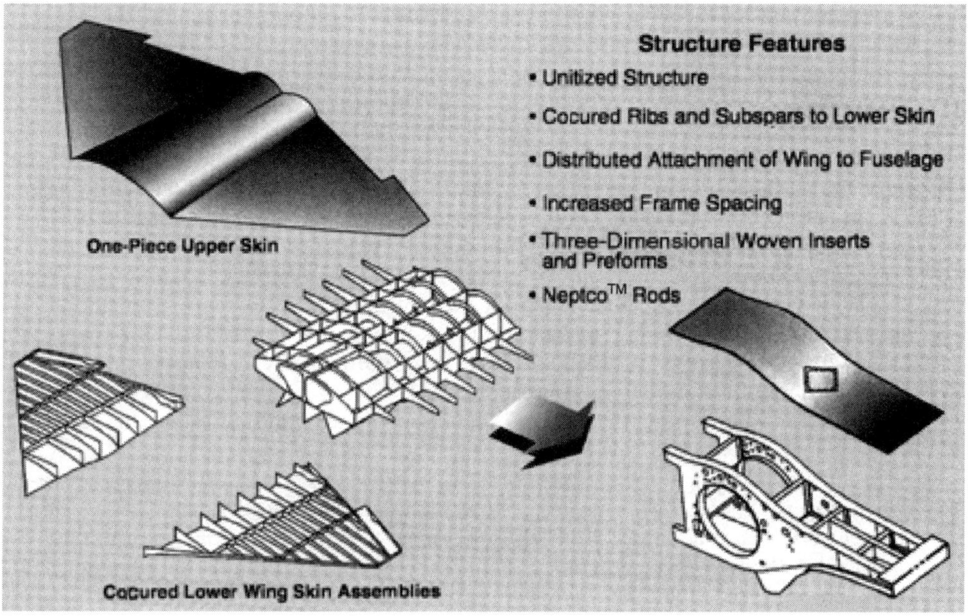

Fig. 2.55 Unitized sandwich assembly.

The airframe used spars that were spaced 15 in. apart to reduce the parts count. The illustration shows a typical wing joint in which the sandwich was full depth through the joint with a metallic insert (internal slug) for joining the skin to the spars. This insert provided a foundation for mechanically fastened skins, similar to the function provided by textile preforms in advanced composite sandwich configurations previously discussed. The B-58 structure experienced high maintenance costs due to moisture retention, severe corrosion, and damage-tolerance problems.

The unitized center fuselage, shown in Fig. 2.55 makes extensive use of noncorrosive advanced composite foam-filled sandwich structure, three-dimensional composite preforms, and Z-Fiber™ in joint intersections. These technologies will enable sandwich composite in future aircraft since they eliminate many of the historic problems that have been encountered.

Investigation to characterize the benefits of modern sandwich technology has been conducted.[37] This effort was directed at an all-composite section of a blended wing-body fighter aircraft and was compared to a baseline structural configuration representative of current day technology. The generic baseline vehicle configuration consisted of composite skins mechanically fastened to titanium/aluminum substructure. The wing of the baseline structure was designed with titanium attachment lugs typical of fighter aircraft to react the large loads induced from shear and bending moments of the wing. The substructure of the baseline is fabricated using advanced machining techniques that represent the state of the art for the present day manufacturing environment. The baseline offered a high level of detail, including cutouts for major subsystems and routings and a pylon fitting integrated to the wing. The sandwich configuration was developed with a similar level of fidelity, also including the use of a pylon fitting, and featured cutouts for major routings. The construction of the sandwich structure involved establishing a representative section that considered the benefit due to increased bulkhead/spar spacing and eliminated the wing attach root fittings by incorporating a blended body design. The lower skin of the structure was sized for a cobonded Z-Fiber™ configuration and the upper skin was mechanically fastened with embedded monolithic stiffeners in the cap structure, so no fasteners extend through the sandwich core. The sandwich design used to size the skins and substructure was mostly foam-filled HC core. The upper and lower skins used a titanium HC core filled with light-weight Rohacell™ foam, and the majority of the substructure panels used a Nomex™ core with the same foam. Mechanical testing of these core systems subjected to impact events under elevated temperature, and moisture conditioning has demonstrated a high level of durability. The foam in combination with the titanium core was shown to inhibit moisture progression and is not susceptible to corrosion. The conceptual configurations of the baseline and advanced configurations that were evaluated are shown in Fig. 2.56.

The weight savings calculated are shown to be approximately 36.2% when comparing the advanced design with baseline configuration. Primary sources of weight savings are achieved from elimination of the metallic load reaction lugs, skins and the benefit derived from the increased bulkhead spacing. Some benefit has been achieved from the advanced design, offering increased structural efficiency, but the primary contributor is from the reduced density of composite material. The overall cost reduction benefit is approximately 20%, which can be attributed largely to the reduced part count achieved through unitization. It is interesting to note that the actual cost per pound of the structure actually increased in the advanced configuration. This fact should not be overlooked because, in general, alternative designs are evaluated on their ability to reduce the cost per pound of structure ratio. Therefore, when evaluating composite design configurations, caution must be exercised when comparing the potential benefits that can be achieved.

BASELINE

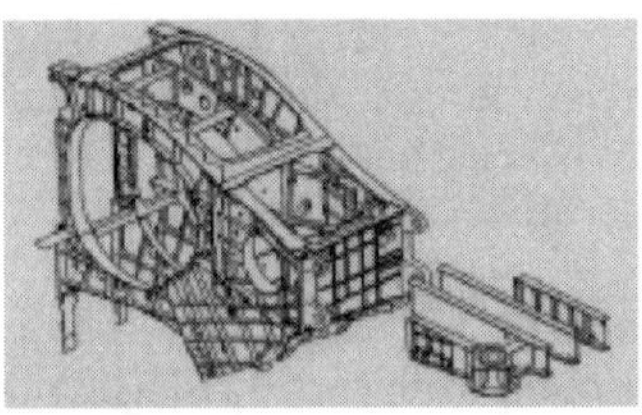

Parts Count	24 Parts
Fasteners	20.1 lb
Sealant / Shims / Radius Blocks	7.1 lb
Composite Parts Weight	126.1 lb
Metallic Parts Weight	475.4 lb
Total Weight	628.7 lb

SANDWICH

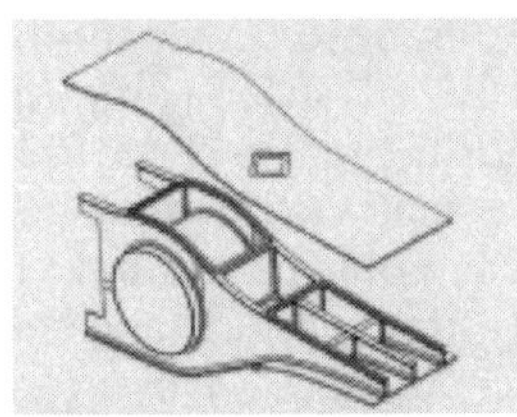

Parts Count	13 Parts
Fasteners	15.1 lb
Sealant / Shims / Radius Blocks	0.6 lb
Composite Parts Weight	385.4 lb
Metallic Parts Weight	0.0 lb
Total Weight	401.1 lb

WEIGHT SAVINGS 36.2%

Fig. 2.56 Weight payoff of sandwich structure.

D. Innovative Structural Layouts

Many vehicle designs have been conceived to enable affordable composites. The following discussion offers some examples[6] of typical considerations when laying out an airframe to meet specific performance and operational requirements with cost as a primary consideration. A design approach is shown in Fig. 2.57 that incorporates a split skin at the center-line of the vehicle and continuous carry-through bulkheads. The key benefit to this design approach is the accessibility to internal systems by having smaller removable skin panels. The challenge in making this design functional is to shear axial loads from the skin into the substructure bulkhead caps so that bending loads can be reacted. This approach does not directly offer the greatest weight-savings method due to the associated inefficiency of the bulkhead cap section properties relative to the skin but does provide upper skin panels of reduced size and some accessibility to subsystems.

An alternative design approach used a continuous skin from fold to fold and a set of discontinuous bulkheads. This approach captured the structural efficiency of continuous skin and provided for simplified vehicle assembly. The primary challenge associated with this design is getting the load transferred from the bulkheads to the skin and maintaining skin stability. The shear that would need to be reacted at the interface of the discontinuous bulkhead would create a significant

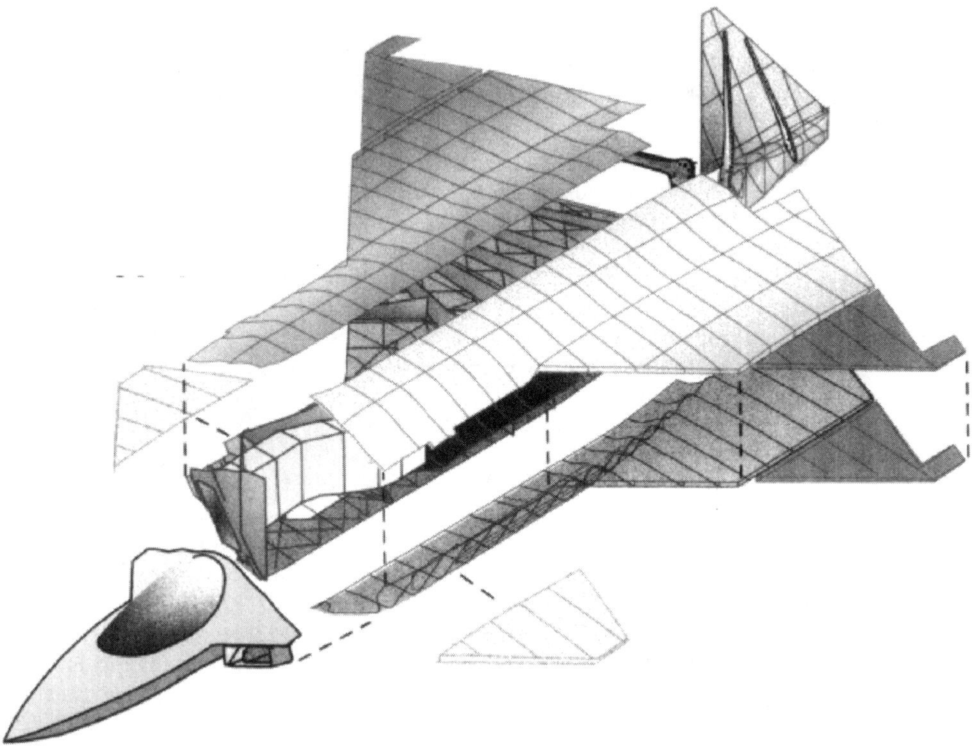

Fig. 2.57 Split skin—continuous bulkhead design. (See color section.)

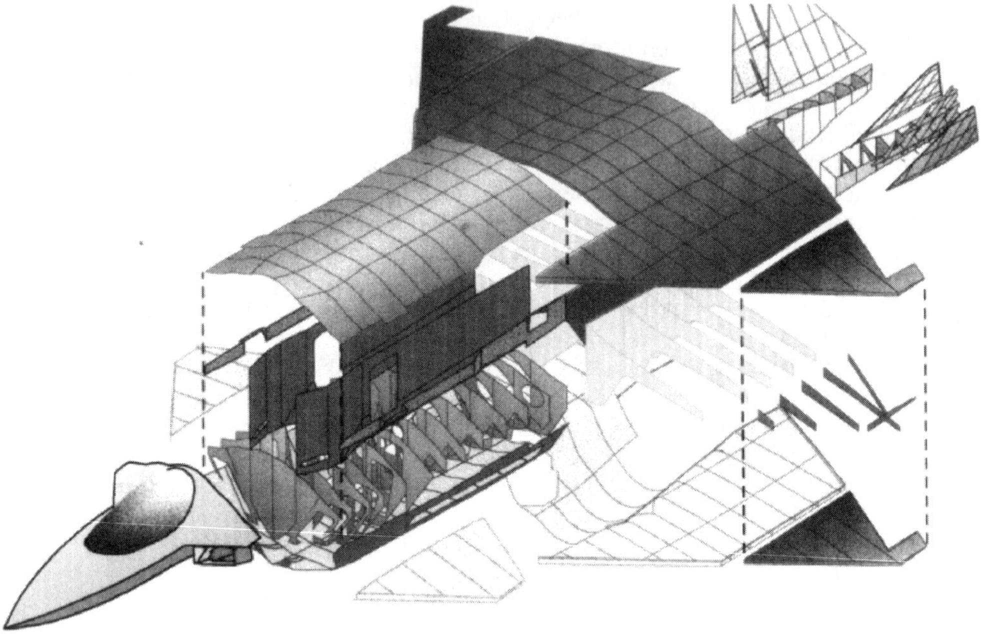

Fig. 2.58 Continuous skin—discontinuous bulkheads. (See color section.)

load path disruption unless the substructure was gradually tailored to minimize its load-carrying capability far enough outboard from the interface to fully transfer cross-ship loads into the skin. This conceptual design is shown in Fig. 2.58.

Additional structural layout configurations can be envisioned. By combining attributes of both of these structural configurations and given the needed requirements, current day designers feel that structural cost and weight can be optimized. Vectoring loads outboard in the skin by blending them into key carry-through bulkheads designed to maximize spar spacing will provide major wing/fuselage cost and weight reductions.

IX. Future Outlook for Composite Airframe Structures

Application of composite materials will continue to increase in future military and commercial aircraft structures. To extend the use of composite material beyond current application in fighter aircraft of approximately 25%, several key developments that have been discussed must be further understood. This specifically includes the application three-dimensional concepts that will enable the integration of composites to loading regimes that have not been possible in historical applications of laminated structures. This necessitates the development of guidelines and procedures that can be used for analysis, definition of failure, characterization of effects on structural durability, and establishment of supportability guidelines for unitized structure. The robust damage-tolerant design concepts discussed will most likely expand the designer's ability to accept composite structures with manufacturing defects beyond what is allowed today. In addition, in-service repair requirements will decrease as well, further reducing the cost of operations. This will come from the fail safety offered from these concepts and from a better understanding of composite material durability in general.

The development of affordable structure must involve an approach that exploits design for manufacturing. Low-cost structure starts with the design; however, to achieve the benefits offered by composites, enabling processes and low-cost manufacturing must be integrated to the development of the concept. For high-volume aircraft manufacture, the cost of hard tooling and the implementation of factory automation can be absorbed; however, in low-volume production, alternative processes must be considered. The evolution of low-temperature curable composite materials will continue to improve in performance capability. As these material developments become available, the amount of composite in aircraft will increase and associated costs will decrease. The aircraft industry is unique in the sense that high volume is typically measured in the thousands, whereas in other industries it is in the millions. Design for high volume and tooling for high-volume production will lead to increased automation, expensive (but durable) tools, and self-tooling design features. Design for low volume and tooling for low volume will lead to increased use of hand layup, throw-away tools, and self- tooling features. Major opportunities exist for cost savings by embracing design/manufacturing approaches that will not require the same level of quality control during manufacturing and from concepts that offer inherent tolerance flexibility to ease assembly complexity.

The requirements suite associated with fighters must be continually assessed to completely ascertain the tradeoff between necessary war fighting capability and low cost. Design approaches to enable the composite as part of a multifunctional system will bring consideration of the structural performance to par with other functional requirements of the system. Multifunctional structural systems will offer load-bearing capability and feature integrated subsystems. In turn, this will result in vehicles that offer improved overall efficiency and affordability.

References

[1]Talwar, R., "Supportable Technology for Affordable Fighter Structures (STAFS)," WL-TR-97-3024, 2 June 1997.

[2]Taggart, D., "Integrated Airframe Technology for Affordability," First Semiannual Rep., April 1995.

[3]Paul, D., Hess, T., and Kautz, E., "Evolution of US Military Aircraft Structure Technology," 37th Structures, Structural Dynamics and Materials Conference, 15–17 April 1996, AIAA-96-1571-CP, pp. 2232–2240.

[4]Raymer, D. P., "Aircraft Design: A Conceptual Approach," AIAA Education Series, Washington, DC, 1989.

[5]Kane, D. M., and Hangen, J. A., "Ultralightweight Structures," Oct. 1991, WL-TR-91-3034.

[6]Baron, W., and Leger, K., "Composites Affordability Initiative, Phase I: Concept Design Maturation," Structures, Structural Dynamics and Materials Conference, April 1998, AIAA-98-1874.

[7]Vosteen, L. F., and Hadcock, R. N., "Composite Chronicles: A Study of the Lessons Learned in the Development, Production, and Service of Composite Structures," NASA Contractor Rep. 4620, Nov. 1994.

[8]Doerner, F., et al., "Low Cost Composite Processing," AFRL-ML-WP-TR-1998-4005, Oct. 1997.

[9]Strong, K., "Electron Beam Curing of Composite Materials," WL/MLBC Study Final Rep., June 1995.

[10]Abrams, F., and Benson, T. T., "An Analysis of E-Beam Potential in Aerospace Composite Manufacturing", 42nd International SAMPE Symposium, 1997, pp. 548–557.

[11]Goodman, et al., "Automated Tape Placement with In-situ Electron Beam Cure," 44th International SAMPE Symposium, 23–27 May 1999, p. 269.

[12]Taggart, D., et al., "Integrated Airframe Technology for Affordability," Final Draft Rep., AFRL, 1999.

[13]Vastava, R. B., "An Assessment of E-Beam Technology for Aircraft Applications", 43rd International SAMPE Symposium, May 1998, pp. 1681–1690.

[14]Leisy, R. A., Pannell, D., and Matsen, M., "Induction Heating for Affordable Polymer Composite Structures," WL-TR-97-4093, July 1997.

[15]Wedgewood, A. R., and Hardy, P. E., "Induction Welding of Thermoset Composite Adherends Using Thermoplastic Susceptors," Third International Conference on Composites Engineering, 1996, p. 915.

[16]Cochran, R., et al., "Advanced Composite Processes for Aerospace Applications," 42nd International SAMPE Symposium, 1997, pp. 653–540.

[17]Goetz, F., and Walsh, S., "Monitoring Resin Impregnation in Graphite Reinforcement," 4th International SAMPE Symposium, 23–27 May 1999, p. 670.

[18]SAMPE Seminar: Path to Affordable Composite Structures, "Myth or Reality," 14 August 1996.

[19]Dexter, B., "Innovative Textile Reinforced Composite Materials for Aircraft Structures," 28th International SAMPE Technical Conference, 4–7 November 1996, pp. 404–416.

[20]Clarke, S., "Engineered Textile Preforms for RTM: A Comparison of Braiding, Knitting and Weaving Technologies," 29th International SAMPE Technical Conference, 28–November 1 October 1997, pp. 21–31.

[21]Chavka, N., and Dahl, J., "P4: Glass Fiber Preforming Technology for Automotive Applications," 44th International SAMPE Symposium, 23–27 May 1999, pp. 281–292.

[22]Rondeau, R., Reeve, S., and Bond, G., "The Effect of Tows and Filament Groups on the Properties of Discontinuous Fiber Composites," 44th International SAMPE Symposium, 23–27 May 1999, pp. 1449–1460.

[23]Goodman, D., et al., "Automated Tape Placement with In-Situ Electron Beam Cure," 44th International SAMPE Symposium, 23–27 May 1999, pp. 269–280.

[24]Harper-Tervet, J., Reeve, S., Schwarz, J., Stratton, R., Tervet, F., Barnes, A., and Benson Tolle, T., "Superplastic Diaphragm Forming of Thermoplastic Complex Shaped Composites as a Cost Effective Manufacturing Process," 29th International SAMPE Technical Conference, 28 Oct–28 Nov 1997, pp. 383–394.

[25]Costen, R. C., and Marchello, J. M., "High Productivity Model for In-Situ Tow Placement of Thermoplastic Composites," 41st International SAMPE Symposium, 24–28 March 1996, pp. 1346–1360.

[26]Burgess, K., and Paradis, S., "Application of 3-D Woven Preforms to Laminated Composites," Paper No. 95-3890, *1st Aircraft Engineering, Technology, and Operations Congress*, AIAA, Sept. 1995.

[27]Fisher, K., "3D Reinforcements: on the Verge?" *High Performance Composites Magazine*, Mar./Apr. 1996.

[28]Bersuch, L., Hunten, K., Baron, W., and Tuss, J., "Three Dimensional Reinforcement in Composite Structures," AGARD Conference on Bolted/Bonded Joints in Polymeric Composites, Italy, 1996.

[29]Northrop Grumman, "BMI Composite Sandwich Box Beam Briefing," HSR/HSCT Annual Meeting, 17–18 Nov 1998.

[30]Gause, L., and Alper, J., "Braided to Net Section Graphite/Epoxy Composite Shapes," *Journal of Composites Technology & Research*, 1988.

[31]Childress, J., et al., "Structural Technology and Analysis Program (STAP): Hydrodynamic Ram Decoupled Structural Fuel Cells," WL-TR-97-3118, July 1997.

[32]Campell, T., and Richard, J., "Application of Advanced Intersecting Structure Technology to Reduce Cost and Weight of Airframe Composite Components," Air Force Contract F33615-94-C-3207, May 1997.

[33]Anderson, T., and Holzwarth, R., "Design and Manufacture of Low Cost Composite-Bonded Wing," 39th Structures, Structural Dynamics and Materials Conference, AIAA 98-1870, April 1998.

[34]Nunn, K., Dompka, R., and Jones, L., "Design and Manufacture of Low Cost Composites, Bonded Wing," WL-TR-94-8007, Dec. 1993.

[35]TOLO, Inc., "Composite Grid-Lock® Wing Box Assembly," Final Rep., Air Force Contract F33615-96-C-3212, April 1997.

[36]Leissa, A., "Buckling of Laminated Composite Plates and Shell Panels," AFWAL-TR-85-3069, June 1985.

[37]Sheahen, P., Lockheed Martin Company, "Robust Composite Sandwich Structures (ROCSS)," Semi-annual Rep., Air Force Contract F33615-94-C-3210, April 1997.

[38]Jacobsen, M. J., Heitz, R. M., and Yamane, J. R., "Survivable Composite Integral Fuel Tanks, Vol. I. Manufacturing, Testing, and Analysis," AFWAL-TR-85-3085, Jan. 1986.

[39]Czarnecki, G., "High Speed Airflow Effects on Battle-Damaged Composite Surfaces," AFWAL-TR-87-3090, April 1987.

[40]Baron, W., "Survivability of Multispar Cocured Composite Wing Structure," 37th Structures, Structural Dynamics and Materials Conference, AIAA-96-26808, April 1996.

[41]Allen, M. R., Sawdy, D. T., Bradley, S. J., and Avery, J. G., "Survivability Characteristics of Composite Compression Structures," AFWAL-TR-88-3014, JTCG/AS-87-T-005, May 1988.

[42]Holzwarth, R., "The Structural Cost and Weight Reduction Potential of More Unitized Aircraft Structure," 39th Structures, Structural Dynamics and Materials Conference, AIAA 98-1872, April 1998.

Chapter 3

Aging Systems and Sustainment Technology

John W. Lincoln
Aeronautical Systems Center, Wright-Patterson Air Force Base, Ohio

I. Historical Background

A. Introduction

MOST aircraft operated commercially or in the military reach a state referred to as aging sometime after entering operational service. Aging of an aircraft is not the same as becoming obsolete. An aircraft may be obsolete before it reaches the aging state or, more typically, it reaches the state of aging before it is obsolete. A commercial aircraft is obsolete when it is no longer economically viable to keep it operational. A military aircraft is obsolete when its capabilities are no longer competitive with potential adversaries. The time when an aircraft reaches the aging state is usually much more difficult to determine. It is important to distinguish between characteristics of the structure of a young aircraft and an aging aircraft. A young aircraft is one that continues to be airworthy with the maintenance program prescribed at the time of manufacture. The primary concern with a young aircraft is the potential for design errors that introduce unintentional high stresses in the structure that could lead to premature fatigue-cracking incidents. When these are discovered, the structure is modified to eliminate the problem. An aging aircraft may be characterized as one where the effects of corrosion and cracking from fatigue require modification of the maintenance program to retain adequate structural integrity. The word adequate here means that the expected number of failures would be less than one in a given fleet of aircraft. As an aircraft accumulates calendar time and flight time, the effects of corrosion and cracking from fatigue as well as accidental damage lead to repairs on the aircraft. Cracking from fatigue can be so widespread that it degrades the integrity of the structure. When this occurs, the structure is said to be in a state of widespread fatigue damage (WFD) and must undergo modifications to remove this problem. In addition, as an aircraft accumulates flight time, it may exceed its design life goal. Therefore, the maintenance program will require modification to include additional structural inspections. If the initial maintenance program requires modification from any of these events, then the aircraft may be considered to be in a state of aging.

No one should be surprised that there are aircraft all over the world today in a state of aging. Economic considerations demand that aircraft be operated long beyond originally identified retirement times. One reason for keeping aircraft in the inventory is that technological advances allow currently designed aircraft to effectively perform their mission for much longer than previously possible. An aircraft, even when sold by one airline, sees extended life in another airline's operations. In the commercial sector, new aircraft tend to be evolutionary in their designs. Consequently, they are maintained in service until they are not economically viable to operate. The cost of new aircraft, particularly for the military, is enormous. Each new military aircraft is a revolutionary change from the previous model since the services must maintain combat effectiveness in an environment of ever-changing threats. Therefore, military aircraft stay in inventory until they are operationally obsolete or they are no longer economically viable to operate. The United States Air Force (USAF) retired many F-4 aircraft because they were obsolete as a weapon system rather than being economically nonviable. In the case of the KC-135, the USAF plans to keep these aircraft operational to the year 2040 since they believe it will be economical to operate them until that time. They would likely not be obsolete in the year 2040. If the USAF can maintain these aircraft operationally until 2040, their service life will be approximately 80 years. When the USAF procured these aircraft, they planned for a service life of about 20 years.

Sustainment of an aircraft is the act of keeping it operational (i.e., airworthy). Maintenance of an aircraft (that is, the work done by mechanics in keeping it airworthy) is one aspect of sustainment. However, sustainment also includes engineering analyses and tests needed to determine an adequate maintenance plan for the aircraft. Sustainment is life management. One task of sustainment is the determination of structural inspections based on damage-tolerance principles. These inspections protect against failure from defects that could be in the structure because of manufacturing or operational service. The approach for developing a damage-tolerance-derived inspection program is discussed later. Another task of sustainment is the determination of the time of onset of fatigue cracking in the structure so widespread that the structural integrity of the aircraft could be compromised. Experimental evidence[1] shows that fatigue cracks smaller than those that could be easily detected by current inspection methods could constitute WFD.

Today, the primary concern with aging aircraft is the cost of their sustainment. Commercial operators buy new aircraft when it becomes economically viable for them to do so. The aging aircraft problem, however, has often made itself known to both commercial and military operators through failures of in-service aircraft. Both operators found that maintenance programs did not adequately protect aircraft as they progressed through their service lives. Failures in both commercial and military aircraft have been the primary factor that has changed rules and specifications that are the basis of their design. In many cases, failures have identified threats to structural integrity that were not previously identified by certification authorities. In many cases, commercial failures have influenced military specifications and military failures have influenced commercial aircraft rules. New rules and specifications have lead to better maintenance programs that help alleviate many of the threats to failure that previously existed. However, the economic demand to fly these aircraft longer has emphasized the need to reexamine these aircraft for the possibility of WFD, corrosion damage, and loss of damage-tolerance capability through repairs.

It is difficult to determine the exact moment in time when an aircraft has reached the state of aging. However, all would agree that costs associated with repairs or

Table 3.1 USAF fleet

	Date of combined fleet				
Summary	9/93	9/94	9/95	9/96	9/97
Fighters and attack	3410	3140	2967	2740	2968
Bombers	227	198	203	208	208
Transports and tankers	1975	1971	1912	1921	1864
Trainers	1234	1214	1202	1192	1232
Reconnaissance	92	85	85	80	72
Helicopters	235	227	226	215	215
Grand total	7153	6835	6595	6536	6289
Average age in years	16.9	17.1	17.3	17.6	18.6

modifications from corrosion or cracking from fatigue would be an indicator of this condition. When these costs rise significantly, then the aircraft are certainly in that state. Chronological age is not always a good indicator of aging. However, because corrosion and fatigue are somewhat related to time in service, it does give some insight for the potential of this problem. A summary presented in July of 1997[2] detailed reasons for the increase in chronological age of aircraft. In 1996, the total U.S. military fleet consisted of 20,400 aircraft. Reduced procurement budgets will prevent these aircraft from being retired because, at the current rate of procurement, the time required to replace these aircraft would be approximately 100 years. The basis for this conclusion is that, in 1996, the U.S. Department of Defense budget procured only 159 new aircraft for $5.4 billion. The estimate of the cost of replacing the entire fleet of U.S. military aircraft is $530 billion in 1995 dollars. Many of the aircraft in the population today are currently more than 30 years old. Table 3.1[2] derived from the open literature shows the ages of USAF systems and the decline in the total number of aircraft.

The situation for the U.S. airlines is not quite as grim. However, at the 1995 rate of procurement, replacement of the current airline fleet would require approximately 38 years. This situation has improved since the U.S. economy supported procurement of new commercial aircraft earlier than expected. The more prosperous airlines procure new aircraft and sell the ones they retire to less prosperous operators around the world. Therefore, the problem with the sustainment of aging aircraft is not relieved.

Hadcock[2] studied the cost of maintaining both military and commercial aircraft in operational service. Table 3.2 presents some of his findings. The table shows that there is a considerable difference in the ratio of the flyaway to the maintenance cost (number of flight-hours for the total maintenance cost to equal the purchase price) between military and commercial aircraft. Part of this difference is because of the relatively low utilization rate for military aircraft compared to commercial aircraft. Consequently, the cost of corrosion-maintenance per flight-hour, which is more time dependent than flight-hour dependent, is much larger for military aircraft. In addition, maintenance costs identified include all modifications and avionics. These costs are typically much larger for military aircraft than for commercial

Table 3.2 USAF and commercial aircraft costs in 1995 dollars

Aircraft	Operational year	Flyaway cost, $M	Maintenance cost, $/flight-hour	Flyaway/ maintenance cost
F-4E	1966	8	2,560	3,130
F-16A	1978	25	6,200	4,030
C-130E	1963	32	3,440	9,160
C-5A	1969	163	8,800	18,500
DC-9	1965	23	560	41,000
B737-200	1968	26	440	59,000
DC-10	1971	101	1,500	67,000
B767-200	1982	89	690	129,000
B747-100	1970	110	1,800	61,200
B747-400	1989	155	1,260	123,000

aircraft since military aircraft require modernization for changing missions as the military threat changes.

Table 3.2 also shows the difference between the B747-100 and B747-400. Although the B747-400 initial cost is considerably higher, the maintenance cost per flight-hour for the B747-400 is considerably less. One may attribute at least a part of this to the significant improvement in the corrosion protection system enjoyed by the B747-400. These kinds of differences also help explain why operators try to remove older aircraft from inventory as soon as possible.

This chapter will concentrate on issues with the aging airframe on fixed-wing aircraft. There are many aging mechanical subsystems on aircraft as well. Many of the issues are similar to those found in the airframe. However, one of the main concerns with mechanical subsystems is obsolescence. Many of these systems cannot be maintained because manufacturers are no longer in business or do not make parts any longer. The obsolescence factor is particularly acute for avionics. Both commercial and military operators are in the process of rethinking the architecture of avionics systems to combat obsolescence. Engines have some of the aging problems as the airframe. However, thermal and high cycle loading create environments unique to engines.

B. Aging Commercial Aircraft Research at the Federal Aviation Administration Technical Center

Commercial aircraft failures in the 1940s and 1950s caused significant modifications of the approach to achieve structural integrity. One of these fatigue failures was in the wing of the Martin 202 on 29 August 1948. This failure demonstrated that the wing design was such that the failure of a single member of the wing structure could lead to catastrophic loss of the aircraft. Possibly even more significant were the deHaviland Comet fuselage failures. The first one of these occurred on 10 January 1954 and the second occurred on 8 April 1954. The Comet failures showed that a failure of a single member in the fuselage could lead to catastrophic loss of the aircraft. These failures demonstrated to authorities that an alternate load path was essential to help ensure the structural integrity of large commercial transport aircraft. That is, designs needed to be fail-safe. Swift[3], in his comprehensive report on the Comet failures, also describes fail-safe design practices used in the design of the McDonnell Douglas DC-10 aircraft. Swift recommends that, to account for the possibility of WFD occurring in one projected lifetime, at least two lifetimes of testing should be completed on a full-scale representative section of the aircraft.

In 1978, the Federal Aviation Administration (FAA) significantly changed their program for aging aircraft. Richards[4] from the CAA in the United Kingdom guided this transition. It was at this time that the FAA adopted the use of damage-tolerance-derived inspections of high time aircraft. The primary motivation for this change was concern over continued structural integrity of aging aircraft. On 14 May 1977, a catastrophic failure due to fatigue in a Boeing 707 horizontal tail[5] brought this concern into focus. Boeing designed this structure to be fail-safe; however, the certification authority did not require an inspection program to examine the structure for a failed element. This incident proved there was a need for an inspection program to augment the fail-safe design approach. An earlier failure of an F-4 on 23 January 1973 at Nellis Air Force Base convinced the USAF that a structure was not fail-safe unless the inspection program could find the first failure before the structure failed catastrophically.

Because the technology for performing these damage-tolerance assessments (DTAs) was available to manufacturers in the 1970s and early 1980s, there was little emphasis by the FAA on supporting structurally related research and development activities. This changed dramatically after the 1988 failure of a Boeing 737 due to WFD. This event provided the motivation for the considerable emphasis by the FAA on structural issues associated with aging aircraft. This event occurred on Boeing 737 (N73711) on 28 April 1988. On this date, cracks in the fuselage lap-splices coalesced, resulting in loss of the upper fuselage from just aft of the pilot's cockpit to the wing leading edge.

Boeing 737 (N73711) was number 152 in the production line for this aircraft. The early 737s (line numbers 1–291) were manufactured with cold-bonded lap-splices. The service history indicated quality control problems. The bond quality could also be degraded by moisture absorption by a scrim cloth that was an integral part of the joint. These difficulties lead to bonds with low environmental durability with susceptibility to corrosion and with some areas of the lap-splices that may not have bonded at all. In service, moisture could enter the joint in areas of disbond and corrosion could occur. Moisture and corrosion, in some cases, contributed to further disbonding of the joint because of the accumulation of oxides, water wicking in the joints, and freeze–thaw cycles.

Boeing designed the lap-splice in the 737 such that there was adequate strength in case of a bond line failure in the pristine structure. However, the skin thickness was 0.036 in. in the lap-splice where failure occurred. This meant that countersunk rivet hole preparation left a knife-edge in the skin. Therefore, in case of bond line failure, this feature led to high bearing stresses and early cracking in the skins.

The National Transportation Safety Board[6] cited the Aloha Airlines maintenance program, Aloha Airlines management, Boeing, and the FAA for deficiencies relating to this failure. This event also spawned a major effort in aging aircraft research and development. The FAA brought together most of the aviation community at

Table 3.3 Impact of ATA actions

Aircraft designation	Number of new airworthiness directives	Number of aircraft affected	Cost per aircraft in millions of dollars
Boeing 727	76	1,710	1.0
Boeing 737	58	1,500	0.9
Boeing 747	31	680	2.3
DC-9	52	919	0.6

the First International Conference on Aging Aircraft held in Washington, D.C., 1–3 June 1988 to discuss issues relating to airframes, engines, nondestructive inspections, and human factors of inspection. Numerous actions resulted from this meeting, many of which involved the Air Transport Association (ATA). The ATA formed a joint ATA, Original Equipment Manufacturer (OEM), and FAA group. This group became the Airworthiness Assurance Working Group (AAWG) and is now a part of the FAA Aviation Rulemaking Advisory Committee (ARAC). They formed three subgroups composed of United Airlines, Boeing, and the FAA; Delta Airlines, Lockheed, and the FAA; American Airlines, McDonnell Douglas, and the FAA.

The results of this group's actions are shown in Table 3.3. This table shows the effect on a sample of aircraft. New airworthiness directives made actions mandatory that had previously been service bulletins. It is seen that their actions were significant in terms of economic impact. It is easy to conclude that these actions had a severe impact on the market value of used aircraft.

These model-specific subgroups have demonstrated a cooperative determination to enhance the safety of large category transport aircraft. Achievements have been impressive in accomplishing five of the original tasks identified by the AAWG. The five tasks were 1) review service bulletins, 2) develop a corrosion control program, 3) review the basic maintenance program, 4) review the supplementary structural inspection document programs, and 5) assess repairs. In 1991, the AAWG added an additional task of assessing the WFD situation.

The Aloha incident also had a significant impact on the structural research efforts of the William J. Hughes FAA Technical Center at Atlantic City, New Jersey.[7] They adopted an aggressive program in many aspects relating to aging aircraft, including the following:

1) Fatigue/fracture,
2) Corrosion,
3) Flight loads,
4) Nondestructive evaluation/inspection,
5) Human factors, and
6) Maintenance and repair.

However, the main thrust of their efforts was the development of technology to assess WFD.

The Aloha incident was not the only operational failure caused by WFD. Another one in Japan was much more serious but did not receive the notoriety in the United States that the dramatic Aloha incident received. WFD resulting from a faulty repair of the aft pressure bulkhead of a JAL 747 caused the failure of JAL Flight 123 from Tokyo to Osaka. Total casualties resulting from this accident were 520.

The FAA in the United States has focused much of their attention on aging aircraft issues in the commercial fleet of aircraft in both regulatory and research arenas. There are close parallels, however, between military and commercial approaches for damage-tolerance-derived inspection programs. There has also been a mutual need to develop the capability to estimate the time of onset of WFD that could degrade the damage-tolerance capability of aircraft.

Congress supported the FAA efforts by passing the Aviation Safety Research Act of 1988, or Public Law 100–591. This statute enabled the FAA to initiate an extensive research program addressing nondestructive inspection, analysis of fracture and fatigue mechanisms, corrosion, and issues relating to human factors in the maintenance environment. The Aging Airplane Safety Act of 1991 directed the FAA to initiate rulemaking that would ensure the continuing airworthiness of aging aircraft. In response to this Act, the FAA prepared a rule that will require FAA inspectors or their designated airworthiness representatives to periodically inspect each aging aircraft and review the records of that aircraft.

There have been many highlights in the FAA program for aging aircraft.[8] The First International Conference on Aging Aircraft was held in Washington, D.C., in 1988. It was also in this year that the FAA began the Technical Oversight Group for Aging Aircraft (TOGAA). This group has advised the FAA on many subjects, including new rules for both aircraft and engines, including rotorcraft. In addition, they have advised the FAA Technical Center on research and development activities for aging aircraft. The group is composed of experts in the disciplines of aeronautical structural engineering and maintenance.

In 1989, the FAA issued the airworthiness directive (AD) action for corrective action for the Boeing model aircraft with cold-bonded lap-splices. This was significant in that the corrective action was to start at 40,000 flight cycles although the Aloha incident did not occur until the aircraft had 89,680 flight cycles. The FAA recognized the aircraft at 40,000 flights was at the point of onset of WFD. They intended that the modification to lap-splices would restore the fail-safety the aircraft possessed when originally certified.

The FAA implemented an Aging Aircraft Modification Program to reduce reliance on intensive inspections for known problems.

The FAA recognized that nondestructive inspections would play a significant role in the maintenance of safety of aging aircraft. This led them to establish the Center for Aviation System Reliability Research led by Iowa State University with Northwestern University, Wayne State University, and Tuskegee University in 1990.

The FAA formed the Titanium Rotating Components Review Team in 1989. This action was the result of the DC-10 accident in Sioux City, Iowa, on 19 July 1989, where a hard alpha inclusion in the titanium stage one fan disk in the center engine caused the failure. This accident took the lives of 110 passengers and a flight attendant. The team provided their findings in an internal FAA report on 14 December 1990, in which they recommended the use of inspections of engines based on the damage-tolerance philosophy.

The FAA recognized that they must validate nondestructive inspection (NDI) methods that operators planned to use on commercial aircraft. For this purpose, they established the Aging Aircraft NDI Validation Center (AANC) at Sandia

National Laboratories in 1991. In addition to commercial activities at this installation, the USAF uses this facility for ensuring planned inspections will have desired reliability. The FAA–AANC has created an environment for inspection instrument manufacturers to test their products under conditions simulating operational conditions. The AANC has also been active in probability of detection determination for various inspection techniques.

In October of 1992, the FAA established their first Center of Excellence. They located this center jointly at Rutgers, the State University of New Jersey, and Georgia Institute of Technology,[9] the FAA Center of Excellence for Computational Modeling of Aircraft Structures. FAA research and development programs in analyses methodology to predict the point of onset of WFD relied heavily on the work done by this Center of Excellence.

In addition, in 1992, they issued ADs to mandate the modification and inspection program on the original 11 models of commercial aging aircraft. This action directed manufacturers to identify modifications needed to maintain fail-safety required by original certification.

In 1993, they published Notice of Proposed Rulemaking for revising 14 CFR Part 25.571 to establish minimum full-scale fatigue testing for new aircraft models. This rule established that the manufacturer would test a new aircraft for two lifetimes to ensure that WFD would not occur in its design lifetime. This was not specifically an aging aircraft rule but it serves as a standard for determining safe operating periods for aging aircraft.

In 1993, the FAA established a consortium consisting of Iowa State University and various turbine engine manufacturers to develop reliable and cost-effective methods or improvements to existing methods for in-service and production inspection of titanium safety critical rotating components. The Sioux City accident was the motivation for this activity.

They issued selected airworthiness directives to mandate incorporation of corrosion prevention and control programs in the maintenance of the original 11 models of the transport aging fleet in 1994. This AD action, when followed by operators, should preclude failure as the result of corrosion damage.

In 1995, they formed a Rotorcraft Community Working Group made up of representatives from manufacturers, operators, the research community, and the FAA Southwest Directorate. The group's purpose is to define, address, and reach consensus on aging issues. This group has laid the foundation for the use of damage tolerance in rotating parts of helicopters.

The FAA William J. Hughes Technical Center in Atlantic City, New Jersey, collaborated with the Naval Air Warfare Center, Aircraft Division, Patuxant River, Maryland, to leverage the U.S. Navy's sophisticated landing data acquisition and analysis technology. The Navy developed this technology specifically for naval carrier aircraft. The FAA conducted a video landing parameter survey at Ronald Reagan Washington National Airport in Washington, D.C., and recorded 1070 landings. The results of this study have shown that current sink speed distribution for wide-body aircraft may be unconservative.

In 1996, the FAA revised operating rules requiring 14 CFR Part 135 aircraft having seating of ten passengers or more used in air transportation to meet 14 CFR Part 25.571 damage-tolerance requirements. This includes establishing a schedule to phase these requirements into existing aircraft having seating for 10 to 30 passengers.

The FAA reaffirmed their commitment to aircraft safety by establishing the Damage Tolerance ARAC Working Group in 1997. In 1998, they tasked ARAC to recommend actions required for completion of WFD evaluations on all transport aircraft in a timely manner. In 1998, they established a testing capability at the Technical Center. This facility is capable of testing full-scale curved panels and researchers to study fatigue cracking in fuselage pressure shells through simulation of the effects of pressure, tension, and shear. This capability will provide validation of numerical processes developed by the FAA as well as provide valuable test data to understand the behavior of these structures.

In 1999, they published AC 91-MA for damage-tolerance analysis of commuter aircraft. This action was significant in that they required an inspection program on commuter aircraft based on damage tolerance comparable to that performed on large category transport aircraft.

In October 1998, the FAA expanded the scope of their aging aircraft program to include the aging of nonstructural systems with the release of the FAA aging transport nonstructural systems plan. This plan was put in place to address concerns with potential safety issues associated with problems that may develop in transport category airplane systems as a result of wear and degradation in service. This plan involves study of the existing transport airplane fleet and development of appropriate approaches to mitigation of the effects of system aging, if such action is determined to be necessary.

One important component of the aging transport nonstructural systems plan is research being conducted at the FAA Technical Center as described in the aging nonstructural systems research program (NSRP). The objectives of the NSRP are to develop technologies and techniques to ensure continued safe operation of aircraft electrical and mechanical systems. Research results will be used in support of new and pending regulatory action and to facilitate compliance with new and existing regulations. NSRP research initiatives on aging include development of nonstructural systems test and validation infrastructure, development of wire testing equipment, assessment of visual inspection, and the development of aircraft arc fault circuit breakers. The FAA is working closely with both the USAF and the Navy to conduct research in nonstructural systems.

C. Aging Commercial Aircraft Research at National Aeronautics and Space Administration Langley Research Center

The Langley Research Center (LaRC) recognized that they could make a major contribution to the solution of aging concerns identified by the AAWG. They examined their in-house talents and decided that they could contribute to the solution of the WFD problem. They decided to emphasize WFD in fuselages, which is the most difficult area of the aircraft to analyze because of the effect of pressure on the thin fuselage skins and other factors.[10] They also recognized that they could contribute to the development of NDI technology that is so important in the proper maintenance of an aging aircraft.

LaRC initiated the National Aeronautics and Space Administration (NASA) Aircraft Structural Integrity Program (NASIP) in 1992 and completed that phase in 1998. The goal of their effort was to advance technologies to safely extend the operational life of commercial transport aircraft. Objectives and benefits from this effort provided reliable NDI technology for crack and corrosion detection to replace costly and subjective techniques. In addition, they developed rigorous durability and damage-tolerance analyses to supplement current practices and transferred the technology through effective government/industry partnerships

Their aging aircraft program for structures demonstrated advancement in four areas. First, they continued their development of nonlinear structural analysis methods to replace linear methods commonly in use. Second, they developed advanced fracture mechanics analysis tools based on elastoplastic methods that used the crack tip opening angle (CTOA) fracture criterion. Third, they participated in the development of advanced stress intensity calculation methods with automatic remeshing to replace methods requiring manual remeshing and superposition. Fourth, they advanced the use of crack closure methods to replace empirically based fatigue crack growth life predictions.

Their aging aircraft program for NDI also yielded many significant results. They replaced the time-consuming handheld small probe for disbond detection with the rapid large-area thermal inspection system for disbond detection. They further reduced the existing 10% accuracy in determination of material thinning due to corrosion to 3% accuracy. In addition, they developed a procedure that improved the existing capability of obtaining a 0.9 probability of detection for 2.54-mm-length cracks in the first layer around rivet heads and 10.1-mm cracks in the third layer around rivet heads. The new procedure obtains a 0.9 probability of detection of 0.81-mm cracks in the first layer around rivet heads and a 0.9 probability of detection for 3.0-mm cracks in the third layer around rivet heads.

In 1993, they completed a procedure using thermal nondestructive instruments for rapid detection of disbonding. The motivation for this effort was that, if the bond failed, the joint would be susceptible to crack initiation though fretting damage and corrosion. They transferred this technology with beta site testing of a thermographic inspection system at the Boeing Company and at Warner Robins Air Force Base, Georgia.

In 1994, NASA completed an experimental study[11] of small crack growth in 2024-T3 and 7075-T6 aluminum in corrosive environments. Fatigue crack growth tests with naturally occurring cracks were conducted in laboratory air and salt water environments. Crack length measurements were made using an optical microscope and a replica technique, with detectable crack lengths of 0.03 mm and 0.01 mm, respectively. The study concluded with two major conclusions. The first was that small cracks exposed to salt water exhibited a threefold increase in fatigue crack growth rates compared to growth rates in air, similar to the increase observed for long cracks. The second conclusion was that small fatigue cracks with corrosion exhibited a much lower threshold of crack growth than small fatigue cracks without corrosion. This is an enabling technology for applications of damage tolerance to engines and helicopters. In addition, it provides a tool for the prediction of the onset of WFD. For this purpose, they participated in the development FRANC2D, FADD, and FRANC3D to calculate stress-intensity factors and they developed FASTRAN to predict fatigue crack growth behavior. The codes are currently in use by industry, government, and academic institutions.

In 1995, NASA developed methods to detect corrosion. Multiple-layer construction often hides corrosion damage, making inspections difficult and costly. To solve this problem, they developed thermal and ultrasonic nondestructive instruments for quantitative evaluation of corrosion damage. For transfer of this technology, they performed beta site testing of the thermographic inspection system. They did this at the Boeing Company and Basic Research Aircraft and beta site testing of ultrasonic inspection system at Basic Aircraft Research and Miami NDT (general aviation and commuter aircraft).

In addition, in 1995, they developed methods for small crack detection. The motivation for this effort was cracking at rivet holes. These cracks initiate under the rivet head or at the faying surface and thus may not be visible from the surface. To solve this dilemma, they developed the rotating self-nulling eddy current probe to detect small, subsurface cracks. They transferred the technology by commercialization through Krautkramer Branson and Foerster.

In 1996, NASA completed the experimental verification of analysis methodology. The acceptance of methodologies for residual strength prediction requires extensive experimental verification. They conducted over 200 fracture tests on different coupon sizes and configurations and compared the results with analytical predictions. They developed innovative measurement techniques to verify that models were accurately describing the physical behavior observed in the tests. The technology for the CTOA criterion and fracture data are in use by industry, government, and academic institutions.

In 1998, they verified analysis methods for residual strength, remaining life, and inspection intervals. Residual strength methodology (previously verified by laboratory tests) needed to be verified for fuselage structures. This was done by the residual strength capability to the FRANC3D/STAGS nonlinear shell code and performing verification tests on curved stiffened fuselage panels with pressurization. They completed this effort by transferring FRANC3D/STAGS and FRANC2D codes, residual strength methodology, and fracture data to industry, the government, and academic institutions.

In addition, in 1998, they completed a detailed destructive examination of horizontal riveted lap joint structure[12] removed from a fuselage test article that had received 60,000 pressurization cycles. Three 6-ft-long lap joint panels containing visible outer skin fatigue cracks were removed from different regions of the fuselage. Visual and nondestructive examinations were first performed to characterize the extent of cracking present in the structure, followed by a destructive examination of more than 2,500 rivet holes. Visual and nondestructive examinations revealed cracks as small as 1.5 mm, while the destructive examination yielded cracks as small as 0.1 mm. Destructive examinations identified 1) damage mechanisms contributing to fatigue crack initiation in riveted structures, 2) the physical behavior of fatigue crack propagation from rivet holes, and 3) characteristics of fatigue crack link-up in a riveted lap-splice joint. This research has established a comprehensive database that fully characterizes fuselage riveted lap joint multisite damage. Results showed that fatigue crack growth from rivet holes is well behaved and predictable. FASTRAN code predictions of fatigue crack growth from rivet holes correlated with the destructive examination database, thus showing that fracture mechanics based methods predict the growth of multisite damage in the lap joint.

They also completed commercial product licensing agreements to facilitate the transfer of NDI technology developed under the NASIP program to industry. They established commercial product licensing agreements with companies interested in marketing advanced NDI technology

Finally, in 1998 they completed an Engineering Handbook for Analysis Methodology for facilitating the transfer of structural integrity technology developed under the NASIP program to industry. The handbook contains all of the experimental data generated under the program and provides examples for all of the codes developed.

In summary, they completed the following projects under NASIP for thin-sheet metallic fuselage components:

1) Nondestructive evaluation for disbond, corrosion, and cracking;
2) Riveted lap-splice joints;
3) Riveted thin skin structure;
4) Structural integrity analysis methodology;
5) Nonlinear effects due to combined loads;
6) WFD, including multiple-layer components; and
7) Corrosion fatigue effects.

For the Airframe Airworthiness Assurance (AAA) program, NASA's goal is to develop technologies to ensure continued airworthiness of the commercial transport fleet and enhance its survivability.

Under the AAA program planned for 1998 to 2003, they will expend their efforts on the following activities for the heavily loaded thick-section wing components that would include 7000 series aluminum alloys in plate, extrusion, and forging forms:

1) NDE for corrosion and cracking,
2) Hidden components, and
3) Spectrum loading effects on fatigue behavior.

In addition, LaRC has supported the development of the fracture mechanics program, NASGRO. This program was initially developed for applications to space vehicles however, is also ideally suited to the aging aircraft program. The program as it exists today has an extensive database of crack growth properties as well as solutions for most of the problems encountered in aging aircraft. Its future development will further enhance this capability.

D. Aging Military Aircraft Research in the United States Air Force

Except for those failures in the USAF resulting from exceeding the operational envelope of the aircraft, structural failures before the mid-1940s were rare. One reason for this is that, before the mid-1940s, the USAF rarely accumulated sufficient flight time on their aircraft to suffer fatigue failures. Further, ductile materials and conservative methods used for analysis tended to preclude failures. Experience has shown that early aircraft manufactured with ductile materials and designed based on static strength only are typically safe from failure caused by fatigue for at least 1000 flight-hours. It was rare for a combat aircraft in World War II to remain operational for more than 1000 hours. The demand for improvements in performance in the late 1940s, however, introduced new materials with high strength but few other virtues. Further, the demand for performance improvements reduced analytical conservatism and introduced designs that were to operate at high altitudes. The design community appeared oblivious to the consequences of their actions. Even before the time of the first flight by the Wright Brothers, fatigue was a major issue in many industries.[13] In the railway industry alone, fatigue failures of wheels caused numerous deaths. These failures seemed to make no impression on aircraft designers. The success they had stemming from the days of the Wright Brothers appeared to continue without interruption, although fatigue failures in aircraft can be traced back to the late 1920s.

The reality of the consequences of aging came sharply into view for the USAF on March 13, 1958, when the USAF lost two B-47 aircraft because of fatigue cracking in the wing. It was on this day that the aging aircraft research effort started for the USAF. The USAF did not specify a service life for the B-47. Consequently, they based the design on the assumption that failure from overload was the only threat to its structural integrity. This was common practice for aircraft designed in the late 1940s such as the B-47. Review of the then current literature on structural design provided no hint that fatigue was a serious consideration. However, the USAF intended to maintain the aircraft in service until 1965. The technical basis for maintaining the aircraft in service for that length of time did not exist.

Near Homestead Air Force Base, Florida, a B-47B disintegrated at 15,000 ft, 3 minutes after takeoff. Its center wing section failed at approximately butt line 45. The aircraft had a total flight time of 2,077 hours at the time of failure. The same day, a TB-47B failed at 23,000 ft over Tulsa, Oklahoma, after the bottom wing skin of the left wing failed at butt line 35. This aircraft had flown 2,418 hours. While the USAF and contractor were investigating these accidents, three more occurred, indicating that the March that 13 failures were not isolated events. The USAF was able to attribute one of these failures to overload. The other two, however, they believed to have been caused by fatigue. One of these aircraft appeared to explode at 13,000 ft just before a refueling rendezvous near Langford, New York, on 10 April. Unlike the other accidents, the cause of the accident was a failure in the fuselage. On 15 April, 1958, a B-47E with a total flight time of 1,419 hours took off into a storm from McDill Air Force Base, Florida, and disintegrated shortly afterwards.

The 1958 failures motivated the USAF to establish the USAF Aircraft Structural Integrity Program (ASIP).[14] The USAF designed this program for use with new weapon systems acquisition for their inventory. The program, as originally conceived, defined a sequence of tasks that progressively reduced risk. These tasks, composed of analyses and tests, included all efforts for the qualification of USAF aircraft. This concept is just as valid today as it was in 1958. The approach, although sound in its concept, had a fatal defect. The original program incorporated a reliability concept called safe life to qualify the structure for the loads environment expected in operational service. The USAF determined the safe operational life from the results of a full-scale fatigue test of the structure. They conducted this test in a laboratory environment. They divided the number of successfully tested flight-hours by a factor called the "scatter factor." The scatter factor (usually in the interval from two to four) supposedly accounted for material property and fabrication variations in the population of aircraft. The trouble with this approach was that it did not preclude the use of low ductility materials operating at high stress. Unfortunately, it was at this time that aluminum companies were introducing high-strength alloys in response to the insatiable desire for improved aerodynamic performance. Consequently, the "safe life" concept did not eliminate in-service failures the USAF designed it to prevent. The "safe life" approach adopted by the USAF in 1958 proved to be ineffective in eliminating fatigue cracking as evidenced by failures in operational aircraft.

Probably the most significant in-service event since 1958 that changed the original version of the ASIP was the failure of an F-111 in December1969. F-111A number 94 (SN 67-049) failed on 22 December 1969 as a result of a wing failure in the lower plate of the left wing pivot fitting. At the time of failure, the aircraft had approximately 100 hours of flight time. It occurred during a pull-up from a rocket-firing pass at the Nellis Air Force Base, Nevada, range. Catastrophic loss of this F-111 demonstrated the fatal defect in the "safe life" method. That is, the safe life method did not preclude designs that were intolerant to manufacturing and service-induced defects. Other losses (e.g., F-5, B-52, and T-38) and incidents of serious cracking (e.g., KC-135) during this period confirmed this shortcoming. These failures led to a new approach for the protection of USAF aircraft safety,[15] a damage-tolerance approach.[16] The basis for the process is to assume that the structure has a flaw, a sharp crack, that is the least upper bound of the expected flaw distribution.[17] The operator makes inspections such that the crack does not reach the point of rapid propagation before it is detectable. The damage-tolerance approach is in a state of continual improvement because research and development have led to better methods in fracture mechanics methods and stress analysis over the last 30 years. The introduction of damage-tolerance principles by the USAF in their structural inspection program in the early 1970s virtually eliminated fatigue as a safety issue in their aircraft.

The USAF incorporated the damage-tolerance approach in the ASIP and, in 1975, they published the process.[18] This program, for a new acquisition, provides a series of time-related tasks that will provide progressive risk reduction in the progression of the engineering and manufacturing development phase of procurement. The current version of the ASIP includes five separate tasks that cover all aspects of the development and support of an aircraft structure. For any given program, if the USAF does not plan to include a specific element, then they must establish the rationale and potential impact on the structural integrity of the weapon system for the exclusion. The main tasks of ASIP are as follows:

1) Design information,
2) Design analyses and development tests,
3) Full-scale testing,
4) Force management data package, and
5) Force management.

The original goals of ASIP were to 1) control structural failure in operational aircraft, 2) devise methods of accurately predicting service life, and 3) provide design and test approach that will avoid structural fatigue problems in future weapon systems. The USAF achieved the first and third of these goals with damage-tolerance principles. The second goal is much more difficult. A more realistic goal is to determine the sustainment cost in future years. This may be done with probabilistic methods.[19]

The ASIP is also the standard by which the USAF can evaluate aging aircraft issues for structural components. For this purpose, the USAF normally emphasizes a subset of the elements of ASIP. For example, they extracted appropriate elements of this program to perform (DTAs) during the 1970s and 1980s. The USAF invested approximately 1 million hours in that effort to provide an inspection and modification program that greatly enhanced the safety of aging aircraft. As discussed below, in addition to DTAs, the impact of corrosion, WFD, and repairs must also be determined. Aging aircraft for many years have had a significant influence on USAF research and development programs and have been a major driving influence on the elements of the ASIP.

Two of the main products of the ASIP process are development of the report on strength and operating restrictions and the development of the Force Structural

Maintenance Plan (FSMP). If there is a need to change either of these documents because of flight beyond design usage that could introduce new critical areas, corrosion, WFD, or repairs, then the aircraft is said to be in a state of aging.

Experience with operational aircraft has shown that they rarely fly according to their design spectrum of loads. Data from flight load recorders have typically shown that there are considerable differences in usage severity among aircraft with the same designation. The USAF often finds that the average aircraft usage is more severe than originally perceived early in the design process. This finding is made more significant by the fact that the DTA may not have identified an area that would be a concern for aircraft with usage more severe than that assumed for design. Experience has shown that the mass of an aircraft increases because of additional equipment or modification after an aircraft enters operational service. In addition, there are differences because there are changes in pilot techniques as pilots become more familiar with the aircraft and mission changes because of new weapons and tactics. Aircraft-to-aircraft variability comes from several sources such as base-to-base variations in distance-to-test ranges and training. These experiences tend to degrade the capability of the full-scale durability test that consisted of two lifetimes of average usage to identify all of the areas of the aircraft that could potentially cause a loss of safety. In most cases, an update of the DTA can account for any change needed in the inspection or modification program.

For the past 40 years, the USAF has used the ASIP to maintain safe and economical operation of aging aircraft. This program has been supported over the years by USAF laboratory programs in the areas of fracture mechanics, corrosion prevention, flight loads, nondestructive evaluation, human factors, and maintenance and repair. These efforts provided the USAF with the technology required to support operational aircraft maintenance programs based on damage tolerance.

As indicated above, the USAF significantly changed this program because of the failure of an F-111 in 1969. This event ushered in the era of damage tolerance in the USAF. The first assessments performed on the C-5A and the B-1A in 1971 and 1972 help derive original DTA requirements for the USAF. These requirements were derived for monolithic (i.e., slow crack growth) structures. The failure of an F-4 wing on 23 January 1973 in a structural location the USAF considered fail-safe demonstrated to them that a structure could not be fail-safe without an inspection program. This failure strongly influenced damage-tolerance requirements as initially established first in MIL-A-83444 and subsequently in AFGS-87221A. The technology for the analysis of fail-safe designs has evolved slowly, primarily because of the need for extensive finite element programs supported by expensive test programs. The change to a damage-tolerance approach prompted considerable research and development in the area of fracture mechanics. The then Air Force Flight Dynamics Laboratory was the focal point for much of this research. In the 1960s and 1970s, they developed much of the fracture technology that is still in use today. In addition, because damage-tolerance approach forced the engineer to better understand the stresses in the structure, finite element techniques emerged as the methods of choice for the stress analysis. These capabilities permitted the USAF to perform a DTA of all of the major weapon systems in inventory in the 1970s and 1980s. This effort required over 1 million hours to complete and every major manufacturer was involved with this activity. Because of this activity, industry was able to develop the technology required for this type of analysis. This technology is also suitable for application to new aircraft developments. Consequently, the USAF was able to include damage-tolerance requirements in the specification for new aircraft procurement.

After completion of the DTA on every major weapon system, USAF laboratories continued research in other areas associated with aging aircraft. One of these was to make a better determination of the durability of aircraft. For this purpose, they sponsored research in the determination of initial crack distributions in aircraft structures. Much of this effort was concentrated on the interpretation of cracks found in the teardown inspection of the F-16 wing after completion of the durability test.[20] Another effort related to aging aircraft was the development of the procedure for evaluating the probability of failure for a population of aging aircraft.

Later, it became clear there would be a sharp reduction in new aircraft procurement and the USAF would need to maintain their aircraft much longer than previously believed. In 1991, the then Wright Laboratory named J. Rudd as their aging aircraft focal point. Rudd assessed the work done by the FAA Technical Center and NASA LaRC and made the decision to tailor his program to complement the work of the other two organizations. For example, he emphasized work on wing structures since both the FAA and NASA were expending most of their efforts on fuselages. This decision was consistent with the reality that most of the commercial problems are in fuselages because of the large number of pressure cycles. Most of the USAF problems are in wings because they are subject to increased turbulence associated with flying at low altitudes.

Another major research initiative in the USAF is WFD in primary structural elements. The research will try to reduce the costs incurred to establish an estimate of the time of onset of WFD. The analysis for the time of the onset of WFD depends in the near term on data derived from teardown inspections of fatigue test articles and/or of operational aircraft. Corroboration of these estimates occurs through detailed inspections of suspect structural elements.

Both corrosion and WFD are significant today because of the lack of adequate NDI techniques. The shortcoming in NDI is the ability to find corrosion damage in structural joints and small cracks that would be the indicator of the onset of WFD. It appears the current efforts in research NDI will produce the technology for this capability. It remains to be seen if there is an economic motivation to transition this technology from the laboratory to inspections of operational aircraft.

The need for NDI technology to enable damage-tolerance-driven inspections has been a major thrust of the USAF for many years. Among these technology programs was a major effort to determine the probability of crack detection in an operational environment.[21] Both the USAF and the FAA recognize the need for continuing the effort to quantify the capability of inspection techniques since this capability is critical to flight safety.

The USAF sponsored a very successful program for the eddy current inspection of engines.[22] This effort made it feasible to perform damage-tolerance-derived inspections for engines and significantly reduced the costs associated with the previously used safe life approach. This technology is not restricted to engines. It has the potential for making inspections on aging helicopters and reducing their cost of ownership. The USAF documented[23] the potential for using the damage-tolerance approach for helicopters in a study on HH-53 helicopter.

There are significant research and development efforts currently underway in the area of nondestructive evaluation of aging aircraft. NASA LaRC and several academic institutions, including Iowa State University and Johns Hopkins University, are doing much of this work. The USAF is working with these institutions and the FAA Technical Center to ensure that these efforts meet their requirements.

Three areas of nondestructive evaluation are particularly relevant to aging aircraft. The first and most important is the detection of cracks that could be significant for determination of the onset of WFD. As indicated above, there is a need to make an estimate of this onset based on probabilistic assessment of cracking data derived from the teardown inspection of fatigue test articles or operational aircraft. The analyst must recognize, however, that this is only an estimate. The actual time may be either somewhat earlier or later than this estimate. It is important, therefore, to be able to validate this prediction with nondestructive evaluation. The size of the defect makes this task difficult since it is quite small. Experimental evidence to date indicates that cracks of the order of 1 mm can significantly lower the fail safety capability of certain structural configurations. The second area of importance in nondestructive evaluation is the identification of cracks in the second and third layer of the structure. Typically, operators have used low-frequency eddy current for this purpose. However, the detectable flaw size using eddy current is too large to enable long inspection intervals. The final NDI area of interest for aging aircraft is the detection of corrosion hidden in the joints of the structure. The usual method of corrosion detection is visual. Maintenance personnel rely on evidence of white powder or bulging to detect the presence of corrosion. This process is far from satisfactory based on evidence of corrosion found in teardown inspections and modification or repair programs. Another factor is that NDI methods are normally optimal for the detection of cracks, not for the detection of corrosion. Methods using multifrequency eddy current show significant potential for detecting corrosion throughout multiple layers of thin aluminum skins.

The primary effort in fracture mechanics with respect to aging aircraft is in improving the capability for assessment of the time of onset of WFD. The USAF has done this for aircraft using probabilistic approaches in combination with fracture mechanics. As indicated, the USAF has performed teardown inspection of full-scale fatigue test articles or operational aircraft. They believe this is the best method currently available to obtain data required to derive the probability distribution function for equivalent initial cracks in each of the structurally critical areas. The word "critical" here refers to an area that could significantly contribute to the probability of failure.

The USAF research and development program for aging aircraft has provided the technology base for safe and economic operation of military aircraft through the ASIP. As an indicator of this success, the failure rate for all systems designed to and/or maintained to the current policy is one aircraft lost due to structural reasons in more than 10 million flight-hours. This is significantly less than the overall rate of aircraft losses from all causes by two orders of magnitude. The success of the program has made it more difficult to obtain funding for research and development in the structural area. It has also, at times, given program managers a false sense about the remoteness of structural failures. This success, however, should not be used to indicate that there is no need for continued research on the structure of aging aircraft. The return on the investment in this research is reduced cost and downtime with inevitable structural problems.

The USAF believes that these are the major issues associated with the aging aircraft. The USAF also believes that their work is complementary to FAA programs.

Therefore, the USAF has entered a memorandum of understanding with the FAA that has the following objectives:

1) Encourage coordinated efforts to provide the technology required for ensuring the continued safe and economical operation of aging U.S. commercial and USAF aircraft.

2) Define aging aircraft requirements and coordinate technical programs and plans.

3) Provide a forum to enable the FAA and the USAF to focus each other's strengths to improve capability.

Many of the problems with aging military aircraft find an exact parallel in aging commercial aircraft. It is prudent, therefore, to find solutions through the combined talents and resources of the FAA, NASA, and the USAF. Efforts to date indicate that this approach is viable and productive.

As indicated above, materials in many aircraft were the result of the desire for improved performance, with little attention given to the potential for corrosion and stress corrosion cracking damage. Further, at the time of manufacture of many of these aircraft, focus on corrosion protection was not what it is today. Many of these early corrosion protection systems have broken down. In the open areas, the operator can readily renew them. There is, however, no easy way to renew the corrosion protection system in the numerous joints. Experience with modification and repair of aging aircraft has revealed that joints without proper protection experience significant damage that results in costly part replacements.

The corrosion concern is now becoming more acute in that environmental protection laws have eliminated the use of some standard corrosion inhibitors. Another issue is that nondestructive evaluation techniques are marginal. Standards for corrosion damage are so poorly defined that it is difficult to properly characterize the damage found. This deficiency creates a real problem in the future years cost projection for structural maintenance.

Because the predictive capability for future corrosion damage in a given aircraft model is non-existent, the research community must emphasize providing tools for NDI for corrosion in hidden areas of the structure. There is a need to enforce the policy in the USAF to repair corrosion damage and not allow it to jeopardize safe and economical operation. In addition, emphasis must be placed on developing new corrosion protection systems that are environmentally safe. Thus, the focus must be on detection and protection.

E. Aging Military Aircraft Research in the United States Navy

The U.S. Navy has an approach to aging aircraft that is somewhat different from the USAF. As with the USAF, the Navy recognized that degradation of the structure from fatigue cracking was an important aging consideration. In addition, as with the USAF, the Navy adopted the safe life approach for maintaining safety of their aircraft. However, the implementation of the safe life method was considerably different from the USAF. The USAF approach determines the stress environment for an aircraft through the design mission usage for the aircraft, with a load factor distribution derived from experience with their earlier aircraft. The Navy, however, for the design load factor exceedance, determines the greatest possible fatigue damage that may occur within the operational envelope of the aircraft. This results in more conservative design than the previously used USAF safe life approach.

Studies on the Navy approach reveal that the sizing of the aircraft would be quite similar to that associated with the currently used damage-tolerance approach of the USAF. From a management perspective, the Navy approach has considerable virtue. They can determine, through their tracking program for individual aircraft, the fatigue life expended and thereby provide a clear picture of how long the aircraft could remain in service. The USAF, on the other hand, determines from their tracking program when to inspect their aircraft to determine if modifications or repairs are needed to maintain it in the inventory.

The Navy's motivation to eliminate structural inspections as an alternative to maintaining aircraft safety is because of operations aboard aircraft carriers. Space is very limited on these ships and consequently there is no room or the capability to perform detailed structural inspections of aircraft. Consequently, the safe life approach is well suited to their operational restrictions.

It is difficult to imagine a more severe corrosion environment than that found aboard an aircraft carrier. The salt environment penetrates to the interior of the structure as well as coating the exterior of the aircraft. The Navy has found that frequent washing of the aircraft and corrosion prevention compounds are essential to their corrosion protection program. Aircraft that have enjoyed relative minor problems with corrosion such as the A-7D in the USAF have shown significant corrosion problems when operated as A-7Es in the marine environment. Consequently, the major effort for aging aircraft by the Navy is research in corrosion prevention, prediction, and detection. Unfortunately, the Navy's aging aircraft have many of the same materials, which are corrosion prone, as the USAF's aging aircraft.

II. Widespread Fatigue Damage

A. Introduction

For some of the older aircraft designed to be fail-safe, there is a potential for WFD to occur. That is, cracking may be so widespread in the structure that even the application of the deterministic damage-tolerance process may not protect its safety. The USAF has experienced WFD on the KC-135, C-5A, and the C-141 aircraft.[24] They have also experienced it on the 707 aircraft they procured off the shelf.[25] Commercial aircraft have experienced many cases of WFD the Aloha incident being one of many. It is helpful for the understanding of the issues with WFD to understand the major aspects of the damage-tolerance process. Much of the technology used in the assessment of WFD is an integral part of the damage-tolerance process. However, the analyst must consider some important differences. One of these differences is that the inspection program derived from the DTA is usually ineffective in finding WFD since cracks of the order of 1 mm are large enough to cause WFD. Typically, the detection of cracks of this size requires the removal of fasteners. Another is that the damage-tolerance process is deterministic and the assessment for WFD usually requires the use of probabilistic methods for solution.

B. Damage Tolerance Assessment Process

Figure 3.1 shows the steps in the DTA process. This description applies primarily to the process used by the USAF. The procedure used by commercial operators is quite similar. The DTA is an integral part of the aging aircraft program for both

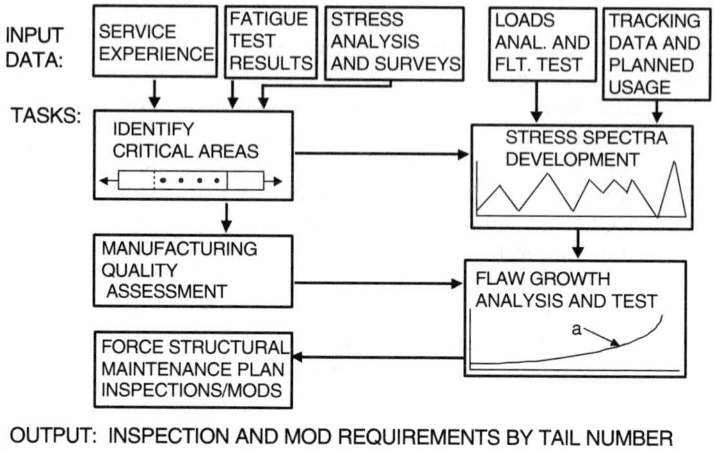

Fig. 3.1 Damage-tolerance process.

military and commercial aircraft. The concept is simple. The flight time to the first inspection is based on the time required for the largest defect expected in a fleet of aircraft from manufacturing or in-service damage to grow to critical crack length. Subsequent inspections are based on flight time for the NDI detectable defect size to grow to critical crack length. A crack growth function illustrating this process is shown in Fig. 3.2.

The ordinate of the point A is the initial flaw assumed for the analysis. The abscissa of the point B is one-half of the time needed for the initial flaw to grow to critical. The ordinate of the point C is the NDI detectable crack length. This crack is then grown to the point D whose abscissa is one-half of the time required for the crack to grow from B to critical crack length. The process is repeated until inspections reveal an actual crack or the structure needs modification for WFD.

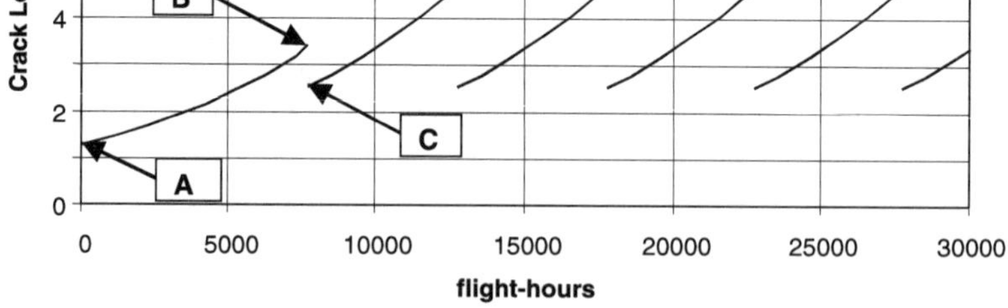

Fig. 3.2 Damage-tolerance crack growth function.

The process evolved over a period of several years after the USAF applied it initially to the B-1A and the C-5A. Its successes include the F-4, an aircraft that did not have a requirement for life when the Navy procured it. The USAF purchased this aircraft in large quantities, and it became an essential ingredient of their fighter fleet. After a crash at Nellis Air Force Base in 1973 caused by fatigue, the USAF found themselves in a difficult situation. They initiated a recovery program that included a DTA and fatigue test conducted in their laboratory at Wright–Patterson Air Force Base. Because of this effort, the F-4 remained in operational service until the 1990s without further incident. During the 1970s and 1980s, the USAF performed a DTA of every major weapon system in their inventory.[17] These successes motivated the USAF to apply this technology to engine structures with similar results. The discussion below describes the method used for damage tolerance with examples on how the USAF applied it.

The first task of the DTA is the identification of critical areas of the structure. A critical area is a location or part of the structure that could affect flight safety and may need maintenance in the form of an inspection or modification during the life of the aircraft. There are several techniques for identification of these areas. Actual cracking experience through service operations or durability testing is usually the most important consideration. Areas that have high predicted or measured stress and details that make them prone to cracking are, of course, prime candidates for the assessment. Another consideration in the selection of critical areas is ease of inspection. In general, the analyst gives higher priority for selection on critical areas difficult to inspect. It has been extremely helpful to use the accrued knowledge of the original aircraft contractor in identifying potentially critical areas. On some of the assessments, preliminary estimates were made of flaw growth in candidate critical areas. For small aircraft, the number of candidate areas generally was of the order of 40 to 70. The analyst would normally be able to screen these down to 10 to 30 for final analysis. For larger aircraft, the number of candidate areas generally was of the order of 60 to 150. The analyst would screen these down to 30 to 60 for final analysis.

The second task of the DTA is the development of the stress spectrum for each area identified for a final analysis. This is one of the more demanding aspects of the DTA process. The reason is that there are significant changes in the rate of flaw growth due to relatively small changes in the cyclic stresses. To perform this task properly, three data items must be available to the analyst. First, the analyst must have operational experience available in a usable form. These operational data must provide a basis for establishing a flight-by-flight sequence of points in the sky (i.e., altitude, mass, and aircraft motion parameters). This is usually available from multichannel data on fighter or attack aircraft. For transport category aircraft, the USAF usually derived the sequence from flight log information supported by multichannel data to define the maneuver and gust environment. In all cases except one, there was a sufficient database from which to derive the sequence. This exception was the A-7D, which was equipped with counting accelerometers only. Consequently, as part of the A-7D DTA, an operational database was derived from collecting 1,250 hours of multichannel data from aircraft located at two bases. The second data item necessary for the derivation of stress spectra is the set of equations needed to determine external loads (i.e., shear, bending moment, and torsion) for a given point in the sky. For USAF aircraft, the manufacturer usually determined the external loads through analyses, wind tunnel testing, and in-flight strain surveys.

For all aircraft except the F-4, there was sufficient confidence in existing data to perform this task. The USAF elected to perform a flight load survey on this aircraft during the course of the DTA. This turned out to be very beneficial because preexisting data would have produced a pessimistic view of the maintenance burden for this aircraft. The final data item needed for the generation of stress spectra is the external load to stress transformation. For all of the aircraft studied, there were at least some experimentally derived stresses from previous static and durability tests. However, in all cases, it was necessary to conduct additional stress analyses. The contractor performed these additional analyses, typically using the finite element method. The scope of this finite element effort ranged from evaluating stresses at local details to finite element models of the complete airframe. The finite element effort varied significantly from aircraft to aircraft because of differences in the test database and the complexity of critical details. Simplification of the stress spectra effort would have been possible if direct strain measurements had been available. In general, these data were not available. In a few cases, such as the C-5, this kind of information was available and was invaluable for determining the environment from maneuver, turbulence, and aerial refueling.

Techniques used in deriving stress spectra for the assessments varied quite widely from aircraft to aircraft. Part of the reason for this difference was due to the available database. For example, for the F-4, data collected from the VGH recorder provided the number of occurrences of combinations of Mach number, load factor, and altitude in predetermined bands. Consequently, the assessment of areas of the aircraft sensitive to asymmetrical loading required data from other aircraft or from pilot interviews. For the F-15, however, the signal data recorder provided a time history of both symmetrical and unsymmetrical parameters for use in developing the stress spectrum. The F-15 database more accurately accounted for unsymmetrical loading. Moreover, it permitted a more realistic assessment of the minimum stress excursion that followed a maximum stress excursion. For the F-4, the conservative assumption had to be made that, after a maximum stress, there followed either a stress corresponding to 1-*g* flight or a stress corresponding to less than 1-*g* flight. The database on the F-15 enabled the analyst to remove this conservatism.

The USAF performed fighter and attack aircraft assessments by reconstituting individual flights from databases except for the F-111. For this aircraft, the multichannel recorder data was used directly to randomly generate a "block" of flights of approximately 500 hours. This is a very effective approach if one can be sure that selected flights are representative of aircraft usage. For the F-111, they used counter data for load factor as a guide for this selection. There was no attempt made to maintain the original order of individual flights since previous studies for the F-4 and other aircraft showed sequence effects were insignificant if flights were randomly selected.

As indicated earlier, the VGH recorder was the basis for the F-4 usage database. A sampling technique based on VGH recorder data provided the approach for derivation of the stress exceedance function. In this method, the analyst computed the stress for a representative set of Mach number, load factor, mass, and altitude combinations. The surface derived from the representative points provided the means to determine the stress at flight-measured points. Thus, recorder data determined the stress exceedance function accounting for "all points in sky." The USAF used a modification of this approach in the development of the stress exceedance function for the A-7D DTA. For the A-7D, the approach involved a regression equation to interpolate based on stresses computed for a representative set of aircraft flight conditions.

Environmental data that augmented flight log data for large aircraft are extremely important. The USAF refers to these data, used in the ASIP, as the loads/environmental spectral survey data. It provides the means to quantify the three-dimensional nature of wing gust loads, the phasing of shear and bending moment, and aerial refueling loads on the C-5. These data were also very helpful for evaluating the low-level turbulence on the B-52, C-141, and C-130. In many cases, such as the fire-fighting mission for the C-130, special mission maneuver data needed quantification. It is the intent in the derivation of stress spectra to determine the "baseline usage" as an average usage for the force. For aircraft where there were significant usage changes during their life or there were possible changes in their future usage, the baseline usage reflected these changes. For some aircraft, such as the F-111, with different mission design series (MDS), the USAF derived a separate baseline usage for each MDS. In addition to the baseline usage, there is a need to derive stress spectra that represent potential variations from the baseline. Testing of these variations develops confidence that the procedure for tail number tracking by fracture mechanics methods is valid. For older aircraft assessments, the usual procedure was to define a spectrum more severe than the baseline and a spectrum less severe than the baseline. Changing the baseline mission mix generally accomplished this.

For larger tanker, transport, and bomber aircraft, the main source of data was flight logs. In general, these logs had sufficient detail such that engineers could divide the usage among a relatively few missions (of the order of ten). Typically, the assessment had to include two or more distinct usage changes. For example, for B-52Gs, there were differences in usage before, during, and after their Southeast Asia operations. In addition, the USAF anticipated the usage of the aircraft in the future to be different from all previous usage.

The third task of the DTA is to establish the initial flaw size for the fracture analysis. Because of their inherent stress concentration, fastener holes were predominant as candidates for critical areas of the airframe. The USAF noted that there had not been a structural failure in the number of flights it takes for a 1.27-mm corner flaw in a fastener hole to grow to critical crack length. By 1975, they believed there was sufficient data to make the judgment that this size was sufficient to ensure aircraft safety. They derived this belief partially from teardown inspections of full-scale fatigue test aircraft, but primarily from observing operational aircraft such as the F-4, C-5A, and the KC-135. There has never been a rigorous substantiation of this belief. However, experience in subsequent years supports use of this size defect as being adequate to protect flight safety. The remaining task then was to determine the flaw size for holes that were cold worked or filled with an interference fit pin. The USAF determined this flaw size on an ad hoc basis. In some cases, where there was a question of the adequacy of the installation of interference fit pins, there was no reduction allowed. In other cases, where there was confidence that the installation was proper, the USAF reduced the initial flaw size to 0.127 mm. Primary considerations in making this judgment were durability test performance and manufacturing procedures. For example, the C-141 durability test showed that tapered fasteners did extend the life of that aircraft. However, there was some concern about the quality of handheld drilling operations in some areas of the wing. Consequently, the DTA did not account for the benefit of tapered pins. However, machines with controlled feed and speed drilled many wing fastener holes. For these holes, the USAF made a decision to reduce the initial flaw by approximately a factor of two.

The identification of stress spectra for each critical area and the initial flaw permits the initiation of the task of establishing operational limits. This combined analysis and test effort uses the disciplines of fracture mechanics to find the safety limit for each critical area. Fracture mechanics technology has improved significantly from the early 1970 period. However, even with the analytical capability available today, the process would be enhanced with test substantiation.

There are two main reasons for fracture testing. First is analysis verification. The aim is to accomplish this with the least specimen complexity possible to isolate the local detail (e.g., a fastener hole) and evaluate spectrum retardation. Specimens used for this purpose were generally dog bone specimens with the proper material, thickness, size of fastener, and load transfer. The second reason for testing is to establish high and low side truncation levels. The low side truncation is primarily an economic consideration. The object is to eliminate as many cycles as possible with a small stress range without significant change in the crack growth. The removal of the high stresses (or clipping) eliminates those cycles that make a crack grow slower and retain those that make a crack grow faster.

In the early 1970 period, there was a belief that crack growth was quite sensitive to the loading sequence within a flight. Therefore, as part of the DTA, the USAF required tests to evaluate these effects. It was learned that, if they simulated loading on a flight-by-flight basis, then the ordering of loads within a flight was of secondary importance. Consequently, they discontinued this type of testing. For almost all of the aircraft, the fact that critical crack sizes were sufficiently small such that there was little if any redistribution of stresses during crack growth simplified the fracture analysis. Further, the primary structural issue was crack growth from a fastener hole or an open hole. Therefore, the analyst needed to concentrate on the flaw in mode I cracking from both filled and unfilled holes, load transfer on the fasteners, and retardation effects. For simulating retardation effects, analysts generally used the Wheeler,[26] the Willenborg,[27] the modified Willenborg,[28] or some form of a contact stress model.[29] The T-38 analysis used the Vroman model for part of its DTA. This model was not used for any other assessment.

Analysts learned that proper counting of stress cycles in the spectrum was essential for obtaining accuracy. The so-called rain flow procedure is now commonly accepted as an adequate procedure for counting the stress cycles in the spectrum.[30]

After verification of the crack growth analytical model through coupon and, in some cases, component testing, the effects of the chemical environment entered the analysis process. For early assessments, the tendency was to take a conservative view of the environment. That is, the USAF required the selection of an environment more aggressive with respect to crack growth than actually expected. This position was relaxed in the late 1970s and they placed emphasis on selecting a realistic environment. Constant amplitude crack growth tests performed in the desired environment provide the basis for quantification of these effects. Crack growth analysis includes the environmental effect in the data used for crack growth rates. The procedure is subject to criticism because it may not accurately account for the effects of cyclic frequency and load interaction effects with the environment.

There is no indication from inspections performed on operational aircraft that the error is significant.

Crack growth analysis plays a dominant role in the damage-tolerance approach. The tool must be usable for different chemical as well as loading environments. In other words, it is the mechanism for tracking the crack growth on each tail number in the force, thereby ensuring aircraft safety. Therefore, it is extremely important to validate the analysis for the expected range of service operations. After the analyst establishes safety limits for all of the critical areas of the structure, the development of the FSMP can proceed. The FSMP provides the how, when, and where for structural inspections or the when and where for modifications. In many cases, it was found, based on either economic or safety considerations, that modifications were preferable to continued inspections. This situation existed for the C-5, T-38, F-4, and KC-135, for example. Of course, the DTA process should include the modifications.

One of the more important tasks in the damage-tolerance approach was to establish the NDI capability. This was done with the help of NDI experts from the now Air Force Research Laboratory (AFRL) Materials Directorate, the appropriate Air Logistics Center (ALC), and the contractor. In some cases, such as with the EF-111, they conducted an NDI reliability program to determine the flaw size corresponding to 90% probability of detection with 95% confidence. However, these cases were in the minority and, consequently, the USAF based most of the NDI detection capability on judgment. When possible, they avoided inspections that involved removal of fasteners. In addition, the USAF rejected the concept of sampling inspections rather than inspecting 100% of the force.

The FSMP covered the period of the planned operational usage of the aircraft. Thus, the FSMP permitted the ASIP manager at the ALC to determine the out years maintenance cost. The accuracy of these costs was suitable for budgetary estimates. The accuracy for any given tail number is, of course, dependent on how closely that aircraft flies to the baseline.

USAF structural engineers have long recognized the need for tail number tracking of aircraft. This is evident from the emphasis given to it in the 1959 version of ASIP. The only significant change from the original version in the tracking process is for crack growth rather than fatigue damage. The USAF developed the first tracking program based on fracture for the F-4 during its DTA. Now all aircraft that have had a DTA have a tracking program based on fracture mechanics. For many of these aircraft, the ASIP manager has the computer programs to provide an immediate view of the maintenance status of the aircraft. This provides both near- and far-term planning and decisionmaking capability. It provides the capability to determine the consequences of a mission change. There is also a need for commercial aircraft to have a periodic reassessment of their usage. The availability today of excellent digital recording devices has made this task considerably more manageable than in the past.

The damage-tolerance approach has led to a greatly improved understanding of aircraft structures and their performance. It is the foundation for maintaining flight safety in aging aircraft. It has also led to a greater recognition that additional research and development in the areas of materials, structures, and nondestructive evaluation were not only needed, but could further increase the reliability of systems. Consequently, over the last several years, many programs have focused themselves on increasing the knowledge base available to enable longer lives and

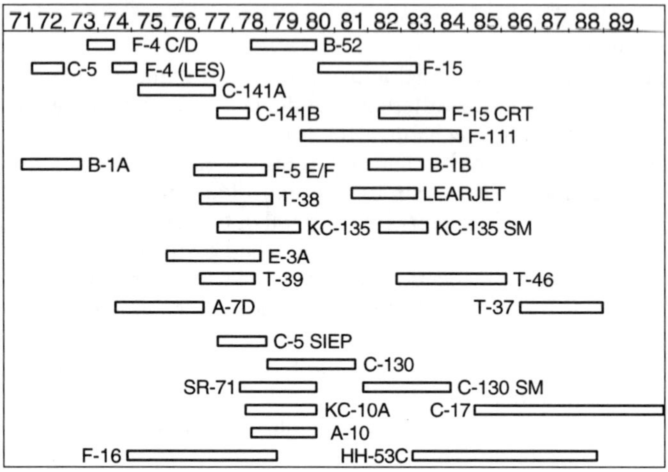

Fig. 3.3 Damage-tolerance experience in the USAF.

more reliability from airframes and engines. Overall, the damage-tolerance experience has been good. Criticism, which is rare, has come from people who believe the approach is too conservative when they perform an inspection and find no cracks. On the other hand, the DTA process has correctly directed inspections to areas that full-scale testing did not indicate were critical. Figure 3.3 shows the DTAs performed by the USAF during the 1970s and 1980s.

However, fatigue cracking of operational aircraft in the USAF is still a significant economical factor. The USAF estimates that fatigue cracking cost approximately $250 million in 1997. The USAF attributes much of this burden to operational usage being more severe than the usage assumed for design.

C. Technology for Assessment of Widespread Fatigue Damage

As indicated previously, the cause of the Aloha accident was WFD. However, this aircraft had flown many flights with its fail-safety significantly degraded. To understand how this is possible, it is helpful to define WFD. The definition adopted by the TOGAA for WFD is the following:

The simultaneous presence of cracks at multiple structural details characterizes the onset of WFD. These cracks are of sufficient size and density whereby the structure will no longer meet its damage tolerance requirement (e.g., maintaining required residual strength[16] after partial structural failure).

Where damage tolerance is defined as follows:

Damage tolerance is the attribute of a structure that permits it to retain its required residual strength for a period of unrepaired usage. It must be able to do this after it has sustained specified levels of fatigue, corrosion, accidental, or discrete source damage. Examples of such damage are (a) unstable propagation of fatigue cracks, (b) unstable propagation of initial or service induced damage, and/or (c) impact damage from a discrete source.

The onset of WFD is the time in flight-hours or flights when there is a degradation of the design fail-safety of the aircraft caused by cracking. The cause of the

April 1988 failure of the Aloha Airlines 737 was WFD. However, the situation with this particular aircraft needs explanation. The aircraft at the time of the incident had 89,680 flights. The cracks out of the fastener holes in the fuselage lap-splices had grown to such an extent they had joined together (i.e., linked up) to form a large crack and subsequent failure. This incident motivated many researchers to expend considerable effort to analyze this particular scenario of cracking. However, this particular situation was misleading in that the onset of WFD occurred long before the time of failure. This aircraft did not experience discrete source damage before it failed. If it had, analyses show that fatigue cracking degraded the fail-safety of the aircraft at 40,000 flights and failure could have occurred then. Therefore, the FAA should have mandated that the operators modify their aircraft to remove this problem at 40,000 flights. The subsequent AD did this by declaring that aircraft modifications take place starting at 40,000 flights. There is no reason to assess the time of linkup of the cracks in the lap-splices since this time is unimportant for the maintenance of safety.

The effect of WFD on flight safety has long been a concern of many researchers. In particular, T. Swift, formerly a National Resource Specialist for Fracture Mechanics for the FAA, has written many papers and given many lectures on this subject.[3,31,32] He motivated an independent review of WFD in 1986. This team wrote a report citing concerns about WFD, but neither the FAA nor the manufacturers acted on their findings. It is interesting that, only 1 year after release of the report, the Aloha accident occurred.

Most of the older USAF aircraft designs did not comply with the modern DTA requirements as described above. Consequently, there is a potential for cracking to be so widespread in the structure that the application of the deterministic damage-tolerance process may not protect the safety of these aircraft. Widespread cracking could exist in monolithic structures such as the T-38 aircraft, which the USAF analyzed using probabilistic methods.[33] The USAF refers to cracking as found in the T-38 aircraft as generalized cracking rather than WFD. However, the primary concern is for structures initially designed to be fail-safe. For these structures, the occurrence of WFD can significantly degrade the fail safety of the structure. This problem has presented itself on the KC-135, C-5A, and C-141 aircraft. The USAF subjected these aircraft to teardown inspections. They incorporated the results of these inspections in a risk assessment to quantify the time when the probability of failure, conditioned by the fact there had been discrete source damage, becomes unacceptable.

In the case of the C-5A, it was judged that fail safety had been compromised if the conditional failure probability on a single flight was greater than 10^{-4}. For that aircraft, the critical area of the wing was the inner wing surface—constructed with individual panels that incorporated integral stringers. For this location, the USAF assumed that the threat was a loss of a wing panel from engine disintegration or from battle damage. For a given aircraft, it is necessary that the conditional failure probability and the threat be determined for each area. For example, for areas that are in the path of failed engine parts, penetrations of these parts would be the threat. The likelihood of occurrence may be determined from available statistics on in-flight engine failures. In other parts of the aircraft, an evaluation must be made for the possibility of other threats, such as accidental damage from manufacturing, maintenance, or ground handling. Threat assessments must be made to ensure that the probability of catastrophic failure is maintained within an acceptable bound.

For the USAF, the acceptable bound is a catastrophic failure probability of 10^{-7}. The basis for this number is that, for a failure probability as remote as this, the expected number of failures in service for a given MDS would be less than one. In the case of the C-5A, the time at which the joint probability of discrete source damage and fatigue cracking exceeded 10^{-7} was less than the time at which fatigue cracking alone of the intact structure exceeded 10^{-7}. However, in some cases, the threat of fatigue cracking may be the dominant threat. This was the case for the C-141 inner to outer wing splice, as demonstrated by Lockheed.[34] In this case, there was progressive failure of the structure because of fatigue cracking. Both the discrete source damage case and the intact structure case should be examined for any structure designed to be fail-safe.

Damage scenarios in an airplane that could constitute WFD differ depending on location in the aircraft. However, typically, they fall into two categories. The first of these is multiple site damage that is characterized by cracks in multiple details in the same structural element. The second is multiple element damage where there are cracks in multiple structural elements. For the case of fail-safe structures, design discrete source damage elevates stresses in the remaining structure. These elevated stresses have been shown by test data and analyses to significantly lower the crack size at which the structure will fail by rapid fracture. If these failures cannot be arrested, then the aircraft will fail catastrophically. Each of the potential failure scenarios must be evaluated to ensure it will not lead to total structural failure.

Previous efforts on this type of analysis have shown that it can readily be applied to structures where the concern is multiple element damage. This was the case, for example, for the KC-135 and C-5A. The application of risk assessment technology to the case of multiple site damage is very much the same as it is for the case of multiple element damage. In the case of damage at multiple sites, there will typically be a "structural boundary" that will determine if cracking has the potential to become catastrophic. For example, in the case of the fuselage lap-splice, the boundary would be the crack stopper built into the structure at the frame or between the frames and its surrounding structure. This crack arrest feature protects the integrity of the structure. The condition of the crack stopper and its surrounding structure (i.e., the boundary) will determine if the damage could propagate to catastrophic failure. Therefore, the interest is primarily in the degradation of the boundary with time and not the growth of the cracks from the holes in the lap-splice to linkup. Consequently, the problem is mathematically similarly to multiple element damage. Alford et al.[34] demonstrated this in their work on the risk assessment of the inner to outer wing joint of the C-141 aircraft.

There must be an emphasis placed on detection, through nondestructive evaluation, of cracks that could be significant for determination of the onset of WFD. As indicated above, there is a need to make an estimate of this onset based on probabilistic assessment of cracking data derived from the teardown inspection of fatigue test articles or operational aircraft. It must be recognized, however, that this is only an estimate. It is not realistic to expect that this time could be determined with great accuracy, even with the most sophisticated fracture mechanics programs. The actual time may be either somewhat earlier or later than this estimate. It is important, therefore to be able to validate this prediction with nondestructive evaluation. This task is made difficult by the fact that the size of defect to be found is quite small. Experimental evidence to date indicates that cracks of the order of 1 mm can significantly lower the fail-safety capability of certain structural configurations.

The USAF recognized the need for additional work in the area of risk assessment technology for maintaining safety of operational aircraft. Consequently, the Wright Laboratories sponsored a program in this area.[35] In this effort, the University of Dayton Research Institute developed the computer program PROF (PRobability Of Fracture). The program, although originally designed to evaluate the risk of failure, can perform studies on the economics of maintenance.

The probabilistic approach also requires the analyst to determine the stress probability density function for each critical area. Usage information generated by individual aircraft tracking programs that is part of ASIP for USAF aircraft data is the basis for the stress probability density function. The desired stress density function is for a single flight of an aircraft selected at random. The stress exceedance function, used in developing the flight-by-flight spectrum for the deterministic DTA, is the basis for generating this function. The assumption that crack distribution and stress distribution functions are independent permits a simple calculation of the joint probability density function of cracks and stress. This assumption is valid since the calculation determines the probability of failure for a single flight of an aircraft selected randomly from the population. The integration of this joint probability density function over the point set where the crack size has reached critical length determines the single flight probability of failure. The time at which the probability of failure is unacceptable is the onset of WFD.

Damage scenarios in an airplane that could constitute WFD differ depending on location in the aircraft. However, typically, they fall into two categories. Cracks in multiple details in the same structural element characterize the first of these multiple site damage categories. The second is multiple element damage where there are cracks in multiple structural elements. For the case of fail-safe structures, the occurrence of discrete source damage typically results in significantly increased stresses. These elevated stresses will lower the crack size at which the structure will fail by rapid fracture. When arrest of these failures is not possible, the aircraft will fail catastrophically. An evaluation of each of the potential failure scenarios reveals whether any one of them will lead to total structural failure. Previous efforts on this type of analysis have shown that it is readily applicable to structures where the concern is multiple element damage. This was the case, for example, for the KC-135 and the C-5A. The solution of the problem for multiple site damage follows the same logic as that used for multiple element damage.

D. Case Study on Aging Aircraft and Widespread Fatigue Damage

The commercial Boeing 707 (B707) has been successfully transitioned to USAF usage in several important programs. These programs relied heavily on Boeing structural integrity programs that had been initiated by the FAA. Further, the programs utilized aircraft that were relatively young compared to aircraft that were being operated in commercial operation. Some of the aircraft the USAF chose for their program called Joint Stars (Joint Surveillance Target Attack Radar System) are exceptions to this.[36] About one-third of these aircraft are close to or above the original 60,000 hour life goal established by Boeing for these aircraft. Although, these aircraft may be the best available from the B707 population, this amount of time in operational service has reflected itself in significant corrosion damage and fatigue cracking. When these aircraft were selected for the Joint Stars mission, the nondestructive capability for the detection of corrosion damage was marginal at best. Consequently, some of the aircraft procured for this activity were found to have corroded beyond what was initially believed to be present. For the fuselage, the corrosion was quite extensive. The contractor aggressively removed the corrosion by removing the fasteners from every longitudinal lap to enable detection. Many of the fuselage skins and stringers required complete replacement because of corrosion damage. For the wings, there was corrosion around the fasteners attaching the wing skins to the wing substructure. These areas have been ground out to remove corrosion damage. In some cases, the number that had been subject to this grinding had been so extensive there had been a concern about retention of the static strength of the wing. Boeing has been working with the contractor to help them determine what should be done to restore the lost strength. On some aircraft, Boeing has recommended the cold expansion of the order of 15,000 fastener holes that have been subject to excessive grinding. The USAF has given the contractor freedom to perform the modifications needed to remove corrosion. Typically, there are from 100 to 200 repairs on the aircraft that are not included in the Boeing repair manual. Some of these are in structure that is flight safety critical. To make these repairs damage tolerant, the contractor has had to perform extensive in-house external and internal load assessments to generate a basis for the fracture analysis of these details.

The largest concern about this effort, however, is the potential for the degradation of fail-safety because of WFD in the wing. The wing of the B707 is a single cell box constructed such that the wing reinforcing members, or stringers, are mechanically fastened to the skins (i.e., the outer surfaces of the box). The skins are composed of multiple panels spliced together by members called splicing stringers. There are manholes in the lower surface reinforced by stringers that are larger than those used to reinforce the skin between splice joints. Previous teardown inspections of wings from B707 aircraft have shown evidence of WFD. The cracks found, although generally small (in the order of 0.05 to 0.10 in.), could affect the fail-safety of this aircraft. These teardown inspections were performed on aircraft that had less time than the Joint Stars aircraft. The database, however, was not definitive enough to be usable in an assessment of the risk of failure. Consequently, the USAF contracted with Boeing to examine aircraft parts taken from a retired aircraft. Boeing performed a teardown inspection on a 707 wing from aircraft taken from Davis Monthan Air Force Base. This airplane was a 300 series aircraft, representative of the Joint Stars aircraft, had experienced 57,382 flight-hours and 22,533 flights. Boeing delivered this aircraft to Pan American in 1966. They operated it until 1977 when they sold it. The aircraft operated in Africa until Boeing purchased it for the USAF in 1990. The USAF could use these data to make projections for cracking for future Joint Stars operations. They performed the teardown inspection on the wing lower surface and the wing stringers. Stringers and skins where Boeing used steel fasteners contained most of the cracks found. This was typically in the area of the wing skin splices and large adjacent stringers. Large adjacent stringers rest on lands machined into the skins. The skin at the lands is typically approximately 9.5 mm thick and the wing skin under the splicing stringers is approximately 4.0 mm thick. The beneficial effects of aluminum rivets attaching the other stringers to the wing skins apparently reduced the amount of cracking there. There was, however, some cracking found in these locations.

Boeing found that cracking in the aircraft in the area of the steel fasteners was quite extensive. In these aircraft, there was a total 1915 cracks found in five sections removed from the aircraft. Most of the cracks found in the aircraft were small. However, they found that a significant population was cracked to the point of considerable concern. They found that increasing the size of the holes in the splicing stringer and the large adjacent stringer would not remove all of the cracks. About 20% of these holes would still have stringer cracks. Further, they found significant cracking outboard of the Wing Station 360 production joint. Therefore, the problem involved most of the wing. Typically, the large adjacent stringers had more large cracks than the splicing stringers. The largest crack found in Stringer 7 was approximately 38 mm in length. It was near the point of rapid fracture. Such a crack in Stringer 7, since it is a large adjacent stringer with a thick land, would be difficult to detect with low-frequency eddy current from the outside of the wing. An inspector, however, would likely detect a completely severed stringer by low-frequency eddy current. It would not be readily detectable from the inside of the wing by high-frequency eddy current since the crack initiated on the faying surface. There were, however, many cracks found would have led to failure in the planned life span of the Joint Stars aircraft. There was a concern about cracking that would degrade the capability of the structure to sustain discrete source damage. There was also a concern about the fatigue failure of the stringers and subsequent catastrophic loss of the aircraft after a skin failure.

The USAF made the decision to assess the Joint Stars structure in the same manner as they used in addressing WFD on the C-5A and C-141. For these airplanes, the USAF used probabilistic methods to determine their risk of catastrophic failure. The procedure for accomplishing this[37] considers crack length distribution and stress distribution as random number sets. The procedure assumes that the crack growth function and the stress that would result in rapid fracture for a given crack length are deterministic.

Boeing determined an approximate stress intensity of each of the cracks found. They then determined, for each of them, the size of the corner crack with the same stress intensity. From these cracks, the USAF derived the crack distribution function. They used a population taken from the largest of them to approximate the crack distribution with a Weibull distribution function with two parameters. It is typical that a single Weibull distribution function will not approximate longer cracks as well as shorter cracks. This is not a problem since the shorter cracks were too small to have a significant effect on the risk of failure. The USAF needed two stress distribution functions for the assessment. The first is the stress distribution function for the intact structure. Boeing derived this in the usual manner from the intended usage of the aircraft, the external load analysis, and the stress analysis of the wing. Second, for the cases where discrete source damage was possible, they determined the local stress increase from that damage. In many cases, the local stress increased to the point where there was significant plastic deformation of the structure. When this occurs, it is essential that the analyst include plastic deformation to obtain accurate stresses. A linear analysis in these cases would likely lead to serious errors in the determination of risk. For this study, the USAF considered two discrete source damage scenarios. The first was the assumed failure of a splicing stringer and failure of the skin on each side to the next stringer. The second was the failure of the large adjacent stringer and the skin on each side to the next stringer. For these cases, the resulting stresses were so high that the finite element program had to account for plastic deformations in the skin and stringers as well as the fasteners. Boeing performed this nonlinear analysis by using MSC/NASTRAN solution 66 with manual restarts for changing joint and fastener stiffness. They did this analysis in approximately 14 steps from a zero intact structure stress to the 255 MPA limit stress. The Boeing analysis showed that the structure, with the discrete source damage and no cracking in the structure, was barely able to sustain limit load.

One aspect of the analysis that is somewhat subjective is the determination of the threat of discrete source damage. The threat for a military aircraft must include battle damage as well as other sources such as engine disintegration. The USAF performed a risk assessment for the C-5A in the late 1970s. For that assessment, they made a judgment that the risk of discrete source damage was 10^{-3} in a random single flight of the aircraft.[24] For the Joint Stars, they assumed the threat from battle damage no worse than that assumed for the C-5. However, for the 707, they reasoned this risk should be somewhat less because the portion of the wing that would be susceptible to damage was smaller. Consequently, they selected 10^{-4} for this analysis as the discrete source damage probability of occurrence. The risk assessment also requires that a decision be made on an acceptable probability of structural failure. The USAF provides guidance for this.[38] In this document, they considered a failure probability in a single flight of 10^{-7} as being acceptable. The USAF would not expect an aircraft failure in a given population in their planned lifetime with this probability of failure in a single flight. They used this failure rate in the assessment of the 707 for the Joint Stars mission. Therefore, for the cases of discrete source damage, the maximum failure probability allowed for a single flight was 10^{-3} and, for the intact structure, the maximum failure probability allowed for a single flight was 10^{-7}. The USAF performed the risk assessment initially for an aircraft that had the same number of flight-hours, 57,382, as the aircraft subjected to the teardown inspection. They found, based on the criteria identified above, that the risk for discrete damage cases was unacceptable. To make a determination when the risk initially became unacceptable, they transformed the crack distribution. They mapped the cracks found in a 57,382-flight-hour aircraft to an aircraft that had 40,000 flight-hours in commercial operation. They accomplished this with a crack growth function for an initial crack that was smaller than any member of the crack distribution. They then found the abscissa of a point whose ordinate is a member of the crack distribution. They reduced this abscissa 17,382 hours (57,382 – 40,000). The ordinate of the point with the reduced abscissa is the size of the crack at 40,000 flight-hours that was the chosen member of the crack distribution at 57,382 hours. Continuing this process for all cracks in the distribution for an aircraft with 57,382 flight-hours produces the desired result.

The discrete source damage case for the splicing stringer and neighboring skins is slightly less critical. The USAF found that the aircraft would have failed the given criterion when it had the equivalent of 50,000 hours of commercial usage. For the intact structure risk assessment, the more critical case was the failure from fatigue of the large adjacent stringer.

As indicated above, Boeing found significant cracking outboard the production joint at Wing Station 360. Therefore, any modification action must include the structure up to the production joint at Wing Station 733.

For the stated criteria for discrete source damage, the USAF found significant degradation of fail-safety beyond 50,000 flight-hours of commercial usage. Therefore, for some aircraft, there will be unacceptable fail-safety degradation before the end of the planned 20,000 hours of Joint Stars usage. This will occur for Joint Stars aircraft with more than 36,000 commercial usage hours. Further, for the case of no discrete source damage, there will be safety degradation beyond 58,000 hours of commercial usage. Therefore, aircraft with initially more than 44,000 hours of commercial usage will have a high probability of failure before operationally flying 20,000 hours.

The USAF chose to modify the wings of the high time Joint Stars aircraft to eliminate WFD. This modification coupled with the unexpected costs because of extensive corrosion significantly increased the acquisition costs of this aircraft. The USAF will have increasing pressure to buy aircraft that have been certified by an agency outside of the USAF since this would eliminate some of the acquisition cost. It is the intent to judge these aircraft based on the content of the ASIP. As has been done in the past, the USAF will assess these aircraft procured in this manner to determine the requirements under which they were certified. In addition, they will assess them to determine whether they meet military requirements with their design missions and service experience. These assessments will determine the need for additional structural testing and analysis.

III. Corrosion

One of the major problems found in operations with aging aircraft is the cost associated with corrosion damage. Unfortunately, the progress made in the recent past in the control of corrosion does not bode well for the future. This is especially true when one considers the impact of new environmental laws that remove many of the corrosion-fighting chemicals currently used. Continued emphasis in research in the area of corrosion control is certainly one area that could have significant benefit. The USAF sponsored a contract in 1997 to determine the cost of corrosion in USAF aircraft and found in this study as in a previous study performed in 1990 that the cost of corrosion prevention and repair is significant. They found the total cost from corrosion in 1997 was approximately $795 million dollars. This was an increase of 4% over the 1990 costs, although the USAF reduced the force structure by 28%. The C-5, KC-135, and C-141 account for 50% of all direct corrosion maintenance costs. Of the $795 million spent, painting the aircraft cost $136 million. The USAF spent approximately $425 million of the $795 million specifically on corrosion repairs. The survey highlighted the A-10 as a success story in defeating a serious corrosion problem.

In many cases, the desire for improved performance motivated the use of the materials in these aircraft and, consequently, there was little attention given to the potential for corrosion and stress corrosion cracking damage. Further, at the time of manufacture of these aircraft, corrosion protection was much poorer than that which is available today. Consequently, many of these corrosion protection systems have broken down. In open areas, maintenance personnel can readily renew protection systems. There is, however, no easy way to renew corrosion protection system in joints. Experience with modification of aging aircraft has revealed that joints without proper protection can experience significant damage.

The USAF has undertaken a teardown of a KC-135 removed from service to assess the extent of corrosion on these aircraft. The USAF owns over 700 of these tanker aircraft and they consider them a national asset. It is essential that they remain in inventory well into the next century. Another goal of the program is to evaluate the potential of nondestructive evaluation to detect hidden corrosion in the joints of aircraft. Preliminary evaluations have indicated that eddy current methods using multifrequencies have the best potential for detecting and quantifying corrosion in the thin aluminum skins used in fuselage lap joints. One of the products of this program is an assessment of the costs to maintain these aircraft in future years. Another product is a determination of the approaches to accomplish this maintenance.

Pitting corrosion has long been a concern of the USAF. This form of corrosion attacks intrinsic particles in aluminum structures, enlarges them, and exacerbates the growth of cracks from these origins. This is a potential safety problem because pitting corrosion can initiate WFD before laboratory durability test would indicate its occurrence. It is, however, more of a durability problem and can significantly increase the maintenance burden beyond that expected based on the laboratory durability test. In addition, pitting corrosion is typically difficult to find in its early stages by NDI. There are three phases in a research and development program for pitting corrosion. First is establishing a process for reproducing pitting corrosion that is typical of what is being found in aging aircraft fuselage structures. Second is investigating the effects of this corrosion on the life of test specimens. Third is the establishment of an empirical model that will enable the prediction of crack growth behavior in corroded aluminum parts. The third phase is particularly difficult and final success will be hard to achieve.

A major concern to the corrosion prevention community is the impact of environmental laws that restrict the use of material with chromates, for example. The industry is working on substitutes, but this work is far from being completed. In the meantime, the solution will be to expend more hours at depots in corrosion removal and repair.

Corrosion and fatigue separately have both led to serious safety as well as economic problems. Corrosion alone, in forms such as uniform corrosion or exfoliation, may reduce the strength of aircraft and lead to failure. Both of these forms of corrosion may lead also to expensive component repair or replacement. There are many cases where corrosion alone is not significant from a safety consideration, but is a very significant economic problem. In the case of corrosion alone, one must judge the seriousness of this problem on an individual basis. NDIs have found fatigue cracking where there is essentially no influence from corrosion. Researchers have documented many cases over the years where consequences were catastrophic. The results of fatigue cracking have caused many expensive repairs and modifications to the structure, including component replacement. Fatigue often combines synergistically with corrosion and, in these cases, the term corrosion fatigue is more appropriate. In most cases, corrosion, fatigue, or corrosion fatigue becomes a safety consideration only when either maintenance is not performed properly or the maintenance program is inappropriate. Experience derived from diligent maintenance has repeatedly shown that the operator need not compromise safety resulting from these problems.

Based on historical evidence, there can be no question that corrosion is a major economic factor with both military and commercial aircraft. Many of the older military aircraft that are currently operating were victims of the zeal for increased performance that prompted the development of high-strength alloys without any consideration of the threats that they could have on structural integrity. One such development was the 7178-T6 that the USAF used for the lower wing skins of KC-135. They selected this material rather than the 2024-T3 material used in the Boeing

707 to provide additional tanker capacity for aerial refueling. Experience has shown that high strength was the only virtue of the material. That material exhibited low fracture toughness, fast crack growth rates, susceptibility to corrosion, and low resistance to stress corrosion cracking. To make matters worse, the USAF elected to save money by omitting corrosion protection of the material by faying surface sealing or wet installation of fasteners. The result was that the USAF had to replace the lower wing skins at 8,500 flight-hours. By that time, the structure had lost its fail-safe capability because of WFD. Recent teardown inspections of a high time 707 wing with 2024-T3 indicate that the onset of WFD occurs at about 50,000 flight-hours of commercial usage.[25] The results of previous laboratory full-scale durability testing would not have indicated that the KC-135 wing structure would be in a state of WFD at 8,500 flight-hours. Corrosion fatigue undoubtedly played an important role in the demise of this structure. The economic impact of this problem was enormous. However, the alertness of the engineers at the Oklahoma Air Logistics Center (OC-ALC) at Tinker Air Force Base prevented this problem from causing loss of an aircraft. This same material, 7178-T6, is used in the upper wing skins of the KC-135. It is now causing a significant problem in that there is considerable exfoliation corrosion around fastener holes. The OC-ALC finds many areas of corrosion through the process called search peening. In this process, maintenance personnel shotpeen the upper wing skins with glass beads that cause corroded areas to reveal themselves through local deformation around fastener holes. In some cases, the problem is severe enough to cause panel replacement. When the OC-ALC replaces upper wing panels, they select a replacement material much more resistant to corrosion than the original material. The troublesome part is that the USAF does not have a solution that would preclude replacements in the future. The use of peening, however, does appear to extend their lives. There is evidence of beneficial effects of shotpeening from the B-52H upper wing. These wings, which had the upper surface shotpeened at the time of fabrication, are showing only minor corrosion damage although they have been in service for many years.

The USAF's C-141 is another aircraft that has experienced fatigue cracking in operational service sooner than indicated by the results of full-scale durability testing. The USAF believes that at least part of this difference could be explained by the environmental effect found in operational usage. However, at least part of this difference could be the result of variations in quality that could be expected from aircraft to aircraft. The manufacturer sealed the faying surfaces and wet installed the steel fasteners. These actions appear to have diminished corrosion significantly.

The AGARD Corrosion Handbook[39] discusses problems found on these aircraft and many others through case studies. This document places the cost of corrosion in the United States alone in 1978 at $70 billion overall. The cost today is likely even greater. The USAF can account for approximately one of those billions for their airplanes.[40] Case studies in this handbook show the results of corrosion that remained undiscovered by maintenance personnel until it was a large economic or safety problem.

Corrosion is found on both military and commercial aircraft. As evidenced by the currently documented cost of major maintenance visits by large category transport aircraft, the cost of corrosion is a major factor in commercial maintenance budgets. However, the use of corrosion prevention compounds in commercial aircraft has significantly reduced the burden. It is encouraging that airframe manufacturers are doing a much better job of applying corrosion protection during fabrication. One would anticipate these improvements would relieve some of the maintenance cost burden when operators bring these aircraft into inventory.

The commercial approach of designing structures to be fail-safe and then augmenting this design principle with damage-tolerance-driven inspections has virtually eliminated fatigue as a safety problem in large category transport aircraft. The remaining issue with some of these commercial aircraft is to determine the time of onset of WFD. That is, the time when their fail-safe capability is degraded. The work done by the FAA Technical Center and NASA Langley after the 1988 Aloha Airlines incident has established much of the methodology needed for the determination of the time of onset of WFD. The commercial aircraft manufacturing community can now use this technology to evaluate their aircraft for the occurrence of this problem.

The synergistic effect of corrosion and fatigue together, called corrosion fatigue, is also a safety and economic problem. The author of the ICAF Plantema lecture of 1995[41] discusses corrosion fatigue in detail and provides a large bibliography for further study. The author suggests that a factor of two should be acceptable as a life reduction factor to account for the effect of corrosion on fatigue. This appears to be larger than would be expected from USAF experience. The specific reduction factor for damage-tolerance-driven inspections would need to be determined on a case-by-case basis. Because pitting corrosion could affect the total life, it is likely that the factor would be significantly less than two. The reason is that the size of defects derived from pitting corrosion is considerably less than that assumed as initial flaw size for damage tolerance. Consequently, the crack nucleation time from corrosion pits would not affect the inspection interval. This situation could change for an aircraft that has a long history of unidentified pitting corrosion. In this case, multiple cracks from corrosion pits could require the operator to shorten inspection intervals. Unidentified pitting corrosion could constitute a loss of safety. One author indicates that the result may be different for constant amplitude loading and spectrum loading.[41] The effect of pitting corrosion on life is another shortcoming of the safe life approach. The safe life approach may not adequately account for the reduction in crack nucleation time from corrosion pits. Consequently, the safe life derived from laboratory durability testing could be unconservative.

The fail-safe design augmented by damage-tolerance-derived inspections is largely responsible for preventing the combined effect of corrosion and fatigue from becoming a major factor on large category transport aircraft. For the USAF, the inspection program derived from damage-tolerance principles is certainly the primary means of maintaining safety from corrosion fatigue. The USAF also uses another inspection process called the analytical condition inspection (ACI) to augment damage-tolerance inspections. The USAF initiated the ACI program many years ago to help maintenance personnel discover distressed areas of the aircraft not identified by design analyses or testing. For this process, the USAF selects a small sample of aircraft from the inventory and inspects the entire aircraft thoroughly. To gain better access to concealed areas, they remove fasteners and panels to inspect the structure. This is done as completely as possible nondestructively to ensure that the structure is not experiencing corrosion or fatigue cracking that could jeopardize continued flight safety or economic operation. When they find an area that does not correlate with design experience, they make appropriate changes to the FSMP, which is an integral part of the ASIP. The FSMP tells the maintenance personnel how, when, and where to inspect the structure to maintain safety.

A paper written in 1997[42] describes the forms of corrosion that could compromise flight safety. The authors of this paper list pitting corrosion, intergranular corrosion, exfoliation corrosion, stress corrosion cracking, corrosion fatigue, and uniform corrosion as safety issues. In this document, they suggest a number of research and development activities that should enhance the state of knowledge about the effects of corrosion. They suggest many activities, including teardown inspections of exservice aircraft and scheduling of maintenance on a time basis as well as a usage basis.

The USAF ASIP initiated in 1958 and modified in 1975 makes an assumption about the effects of some of the forms of corrosion listed above. It presupposes that corrosion damage will not raise the stress sufficiently to change the inspection program that ensures flight safety. This infers that the maintenance program is able to control corrosion damage through inspections and preventative measures. Experience has shown that this assumption has not led to serious consequences. The USAF maintenance program is outstanding. One reason for its success is the advice given to them by resident ASIP managers and other engineers at air logistics centers. These engineers provide them with a tailored inspection program for each individual aircraft from data collected from operational aircraft. The USAF experience, since they introduced damage tolerance, has shown the failure rate from structural causes is less than one in 15 million flight-hours. The USAF cannot attribute any catastrophic failure since the early 1970s to corrosion or corrosion-related phenomena. There have been, however, numerous local failures from corrosion, particularly stress corrosion cracking. This has been a chronic problem in many landing gear systems and in large bulkhead forgings.

Applications of the 1975 version of ASIP addressed corrosion fatigue by modifying the crack growth rates in an attempt to account for the corrosion environment such as moisture, salt, or sump tank water as appropriate. The validity of these corrections is subject to question because of such factors as loading frequency, temperature, and the use of crack growth rate data from constant amplitude testing. However, the USAF has found few errors in DTAs that they can directly attribute to the manner in which they accounted for the effects of the environment. Most often, the increase in crack growth rates in critical locations in the structure is associated with operational usage spectra that are more severe than the design spectrum of loading. As indicated above, the USAF accounts for these differences through individual aircraft tracking for loads.

There are serious concerns about the impact of corrosion on structural integrity. The first and most obvious is the effect of lapses in proper maintenance that have led to significant loss of structural strength. Thinning of the structure will increase the stress in the structure such that the inspection intervals required for flight safety will need to be increased. There is limited NDI capability to accurately quantify the amount of material loss due to corrosion. A much easier task is to inspect the structure for the existence of corrosion. Consequently, at this time, the safest approach is to remove corrosion indications when found. As the capability for NDI is improved, then the opportunity for corrosion management is improved. Further, it is essential that those areas that are damage-tolerance critical be an integral part of the corrosion detection program.

As an example of the potential danger of damage from corrosion, a Far East Transport (FAT) 737 experienced a fuselage failure by explosive decompression in Sanyi, Taiwan, on 22 August 1981. The operator had been using the aircraft to transport fish in open barrels in the cargo compartment. The failure was due to severe undetected skin corrosion over a length of approximately 2.5 m in the lower lobe of the fuselage. Corrosion caused significant thinning in the skin causing the hoop stresses due to cabin pressure to increase. There was some evidence of fatigue in some areas on the final fracture surface. This occurred because of the increased stress. The skin must have eventually become thin enough to cause the stress to exceed the tension capability in at least one area. This aircraft had received a D check 1 month before the accident. The inspection instruction required outboard cabin floor panel removal. A proper inspection would most likely have found the corrosion before it became catastrophic.

Corrosion was also present in the lap-splices of the April 1988 Aloha Airlines aircraft. This failure, however, was not in the lower lobe of the fuselage as was the August 1981 failure. The aircraft suffered loss of adhesion, causing the stresses to increase at the fastener holes. The knife-edge countersink caused the local stress at the fasteners to further increase. What remained of the skins themselves indicated that the corrosion had not thinned the skins below normal tolerances. Subsequent tests on simulated lap-splices without corrosion indicated that the knife-edge was the major influence in the formation of cracks. Corrosion was only a minor factor, if any, in this accident since the environmental effects on the metallic material used for the fuselage skins are small. As indicated previously, the accident report[6] issued on 14 June 1989 stated that the probable cause was the failure of the maintenance program to detect the presence of significant disbonding and fatigue damage. This ultimately led to failure of the lap joint.

The USAF has had poor experience in the purchase of older aircraft from the commercial market. Some of these were so badly corroded that they could not be economically repaired. The USAF believes that the decision by the owners to sell the aircraft exacerbated the extent of corrosion damage in these aircraft. After the decision to sell was made, the aircraft evidently received very little maintenance until the owners could sell them. Under these conditions, it did not take long before the damage from corrosion was so extensive that the USAF had to condemn the aircraft immediately after they purchased them. If the USAF had used the current NDI capability at the time of purchase, they would not have purchased these aircraft. Examples such as this highlight the diligence needed in a maintenance program. The owners of these aircraft, up to few years before they sold them, maintained them properly and they were flying in an airworthy condition.

Another major concern is pitting corrosion may accelerate the onset of WFD. This is a safety concern since WFD is a condition where the cracks degrade the fail-safe capability of the structure from its original design requirement. This could be a serious concern unless operators take proper care to use teardown inspections and NDIs to reveal that the aircraft has reached the onset of WFD. The damage-tolerance initial flaw requirements as adopted by the USAF in the early 1970s are well in excess of the size of defects associated with pitting corrosion. The criterion used by the USAF for damage-tolerance design of new aircraft is that the initial flaw will not grow to critical in two design lifetimes. Further, damage-tolerance requirements force the selection of ductile materials that are tolerant to defects. As a further safety measure, the USAF guidance to the contractor is that they design the structure such that it is inspectable. This is an advantage, not only for crack detection, but also for corrosion detection. The effects of pitting corrosion, however, could affect the safety of older aircraft that were not designed to current damage-tolerance requirements. Therefore, for these aircraft, it is likely that cracks derived from pitting corrosion will need to be found by the inspection program. The other concern with pitting corrosion is it could result in a significant degradation of the durability of the structure and hence shorten its useful life in service. Experiments with specimens exhibiting pitting corrosion show that cracks appear much sooner than otherwise expected. These cracks could degrade the fail-safe capability of the aircraft and consequently precipitate the need for structural modifications. A teardown inspection of high time Boeing 707 wing revealed many significant cracks. These cracks appeared to be predominantly in holes where the manufacturer used steel fasteners. There was essentially no protective barrier between the steel and the aluminum wing skins and stringers. Consequently, it is likely pitting corrosion did contribute to the cracking found in this structure. In contrast, the teardown inspection of the C-141 wing structure with wet-installed steel fasteners did not indicate evidence of pitting corrosion. The FAA has sponsored extensive studies in pitting corrosion[43,44] and corrosion fatigue. Studies show that, for bare materials, cracks nucleated from constituent particles, and for clad materials, the pitting promoted cracking on a nearby favorably oriented grain. The work done to date on this subject has led to a good qualitative understanding of the problem and has quantified many aspects of the phenomenon. However, researchers will not be able to provide a predictive method for the effects of pitting corrosion until sometime in the future.

Hidden corrosion is also a major concern for continuing structural integrity. Hidden corrosion is corrosion that remains undetected after inspection with the specified method for the area. Based on this definition, the amount of hidden corrosion is dependent of the equipment used in the nondestructive evaluation. It is essential the maintenance personnel have a clear understanding of where they may find corrosion in the structure so they may use appropriate inspection procedures to find it. It is also incumbent on the laboratory to develop the inspection equipment to improve the likelihood of detecting corrosion. Because corrosion damage appears to reoccur in the same locations in a population of airplanes, the use of teardown inspections is helpful for locating potential damage. The Oklahoma Air Logistics Center performed a teardown inspection on a KC-135 retired to Davis Monthan Air Force Base in 1991. This aircraft, delivered to the USAF in 1962, spent 29 years at Mildenhall Air Base in the United Kingdom. Therefore, the aircraft saw a severe corrosion environment during its life. The inspection interrogated the structure for cracking as well as corrosion. The USAF found little cracking since the aircraft had only 16,521 flight-hours and 2,942 flights. They cleaned the parts and etched them approximately 30 μm to enhance corrosion and crack detection. The USAF, for this study, classified corrosion as light if it was less than 25 μm, moderate if it was between 25 and 250 μm, and severe if it was greater than 250 μm. For the fuselage, there was extensive light corrosion in the skin and doubler faying surfaces. They found limited moderate and severe corrosion below the cargo door, lower bilge, and spotwelds. None of the fuselage corrosion was severe enough to affect flight safety. For the wing, there was extensive moderate and severe pitting at the steel fasteners in the upper surface. Most of these had not progressed to exfoliation and none was severe enough to affect safety of flight. There were several areas of severe corrosion at the upper wing skin and spar interface. The center section of the horizontal tail suffered from severe exfoliation on the lower spar caps. They also found stress corrosion cracking in the horizontal tail. This inspection is significant in that it provided considerable insight to the extent of corrosion. It also served as an excellent representative aircraft for identifying areas of hidden corrosion that the USAF did not address in previous depot maintenance activities. In addition, it assessed the ability of available NDI procedures to locate corrosion. However, the NDI community has made significant progress in detecting corrosion since the completion of this work.

Another concern about the effect of corrosion on structural integrity is the effect of corrosion products on stresses in structural joints. In the longitudinal lap-splices in the fuselage, corrosion products are much less dense than the original material and, consequently, the trapped material causes the joint to expand. Stresses derived from this expansion are significant. At the lower limit of detectability of joint thinning, stresses may reach yield. There have been many cases found where stresses are sufficiently high to cause cracking in the skins. The National Research Council in Canada has studied this problem extensively. They developed an optical device for detecting this problem on operational aircraft. There is no known case where this problem has caused a catastrophic failure or the onset of WFD. However, the potential is there for the cracks to turn into fatigue cracks and propagate from the cyclic flight or pressurization loads. There is also the possibility that, even without cracking, the stresses may degrade the fail-safe capability of the structure. The USAF plans to investigate this phenomenon as part of their aging aircraft program.

The potential for stress corrosion cracking to become a fatigue crack is another concern. This seldom happens but the result can be serious if maintenance does not find the crack. The USAF and FAA records do not reveal that there has ever been a case where a stress corrosion crack in an airframe has resulted in a catastrophic failure. The use of 7079-T6, 7178-T6, and 7075-T6 in aircraft designed in the 1960s and 1970s has led to numerous cases of stress corrosion cracking. In the USAF, the C-130, C-141, C-5A, and T-38 are aircraft for which stress corrosion cracking is a significant economic burden. The Materials Information Analysis Center performed a study[45] in 1996 and found that 70 out of 115, or 60.9%, of the corrosion failures (problems) in C-130 airplanes were stress corrosion related.

The FAA took a firm stand on maintaining safety of corroded aircraft when they issued ADs against Part 25 and some Part 23 aircraft. These ADs provided guidance to commercial operators on the amount of allowable corrosion on their airplanes.

It has been the policy of the USAF that if corrosion was found it would be removed. That is, they have been in a find-it-and-fix-it mode of operation. They have not always followed this policy because of the cost and the pressure to get aircraft back into service operations. However, at least in part, the excellent safety record of USAF aircraft is the result of this policy, costly though it may be. They must consider the safety implications to any change to this policy.

There is at least one other known case where diligent maintenance defeated a corrosion problem. TOGAA visited Petroleum Helicopter, Inc. (PHI), in Layfayette, Louisiana, on 27 March 1996. They found that PHI had defeated corrosion resulting from the harsh environment of warm moist salt air associated with flying from the coast to the Gulf of Mexico oil platforms. PHI had earlier experienced the severe corrosion damage one would expect in this environment. They decided to take matters in their own hands and, through careful washing of the aircraft and the use of corrosion prevention compounds, reduced. corrosion to insignificance.

More recently, a paper[46] given by a member of the AFRL begged the question about when a predictive model would be available for corrosion. There is no doubt that predictive capability would be helpful for making maintenance decisions. However, the ability to predict future corrosion will likely always be limited.

The dangers from corrosion, fatigue, or corrosion fatigue are ever present in operational aircraft. Presently, the largest danger by a considerable margin is economic rather than flight safety. All of the collective experience from both military and commercial operations indicates this to be true. No one can foretell with any degree of accuracy what to expect as both military and commercial aircraft push further into the uncharted waters of aging. Thus far, experience has not indicated that there is a "falling off point," as once imagined when there was widespread belief that the world was flat. It is incumbent, however, for the researcher, the engineer, and maintenance personnel to maintain a diligent approach to the problem. They must use all available techniques such as damage-tolerance scheduled inspections, ACIs, and assessments for the onset of WFD to help ensure that they maintain the safety of future aircraft operations. Diligent use of CPCs and research into better means of corrosion detection and prevention appear to be the most promising ways to reduce the economic burden in the future.

IV. Repair

A. Introduction

Traditionally, the basis for repairs placed on aircraft is replacement of the original static strength. Engineers can accomplish this easily since they do not need to have detailed knowledge of the loads on aircraft. It is a much more difficult task to design a repair that is damage tolerant. That is, one where inspections are identified that maintain the safety of the aircraft throughout its life. A damage-tolerant repair does require that the designer of the repair have knowledge of the loads. Typically, only the OEM has this knowledge and since they see a return on the investment to develop the loads for the life of the aircraft, they do not sell or release them to anyone. In most cases, even the USAF does not have the detail loads on aircraft they procured as new aircraft. On some of the aging aircraft, however, the USAF has funded the effort for the OEM to design the standard repairs to be damage tolerant. This effort was a logical activity after the DTA derived the inspection program for the intact structure.

One of the major concerns with repairs is that they can destroy the inherent damage-tolerant capability of the structure.[47] For example, in lap-splices, they can cover enough of the splice such that the crack would be evident only after the crack would have reached critical size. Repairs can produce areas of stress concentration such that areas may crack that were not previously subject to high stresses.

B. Metallic Repairs

Most of the repairs designed for aging aircraft are metallic. The FAA and the USAF recognized the need for a procedure that would enable operators to rapidly design metallic repairs for metallic aircraft. A software tool called RAPID[48] contains this procedure. The development of RAPID is one good example of cooperation and coordination of the FAA and USAF for aging aircraft research. They shared the funding of the program and the guidance of its development. The program permits operators to design repairs for their aircraft that satisfy both static and damage-tolerance requirements. There are simplifying assumptions made for loads that permit repairs in most of the structure. There are, however, regions of the structure, for example, for the fuselage in the area of the wing attachment where the results of the program would not apply. The program has repair options that will satisfy many repair situations. The software has built-in rules that help prevent the design of a bad repair. There is a need to reevaluate metallic repairs in light of the emerging capability for composite repairs of metallic structure.

C. Composite Repairs

There are numerous applications currently of composite repairs in the USAF, in addition to applications in Australia and Canada. The USAF applications include the C-130, C-141, KC-135, B-1, and F-16. The success of these applications has motivated the USAF to spend the resources to fully exploit this technology. They have established procedures for successfully preparing the surface for the bonding operation. However, they have found that the process is demanding with respect to the mechanics of application of the patch. The primary issue with bonded repairs is there is no NDI procedure to determine the strength of the bond. Innovations such as a self-inspecting smart patch are the subject of research and development activities. The FAA has shown considerable interest in the application of this technology to commercial aircraft. The work by the USAF should establish a sound basis for them to establish a plan to transition this technology to commercial operators.

Aircraft manufacturers have used bonded structures for many years to reduce weight and reduce fabrication costs. Fokker, for example, in The Netherlands, has used bonding technology to great advantage in the manufacture of their aircraft. The laminated skins provide for considerable improvements in durability as well as damage tolerance. Products such as ARALL™ and GLARE™ attest to the benefits that one is able to derive from bonded structures. McDonnell Douglas and Northrop used bonding technology for the design of the step-lap joint in the F-18. There is no known case of failure of this design in operational service. There are several important elements to success of these programs. First, the design must be amenable to analysis and manufacture. Second, the manufacturer must subject these bonded structures to intensive testing and rigorous quality control procedures. Historical evidence indicates that these elements are viable in the manufacturing environment. Are these elements feasible in the environment where technicians make bonded repairs on aircraft? Certification authorities must address questions such as this if they permit the use of these technologies.

The bonding of metal or composites for the repair of cracks derived from fatigue is a very simple concept. The bonded-on patch impedes the opening of the crack and hence slows its growth. There are many aspects of this repair process, however, that are far from simple. The design of the patch such that it does its job efficiently without damaging the surrounding structure is a major challenge. Materials and process issues for surface preparation and application of the patch are extremely critical to its success. Finally, the training of the technician that applies the patch is essential for ensuring the integrity of the patch. It is the view of some experienced installers of bonded repairs that there is a risk not inherent in other types of repairs. The reason is that strength depends on a complex repair process that is not quantifiable after the repair is completed. The repair process if not properly adhered to may produce a repair with no strength at all. Deviations in the cure cycle, handling of the adhesives, or in grit blasting can lead to unexpected and sometimes unknown consequences. To mitigate this risk, the use of an engineer on site is common. Even this practice is no guarantee of a successful repair. Consequently, one sees that issues in the production environment are considerably different than those encountered in field application of the repair. The need for rigorous compliance with process standards has not precluded success in the application of this type of repair. Four hundred repairs made by an installation in the USAF without a failure or known degradation demonstrates this. Australia can identify similar success stories. Fortunately, dedicated teams have made the effort to master the process. This background of experience separates them from the rest of the world.

Analysts have separated the design of patches of cracked structures into three categories. The first is a patch they intend for use in the repair of defects in stress fields that approximate assumptions of uniformity inherent in the derivation of the patch configuration. An example of such a patch would be the repair of a crown skin of the fuselage of the C-5 aircraft. A team applied two Glare patches using the CalcuRep program.[49] The basis of the CalcuRep program is the Rose model. The analyst should determine if cracks are small enough such that stresses do not change significantly in the dimensions of the patch. The second category of patches are those where the stress field is complex enough to warrant the use of finite element methods to accurately determine the effect on the repaired structure. An example of this would be the patch used on the C-141 to repair the cracks in the fuel transfer holes. The Warner Robins Air Logistic Center repaired 140 aircraft using 1,700 boron patches on the skins and integral risers. Here the analyst had to determine the change in the stress field for the fasteners near the patch to establish suitable inspection intervals. Another example in this category is fuel vent-hole repair for the F-16. The analyst used IDEAS to model the wing and the patch. The third category is for cases with complexity intermediate between the first and the second. Designs in this category are closed-form solutions. Not all analysts agree that there is a need for the third category. The reason is that many check closed-form solutions with finite element calculations to test their accuracy.

Bonding technology may also be used to restore the fail-safety for a structure that has suffered the onset of WFD. The USAF examined the possibility of this application in two cases. The first was the C-141 wing structure that had lost its fail-safe capability because of cracking in spanwise splices. Lockheed examined the potential use of longitudinal composite strips to act as tear straps in the case of a failed wing panel. Unfortunately, the C-141 program office was unwilling to fund the effort to complete the work started by Lockheed. The USAF examined a second application of this kind for the aging 707 aircraft they purchased for the Joint Stars program. They terminated further consideration of this approach when the Joint Stars program office made the decision to replace the lower wing skins and stringers.

The damage-tolerance approach is the basis used for ensuring flight safety by many certification authorities in the world. The approach, although different in its details, applies to composite as well as metallic structures. Curiously, many certification authorities that do not use it for metallic structures embrace the concept for composite structures. This approach appears to be ideally suited for ensuring safety of bonded structures used as a repair of cracked structure.

As indicated above, the threat to the integrity of the metallic portion of the structure is fatigue cracking. For the bond itself, there are multiple threats to structural integrity. One is delamination in the bond line. This defect is detectable through NDI and is quantifiable as to its size. In general, however, the growth of this defect is not analytically predictable. The analyst must use testing to determine the growth of this defect and the consequences of growth. Another threat to the bond is environmental degradation. This is not predictable analytically and is quantifiable through testing only with great difficulty. The designer must configure the design of the bond and establish its manufacturing parameters such that this failure mode does not exist. Otherwise, the use of slow crack growth is not a viable alternative.

The preferred approach for the design of composite repairs to metallic components is the "fail-safe" approach. With this approach, the designer maintains structural integrity using an alternate load path in case of loss of integrity of the bond. The FAA and the USAF believe fail-safety is an extremely important asset to any aircraft. The use of the fail-safe approach for bonded structure eliminates essentially all of the concerns with their inspection. The problem associated with inspecting the bond for strength becomes less important.

In addition, the issue of environmental degradation of the bond disappears. However, while the issues of inspection and environmental degradation may not be significant for certification, they must be addressed if an effective repair is to be implemented. In addition to meeting damage-tolerance requirements, the designer must also ensure that the durability of a bonded structure is adequate. Often, but not always, a bonded structure design that meets damage-tolerance requirements also meets durability requirements. This is normally the case if the bonded structure meets the slow crack growth requirement without requiring in-service inspection. In these cases, the initial safety limit is two design lifetimes. For bonded structures, the designer should base the durability of the metallic portion of the structure on a fracture mechanics based approach. The initial defect is one the designer would not expect to occur more than once in a given aircraft from the population. The durability flaw, therefore, is significantly smaller than the initial flaw used for the damage-tolerance analysis. The durability life is the time for this flaw to grow to functional impairment.

The design and manufacturing process must be capable of eliminating environmental degradation as a threat to bond line durability. When the manufacturer controls processes in a production operation, experience indicates that this type of degradation is not a threat. When technicians make the bond for repairing a structure, the same confidence is not achieved. There is no known approach using NDI to determine if the bond will suffer from environmental degradation.

As indicated above, threats to structural integrity of bonded structure are many. One of the main threats is environmental degradation. A bond could initially have apparently adequate strength but could degrade with time. The degradation process is poorly understood but based on successes of many repairs in the field and those made in a factory. The adherence to strict quality controls during the bonding process appears to defeat the problem. Another threat to bond integrity is that the bond did not initially have adequate strength. This is a difficult threat to overcome because there are no known methods to determine the bond strength by NDI. However, there is evidence that a bond made under strict quality control conditions appears to have its intended strength.

It is essential that the bond will have the strict quality control to enable the operator to rely on its strength and its resistance to environmental degradation. The five-element process for technology transition[50] for materials and material processes accomplishes this. In a study of successful transitions of structural technologies from the laboratory to engineering and manufacturing development, the USAF found that five factors constituted a common thread among these successes. These five factors are

1) Stabilized material and/or material processes,
2) Producibility,
3) Characterized mechanical properties,
4) Predictability of structural performance, and
5) Supportability.

The USAF did not attempt to establish a ranking of importance of these factors. A deficiency in any one of the factors could constitute a fatal defect. When the designer cannot ensure that the satisfaction of these factors is achievable then an alternate means of compliance is in order. The easiest approach that eliminates the safety implications of not being able to ensure the quality of the bonded structure is to use the fail-safe approach. The fail-safe approach used for assurance of aircraft structural integrity has saved many aircraft from catastrophic failure. Designers starting using this approach after catastrophic accidents in the mid-1950s. This approach is very simple to apply to bonded structures. One merely assumes that the bond is not effective and performs inspections based on slow crack growth in the remaining structure. The obvious difficulty with this approach is that the inspection burden would be greater than that based on the assumption the bond had its intended integrity. In many cases, this would not be a major problem. For example, for new structures or those where the defect is small relative to critical crack size, the penalty for additional inspections would not be great. The use of the fail-safe approach, however, would preclude the application of patches on areas where the crack was near or exceeded the critical crack length.

Another alternative is available through probabilistic methods. Many aspects of structural certification depend on probabilistic methods.

V. Future Actions

A. Introduction

A particularly difficult problem for an operator is the procurement of aging off-the-shelf aircraft. In the marketplace, there are many used aircraft available at far below the price of a new off-the-shelf aircraft. Procurement of such aircraft is difficult to resist since some of the actions by the FAA have drastically reduced their sales potential. Other reasons for the low price on these aircraft are that they may have flown beyond their design service life. In addition, they may have corrosion in numerous areas, WFD, and numerous repairs, many of which are not damage tolerant. That is, they may possess all of the ingredients to be in the classification of an aging aircraft. Unfortunately, many of these problems are not readily observable and the aircraft appear to be airworthy. Experience has shown that significant problems do exist and the cost of refurbishing these aircraft is considerably above original expectations. The USAF has had numerous bad experiences with the purchase of used aircraft. The purchase of Boeing 727 aircraft that had be scrapped because of corrosion damage and the purchase of Boeing 707 aircraft for use in the Joint Stars program are examples of such experiences. One of the major problems is that it is extremely difficult to determine the extent of corrosion damage from a visual inspection. Even today, the NDI capability is marginal for the detection of hidden corrosion. Making such procurement requires the purchaser follows a disciplined procedure to help ensure that it will be successful.

As indicated previously, another problem with procurement of aging aircraft is the lack of external and internal loads information. This has been a chronic problem in the commercial world because of the need for knowledge of the loads to make damage-tolerant repairs. In most cases, the OEM has not seen that it would be to its benefit to release loads information.

If the aircraft has flown beyond its design lifetime or the operator expects to fly it beyond its design service lifetime, then the inspection program that currently exists must be updated. This update includes an examination of the structure for previously unidentified locations that would need an inspection program to maintain safety. Inspections of individual aircraft must determine the condition of aircraft with respect to corrosion. For those areas found to be corroded, the preferred approach is to eliminate the corrosion by cleaning it out or replacing the structural elements in question. If this is not feasible, then the corrosion must be assessed to determine the change in the inspection program that will account for the influence of corrosion on accelerating crack growth in the structure. Analyses must show that the structure is free of any area with WFD or any area that could expect to experience WFD within the operating life of the aircraft in the USAF inventory. This condition typically will not be found by routine inspections based on the damage-tolerance approach. Consequently, the potential for catastrophic failure is significantly increased. The time of the onset of WFD is normally determined through teardown inspections of the fatigue test article or through teardown inspections of operational aircraft. On aircraft that have been in operational service, there are numerous repairs, many of which are not designed based on damage-tolerance principles. Repairs on each individual aircraft must be examined to determine if they can be safely inspected during the time the aircraft is expected to remain in operational service. Those repairs failing to meet this criterion should be replaced with a properly designed damage-tolerant repair.

B. United States Air Force Actions

The USAF initiated a new effort to reduce the maintenance burden associated with aging aircraft and maintain safe operations. They formed a team called the Aging Aircraft Technologies Team (AATT) whose mission is to provide technical guidance, foster coordination, and contribute programmatic support to reduce the economic burden and maintain long-term safety. This initiative came from recommendations from a recently completed study on USAF aging aircraft conducted by the National Research Council (NRC).[51] The committee's composition included people from universities, industries, and government. The primary motivation for requesting support of the NRC was to ensure that the USAF was funding technology development efforts with high priority to help solve the aging aircraft problem. committee completed this task, but, in addition, identified engineering tasks to update existing DTAs.

The NRC committee recommendation to update DTAs conducted in the 1970s and 1980s was proper. These early assessments have served the USAF well. The USAF aircraft loss rate due to structural failure is less than one in 15 million flight-hours. DTAs and the work done to identify and correct problems associated with WFD are responsible for this remarkable record. The NRC committee briefly reviewed the majority of aging aircraft in the USAF. From this review, they established priorities in three categories on the urgency to conduct an in-depth review of the aircraft. Subsequent reviews by USAF personnel revealed that most of the NRC concerns about specific aircraft were valid. However, the ALCs that manage the maintenance of USAF aircraft had taken the initiative to contract with the manufacturers to update the DTAs. The reviews also showed the close coordination of ALC engineers, maintenance personnel, and the manufacturer.

The updating of DTAs will serve two purposes. First, they will identify any changes to the FSMP, as identified in the ASIP, needed for the airplanes to operate safely in the future. Second, they will identify research and development activities that may have not previously surfaced as a need. The USAF will contract, usually with the original equipment manufacturer, to update DTAs. This effort will meet the intent of the USAF policy and instruction (63-10 and 63-1001) on aircraft structural integrity.

The AATT's concept of operations generally follows the recommendations of the NRC's report. It is consists of the following elements:

1) A single technical leader from the Aeronautical Systems Center (ASC), who provides guidance to the team;

2) A structural assessment group that will determine the structural health of the aging aircraft inventory;

3) A research and development group led by a member of the AFRL;

4) A technology implementation group led by ASC's aging aircraft program office; and

5) A top-level advisory committee to provide advice and council to the single technical leader.

The technical groups identify and define aging aircraft requirements, develop technical programs responsive to customer requirements, coordinate output- and cost-focused research and development, and transition maturing technologies to organizations responsible for maintaining and operating aging aircraft. These functions are performed in an integrated and seamless fashion with input from the aging aircraft community, including the commercial sector.

The AATT wants to enlist the world's leading experts in structural technologies to participate in this activity by offering solutions to USAF problems with aging aircraft. To make their vision as clear as possible, they have developed the following guiding principles to communicate to the world what they expect from research and development activities:

1) Must meet the needs of USAF aircraft;

2) Must be oriented towards flight safety, maintenance cost-reduction, or enhanced availability;

3) Must be product oriented and cost focused;

4) Must augment the highest level of current capability in industry or government; and

5) Must develop technology that can be transitioned.

The AATT, in conjunction with ALCs, will identify problems with aging USAF aircraft that could potentially have a solution through research and development. The problem identification is primarily from visits by the AATT to depots where aircraft are maintained. These visits provide for accurate and concise definitions of the problems. The AATT will communicate these problems to sources worldwide to try to find solutions. The potential solutions will be screened for level of maturity. Those needing additional development will be sent to the AFRL to be funded. Those ready for transition to ALCs will be funded for transfer of technology.

Research and development efforts will initially focus on alleviating corrosion damage, improving NDIs, process, improving repair processes, and reducing the assessment costs of WFD. These efforts will be accomplished through team efforts with the AFRL, ASC, and ALCs working together to identify and focus this activity. The laboratory programs will be transferred to ALCs through a focal point for technology transition located in ASC.

The organizational structure to accomplish the goals of the aging aircraft program follows the guidance of the NRC report.

VI. Conclusions

Aging aircraft will continue to be a problem for operators and certification authorities in the future. The rate of procurement of military aircraft attests to this fact. The current state of technology is adequate for the protection of their safety in the near term. However, it is not evident that the technology is available to protect them in the future unless better methods are found to arrest corrosion. There is still a need for technology developments that will reduce the maintenance burden associated with continued operations. It appears that operators will need significant improvements in NDI to accomplish this goal. Aggressive programs initiated by the FAA, NASA, and the USAF should be able to maintain safe and economical operation of aging fleets. However, to attain this goal, the operator must be vigilant in the search for indications of aging and institute preventive processes in a timely manner.

References

[1]Mclin, J. R., "Performance of Fuselage Pressure Structure," *Proceedings of the 1991 International Conference on Aging Aircraft and Structural Airworthiness*, Washington, DC. Nov. 1991, pp. 67–74.

[2]Hadcock, R. N., "Aging Airframe Structural Life Enhancement: An Overview," *Proceedings of the First Joint DOD/FAA/NASA Conference on Aging Aircraft*, Ogden, UT, July 1997.

[3]Swift, T., "Damage Tolerance in Pressurized Fuselages," 14th Symposium of the International Committee on Aeronautical Fatigue, 11th Plantema Memorial Lecture, Ottawa, Canada, 1987, pp. 1–77.

[4]Richards, P. F., "Continuing Structural Airworthiness of Civil Transport Aircraft," *The Aeronautical Journal of the Royal Aeronautical Society*, Nov. 1980.

[5]Department of Trade, Accidents Investigation Branch, Aircraft Accident Report 9/78, "Boeing 707 321C G-BEBP, Report on the accident near Lusaka International Airport, Zambia, on 14 May 1977," Her Majesty's Stationery Office, London.

[6]NTSB/AAR-89/03, National Transportation Safety Board Aircraft Accident Report, Aloha Airlines, Flight 243, Boeing 737-200, N73711, Near Maui, Hawaii, April 28, 1988.

[7]DOT/FAA/CT 88/32-1, "Program Plan, National Aging Aircraft Research Program," Sept. 1991.

[8]DOT-VNTSC-FA2H2-91-14, Federal Aviation Administration, "The FAA National Aging Aircraft Program Plan," FAA-A:R-98-03, 1998.

[9]Atluri, S. N., *Structural Integrity & Durability*, Tech Science Press, 1997.

[10]Harris, C., Newman, Jr., J., and Starnes, Jr., J., "Analytical Methodology to Predict the Onset of Widespread Fatigue Damage in Airframe Structure," *Proceedings of the 1995 USAF Structural Integrity Conference.*

[11]Piascik, R. S., and Willard, S. A., "The Growth of Small Corrosion Fatigue Cracks In Alloy 2024," *J. of Fatigue Fract. Mater. Struct.*, Vol.17, No.11, 1994, pp. 1247–1259.

[12]Piascik, R. S., and Willard, S. A., "The Characterization of Multi-Site Fatigue Damage in the Fuselage Riveted Lap Spice Joint," ICAF 97 Fatigue in New and Aging Aircraft, EMAS Publishing, Cradley Heath, West Midlands, UK, pp. 93–114, 1997.

[13]Schütz, W., "A History of Fatigue," *Engineering Fracture Mechanics*, Vol. 54, No. 2, 1996, pp. 263–300.

[14]Negaard, G. R., "The History of the Aircraft Structural Integrity Program," Aerospace Structures Information and Analysis Center (ASIAC), Rep. No. 680.1B, 1980.

[15]Coffin, M. D., and Tiffany, C. F., "New Air Force Requirements for Structural Safety, Durability, and Life Management," *Journal of Aircraft*, Feb. 1977.

[16]Department of the Air Force, 1974, "Airplane Damage Tolerance Requirements," Military Specification MIL-A-83444.

[17]Lincoln, J. W., "Damage Tolerance—USAF Experience," *Proceeding of the 13th Symposium of the International Committee on Aeronautical Fatigue*," Pisa, Italy, 1985, pp. 265–295.

[18]Department of the Air Force, "Aircraft Structural Integrity Program, Airplane Requirements," MIL-STD-1530A, 1975.

[19]Lincoln, J. W., and Milliere, R. A., "Economic Life Determination for a Military Aircraft," *Journal of Aircraft*, Vol. 36, No. 5, pp. 737–742.

[20]Manning, S. D., and Yang, J. N., "Advanced Durability Analysis," Vol 1-5, AFWAL-TR-86-3017, Flight Dynamics Laboratory, Wright-Patterson AFB, OH, July 1988.

[21]Lewis, W. G., Dodd, B. D., Sproat, W. H., and Hamilton, J. M., "Reliability of Nondestructive Inspections-Final Report," SA-ALC/MME 76-6-38-1, Dec. 1978.

[22]Hoppe, W. C., "A General Discussion of the Retirement for Cause Inspection System," 1987 Aircraft/Engine Structural Integrity Program (ASIP/ENSIP) Conference, AFWAL-TR-4128.

[23]Lincoln, J. W., "Damage Tolerance for Helicopters," *Proceeding of the 15th Symposium of the International Committee on Aeronautical Fatigue*, Jerusalem, Israel, 1989, pp. 619–629.

[24]Lincoln, J. W., "Risk Assessments-USAF Experience," *Proceedings of the International Workshop on Structural Integrity of Aging Airplanes*, Atlanta, GA, 31 March–2 April 1992.

[25]Lincoln, J. W., "Aging Aircraft—USAF Experience And Actions," *Proceedings of the 19th Symposium of the International Committee on Aeronautical Fatigue*, 16th Plantema Memorial Lecture, Edinburgh, Scotland, 1997, pp. 3–38.

[26]Wheeler, O. E., "Spectrum Loading and Crack Growth," *Journal of Basic Engineering*, 94D, 1972, p. 181.

[27]Gallagher, J. P., and Hughes, T. F., "Influence of the Yield Stress on Overload Affected Fatigue- Crack-Growth Behavior of 4340 Steel," AFFDL-TR-74-27, 1974.

[28]Willenborg, J. D., Engle, R. M., and Wood, H. A., "A Crack-Growth Redardation Model Using an Effective Stress Concept," AFFDL-TM-71-1FBR, 1971.

[29]Dill, H. D., and Saff, C. R., "Analysis of Crack Growth Following Compressive Loads Based on Crack Surface Displacements and Contact Analysis," *Cyclic Stress-Strain and Plastic Deformation Aspects of Fatigue Crack Growth*, ASTM STP 637, Am. Soc. For Testing and Mat'ls, 1997, pp. 141–153.

[30]Dowling, N. E., "Fatigue Failure Predictions for Complicated Stress-Strain Histories," T. & A. M. Rep. No. 337, Univ. of Illinois, Urbana, Jan. 1971.

[31]Swift, T., "Unarrested Fast Fracture," *Durability of Metal Aircraft Structures*, Atlanta Technology Publications, Proceedings of the International Workshop on Structural Integrity of Aging Aircraft, 21 March–2 April 1992.

[32]Swift, T., "Damage Tolerance Capability," *Fatigue of Aircraft Materials*, Delft University Press, Proceedings of the Specialists' Conference dedicated to the 65th birthday of J. Schijve, Delft, The Netherlands, 1992.

[33]Lincoln, J. W., "Risk Assessment of an Aging Military Aircraft," *Journal of Aircraft*, Vol. 22, No. 8, 1985.

[34]Alford, R. E., Bell, R. P., Cochran, J. B., and Hammond, D. O., "C-141 WS 405 Risk Assessment," WL-TR-92-4045, *Proceedings of the 1991 USAF Structural Integrity Program Conference*, San Antonio, TX, 1992.

[35]Berens, A. P., Hovey, P. W., and Skinn, D. A., "Risk Analysis for Aging Aircraft Fleets," WL-TR-91-3066, Wright Laboratory, Wright-Patterson AFB, OH, Oct. 1991.

[36]Carmody, D. P., "Incorporation of an Aircraft Structural Integrity Program into an Aging Commercial Aircraft," *Proceedings of the 1998 USAF Structural Integrity Conference*, San Antonio, TX.

[37]Lincoln, J. W., "Risk Assessments of Aging Aircraft," Proceedings of the First Joint DOD/FAA/NASA Conference on Aging Aircraft, Ogden, UT, July 1997.

[38]Department of the Air Force, Air Force Guide Specification, Aircraft Structures, General Specification for, AFGS-87221A (1990).

[39]Wallace, W., Hoeppner, D. W., and Kandachar, P. V., "AGARD Corrosion Handbook, Volume 1, Aircraft Corrosion: Causes and Case Histories," July 1985.

[40]Cooke, G. E., and Kealy, H. D., "A Study to Determine the Annual Direct Cost of Corrosion Maintenance for Weapon Systems and Equipment in the United States Air Force," Final Report on Contract #F09603-89-C-3016, Dayton, OH: Systems Exploration, Inc (1990).

[41]Schütz, W., "Corrosion Fatigue—The Forgotten Factor in Assessing Durability," Proceedings of the 18th Symposium of the International Committee on Aeronautical Fatigue, 15th Plantema Memorial Lecture, Melbourne, Australia, 1995, pp. 1–51.

[42]Cole, G. K., Clark, G., and Sharp, P. K., "The Implications of Corrosion with Respect to Aircraft Structural Integrity," DSTO-RR-0102, March 1997.

[43]Wei, Robert, P., and Harlow, D. Gary, "Corrosion and Corrosion Fatigue of Airframe Materials," DOT/FAA/AR-95/76, Feb. 1996.

[44]Schmidt, C. G., et al., "Characterization of Early Stages of Corrosion Fatigue in Aircraft Skin," DOT/FAA/AR-95/108, Feb. 1996.

[45]Mindlin, H., Gilp, B. F., and Elliott, L. S., "Corrosion in DoD Systems: Data Collection and Analysis (Phase 1), Feb. 1996.

[46]Paul, C. A., and Mills, T., "Structural Integrity of Aging Aircraft," presented at AeroMat '98, June 1998.

[47]Swift, T., "Repairs to Damage Tolerant Aircraft," International Symposium on Structural Integrity of Aging Aircraft, Atlanta, Georgia, March 1990.

[48]Bakuckas Jr., J., Bigelow, C., Tan, P., Yu, J., and Blauvelt, R., "Enhanced Repair Assessment Procedure and Integrated Design (RAPID)," *Proceedings of the 1996 USAF Aircraft Structural Integrity Conference*, San Antonio, 1996.

[49]Fredell, R., "Bonded Repair of Multiple Site Damage with GLARE Fiber Metal Laminate Patches," *Proceedings of the 1994 USAF Structural Integrity Program Conference*, San Antonio, Dec. 1994.

[50]Lincoln, J. W., "Structural Technology Transition to New Aircraft," *Proceedings of the 14th Symposium of the International Committee on Aeronautical Fatigue*, Ottawa, Ontario, Canada, 1987, pp. 619–629.

[51]National Research Council, "Aging of U.S. Air Force Aircraft," Publication NMAB-488-2, National Academy Press, Washington, DC. 1997.

Chapter 4

Extreme Environment Structures

Donald B. Paul, Christopher L. Clay, Brett Harber, and Harold Croop
Air Force Research Laboratory, Dayton, Ohio

and
David Glass and Steven Scotti
NASA Langley Research Center, Hampton, Virginia

I. Introduction

WITH the renewed interest in high-speed flight, the research and development community is once again focusing on extreme environment structures technology. Extreme environment structures are required technologies where the load environment includes extremes of temperatures (both hot and cold) or extreme temperature combined with significant acoustic loading. These structures technologies are not new. The best examples of these types of structures are usually found within the engines of aircraft. However, in recent years, the airframe has begun to experience similar types of design environments. Thermal protection systems, hot structures, actively cooled structures, heat pipe structures, and cryogenic tanks are key airframe components experiencing extreme environments. The focus of this chapter will be on the current and most recent efforts in the development of fully reusable extreme environment structures for high-speed vehicle applications.

The increase in interest of airframe requirements for this area of structures technology really began to receive prominence due to the advent of stealth technology. Stealth technology typically results in engines being buried deep within the airframe. Often, the exhaust is vented over the aft airframe fuselage, resulting in both high temperatures and high acoustic loads impinging directly on load-carrying structure (see Fig. 4.1).

This environment can cause premature failure of the structure, leading to higher operations and support costs plus more downtime than is desirable. Typical structural temperatures in these aft fuselage areas may approach 1000°F with acoustic load levels of approximately 168 dB.

Vehicle configurations needed for hypersonic flight lead to much larger areas of airframe structure subjected to these types of loads. In most rocket-based systems, such as the X-33, peak temperatures occur out-of-cycle with the peak acoustic loads (see Fig. 4.2).

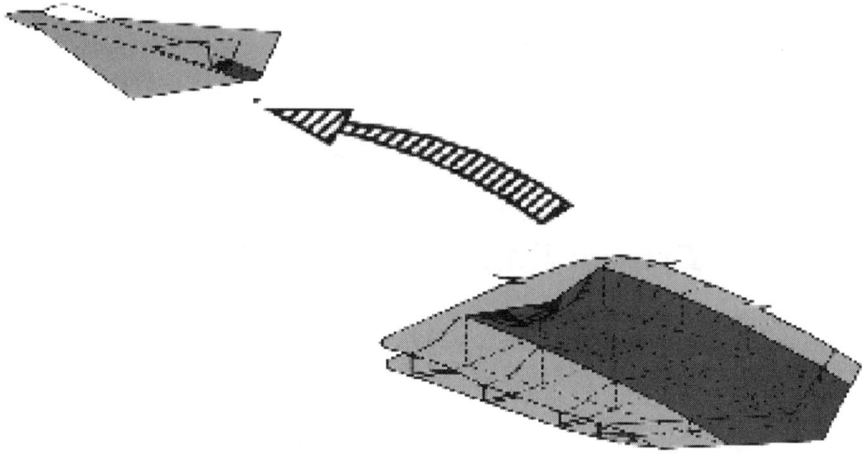

Fig. 4.1 Exhaust washed structure.

Peak acoustic loads typically occur during liftoff and range in magnitude up to 168 dB on the airframe. The structures, including the thermal protection system, are usually at or below the ambient temperature. Peak temperatures, however, will occur during reentry. Peak temperatures may range from 2400° to 2800°F for most concepts under study today.

Airbreathing hypersonic concepts typically operate with temperatures up to 3000°F concurrently with acoustic loads approaching 174 dB on the airframe. Even higher acoustic loading can be found on the engine structure of these airbreathing vehicles (see Fig. 4.3).

Fig. 4.2 X-33 reusable launch vehicle.

Fig. 4.3 High speed vehicle.

Challenges associated with structures for the environments just described are many. Competing requirements for durability and weight are foremost in the mind of the designer. Increased durability can reduce the operations and support cost burden found with these structures, but the added weight could make a hypersonic system nonviable. The following will address several novel technologies that may be required to meet the challenges of future high-speed flight.

II. Actively Cooled Structures

A. Introduction

In the past 15 years or so, a great deal of interest in low-cost access to space and rapid strike/global reach vehicles has developed. This interest has manifested itself in a number of programs ranging from component technology to propulsion and vehicle programs. Among these programs, several have included major activities focused on the development of structures for applications in the most extreme thermal environments of the vehicle. Foremost among the structural concepts developed for these high heat flux environments have been actively cooled structures. Designs using numerous material systems have been developed and tested to address thermal challenges associated with internal ramjet and scramjet engine flow passage surfaces, engine external inlet and nozzle surfaces, and wing and engine leading edges. Some of the recent programs to include development of actively cooled structures have been the National Aerospace Plane (NASP) program, the National Aeronautics and Space Administration (NASA)/Lockheed Martin VentureStar program, and the Air Force Hypersonic Scramjet Engine Technology (HyTec) program. Within these programs, advancements in materials, coatings, and designs have been made that have significantly improved the state of the art in reusable and nonreusable cooled structures for hypersonic vehicle applications.

B. Design Requirements and Conditions

Design requirements and conditions vary widely depending on the application. The NASP propulsion system encompassed probably the most demanding requirements and conditions of any past or present system design. It's engine heat flux levels reached and, in local areas exceeded, values as high as 2500 Btu/ft^2·s while

Table 4.1 NASP design conditions

Condition	Value
Heat flux (face)	>2500 Btu/ft^2·s
Heat flux (gap)	~60% of face
Acoustic pressure	>176 dB
Coolant temperature	Cryo to 1100 R
Coolant pressure	4000 psi

other environmental conditions of its structure also pushed the extremes of materials and design capabilities (Table 4.1). In comparison, heat flux levels for the VentureStar engine reach levels of 1200 Btu/ft^2·s and those in HyTec reach only a few hundred. In addition to the thermal and mechanical environment to which these designs are exposed, design life and fuel compatibility are major considerations in their individual designs.

For those applications involving hydrogen (H_2) coolant, compatibility of the materials with this coolant must be considered. Many common materials that otherwise would be described as initial candidates for a design quickly prove to be adversely affected by hydrogen exposure. For these materials, hydrogen exposure results in loss of ductility, strength, or one or more of a whole range of detrimental effects. As a result, the list of materials that may be used in a reusable H_2 application is severely limited.

In other designs, where alternate coolants such as hydrocarbon fuels are used, other issues become important. For instance, for a number of reasons hydrocarbon-cooled designs require the fuel to be cracked within actively cooled panels. High temperatures lead to coking of the coolant passages, a condition that can be made worse depending on the material selection of the panels.

C. Design Development

Over the past 15 years, a great deal of effort has gone into the development of materials for actively cooled structures. The bulk of this work was accomplished by teams of government and industry engineers working together to support the NASP program. Within the NASP framework, programs such as TechMat and the materials and structures augmentation program (MASAP) and individual contractor programs explored material production, characterization, component fabrication, and process scale-up issues for a wide variety of advanced materials. More recently, programs such as HyTec, VentureStar, and a host of smaller efforts have continued the work started under NASP. These programs have worked to refine the developments started by NASP, expanded the range of useful materials, and explored the use of alternate coolants such as Hydrocarbon (JP) fuels. To date, the main list of materials considered for hydrogen or JP-cooled structures includes those listed in Table 4.2.

The NASP programs explored many advanced materials in support of lightweight, durable heat exchangers. In broad categories, these included metallics, composites, micro-composites, and refractory composites. Over time and the confluence of requirements, the initial range of materials in these studies decreased significantly. Titanium alloys and copper microcomposites were dropped due to

Table 4.2 Material capabilities

	Material temperature range	Coolant temperature range	Heat flux	Hydrogen compatibility
Aluminum	Low	Low	Low	Yes
Beryllium	Low	Low	Low	Yes
Lockalloy	Low	Low	Low	Yes
Titanium	Low	Low	Low	No
Cu	Low	Low	High	Yes
Cu Micro-composites	Low	Low	High	No
Cu/GR	Low	Low	High	Yes
H188	Medium	Medium	Medium	Yes
IN625	Medium	Medium	Medium	?
IN718	Medium	Medium	Medium	No
IN909	Medium	Medium	Medium	Yes
MoRe	High	High	High	Yes
C/C	High	High	Medium	Yes
C/SiC	High	High	Medium	Yes

the adverse affect of hydrogen on their material properties. Beryllium was dropped after the catastrophic failure of a 20 in. $\times$ 20 in. test panel emphasized the need for more ductile heat-exchanger materials. In contrast, materials such as C/C and C/SiC, Mo-Re, Lockalloy, and copper graphite composite demonstrated significant promise for application in actively cooled structures.

D. Refractory and Ceramic Composite Development

During the NASP program, several activities investigated refractory composite materials for use in actively cooled panels. These included C/C, C/SiC, and SiC/C. Of these, only C/C and C/SiC were extensively studied. The temperature limit of SiC/C at 2500°F was felt to be too limiting when compared to the nearly 3000°F limits of the other materials. During the MASAP program, work focused on extensive development of the C/SiC material and heat-exchanger design. Two- and three-dimensional composite panels were procured from several vendors (DuPont, Americom, and Refractory Composites) and subjected to mechanical tests and inspections. Of these designs, the DuPont composite using a T300 fiber/three-dimensional angle interlock material from Textile Technologies, Inc. (TTI), and Chemical Vapor Infiltration of the SiC matrix was found to have excellent mechanical properties. Warp tensile and compressive ultimate strengths were 38 and 54 ksi.

In parallel with the work accomplished for the composite, the MASAP program also worked on the development of embedded coolant tubes. Materials including tungsten, tungsten rhenium, molybdenum, and molybdenum rhenium alloys were studied. Work on the tungsten tubes was cancelled because of the brittleness of the material. Tungsten rhenium tubes were also dropped from the program when the layered approach (alternating layers of deposited tungsten and rhenium that were then annealed to consolidate the structures) by Delta-G was found to be too costly to continue within the budget of the program and Ultramet's deposition

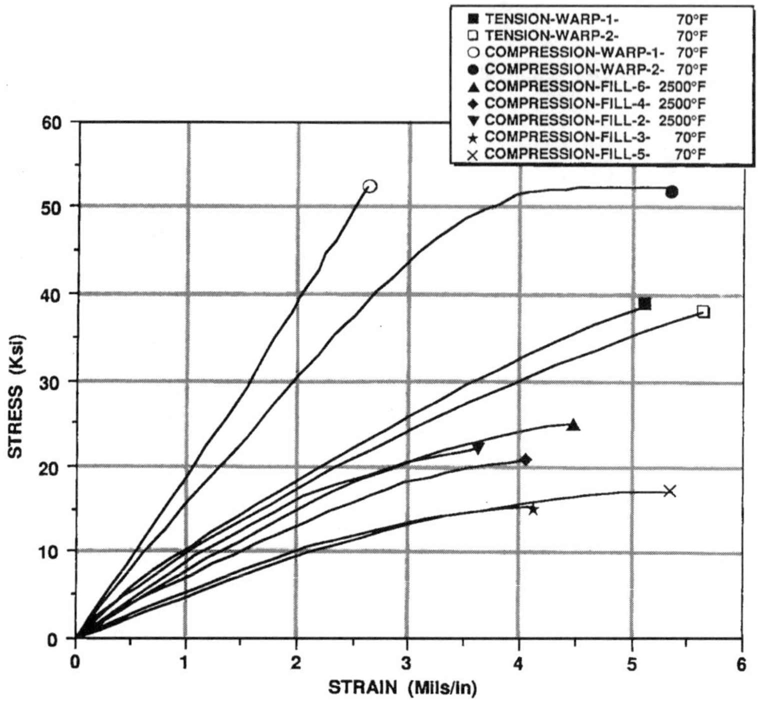

Fig. 4.4 Dupont C/SiC panel tensile stress-strain response.

process resulted in large nonuniform grains. Molybdenum tubing was dropped after tests of material exposed to 2000°F temperatures exhibited a 20% reduction in axial strength. Molybdenum rhenium was the only material system to complete the program. Tests of several alloys from several suppliers were conducted to identify the best alloy system. For these tests, sheet and tubes of Mo-13Re, Mo-41Re, and Mo-47.5Re were procured for characterization. The Mo-13Re was found to be unstable at temperatures up to 2000°F while the Mo-47.5Re properties remained stable after a 2000°F, 8-hour exposure. Based on these results, the Mo-47.5Re material was selected for further development in consolidated composite heat exchangers. Seven heat exchangers were fabricated after the down select to the Mo-47.5Re tube. These panels demonstrated the ability to fabricate larger panels (12 in. × 6.5 in.) with unit weights of 1.5 lb/ft^2.

The final step in this development process included measuring the thermal performance of these designs. During these tests, a proof of concept panel using the TTI three-dimensional weave and molybdenum tubes was exposed to heat fluxes up to 100 Btu/ft^2·s using water as a coolant. At total of three runs were made, with heat flux increasing in 25-Btu/ft^2·s intervals. Results from the tests indicated good heat transfer between the composite and the coolant.

Several years later, the MASAP tests were followed by extensive tests of C/C and C/SiC heat exchangers by Lockheed. For these tests, 26 1 in. × 4 in. C/C and C/SiC specimens of differing configurations were made. The primary variables distinguishing one specimen from the next were material (C/C or C/Sic), composite

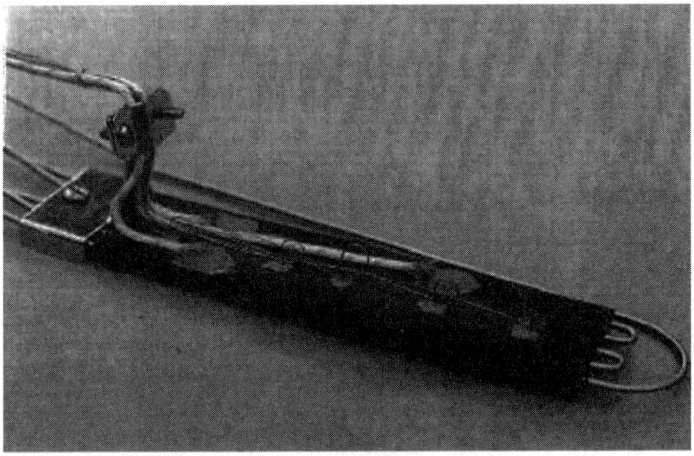

Fig. 4.5 Completed test article (2).

architecture (two- or three-dimensional), SiC oxidation coating and coolant tube spacing. Also, unlike the panels built and tested under the MASAP program, these panels did not incorporate coolant tubes that were intimately bonded to the composite. Instead, the H188 tubes, which were 0.0625 in. in diameter, were merely slipped inside holes in the composite. This was done to eliminate strains between the tube and the composite caused by the large thermal expansion disparity between the two materials.

A total of 11 of these articles were then tested in the Air Force Research Laboratory (AFRL) Hydrogen test facility under of wide range of coolant and heat flux conditions. A total of 58 tests in two test series were run on the 11 specimens. In these tests, coolant pressure mass flow rate and temperature was varied to study the effects of these parameters on heat-exchanger performance. Inlet pressure was adjusted from 1900 psi to 3000 psi, while mass flow rates were from 0.01 to 0.03 lb m·s and coolant temperatures were from $-200°$ to 700°F. Incident radiation heat flux in these tests ranged from 52 Btu/ft^2·s to 894 Btu/ft^2·s. Unfortunately, posttest analysis of the thermal data generated mystifying results. Four of the 11 test articles were selected for detailed posttest analysis with two of the four spanning tests in both series. Results from the first series agreed well with expectations. Absorbed heat flux in these tests reached a high of 250 Btu/ft^2·s, and measured surface temperatures reached as high as 2300°F. However, results from the series two tests were quite different. In these tests, measured heat flux differed significantly from the incident flux. In the four cases studied, absorbed heat flux never exceeded 50% of the incident flux and was as low as 30% in one case. However, if the absorbed heat flux numbers are to be believed, then the test articles demonstrated capabilities up to at least ~400 Btu/ft^2·s, surviving the tests in satisfactory condition.

E. Refractory and Ceramic Composite Development (Recent Activities)

More recently, NASA, Lockheed Martin, and their subcontractors have explored the application of refractory composite heat exchangers for the VentureStar program. These activities have focused on lightweight durable designs that are

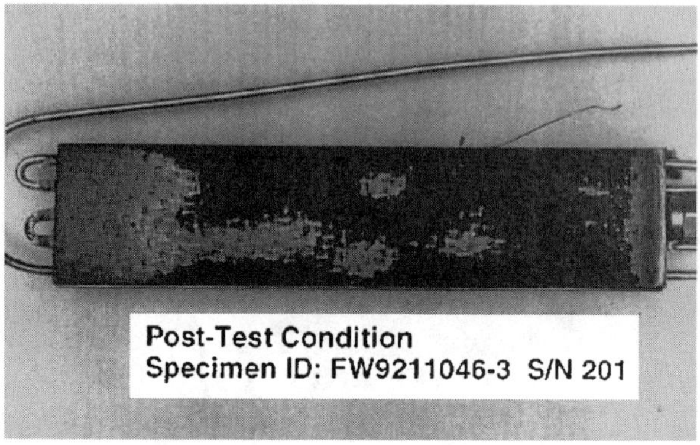

Fig. 4.6 Post test appearance of 2-D C/C specimen front side.

acceptable for reusable and potentially man-rated vehicles. To date, several proprietary designs have been proposed though none have been fabricated.

F. NARloy-Z Development

NARloy-Z is probably the best characterized and most tested material to be used as a flat actively cooled panel. The material has a long history of use by Rocketdyne in rocket engine applications and was a natural candidate for the NASP engine. With the material so well characterized for previous applications, most of the effort in NASP focused on designs and manufacturing processes for flat nonintegral panels and on thermomechanical tests. During this development, several dozen 1 in. × 4 in. specimens were fabricated, of which approximately six were tested in the AFRL Hydrogen test facility or the General Applied Sciences Laboratory panel oxidation and erosion test (GASL POET) facility. In addition, a 12 in. × 12 in. panel was made and tested at Rocketdyne.

For NASP, NARloy-Z was chosen as one of two materials for panels in the engine combustor. Its high conductivity, nearly that of pure copper, made it one of the best possible materials for the high heat flux environment of the combustor where heat fluxes exceed 2000 Btu/ft^2·s. Over the course of several design iterations, a final design tailored to the heat flux, coolant temperature, pressure, and acoustic conditions of the combustor was developed. This design included several key features. To account for gap heating between individual panels, the design incorporated a second channel along the edge. This channel performed several functions: the first and primary function was to control edge temperatures that otherwise would be hundreds of degrees higher than those found away from the edge of the panel. The second function was to reduce the heat load on the edges of the channels; for, without such help, it was possible that severe flow maldistribution would develop between the edge and center channels. In such an event, edge temperatures would have risen further with respect to panel center channels, possibly setting up a self-destructive cycle of increasing edge temperature and flow maldistribution. A second feature of the design was the use of an oxidation protection coating. To get the highest performing design required very high panel surface temperature. To avoid rapid oxidation and erosion of the panel surface at these temperatures, the design incorporated a Cu–Cr coating several mils thick. This surface was shown in a several subsequent test programs to be very durable, with little loss in thickness over many cycles and hours of testing. The final unique feature of the design was the attachment method. Because of the high thermal expansion of NARloy-Z and the high backside temperatures of the panel, an integral structure using lightweight composites was not possible. Such a design would have suffered from high shear stresses; high composite temperatures; and, as a result, very short life. To alleviate these problems, a nonintegral design was picked for this as well as all of the engine panels found in the final engine design. For this particular design, an arrangement called a pin/cup design was developed.

G. Haynes 188 Development

Another material studied extensively for actively cooled engine panel applications in the NASP program was Haynes 188 (H188). This material was selected primarily for its high-temperature capability and its hydrogen compatibility. Designs for the NASP propulsion system incorporating this material focused on midlevel conditions within the engine. These conditions included those where heat flux levels were moderate (approximately 1000 Btu/ft^2·s) and coolant temperatures were moderately high (1500° R). Primary areas of development activity for the H188 designs were in fabrication and design, analysis, and verification. By the conclusion of the program, the preferred method of fabrication was shown to be one incorporating electrical discharge machining (EDM) and diffusion bonding. Using this method, a number of specimens were fabricated. Testing yielded mixed results.

The design for NASP that evolved out of several years of analysis and testing was much like the other cooled panel designs for the engine. To survive over many cycles in the extreme environments found in the engine, the design incorporated a thin heat exchange design with small coolant passages measuring on average 0.042 in. × 0.019 in.

This design by Rockwell International was fabricated by Allison. Their process involved several manufacturing steps, including EDM processes to produce coolant passages and manifolds, diffusion bonding of the heat exchanger closeout to the hot face sheet, and brazing of the completed face sheet to the manifold. Using this process, approximately six 1 in. × 4 in. tests specimens were completed. Of these, five were tested in the AFRL hydrogen and GASL POET facilities.

Tests in the AFRL hydrogen facility began with the premature failure of the first test article. Subsequent investigation of the failure revealed residual braze stop-off material blocking the coolant channels. This material decreased or prevented flow through the channels, leading to localized hot spots that exceeded the design's temperature capability and thus the failure. Inspections of the remaining test articles indicated similar blockages requiring removal before further testing could continue. To that end, a methodology was developed to clean and inspect the remaining test articles. The result of the effort was test articles that were better but not entirely free of restrictions. Despite the less-than-perfect condition of these test articles, they were submitted for testing, in both facilities. Two articles each were tested in the hydrogen and POET facilities. Tests in the hydrogen facility intended to demonstrate the thermomechanical life of the design and tests in the POET facility intended to demonstrate oxidation and erosion resistance of the designs.

Results in the hydrogen facility were very positive, with both test articles easily exceeding the requirement of 80 thermal cycles without failure with demonstrated lives as high as 160 cycles. Conditions during these tests included a peak heat flux in each cycle of 1131 Btu/ft^2·s and coolant outlet temperature averaging 750°F to 800°F.

Results in the POET facility tests were far different. Both test articles placed in the facility failed early and catastrophically. In fact, failure of the second article was so rapid and complete that nearly 75% of the test article was burned away in the few seconds needed to shut the test down. Posttest analyses of these tests failed to determine the cause of the failures though several possible explanations exist and all may have contributed to varying degrees. The primary suspects in the failure were the known presence of residual stop-off in the channels and gap heating. While every effort was made to clear this material from the test articles, some was known to remain in the test articles, a fact posttest investigation confirmed. It was assumed that the presence of this material might have starved selected channels of coolant, leading to localized overtemperature conditions, resulting in channel failures. Gap heating (the gap between panels or, in this case, between the test article and the wall of the test chamber) was also suspected of playing a role in the failure. Unfortunately, these articles were designed and fabricated before gap heating was identified as a major contributor in the POET facility and thus the design did not include features necessary to mitigate the effect of gap heating. If, in fact, gap heating placed a significant additional thermal load on the edge channels of the panel, this condition may have contributed to localized overheating of the panel and the failure. Finally, the third factor that was thought to have contributed to the failure was the possibility that the test was inadvertently run at heat flux levels much higher than intended. Planned conditions for the test were to include a peak heat flux of 780 Btu/ft^2·s, with coolant flow rate and temperature set to produce panel temperatures of 1727°F. Instead, analysis of coolant temperature rise and mass flow rate indicated that heat flux might instead have been as high as 1108 Btu/ft^2·s. If this was the case, surface temperatures would have easily exceeded the capability of the design (see Fig. 4.7).

H. Molybdenum Rhenium Development

Work on molybdenum rhenium alloys for actively cooled panels initially began during the MASAP program, where the effort focused on development of a coolant tube material for refractory and ceramic composites being developed. Later, within

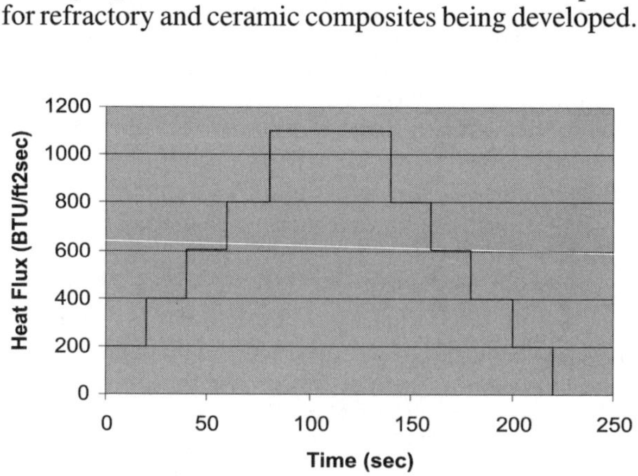

Fig. 4.7 Heat flux cycle.

the mainline NASP program, Mo–47.5%Re was selected as the material of choice for high-temperature engine actively cooled panels where its placement at the end of the coolant circuit provided the coolant/fuel temperature boost required to get peak combustion efficiency. By design, Mo–Re panels were placed in those areas of the propulsion system where heat fluxes were the most extreme and at the end of the engine cooling circuit (after NARloy-Z and H188 panels) where the coolant temperatures were the highest. However, to make this material practical in the application required significant fabrication and coating technology development. Therefore, during this period, major activities focused on fabrication methods development and oxidation protection coating development. In parallel, a moderate design effort progressed to develop designs for larger panels. By the end of the NASP program, these efforts had largely paid off with the identification of promising fabrication and coating technologies and development of a 20 in. × 20 in. panel design.

The bulk of the fabrication effort for Mo–Re revolved around machining methods capable of economically producing coolant channels in large panels. Conventional milling processes were inadequate given the toughness of Mo–Re; therefore, alternate methods such as high-speed, electrochemical, and laser machining processes were explored. These processes were applied to several panel assembly geometries, including a conventional two-sheet assembly, a three-sheet assembly, and a platelet assembly (see Fig. 4.8).

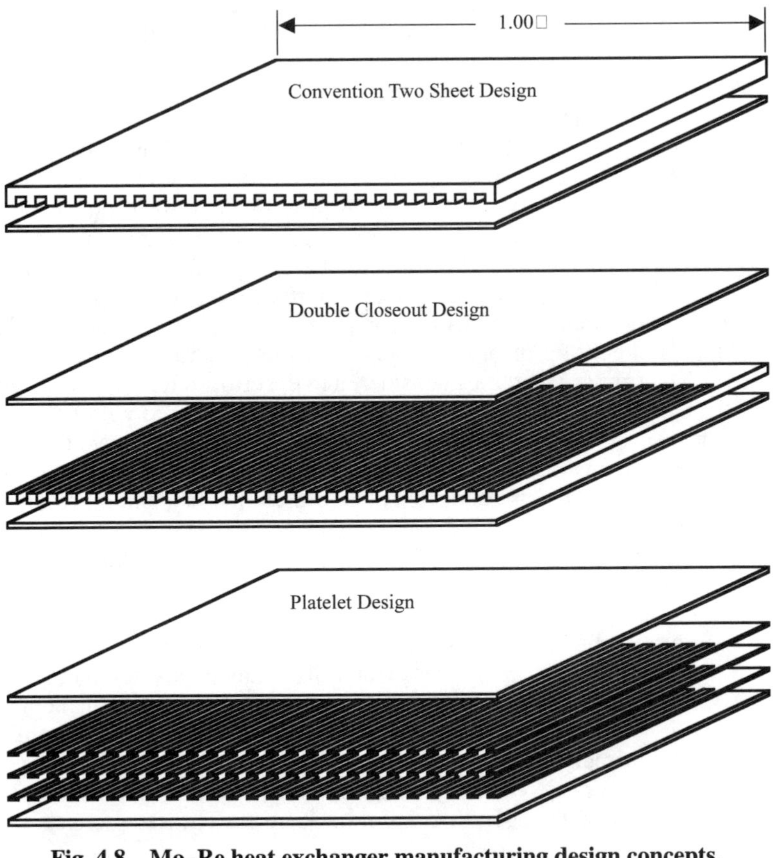

Fig. 4.8 Mo–Re heat exchanger manufacturing design concepts.

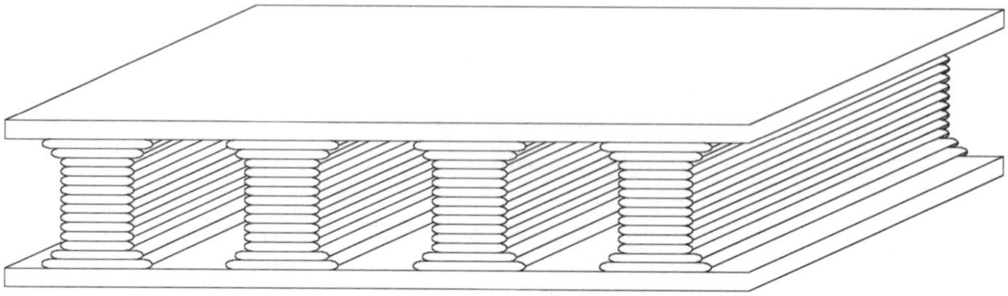

Fig. 4.9 Plate assembly cross-section.

The conventional design was attempted using conventional and high-speed machining techniques before these methods were dropped from consideration. As stated previously, conventional milling was not practical given the toughness of the material. The electrochemical process was used to produce the platelet design. One drawback to this method was the cusping present in the bonded assembly. This cusping was the result of the more extensive loss of material at the surface of a platelet compared to that experienced in the midplane of the platelet. The resulting assembly cross section can be seen in Fig. 4.9. This cusping was and is a concern given that it may lead to premature cracking. As a result, laser machining was explored. The laser process was used to manufacture the double closeout design. In this process, the laser was used to mill the long narrow slots in the thick center sheet, after which thin closeout sheets were diffusion bonded to either side to from the channels of the heat exchange.

Coating development for Mo–Re was probably the most aggressively pursued technology effort for the panel design because, unlike many materials that form a somewhat protective scale as they oxidize, Mo–Re alloys do not. Instead, at temperatures as low as 1250°F, Mo–Re oxides are gaseous and, therefore, provide no protective scale that might slow the rate of material loss. As a result, the requirement to use the alloy at temperatures approaching 2500°F required the development of a dependable, defect-free coating that could be repeatably produced. To that end, a number of protective materials and methods for applying them were considered. The primary materials considered in the effort fell into two categories, glassy coatings and metallic coatings. The glassy coating used with success was R512E. One example of the effectiveness of this coating is seen in weight change. Specimen one demonstrated a 0.4% weight change in 30 hours of static conditions at 2350°F. In other tests conducted at a different facility, similar results were obtained (see Fig. 4.10).

The metallic coatings investigated included platinum, rhodium, and iridium. Experiments with these materials explored the benefits of these materials applied individually and in alternating layers. When applied in alternating layers, platinum was normally the primary (outer) protective material with under coatings of rhodium and/or iridium. To apply these materials, many deposition methods were tried as shown in Table 4.3. By the conclusion of the program, significant progress had been made with the development of a layered metallic coating system using platinum as the primary protective material. An example of the effectiveness of platinum was demonstrated in early oxidation trials on HIP Clad Mo–Re where it demonstrated reasonably good oxidation protection. The sample tested was a

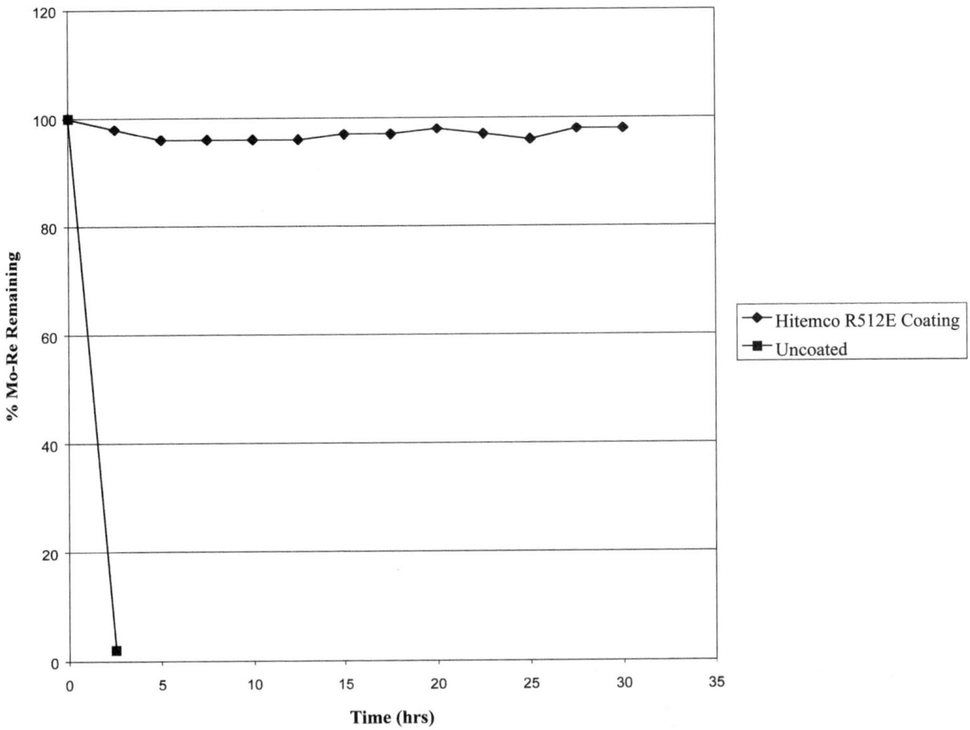

Fig. 4.10 Coating Mo–Re oxidation evaluation @ 2300°F.

thin sheet of Mo–Re HIP clad inside a 0.005-in. thick envelope of platinum. After 5 hours in an oxyacetylene torch flame at 2350°F, the sample experienced a 0.4% weight change. In a complementary test, the same specimen type demonstrated a 0.4% weight change in 30 hours of static conditions at 2350°F (see Fig. 4.11).

In parallel with fabrication and coating development, development of panel designs included a subelement design (1 in. × 4 in.) and a full size (20 in. × 20 in.) flight weight design. The specimen design was sized based on test facility capabilities and facility interface requirements, with attention paid to ensuring that the results obtain in the POET and hydrogen facilities were applicable to full-scale panels exposed to actual flight environments. The resulting design of the 1 in. × 4 in.

Table 4.3 Metallic coating deposition methods

Clad bonding	Aqueous plating
Aqueous brush plating	Magnetron ion deposition
Cathodic arc deposition	Ion-assisted deposition
Ram bonding	Vacuum plasma spray
Low-temperature arc vapor deposition	Molten salt bath
CVD fluidized bed	

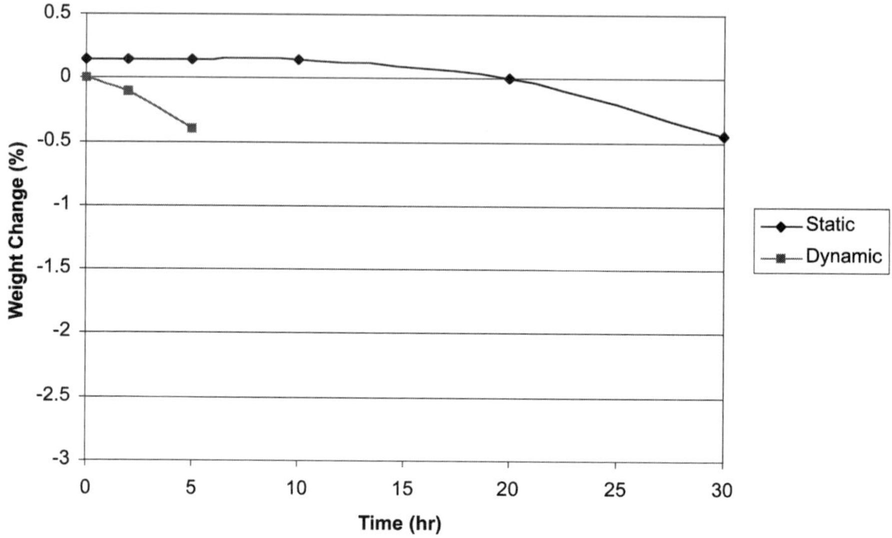

Fig. 4.11 HIP clad Mo–Re oxidation rate.

specimens were similar to other panels in the engine. Material thicknesses and channel dimensions were based on a balance between thermal stresses and coolant pressure drop. The resulting cross section of the design is shown in Fig. 4.12.

The full-scale panel that evolved from this design maintained the same coolant channel geometry but added a manifold layer similar to the heat exchanger to distribute coolant across the panel. It also incorporated an area around the perimeter of the panel against which a seal to prevent hot flow path gases from entering the rear cavity of the panel would be placed. To support out-of-plane pressure loads, the panel relied on pins and an egg crate style backup structure to attach it to and distribute loads to the primary structure. Views of the conceptual design are shown in Fig. 4.13.

In summary, development of a viable Mo–47.5Re actively cooled panel appeared promising. Many of the technical challenges associated with manufacturing and oxidation coatings had been addressed at the specimen level. What remains today is the validation of the design at the subelement and full-scale panel levels in test facilities that can replicate the harsh environments of flight.

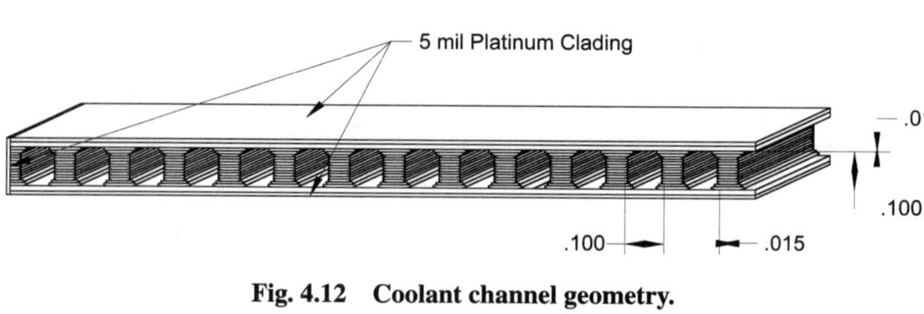

Fig. 4.12 Coolant channel geometry.

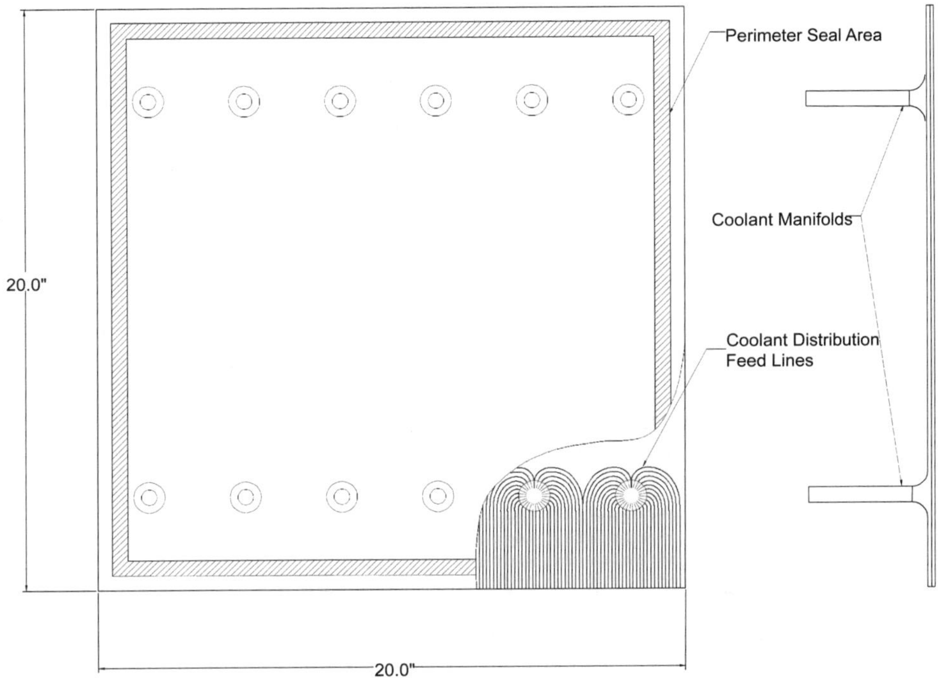

Fig. 4.13 Conceptual full scale Mo–Re panel design.

I. Copper Graphite Development

Copper graphite designs were explored as a means to achieve panels with the high conductivity of copper while avoiding the high weight associated with a monolithic copper alloy panel. Eventually, issues of compatibility between graphite fibers and the copper matrix were overcome as a lightweight panel design was developed and produced in 1 in. × 4 in. specimens for testing; however, these panels were never tested.

J. Copper Microcomposite Development

Copper microcomposites materials were also investigated during the MASAP program. This work sought to develop copper materials strengthened through the addition of such high-temperature materials as niobium, tantalum, chromium, molybdenum, and Rhenium. Processing of these materials yielded copper sheet materials with ribbons of one or more of these materials throughout the sheet intended to provide superior high-temperature strength. Of those material systems investigated, several—including the Cu–Nb system—showed some promise though none was ever pursued far enough to result in panel tests.

K. Test Facilities

Several test facilities were constructed, specifically to test actively cooled panels under simulated flight conditions. These include the AFRL hydrogen facility

Table 4.4 Hydrogen facility capabilities

Test article inlet pressure	0 to 4500 psig
Maximum GH_2 mass flow rate	0.15 lb/m·s
Maximum test article ΔP	250 psig
Test article inlet temperature	−300 to 750°F (GH_2)
	−300 to 500°F (GHe)
Maximum test article outlet	830°F (GH_2)
	500°F (GHe)
Maximum test article ΔT	525°F (GH_2)
	300°F (GHe)
Test article face heating	13 to 2000 Btu/ft^2·s (incident)

(also known as the Vortek facility), GASL POET facility, and Rocketdyne's material and structure thermal validation (MSTV1) rig. In addition to these facilities, NASA Glenn in Cleveland, Ohio, operates the C22 Facility, and the half-completed high heat flux test rig, HHFTR fabricated by Aerojet.

L. Air Force Research Laboratory Hydrogen Facility

The facility was built in the early 1990s to test 1 in. × 4 in. actively cooled subelements under simulated flight thermal conditions (Table 4.4). At its heart is the Vortek III system supplying power to a single parabolic reflector, which focuses the radiant energy of the Vortek III onto a 1in. × 4 in. area. With a test article placed at the focal point of the reflector, incident heat flux levels approaching 2000 Btu/ft^2·s can be achieved over the entire surface of the test article. Another feature of the Vortek system is the capability to vary heat flux with time to simulate the heat flux profile that would be experienced in flight. The second major system within the facility is the coolant system. This closed-loop system is designed to provide continuous coolant flow, usually hydrogen, for as long as required and also allows coolant mass flow rate and temperature to be varied as a function of time to simulate actual coolant conditions experienced in flight.

Since its construction, the hydrogen facility has been used to test a number of test articles, including NARloy-Z, H188, IN909, C/C, and C/SiC panels for NASP and, more recently, a C/C panel for the VentureStar program was tested. Throughout these tests, the facility demonstrated the capability to complete dozens of tests in a single day or run continuously for over an hour.

M. General Applied Sciences Laboratory Panel Oxidation and Erosion Test Facility

The POET facility was constructed at about the same time as the hydrogen facility. Its purpose was to expose 1 in. × 4 in. test articles to enginelike erosion, oxidation, and heat flux conditions for extended duration and over many cycles. The completed facility accomplished these objectives. The POET facility, as it exists today, uses a vitiated air heater to produce a test environment capable of convection heat flux levels up to 2000 Btu/ft^2·s and oxygen concentrations up to 21%. The facility also provides conditioned hydrogen coolant at pressures up to

5000 psi at room temperature or 1000 psi if coolant inlet temperature is increased to 1000°F. Other features of the facility include 1) 12-minute run time with several runs a day possible, 2) a window for viewing the heat face of the test article during tests, and 3) reducing environments rather than oxidizing.

N. Rocketdyne Material and Structure Thermal Validation Rig

Rocketdyne's MSTV1 was constructed to test full-sized engine panels attached to representative engine primary structure under actual engine conditions. Within the rig, internal engine conditions such as heat flux, erosion, acoustics, and pressure could be duplicated. Under these conditions and with a segment of the engine accurately reproduced in the form of a test specimen, testing of actual component performance to the combine environment of the engine was possible. Test conditions within the rig included heat flux levels up to 1400 Btu/ft^2·s and pressures up to 350 psi. Unfortunately, while the rig may still exist in storage somewhere, it is no longer part of an active facility.

III. Thermal Protection Systems

There are many differing types of thermal protection systems for reusable vehicle applications. Probably the one most familiar to us is that in use on the United States space shuttle (see Fig. 4.14).

The space shuttle thermal protection systems employ both fibrous blankets and ceramic tiles bonded to an aluminum structure. The HRSI tiles make up the bulk of the lower surfaces that experience the highest temperatures upon reentry. The

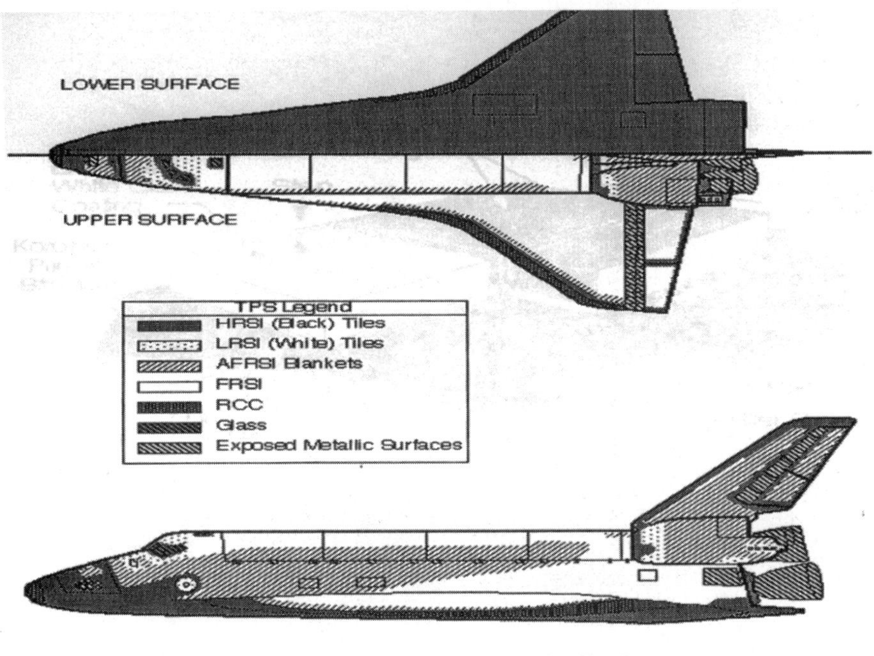

Fig. 4.14 Space shuttle TPS distribution.

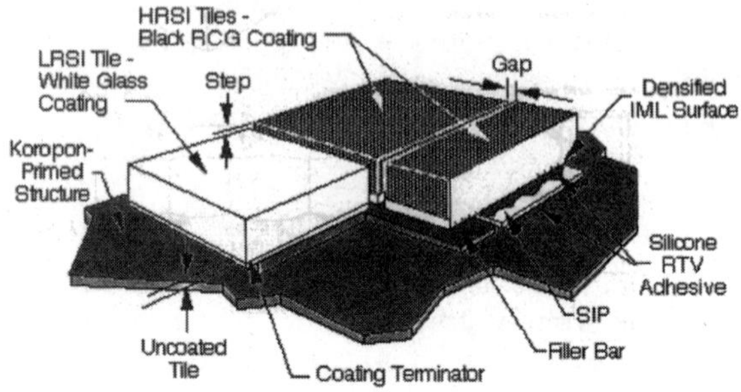

Fig. 4.15 Tile TPS configuration.

lower temperature tiles and the blankets comprise the bulk of the remaining thermal protection system. Although these systems are proven in operation, the cost to operate and maintain these systems are considered too great for many of the planned systems in the future. For the tiles alone, approximately 18,000 hours are needed to ready them for the next flight.

There are over 20,000 tiles on the shuttle orbiter. In addition, there is over 4000 sq ft of blanket TPS on the orbiter. Removal and replacement of these TPS are due to numerous factors. Approximately one-third of remove-and-replace actions are due to flight-induced damage such as those caused by impact. Another third is due to modifications that have to be implemented on the orbiter. The remainder of removal and replacement actions is due to access issues or TPS found to be out of tolerance (see Figs. 4.15 and 4.16).

To achieve the cost and operability goals of future systems will require order-of-magnitude reductions in the maintenance, inspection, repair, and replacement of TPS. Efforts are ongoing to meet this challenge and progress is described below.

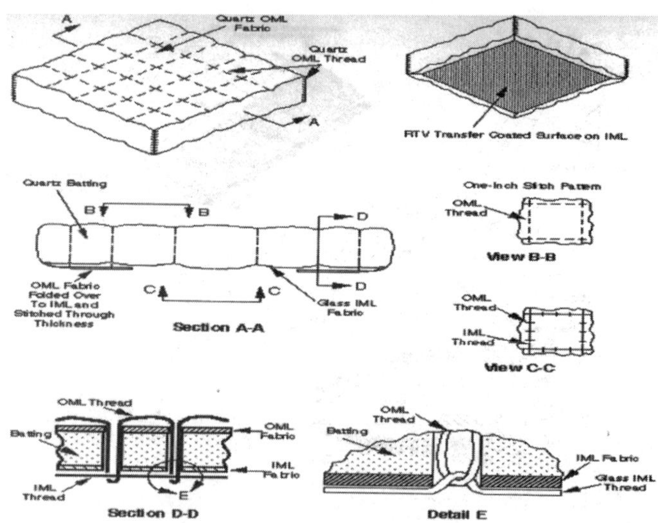

Fig. 4.16 Blanket TPS configuration.

A. Blanket TPS

One approach to address this shortcoming is the development of mechanically attached blanket TPS. Although mechanical attachments for tile TPS has been discussed, very little development has occurred to date. The use of mechanically attached TPS concepts instead of bonded TPS concepts will reduce vehicle operations and support cost, while at the same time reduce the vehicle turnaround time. The ease of removal of the TPS will allow rapid replacement of damaged areas and provide quick access to doors and ease inspection of the structure underneath. In Fig. 4.17, the titanium frame and corner plates can be seen in the x-ray image. The attachment studs to which the fasteners will be attached are also shown. The attachment studs can be adhesively attached to the structure or welded as appropriate. Vibroacoustic and arcjet testing have been performed to verify both structural integrity of the mechanical attachment concept and blanket as well as the thermal performance of the system. Vibroacoustic test conditions were a sound pressure level up to 170 dB with a frequency spectrum of 50 to 500 Hz random. Other improvements to the blanket system include advanced fibers such as Nextel 440 and improved emissivity coatings. A side benefit of the coatings is an improved durability of the surface of the blankets.

In the near future, scaleup to larger blanket sizes will be accomplished. To date, most of the effort has been focused on proof-of-concept activities in blanket sizes up to 12 in. × 18 in. Within the first year or two of the next century, blanket sizes up to 2 ft × 2 ft will be developed. This will pave the way for an eventual flight demonstration and provide additional design data for vehicle designers. Other areas to be addressed include seals between blankets and other TPS concepts (tiles and panels), waterproofing protection, and refinement of the quick-release fastener concept.

Other improvements to blanket TPS are under development as well. These improvements include toughening the blankets for increased robustness and higher

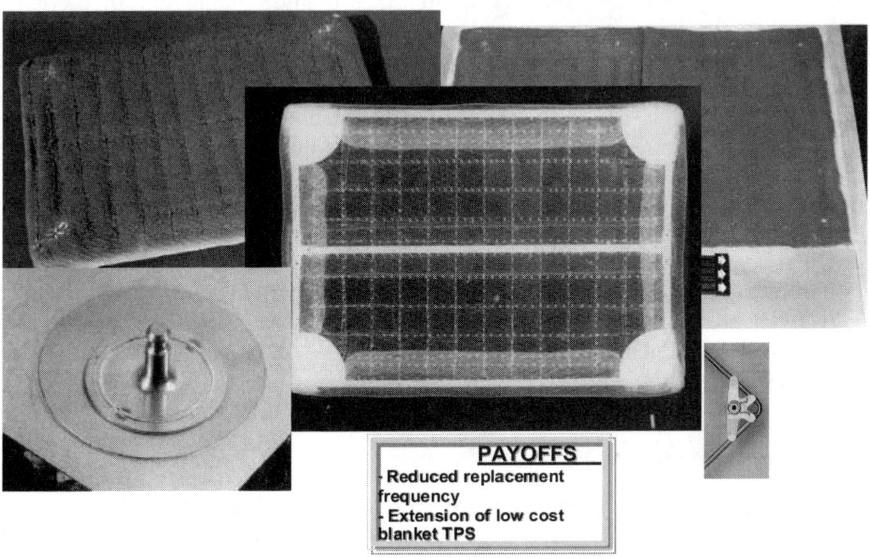

Fig. 4.17 Mechanically attached blanket TPS.

Fig. 4.18 Ceramic matrix composite/tile TPS.

temperature felt blankets (organic, ceramic, and organoceramic) for reduced cost and weight. One such approach to toughening the blanket integrates a metallic foil brazed to the blanket outer surface.

B. Tile TPS

The development of tile TPS has largely been along the same lines as that experienced in the drive to develop blanket TPS. Many of the issues are the same. Tiles experience excessive damage, require waterproofing, and slow vehicle turnaround times. Many approaches to reduce tile damage include toughening of the exterior surfaces to improve durability. Concepts to achieve this toughening include ceramic coatings and encasement of the tile in a thin ceramic matrix composite outer layer. The use of a ceramic matrix composite outer layer is probably the most recent example of tile durability improvements. The use of a ceramic matrix composite outer layer may also provide an avenue to the development of a mechanically attached tile. Shown in Fig. 4.18 is a ceramic tile with a ceramic matrix composite external surface after combined thermal/vibroacoustic testing. The concept successfully survived its design environment.

Other benefits of the ceramic matrix composite wrapped tile include the elimination of rewaterproofing requirements. These systems will begin to make their way into advanced systems in the next century. One example is the X-33 demonstrator, where both toughened tiles and advanced blankets will be featured as part of the TPS package.

C. Stand-Off TPS

For many advanced systems under consideration, such as the space operations vehicle, even greater durability requirements with much lower operations cost will drive TPS systems. For many of these systems, stand-off thermal protection systems with either metallic or ceramic matrix composite designs may be needed.

Fig. 4.19 Metallic TPS concept.

Shown in Fig. 4.19 is a metallic TPS concept being readied for combined thermal and acoustic testing. For lower temperature regimes, titanium is preferred because of its light weight. As the temperature increases, superalloys and other high-temperature metallics may be considered. Typical construction consists of a high-temperature honeycomb core with high-temperature, brazed-on facesheets. Multiwall construction has also been investigated, primarily with titanium materials. Both rectangular and hexagonal panel geometry are typically considered (see Fig. 4.20).

Although not as mature as tile and blanket TPS concepts, stand-off concepts in both metallics and ceramic composites are receiving a growing investment in their development. Several key issues remain before their use in military high-speed

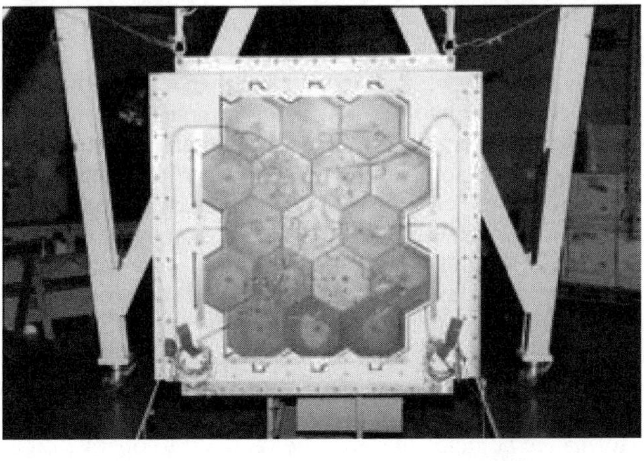

Fig. 4.20 Hexagonal carbon–carbon TPS array.

systems. The metallics are limited in size due to their thermal growth during heating. This growth is usually managed through keeping the designs small (18 to 20 in. typical). The growth is then managed by the gaps between adjacent panels, which reduces thermal stresses experienced during reentry or high-speed cruise. However, this practice increases part count. This increased part count in panels and fasteners may lead to indirect cost and maintenance burdens that are higher than desirable. The coatings needed to ensure high emissivity, noncatalycity, and oxidation protection could also affect turnaround times. In their favor, however, is the light weight that can be attained. Titanium TPS can approach 1 lb/sq ft in weight, which is competitive with blanket concepts, while the higher temperature superalloy concepts are weight competitive with tiles. In addition, waterproofing is not needed.

The ceramic matrix composite TPS is usually significantly heavier than its metallic, blanket, and tile counterparts. However its advantage are much larger part sizes (two to three times tiles or metallics) can be utilized. Also, for the highest temperatures on the vehicle, it may be the only way to ensure adequate life without resorting to active cooling. Issues with ceramic composites include the immaturity of the basic materials, especially in the fiber–matrix interface, the inability to adequately assess life, and the durability of oxidation protection coatings.

Thermal protection systems are absolutely critical to the successful performance of high-speed systems. Compromises between weight and cost are often required. To make the TPS more durable and operable often results in a heavier weight than is desired. Weight reduction can lead to less operable TPS. Clearly, improvements in materials and design will be needed to meet the performance (weight and cost) goals of future systems.

IV. Hot Structure for High-Speed Vehicles

A. Introduction

This section provides several examples of hot structures development programs, which have been pursued under government sponsorship since the mid-1980s. Generally, for a given design application, hot structures must be evaluated along with the competing design approaches of thermally protected structure and actively cooled structure. Normally, actively cooled structure is proposed only for the highest energy points, such as localized stagnation regions, where no other approach is viable. Though it can be considered for broader application, the major deficiency of actively cooled structure is the design complexity, hence cost, associated with the coolant arrangements and the ensuing reliability issues associated with proper operation of the system. Thermally protected structure is a more mature technology. Thermally protected structure is used for acreage that has high energy for thermally protected structure; however, the thermal barrier is usually parasitic, resulting in additional weight beyond that required by the primary load-carrying structure alone. Also, thermally protected structure can have extensive inspection and maintenance actions required to verify integrity and continued safe operation of the thermal shield. Hot structure allows the load-carrying structure to attain the flight environment temperature. The major attribute of hot structures is that, generally, it is the minimum weight design possible for an intended application. For high-performance vehicles that may be required to operate on a global scale both within and outside the atmosphere, minimization of airframe weight is key to achievement of the vehicle's performance objectives. If the structure can be designed to be reliable and robust without sacrificing its weight advantage, then a superior solution will have been attained.

As is often the case in technology advancement, a driving imperative is needed to force the technology investment needed to achieve real advances. For hot structures, the NASP program provided just such a catalyst. Extensive use of hot structures was proposed for the X-30 demonstrator and its postulated derivatives. This chapter discusses several technology development initiatives that were undertaken by the NASP program and in support of NASP to mature hot structure technology for vehicle application.

B. Two-Dimensional Carbon–Carbon Control Surface

Because of its high specific properties at elevated temperature, carbon–carbon (C–C) is an attractive material for hypersonic vehicle application. During NASP, C–C was investigated by NASA for a control surface component for a generic NASP vehicle configuration. Hot structure is particularly appealing for aerodynamic surfaces since it can often more easily meet stringent volume restrictions imposed by thin surfaces heated on both sides.

Early in the design study, actively cooled designs were evaluated along with passively cooled designs. Actively cooled designs, however, were found to be consistently heavier. The passive C–C design illustrated in Fig. 4.21 was eventually chosen based on a combination of minimum weight, low risk and ease of fabrication.

To evaluate the design, a test article was fabricated from two-dimensional ACC-4 C–C. The major detail parts were two rib-stiffened, upper and lower panels; a

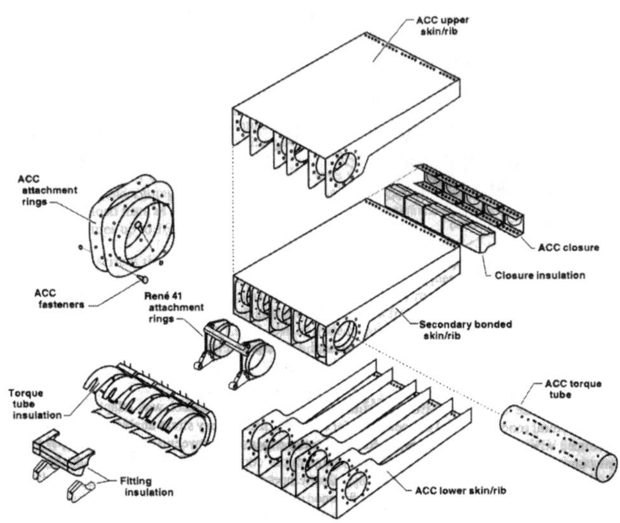

Fig. 4.21 Exploded view of elevon test article.

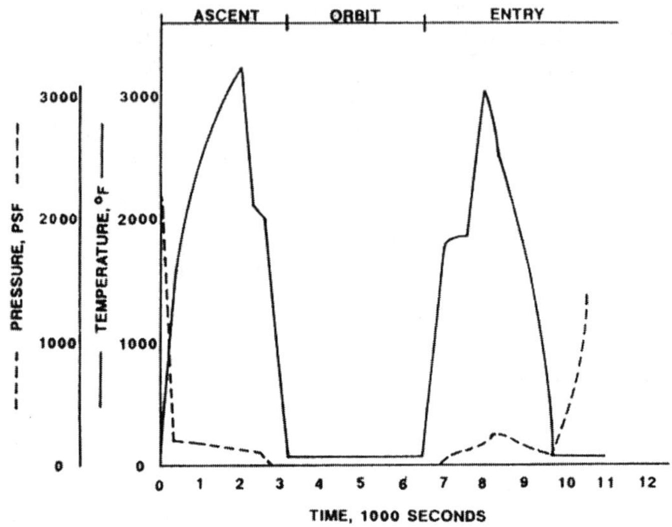

Fig. 4.22 Elevon surface thermal environment.

C–C torque tube; ten C–C attachment rings; a C–C closeout channel, and a Rene' 41 actuator lug fitting. The rib-stiffened panel details were laid up and processed together, eliminating the need for mechanical joining at the rib/skin interfaces. The two panel halves were then joined by secondary bonding with the remaining C–C processing and coating cycles performed on the assembly as a whole. Rene' 41 fasteners and cleats were used to attach the actuator lug fittings to the C–C torque tube. Overall dimensions of the test article were 38.7 in. span, 56 in. chord, and a depth tapering from 14 in. at the forward edge to 6.2 in. at the aft edge.

As fabricated, the test article represented a segment of a full-scale elevon control surface and was designed to sustain a maximum surface temperature of approximately 3000°F. A typical temperature/pressure profile for the elevon for ascent to orbit and reentry is illustrated in Fig. 4.22.

Both room and limited elevated temperature testing of the control surface test article were conducted. For the room temperature testing, loading of the component to 100% design limit load resulted in localized damage at several of the load-transfer points between the actuator fittings and the C–C torque tube. Nonetheless, from the room temperature testing, basic structural viability of the C–C design was established. The component was subsequently repaired and additional experimental evaluation was performed at moderate elevated temperatures. The test article is now being retrofitted with several improved components that will enhance load transfer between the torque tube and actuating fittings. Testing will resume once this activity has been completed.

The development activity to date has focused on static loads evaluation. Before this type of structure can be fully validated for vehicle application, simulated mission cycling will need to be performed in a realistic gas environment to establish the capability of the structure to perform acceptably for its required lifetime.

C. Three-Dimensional Carbon–Carbon Wing Box

Traditional two-dimensional C–C suffers from weak interlaminar material properties. One way of overcoming this weakness is three-dimensional fiber architecture. In this example of C–C hot structure, a three-dimensional angle interlock weave was employed in the design, fabrication, and testing of a three spar, wing torque box structure. In the three-dimensional angle interlock process, most of the fibers remain in the plane of the material. A smaller percentage of the fibers is woven through the thickness to increase the interlaminar tension and shear properties. A side benefit of this fiber architecture is the ability to integrally weave certain design features of the structure as a unitized whole. In this particular example, spar caps and spar webs were integrally woven. Additionally, skins in the area of the spar caps were woven as a bilayer preform. This allowed one layer to be cut and the spar cap inserted to the intervening space, resulting in a coprocessed, fastenerless skin-to-spar joint. For the closure ribs, traditional two-dimensional laminated C–C composite was used.

The baseline for the three-dimensional wing box design demonstration was a single-stage-to-orbit (SSTO) operational derivative of the NASP X-30. Design studies that led to the selection of C–C for the wing structure showed a significant weight savings relative to cooled, titanium- aluminide structure. Maximum temperature of the C–C design was approximately 2250°F. To prevent oxidation, a multilayer, silicon carbide based coating was applied to all surfaces. Elimination of mechanical fasteners through the skins allowed outer surfaces to present an uninterrupted coating interface to erosive and oxidative flow.

A view of the three-dimensional C–C test component as it appeared before thermomechanical evaluation is shown in Fig. 4.23.

Thirty-three simulated mission cycles of heating and mechanical loads were successfully applied to the box. Testing in air was not possible due to a coating anomaly requiring that a nitrogen gas-filled chamber be used instead. A view of the quartz lamp heating array is shown in Fig. 4.24, illustrating the extent of the lamp assembly that was needed to provide proper thermal input to multiple heating zones covering the box.

Figure 4.25 illustrates the relative magnitudes of heating and mechanical loads for the payload-to-orbit mission profile. This profile was applied for one-third of the total number of mission cycles to which the box was subjected. The remaining 22 profiles represented loads for two different cruise missions.

Fig. 4.23 Demonstration C–C wing box test component.

Fig. 4.24 Test article upper and lower lamp arrays.

A final room temperature test to failure was performed with the box successfully exceeding its 150% design limit load objective. A view of the wing box in its load application fixture during the room temperature failure test is shown in Fig. 4.26.

This program was successful in demonstrating the basic structural integrity of three-dimensional woven C–C for use as primary structure. Life testing in an oxidative environment still remains a development issue for this type of structure to be fully validated for reusable vehicle application.

D. Titanium Matrix Composite Hot Structure

One of the significant challenges of the NASP program was development and selection of a material system that would be able to withstand the significant aerodynamic heating encountered by the vehicle's airframe during atmospheric flight. After an extensive development program that examined numerous material

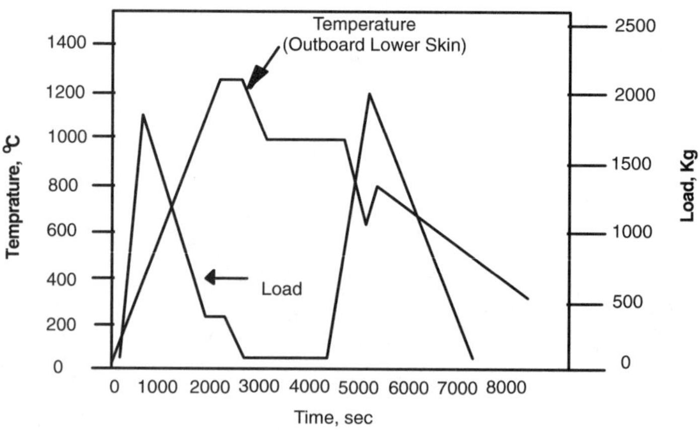

Fig. 4.25 Payload-to-orbit mission load/temperature profile.

Fig. 4.26 C–C wing box configured for RT failure test.

candidates, titanium matrix composites (TMC) were chosen as the principal material system for the X-30 fuselage and empennage structures. Maximum use temperature of the material was to be 1500°F. Use of TMC was shown to offer a weight savings of 50% versus state-of-the-art superalloys and high-temperature steels, a weight advantage that was of critical importance to the vehicle's intended SSTO operation.

The specific TMC material system selected for NASP used coated silicon carbide fibers in a Beta 21S titanium matrix. The matrix alloy, Ti-15Mo-2.8Nb-3Al-0.2Si, was developed by Timet specifically to meet the oxidation-resistance requirements of NASP. The fibers selected included Textron Specialty Material's SCS-6 silicon carbide fiber, chosen as the baseline fiber, and SCS-9. The SCS-9 fiber is essentially the same as SCS-6 except for a smaller diameter. The SCS-9 fiber has lower mechanical properties but its density is lower also. Both fibers were able to meet the acceptance criteria of 50 hours exposure at 1500°F without unacceptable degradation of the fiber-protective coating. This exposure time was greater than what would be encountered for the NASP lifetime.

Titanium was feasible as a matrix material because of its limited exposure at maximum temperatures and because maximum temperatures were to be reached at a high altitude where the availability of oxygen and hydrogen would be reduced. Thermomechanical fatigue (TMF) testing was performed, which demonstrated the ability of the SCS-6/Beta 21S baseline material system to survive more than two NASP lifetimes. Testing also demonstrated that the Beta 21S matrix material was sufficiently oxidatively resistant to survive the NASP environment without a protective coating. Exposure to casual hydrogen in TMF testing at the low anticipated hydrogen pressures additionally demonstrated that the material would not be unacceptably degraded due to hydrogen embrittlement.

Significant progress was achieved by the NASP program in developing processing and manufacturing operations necessary to produce consistently high-quality material and built- up parts using the TMC material. Resistance spot welding, brazing, and weld brazing joining techniques were successfully developed to enable the manufacture of minimum weight X-30 fuselage hat-stiffened skin designs. Mechanical fastening techniques were also developed to enable the joining of structural parts into an assembly. The fastener technology utilized the high-temperature materials, Haynes 230 and Ti-1100.

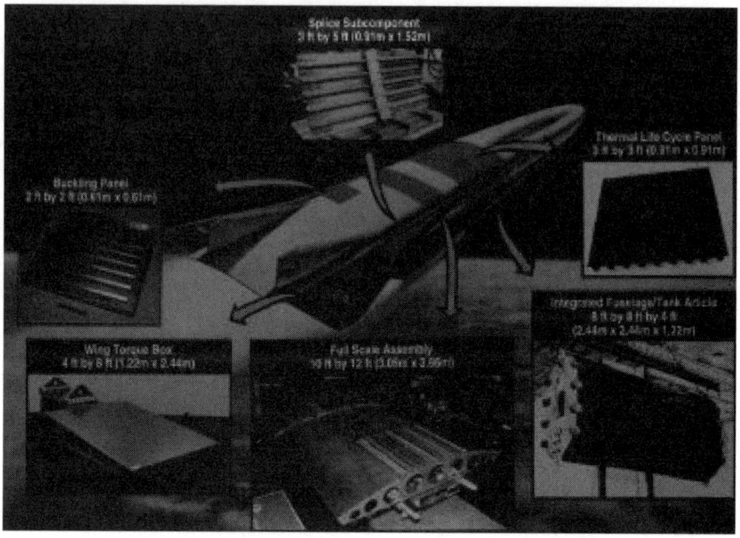

Fig. 4.27 Major TMC components fabricated during NASP.

During the fabrication development, various structural details were produced, demonstrating the ability to fabricate stiffened skin panels. On a larger scale, several shear/compression panels were built and tested to validate the stiffened skin design approach. Fabrication techniques were also developed for a TMC-reinforced sandwich panel taking advantage of the ability to diffusion bond a superplastically formed titanium core to a TMC facesheet. TMC components of multifeet dimensions were also successfully fabricated and tested demonstrating viability of TMC technology for the larger scale structures that would be needed for the X-30. These components, as illustrated in Fig. 4.27, included a splice subcomponent, a stabilator torque box, a full-scale panel assembly, and an integrated fuselage/tank article.

As extensive as the NASP TMC development effort was, additional work would have been required to successfully transition TMC technology to the X-30 airframe. Additional, larger, and more complex structural components needed to be fabricated and tested had X-30 airframe development proceeded. Additional data on fasteners and joints would have been required also, as well as durability and damage-tolerance investigations. Even though TMF testing demonstrated that Beta 21S did not require an oxidation-preventive coating, an emissivity coating would have been needed and a development program would have been required to meet that need. Because of its intended SSTO operation, the X-30 was performance driven. Cost was not a driving parameter. However, the TMC material was expensive and initiation of a cost reduction effort would have been highly desirable had the program gone forward.

V. Heat-Pipe-Cooled Leading Edges for Hypersonic Vehicle Airframes

A. Introduction

Leading edges of hypersonic vehicles experience extreme thermal environments during flight. Materials and cooling techniques for hypersonic vehicle leading edges must thus be chosen carefully to minimize cost and maximize performance. These high heat flux regions can be cooled passively (radiation), actively, or with heat pipes, depending on the magnitude of the heat flux. Passively cooled structures have by far the lowest cost, complexity, and weight but also have the lowest capability to withstand extreme local heating. Once the passive heat flux limit is surpassed, hypersonic vehicles must utilize either heat-pipe or active cooling. Active cooling is well developed, has the highest heat flux capability, and actively cooled structures generally operate at the lowest temperatures for a given heat flux. However, active cooling is also costly, complex, and heavy, requiring numerous auxiliary systems.

Analytical studies, laboratory experiments, and wind tunnel tests indicate that a viable solution to thermal-structural problems associated with stagnation regions of hypersonic aerospace vehicles may be obtained with heat pipes. Heat-pipe-cooled leading edges combine features of both passive and active cooling and generally have a heat flux capability, cost, complexity, and weight somewhere between passive and active cooling. Like a passive leading edge, the maintenance required for a heat-pipe-cooled leading edge is negligible. Because of high localized heating on a leading edge, leading edge geometry and the ability of a heat pipe to passively redistribute the heat, leading edges are an ideal application for heat pipes. The use of heat pipes results in a lighter-weight, lower-cost, and lower-maintenance leading edge than an actively cooled leading edge.

Potential applications for heat pipes include sharp leading edges for hypersonic vehicles and a heat-pipe-cooled leading edge and nosecap for a shuttle upgrade. Heat-pipe-cooled leading edges were estimated to provide a 50% weight savings over active cooling for the NASP. The use of heat-pipe-cooled leading edges and nosecap on the Space Shuttle Orbiter could extend the range of the orbiter up to 400 to 500 miles in the event of an aborted mission by allowing the vehicle to fly a lower angle of attack (AoA). The current C–C passive leading edge maximum temperature limits how low the AoA the orbiter can be during high-speed flight.

Heat-pipe-cooled leading edges are also a critical technology for military systems. Anticipated military requirements for a relatively high lift-to-drag (L/D) configuration suggest that sharp leading-edge heat pipes enable the cross range required to ensure a continental U.S. landing in a once around mission (typically 1200 nautical miles). High L/D requires much sharper leading edges than the current Space Shuttle Orbiter or the planned Venture Star. These sharper leading edges will result in heat fluxes far above the passive limit for most materials.

An additional constraint on future military vehicle designs is the requirement to rapidly service vehicles for further mission activities. Turnaround times of 8 hours or less are desired to minimize fleet size and maintain a high operational tempo during wartime. This service requirement essentially eliminates the use of ablative approaches to achieving sharp leading edges due to the refurbishment time. In addition, ablative concepts may well result in some blunting of the leading edges and a reduction of L/D. One way to reduce support cost for structures is to reduce the number of parts and systems that must be maintained. Actively cooled structures will typically result in an increase in part and system count, while heat-pipe-cooled leading edges are essentially passive, maintenance-free structures with low part count.

Here, we discuss the development of heat-pipe-cooled leading edges and a description of the state of the art. It begins with a brief discussion of the operation of heat-pipe-cooled leading edges. Following this section, a detailed discussion of fabrication and testing experience with heat-pipe-cooled leading-edge test articles is presented.

B. Operation

Heat pipes provide cooling of stagnation regions by transferring heat nearly isothermally to locations aft of the stagnation region, thus raising the temperature aft of the stagnation region above the local radiation equilibrium temperature. When applied to leading-edge cooling, heat pipes operate by accepting heat at a high rate over a small area near the stagnation region and radiating it at a lower rate over a larger surface area, as shown in Fig. 4.28. The use of heat pipes results in a nearly isothermal leading-edge surface, thus reducing temperatures in the stagnation region and raising the temperatures of both the upper and lower aft surfaces. Detailed modeling of heat pipes related to leading edges can be found in the literature.

A heat pipe is a self-contained, two-phase, heat-transfer device that is composed of a container, a wick, and a working fluid. A schematic diagram illustrating the operation of a heat pipe is shown in Fig. 4.29. Heat, input locally to one section of the heat pipe known as the evaporator region, is conducted through the container and into the wick/working-fluid matrix, where it is absorbed by the evaporation of the working fluid. The heated vapor flows to a slightly cooler section of the heat pipe where the working fluid condenses and gives up its stored heat. The heat is then conducted through the wick/working-fluid matrix and container. Ultimately the heat is rejected through the cooler regions of the container. The location of the heat pipe where heat is rejected is called the condenser region. The cycle is completed with the return flow of liquid condensate back to the heated region (evaporator) by the capillary pumping action of the wick. During normal operation, heat pipes operate as devices of very high effective thermal conductance and maintain a nearly uniform temperature over the entire heat-pipe length.

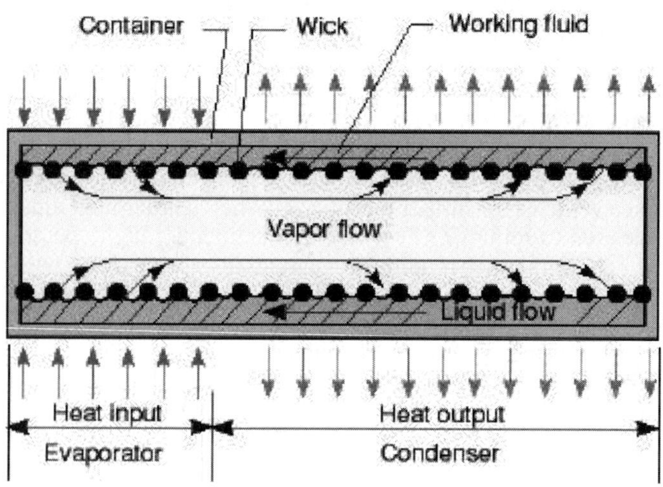

Fig. 4.28 Schematic diagram of a heat-pipe-cooled leading edge showing regions of net heat input (evaporator) and net heat output (condenser).

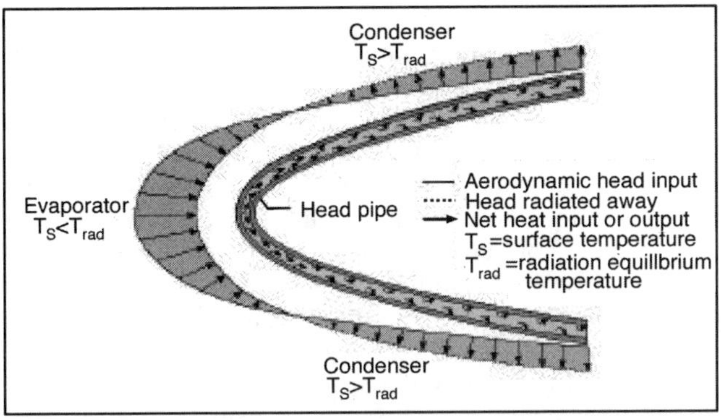

Fig. 4.29 Schematic diagram of the operation of a heat pipe showing the heat-pipe container, working fluid, and wick.

A given heat pipe can operate over a range of temperatures; however, at each temperature, there is a maximum axial heat-transport capacity and surface heat flux above which normal heat-pipe operation is disrupted. Four such operational limits are shown schematically in Fig. 4.30. These limits are the sonic, entrainment, wicking, and boiling limits. The shaded region under each of the limit curves in Fig. 4.30 represents the area of feasible operation. The sonic limit usually occurs at low vapor temperatures when the velocity of the vapor can become sonic and the axial heat transport in the heat pipe is limited by a choked-flow condition. During startup of a liquid metal heat pipe from the frozen state, when the working fluid is solid, the axial heat transfer in the heat pipe is usually limited by sonic flow. Although the sonic limit alone typically does not cause a failure of the heat pipe, it causes the operating temperature of the heat pipe to rise, which may cause

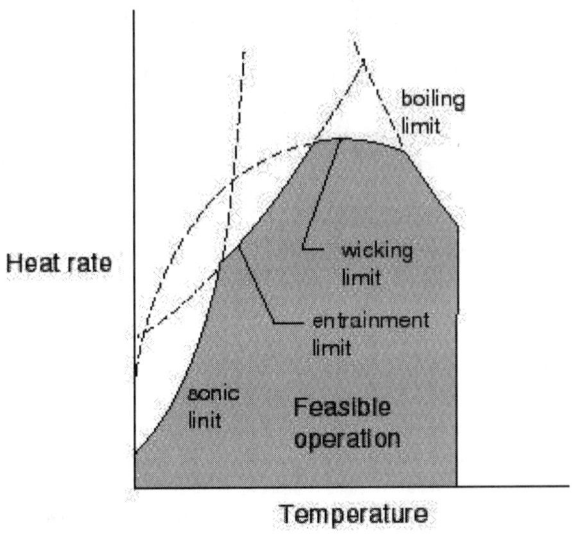

Fig. 4.30 Heat transport limits.

other limiting conditions to be encountered. The entrainment limit occurs when the drag force of the vapor on the liquid condensate returning to the evaporator is sufficient to overcome surface tension forces of the liquid within the wick and entrain liquid droplets in the vapor flow. The entrainment of liquid intended to flow to the evaporator region could lead to dryout of the wick in the evaporator region and result in overheating and excessive local temperatures. The wicking limit is reached when the surface-tension pumping capability of the wick is just sufficient to provide the liquid mass flow rate needed to balance the working fluid evaporation from the applied heating. Liquid and vapor frictional losses tend to decrease the flow rate of liquid condensate to the evaporator region. Inertial forces caused by vehicle acceleration and gravity can tend to reduce or increase the flow of liquid condensate to the evaporator region. If the wicking limit is exceeded, dryout will occur in the evaporator and result in overheating.

Vaporization of the working fluid normally occurs in the evaporator region of the heat pipe at the liquid–vapor interface. At high radial heat input levels, however, the superheated liquid can reach a critical value at which boiling can occur near the heat-pipe wall. The boiling limit is reached when the heating rate is high enough to cause the formation of a continuous vapor film at the container wall that inhibits the flow of heat to the working fluid. The boiling limit is usually not a limiting condition for liquid-metal heat pipes because of the high thermal conductivity of the liquid-metal working fluid and the large amount of superheating needed to initiate boiling. The boiling limit is a limit on the radial heat flux density, whereas sonic, entrainment, and wicking limits are limits on the axial heat flux density.

Heat-pipe-cooled leading edges offer the potential for significant weight savings compared to actively cooled structures for extremely weight-sensitive hypersonic vehicles. As a result of the high weight payoff, development of heat-pipe-cooled leading edges has continued for nearly three decades. In the early 1970s, several feasibility studies were performed to assess the application of heat pipes for cooling leading edges and nosecaps of hypersonic vehicles.[13–16] NASA Langley, through a contractual study, analytically verified the viability of heat pipes for cooling stagnation regions of hypersonic vehicles. In 1972, McDonnell Douglas Astronautics Co. (MDAC) compared four space shuttle orbiter wing-leading-edge concepts: a passive C–C concept, a passive coated-columbium concept, an ablative concept, and a liquid-metal/superalloy heat-pipe-cooled concept. The heat-pipe-cooled concept was determined to be a feasible and durable design concept but was slightly heavier than the other candidate concepts. In 1979, MDAC received a contract from NASA Langley to optimize a heat-pipe-cooled wing leading edge for a single-stage-to-orbit vehicle. Results of the follow-on study indicated that the mass of a shuttle-type, heat-pipe-cooled leading edge could be reduced by over 40% by use of a more efficient structural design.

In addition to the design studies mentioned above, several test articles have been fabricated and tested. The first heat-pipe-cooled leading-edge test article fabricated was a half-scale Hastelloy-X leading edge. The test article was designed and fabricated by MDAC and tested at NASA Langley in both radiant and aerothermal environments. Haynes 188 leading-edge-shaped heat pipes were later fabricated and tested by Boeing to heat fluxes of 220 $Btu/ft^2 \cdot s$ (250 W/cm^2). Boeing and Teledyne fabricated but did not test a leading-edge subcomponent with Nb-alloy heat pipes. Boeing fabricated a niobium vapor chamber for evaluation as a nosecap.

A Hastelloy-X leading-edge-shaped heat pipe was fabricated and tested by MDAC. After testing at MDAC, the heat pipe was sent to Los Alamos where it was tested and characterized by Los Alamos and NASA Langley. Each of these heat pipes is discussed in detail in the following sections.

Based on encouraging results from previous tests and design studies, in the early 1990s the NASP program funded design studies for a metallic vapor-chamber leading edge with Boeing and a refractory-composite/refractory-metal heat-pipe-cooled leading edge with NASA Langley. The Boeing design was based on prior nosecap work and consisted of a Nb–Ta alloy (FS-85) heat-pipe and leading-edge structure with lithium as the working fluid.[18] The NASA Langley effort departed from prior metallic heat-pipe-cooled leading edges and focused on embedding metallic heat pipes in a refractory-composite structure to provide an integrated thermal/structural solution for leading edges. The NASA Langley concept was funded for further development and fabrication and was the impetus for the fabrication and testing of a leading-edge-shaped Mo–Re heat pipe and three straight Mo–Re heat pipes embedded in C–C.

C. Half-Scale Hastelloy-X Leading-Edge Component

In search of alternate shuttle orbiter leading edges, a one-half scale, heat-pipe-cooled wing-leading-edge test model was fabricated by MDAC for a shuttle-type vehicle. The test model, shown in Fig. 4.31, had a 6-in. span, a chord length of 22 in., and was 13 in. high at its base. It consisted of 12 sodium-charged heat pipes that were brazed to each other and the inner surface of a thin (0.02-in.-thick) Hastelloy-X skin. The heat pipes used Hastelloy-X circular tubes that had an outer diameter of 0.5 in., a wall thickness of 0.05 in., and a wick

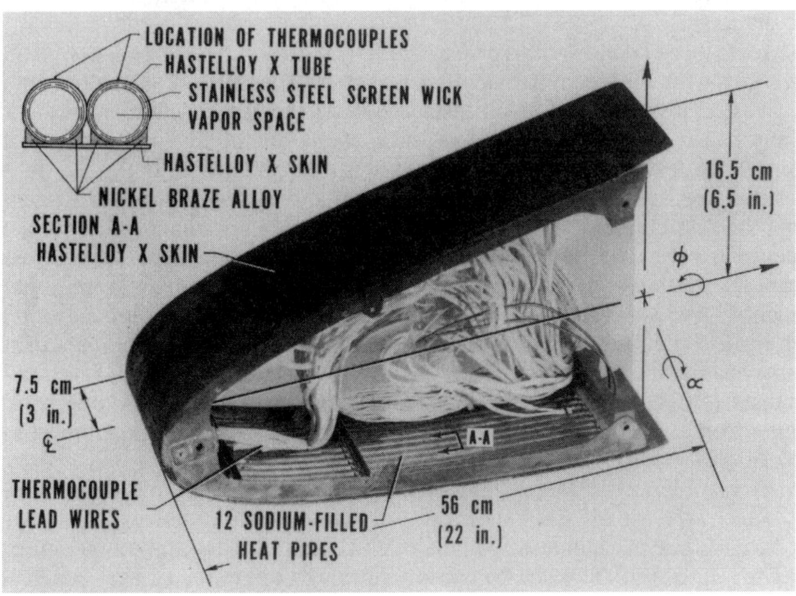

Fig. 4.31 Shuttle-type heat-pipe-cooled wing-leading-edge test model.

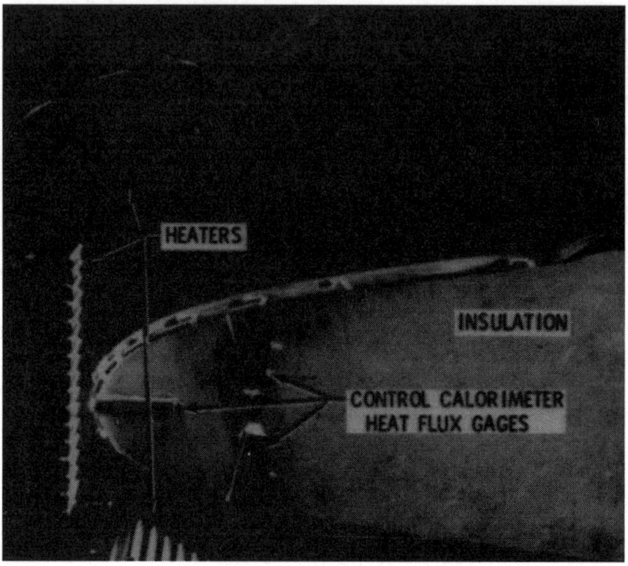

Fig. 4.32 Shuttle-type heat-pipe-cooled wing leading edge during a radiant heat test.

thickness of 0.035 in. The wick consisted of seven alternate layers of 100- and 200-mesh stainless steel screen and was a concentric annulus design. The heat pipes were sized so that one heat pipe could become inoperative without compromising structural integrity, thus providing a fail-safe capability. The skin was coated externally with a high-emissivity ceramic paint to facilitate heat rejection by radiation.

Performance of the shuttle-type, heat-pipe-cooled wing leading edge was studied by a series of radiant heat and aerothermal tests. The purpose of the radiant heat tests was to investigate heat-pipe startup from the frozen state (working fluid initially frozen or solid) and transient and steady-state thermal performance of the leading-edge test article. For the radiant heat tests, shown in Fig. 4.32, an AoA was simulated by orienting the model at an angle with respect to the horizontal to obtain the desired gravity effect and by positioning three radiant heaters about the model to provide a heating distribution representative of the aerodynamic heating distribution at that angle. Three water-cooled calorimeter gauges were used, one in each of three separate feedback control loops, to regulate output from each lamp bank separately. Power to the lamp banks was varied to simulate a shuttle reentry heating trajectory. The test model was subjected to a total of seven radiant heat tests, with cold-wall heating rates ranging from 21.1 to 34.8 Btu/ft^2·s, AoAs ranging from 0 to 20°, and maximum surface temperatures ranging from 1130° to 1380°F.

After completion of the radiant heat tests, the heat-pipe-cooled leading edge was fitted with spanwise extensions and aerodynamic fences as shown in Fig. 4.33 for aerothermal testing in the NASA Langley 8-ft, high-temperature tunnel. The purpose of the extensions and fences was to produce a two-dimensional flow field over the test model and allow measurement of flow conditions over the leading edge. The purpose of the aerothermal tests was to investigate performance

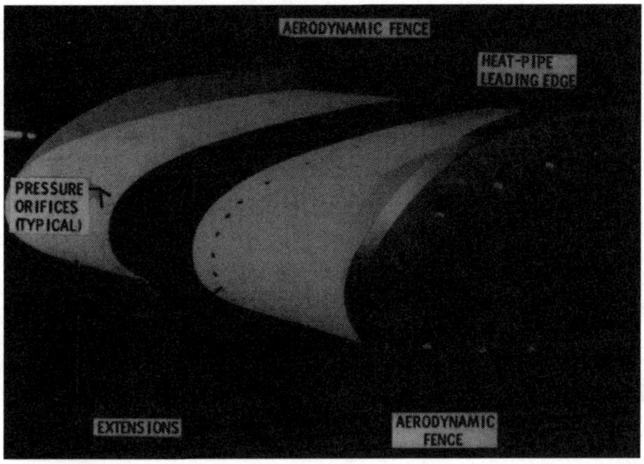

Fig. 4.33 Wind tunnel configuration of heat-pipe-cooled wing leading edge.

of the model subject to realistic aerothermal loads. To avoid unrealistically severe heat pulses and ensure proper startup of the heat pipe, the model was preheated before insertion to the test stream. A simple heater array was used to preheat the model; no attempt was made to simulate an aerodynamic heating distribution during preheating. The model withstood 27 supplemental radiant heat cycles lasting 30 minutes each, during which maximum surface temperatures ranged from 1100° to 1480°F. Cold-wall stagnation heating rates ranged from 21.1 to 34.8 Btu/ft^2·s for radiant preheating cycles and hot-wall stagnation heating rates ranged from 21.5 to 37.2 Btu/ft^2·s for aerothermal tests. The maximum cold-wall heating rate of the aerothermal tests was 56.9 Btu/ft^2·s and maximum heat-pipe operating temperatures ranged from 1090° to 1520°F. The model was subjected to a total of eight aerothermal tests with AoA variations of 0, 10, and 20° and roll variations of 0 and 90°. The 90° roll condition was included to investigate the effect of gravity on heat-pipe performance.

The heat pipes operated isothermally for these test conditions, with maximum operating temperatures ranging from 1250° to 1520°F. Near steady-state temperatures obtained during aerothermal tests were in good agreement with analytical results. Flow pressure and heat-flux distributions simulated by the facility were also in good agreement with design distributions.

D. Haynes 188 Leading Edge "D-Shaped" Heat Pipe

A sodium-working fluid/Haynes 188 leading-edge-shaped heat pipe with a "D-shaped" cross section was designed and tested by Boeing Advanced Systems. The heat pipe, fabricated by Thermacore, utilized a sintered nickel wick, had a length of 11.8 in. (300 mm) and a wall thickness of 0.030 in., with the flat part of the "D-shaped" heat pipe being 1-in. wide. A single screen artery with a diameter of 1.6 mm was utilized and placed away from the heated, flat surface. Thermal performance, vibration sensitivity, and critical heat flux tests were conducted.

Over 20 tests were performed on the leading-edge-shaped heat pipe. The heat flux was increased to 220 Btu/ft^2·s (250 W/cm^2) over 1000 seconds, held for

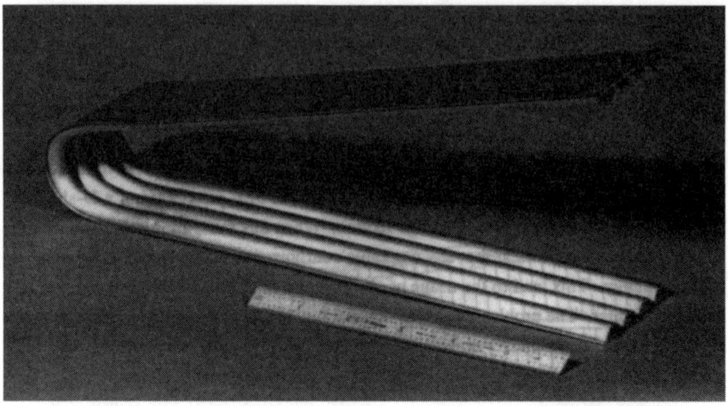

Fig. 4.34 Photograph of C-103 lead edge subcomponent.

800 seconds, and then turned off. No change in the performance of the heat pipe was noticed after several cycles. Tests were also performed to determine the heat flux capability of the heat pipe. Sharp increases in the surface temperature (to over 1830°F) occurred at slightly over 220 Btu/ft^2·s (250 W/cm^2), indicating possible wick dryout.

E. Niobium Heat-Pipe Leading-Edge Subcomponent

A heat-pipe-cooled leading-edge subcomponent was fabricated by Teledyne Wah Chang, and is shown in Fig. 4.34. The heat pipes had a "D-shaped" cross section and were made of C-103 (Nb-10Hf-1Ti) with a porous powder metal wick and included a 0.040-in.-diam (0.016-cm-diam) artery. The heat pipes were joined to C-103 face sheets to form flat panels 2.25 in. × 25 in.

Each of the flat panels was bent around a mandrel to form a leading edge with a 0.875-in. (2.33-cm) radius, as shown in Fig. 4.35. Bending methods demonstrated that the heat-pipe panels could be bent to relatively sharp radii without damaging the porous wick. Successful bending was verified by examining the wick in the

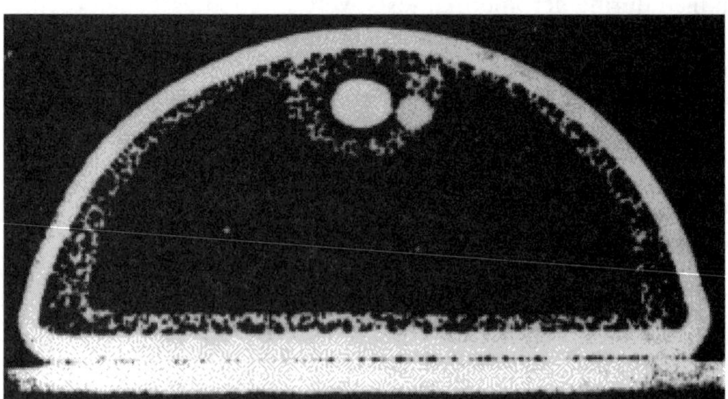

Fig. 4.35 Photograph of heat-pipe cross section in bend section (AIAA-88-2679).

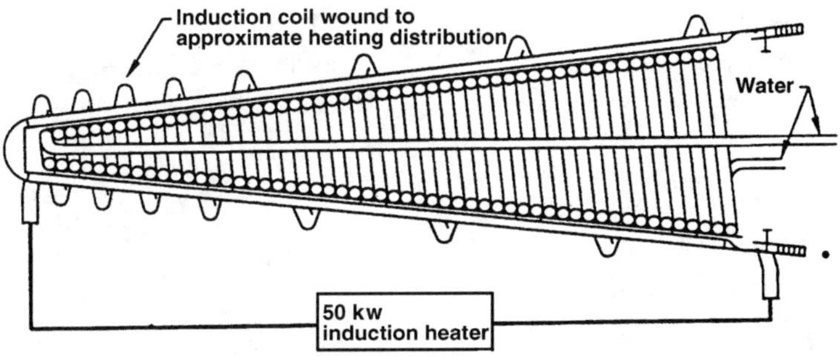

Fig. 4.36 Sketch of test set up at Boeing.

bent section for compaction, voids, or debonding. A photograph of the C-103 container and facesheet, along with the wick and artery are shown in Fig. 4.35. These leading-edge subcomponents were not tested.

F. Niobium Nosecap Vapor Chamber

Boeing fabricated a Nb vapor chamber for evaluation as a nosecap as part of the NASP program. The nosecap was 32 in. long, and had an 11-in. diameter at the aft end. A sintered metal wick, 0.020 in. thick, was used in the vapor chamber to wick the lithium working fluid. The sintered metal wick was interspersed with small arteries to help in the distribution of the working fluid. An oxidation-protection coating was used on the external surface due to the severe oxidation of Nb in an oxidizing environment.

The first series of tests was performed at Boeing. The nosecap was heated by a 50-kW induction heater to approximately 1300°F. A sketch of the test setup is shown in Fig. 4.36 with induction coils wrapped around the outside surface. When the vapor chamber reached approximately 1300°F, an arc jumped to the nosecap, damaging the coating. The coating was repaired at Boeing and the nosecap was sent to the Air Force Research Laboratory (AFRL) for a second series of tests.

AFRL planned to preheat the nosecap with quartz lamps, followed by high heat flux exposure using a Vortek® heater. A sketch of the AFRL test setup is shown in Fig. 4.37 and a photograph of the nosecap mounted in the test assembly with

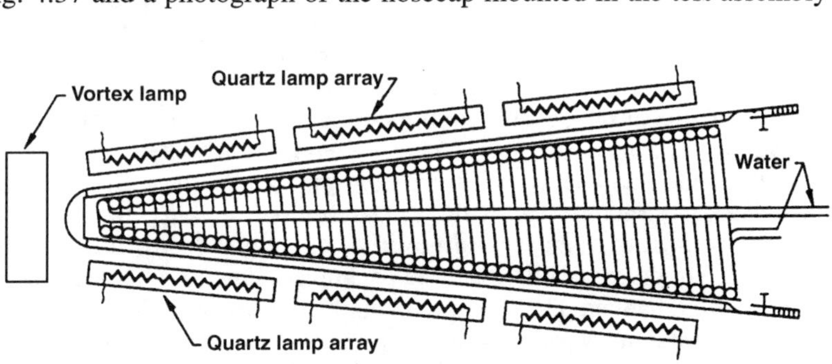

Fig. 4.37 Sketch of set up at WL.

Fig. 4.38 Photograph of Nb vapor chamber in test fixture (Photo courtesy of AFRL).

the quartz lamps surrounding the sides is shown in Fig. 4.37. The repaired coating can be seen in a photograph as Fig. 4.38. The initial heating with quartz lamps was designed to melt the Li working fluid and start the vapor chamber operating. Once the vapor chamber was operational, the Vortek heater would be used to apply the design heat fluxes of up to 1000 Btu/ft^2·s. During initial testing with the quartz lamps, the repaired portion of the coating began to bubble and testing was stopped. Material from the location of the failure was analyzed and found to contain negligible lithium, indicating that the heat pipe did not leak. No further tests were performed on this article.

Boeing utilized their nosecap experience to develop a heat-pipe-cooled wing-leading-edge design for the NASP. The design consisted of a Nb–Ta alloy (FS-85) heat-pipe and leading-edge structure with lithium as the working fluid. They considered four configurations in the study, each consisting of a vapor chamber in the stagnation region. A typical example is shown in Fig. 4.38. Three of the four configurations had a backup structure behind the heat pipe to serve as a ablator in the event of a heat-pipe failure. In the first configuration, all heat was rejected from the outer surfaces. A backup structure was included but no internal cooling. In the second configuration, a convectively cooled heat exchanger was located internal to the leading edge, with the heat pipe radiating energy to the heat exchanger. The third configuration was similar to the second, but a backup structure was included. In the fourth configuration, the convectively cooled heat exchanger was moved to the region between the heat pipe and the backup structure. In this case, the backup structure did not restrict radiation to the heat exchanger. The heat-pipe lengths ranged from 28 to 60 in., with a maximum external surface temperature of ~2500°F. Evaluations of oxidation protection and weight were also conducted. Fabrication of this design was not funded.

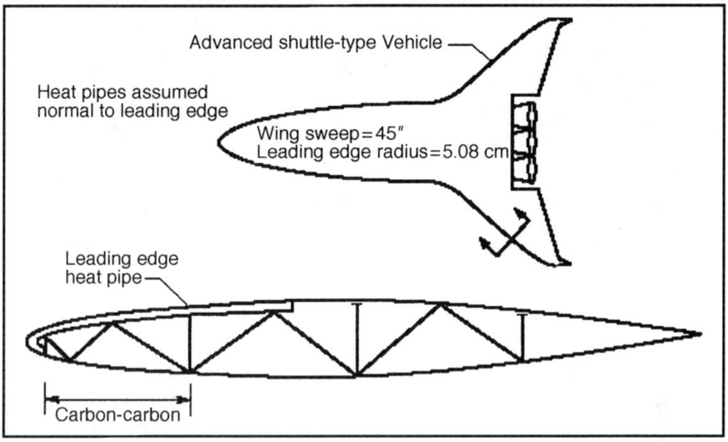

Fig. 4.39 Schematic diagram of a heat pipe for cooling the wing leading edge of an advanced shuttle-type vehicle.

G. Hastelloy-X Leading-Edge-Shaped Heat Pipe

In 1986, MDAC received a contract to fabricate and build a sodium-working-fluid/superalloy heat-pipe-cooled leading-edge component for an advanced shuttle-type vehicle. This advanced shuttle-type heat pipe was 6 ft long and was tested at MDAC by radiant heating and later at Los Alamos by induction heating. The heat pipe was designed to be oriented normal to the leading edge as shown in Fig. 4.39. Aft of the heat pipe, hot-structure designs (e.g., C–C) were assumed on the lower surface. The heat pipe was intended to reduce maximum wing leading-edge temperatures during reentry from 3500° to 1800°F. The total length of the heat pipe necessary to radiate enough heat to reduce maximum temperatures to 1800°F was determined from an energy balance over the entire heat pipe. A large AoA during reentry would result in lower heat loads on the upper surface, thus providing a considerable area on the upper surface to radiate heat away. The heat-pipe length on the lower surface was sized such that the radiation equilibrium temperature of the region beyond the end of the heat pipe was lower than 2800°F. This sizing operation resulted in a total heat-pipe length of 69 in., with 3 in. of that on the lower surface.

After preliminary testing at MDAC, the heat pipe was sent to Los Alamos where several startup tests of the heat pipe were conducted. As tested in the configuration shown Fig. 4.40, it was found that the heat pipe was extremely sensitive to transients, both during and after startup. It was observed that insulating three sides of the condenser assisted in both the startup and steady-state operation of the heat pipe during the tests. The insulation reduced the axial heat flux density of the heat pipe and thus reduced the pumping requirements of the wick.

An initial test was conducted to verify the heat-pipe startup operation from the frozen state and demonstrate freedom from noncondensable gas contamination within the heat pipe. The heat pipe was found to be extremely sensitive to both the magnitude and rate of increase of thermal loading and would not operate when a water-cooled calorimeter was used to remove heat from the condenser region of the heat pipe. The calorimeter had to be removed and one-half of the length of the heat

Fig. 4.40 Photograph of MDC Heat Pipe during test at Los Alamos.

pipe was covered with insulation to ensure successful startup and operation. Thus, it was not possible to determine the heat input to the heat pipe by direct calorimetery and calculated values for heat rejection by radiation were used. Heat-pipe startup was accomplished by increasing the power input to the induction heaters at a slow rate while monitoring evaporator response and recording startup temperatures. After 2 hours, a hot spot was observed in the evaporator region and the test was stopped. Temperatures at the various thermocouple locations are shown in Fig. 4.41 as a function of time. The hot spot in the evaporator region resulted from a lack of working fluid in the evaporator, which caused a dryout of the wick. At the time the hot spot was observed (2 hours into the test), the temperature of the evaporator was 600°C near the stagnation line, 550°C near the evaporator exit, and 450°C at the station $s = 144$ cm.

It was then decided to try a slow startup of the heat pipe with insulation on three sides. The heat pipe started up normally with the operating region of the heat pipe slowly increasing in both temperature and length. The heat flux to the stagnation region was increased in small increments. However, after approximately 2.5 hours, a hot spot was again observed in the evaporator region. The temperature distribution in the heat pipe just before formation of the hot spot is shown in Fig. 4.42. The stagnation temperature was approximately 560°C. At a position

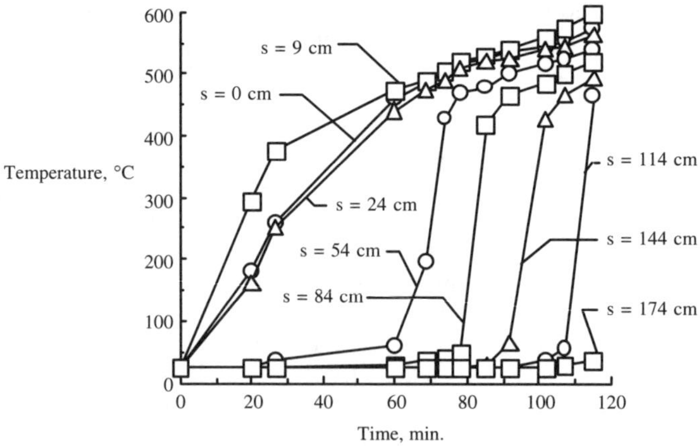

Fig. 4.41 Temperature distribution in the heat pipe as a function of time prior to dryout.

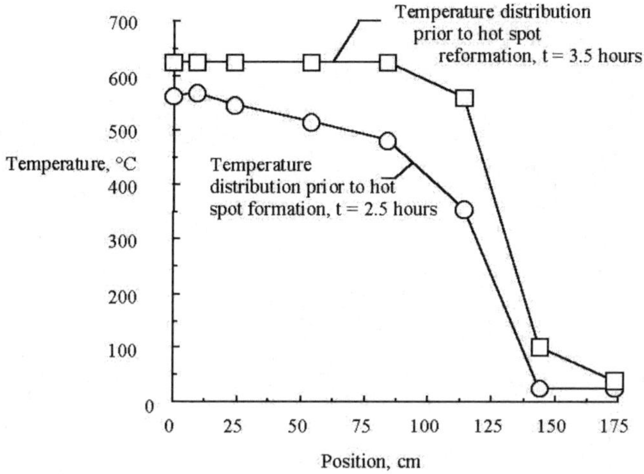

Fig. 4.42 Temperature distribution in the heat pipe just prior to dryout.

of $s = 84$ cm, heat-pipe temperatures were above 475°C. At the time of the hot spot formation, the heat flux was constant, thermocouple No. 6 ($s = 114$ cm) was increasing in temperature, and the first five thermocouples were slowly increasing. Once the hot spot was observed, the heat input was decreased and the wick was rewetted by elevating the condenser end of the heat pipe. Though a slight hot spot remained at the stagnation line, the heat input was slowly increased until the first five thermocouples were nearly isothermal at 625°C. The temperature distribution just before the reformation of the hot spot is also shown in Fig. 4.42. A portion of the insulation was then removed and the hot spot grew in size and temperature. At this time (3.5 hours after startup), power to the induction heaters, and thus heat to the heat pipe, was shut off. The use of insulation on three sides allowed the heat pipe to operate isothermally at 625°C. In an earlier test without insulation on three sides, dryout occurred before isothermal operation, with a maximum stagnation temperature of 600°C. Thus, the insulation on three sides enabled the heat pipe to operate at a higher temperature. It was determined that further testing of the heat pipe would not be productive. Instead, a posttest inspection of the heat pipe was performed.

Posttest permeability measurements revealed that the composite wick, using a 200-mesh screen at the liquid vapor interface for high capillary pressures, did not provide the required pumping pressure. The inability to generate required pumping pressures was probably due to a breach in the 200-mesh screen, shown in Fig. 4.43. With a tear in the 200-mesh screen, the maximum pumping pressures that could be generated resulted from a 50-mesh screen, which was located next to the 200-mesh screen. Estimates of the capillary size that would cause a wicking limit to occur during the induction heating test are consistent with a 50-mesh screen and, thus, tend to verify that a breach in the 200-mesh screen had occurred.

H. Refractory-Composite Heat-Pipe-Cooled Leading Edge

To make heat pipes more mass competitive and structurally efficient, NASA Langley studied embedding individual heat pipes in a refractory-composite structure

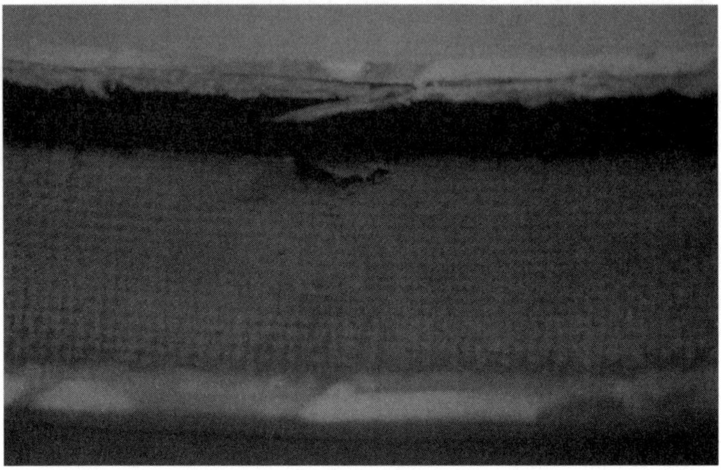

Fig. 4.43 Photograph of the wick in the evaporator region showing breach in the wick.

such as C–C or C–SiC. The maximum operating temperature capability of coated refractory-composite materials for the primary structure of the leading edge is high (~3000°F) relative to refractory metals, which are typically limited to approximately 2400°F. The higher operating temperature increases the radiation heat-rejection efficiency of the heat-pipe-cooled leading edge to permit reductions in the mass of the leading edge for a given leading-edge radius. In addition, the higher operating temperature increases the total heat load that can be accommodated passively by the heat pipe (i.e., no forced convective cooling required). For many space access applications, high operating temperatures help eliminate the need for active cooling and associated mass and operability penalties.

Refractory-metal heat pipes embedded in a refractory-composite structure offer an integrated thermal/structural design solution. As illustrated in Fig. 4.44, heat pipes can be oriented normal to the leading edge and have a "D-shaped" cross section, with the flat part of the "D" forming the wing-leading-edge outer surface.

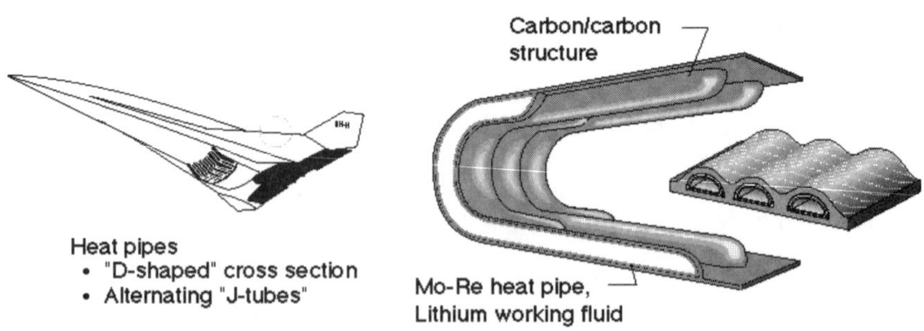

Fig. 4.44 Schematic drawing of a hypersonic vehicle with a diagram of a heat-pipe-cooled wing leading edge.

The flat external surface optimizes heat transfer from the external surface to the heat pipe and also allows for other heat-pipe orientations if desired. The leading edge comprises "J-tube" heat pipes, with a "J-tube" heat pipe being a heat pipe with a long leg on one side of the nose region and a short leg on the other side of the nose region, as shown in Fig. 4.44. An alternating "J-tube" configuration can be utilized to minimize heat-pipe spacing in the nose region where heating is the highest and provide a greater heat-pipe spacing on the upper and lower surfaces where heating is lower, which also serves to reduce mass. The refractory composite structure sustains most of the mechanical structural loads and also offers ablative protection in the event of a heat-pipe failure. Some of the key concepts utilized in the fabrication of the refractory-composite heat-pipe-cooled leading edge, such as a compliant or removable layer to reduce thermal stresses and a slightly convex surface to maintain good thermal contact, have been patented.

A mock-up of the stagnation region of a heat-pipe container embedded in C–C is shown in Fig. 4.45. The photograph should help conceptualize the above leading edge. Only the center tube in the figure is a Mo–Re heat-pipe container. The two adjacent "tubes" are dummy graphite inserts. Graphite and Mo–Re tubes were inserted to a three-dimensional woven carbon preform (T-300 fibers) and the preform was then densified by Carbon–Carbon Advanced Technologies (C–CAT), Fort Worth, Texas. Tooling was used on both the inner and outer surface of the leading-edge component.

NASA Langley's experimental effort to validate the development of a refractory-composite/refractory-metal heat-pipe-cooled leading edge began with the fabrication and testing of a single Mo–Re heat pipe for the validation of the heat-pipe design. After the heat pipe was characterized and performance was verified, three additional heat pipes were fabricated and embedded in C–C. This C–C/heat-pipe subcomponent was tested both at NASA Langley and AFRL. The final test article was a leading-edge-shaped heat pipe. Numerous fabrication issues were resolved in the fabrication of this heat pipe. This test article was tested at Los Alamos.

Fig. 4.45 Photograph of Mo–Re nose section and similarly shaped graphite embedded in C–C.

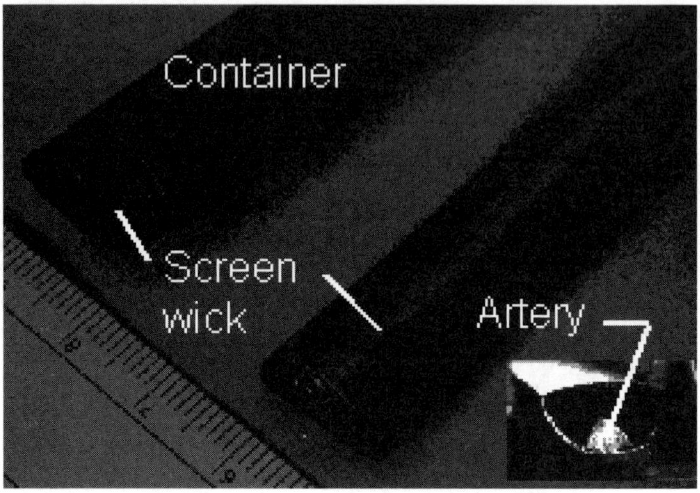

Fig. 4.46 Photograph of a section of "D-tube" and screen wick.

I. Straight Heat Pipes Embedded in Carbon/Carbon

Several straight heat pipes were fabricated and tested in support of a refractory-composite/heat-pipe-cooled wing leading edge that was proposed for the wing leading edge of the NASP. A single 30-in.-long "D-shaped" heat pipe was fabricated and tested at Los Alamos. The heat pipe was fabricated from a Mo-41Re "D-shaped" tube that was drawn from an arc-cast billet. A screen wick was fabricated by wrapping four layers of 400 × 400 mesh Mo-5Re screen around a "D-shaped" mandrel. Figure 4.46 shows a photograph of a section of the "D-tube" and the screen wick. The artery is located in the top center of the curved portion of the heat pipe and has a spring in it to help maintain its shape. Superimposed in Fig. 4.46 is a photograph of the closeout at one end of the wick. One end of the wick is closed out as in the photograph, while the other end is open and is covered by a pool of liquid lithium during operation. Mo-41Re end caps were machined and welded to the ends of the heat pipe.

The heat pipe was started from the frozen state and operated at temperatures up to 2460°F, with an axial heat throughput of 3.1 Btu/s. The maximum heat flux applied, 141 Btu/ft^2·s, was limited by the induction heater capability not by heat-pipe performance. During calibration, the heat pipe was slowly brought to a temperature of 2370°F, and steady-state operation was obtained at various intermediate temperatures, as shown in Fig. 4.47. As can be seen in the figure, temperatures are relatively uniform except at the two ends of the heat pipe. The first thermocouple in the evaporator end was over the endcap and was not cooled by the lithium, thus resulting in slightly higher temperatures. At the condenser end, a lithium pool resulted in cooler thermocouple temperatures. Temperatures were monitored with both an optical pyrometer and thermocouples spot welded on the heat pipe. Knowing the steady-state temperature of the heat pipe, the heat radiated away (the throughput) could be related to the setting on the power supply. Throughputs for steady-state conditions are listed on the figure. However, due to the temperature dependence of the electrical resistance of the Mo–Re, heat input at a given power

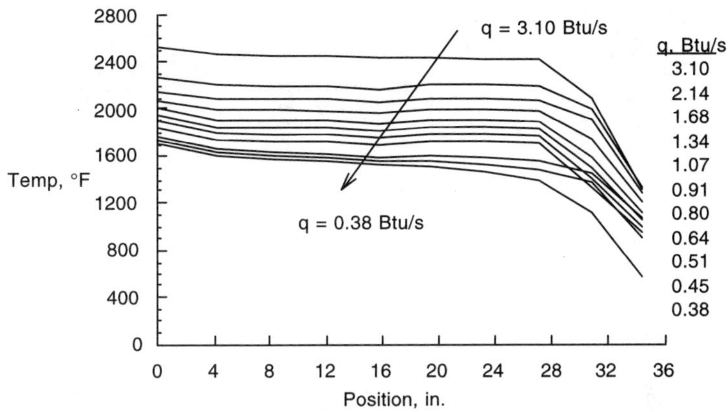

Fig. 4.47 Steady rate temperature distributions for a range of heat fluxes (T = ±1%).

setting of the induction heater varied with heat-pipe temperature. Thus, the heat input determined from the measured throughput is not as accurate during transient conditions as during steady-state operation. In general, this testing of the heat pipe validated the operating characteristics of the design.

After the design of the heat pipe was validated, it was decided to fabricate three additional heat pipes with the same design. During testing of these heat pipes, the performance of each heat pipe varied slightly (hot spots, noncondensable gas, etc.), even though all were fabricated in a similar manner and in parallel. The variability in performance of the heat pipes emphasizes that there is still much art to fabricating heat pipes.

After the performance of the heat pipes was verified, all three were embedded in C–C by C–CAT to form a multiple heat-pipe subcomponent. A carbon preform was woven by Fiberite, with "D-shaped" channels using T-300 fibers. The fabric was then heat treated at 3000°F. The heat treatment was not performed at as high a temperature as is often done because the temperature of the heat-pipe test article was less than 3000°F and higher heat treatment temperatures increase the fiber modulus, making processing (handling) more difficult. For the application here, the lower modulus fibers resulting from the lower temperature heat treatment were preferable. A 0.005-in.-thick layer of Grafoil® was placed between the curved portion of the heat pipe and the carbon fibers before densification. This was done to provide a compliant layer to relieve stresses due to the CTE mismatch between Mo–Re and C–C. A graphite tool was used in the densification process to ensure proper positioning of heat pipes. The C–C/heat-pipe subcomponent is shown in Fig. 4.48.

The C–C/heat-pipe subcomponent was tested at NASA Langley in a vacuum chamber with quartz lamps. The quartz lamps were unable to provide enough heat to bring the heat pipes to fully operational conditions. The condenser section of the C–C/heat-pipe subcomponent was then insulated to bring the test article to temperature. Though isothermal operation was obtained over a significant length of the heat pipes, the throughput of the heat pipes was negligible since the test article was insulated. The test article was operated in both the horizontal and vertical orientations, with only minor differences between the two.

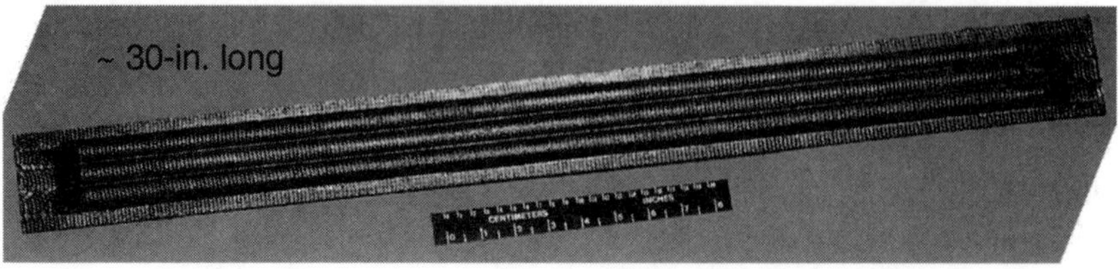

Fig. 4.48 Photograph of C–C/heat-pipe subcomponent.

The C–C/heat-pipe subcomponent was then tested in the Vortek facility at AFRL, where heat fluxes far in excess of design conditions could be obtained. In operation as a leading edge, the outer surface of the leading edge would radiate heat away, while the inner surface would be insulated. Therefore, for the test at AFRL, the flat (outer) surface radiated to a calorimeter, while the inner surface (visible in Fig. 4.49) radiated to a ceramic-board box. Insulation characteristics of the box provided an adiabatic environment. A water-cooled shield was placed between the test article and the Vortek heater to adjust the heated surface area. The components utilized for the test assembly are shown in the assembled condition in Fig. 4.49. Both the shield and calorimeter were painted with a high-temperature, high-emissivity coating. Type K thermocouples were bonded to the C–C with graphite cement and can be seen in the photograph.

Not shown in the photographs are quartz lamps that were used to preheat the heat pipe before initiation of the Vortek power. Quartz lamps heated the heat pipes through a hole in the ceramic box on the side opposite the shield. Because the C–C was uncoated, an inert environment was required. The entire facility was purged with nitrogen and, as added protection, the ceramic box was purged with argon.

Both slow and fast startups of the heat pipes were performed. The slow startup utilized approximately 1.5 hours to reach the maximum heat flux of 500 Btu/ft^2·s, while the fast startup (including quartz lamp preheating) reached the maximum heat flux in approximately 30 minutes. A plot of C–C surface temperatures with time is shown in Fig. 4.50. Eight thermocouples were spaced approximately 4 in. apart down the length of the C–C. The first thermocouple was on the edge of the heated zone and, as a result, recorded higher surface temperatures. After the maximum heat flux was obtained ($\sim$1800 seconds), power settings were held constant until nearly steady-state conditions were obtained. Thermocouples 3 to 6 can be seen to be isothermal, with a C–C surface temperature of approximately 1500°F. The heat flux was later decreased in two steps and it can be seen that the

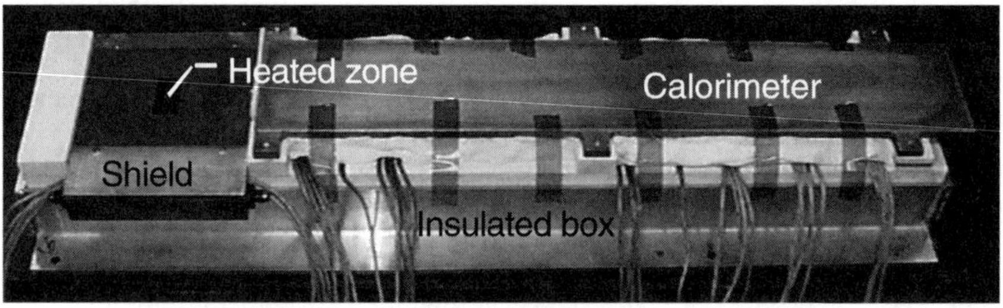

Fig. 4.49 Photograph of the assembled test article.

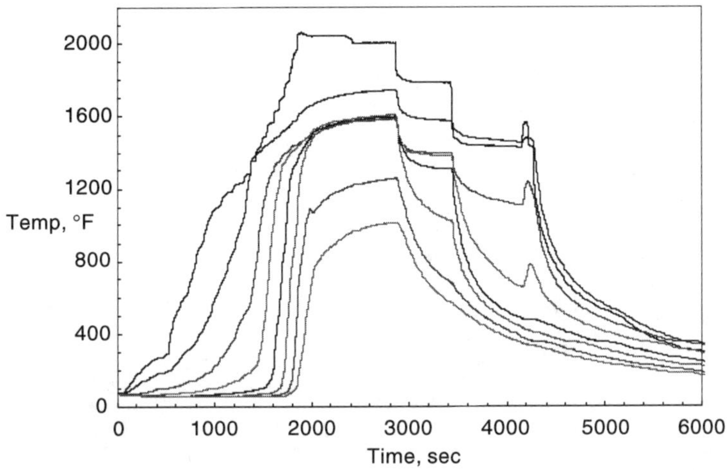

Fig. 4.50 C–C temperatures as a function of time during fast startup of heat pipes.

C–C surface temperature at the opposite end of the test article responded nearly instantaneously, which is a key characteristic of a heat pipe. A short step increase in surface heat flux also resulted in nearly the same instantaneous response.

After the test, the heat pipe was visually inspected. The C–C in the heated region only (1 in. wide) was bowed outward slightly. It is anticipated that the bowing was due to either 1) the decreased strength of the Mo–Re on the flat part of the heat pipes coupled with the increased vapor pressure of the Li temperature in the heat pipe or 2) a thermal expansion mismatch between the Mo–Re and C–C in the heated zone that was relieved by bowing of the Mo–Re. Heat pipes were fabricated with a slight convex shape on the flat surface (see inset in Fig. 4.51) to direct any

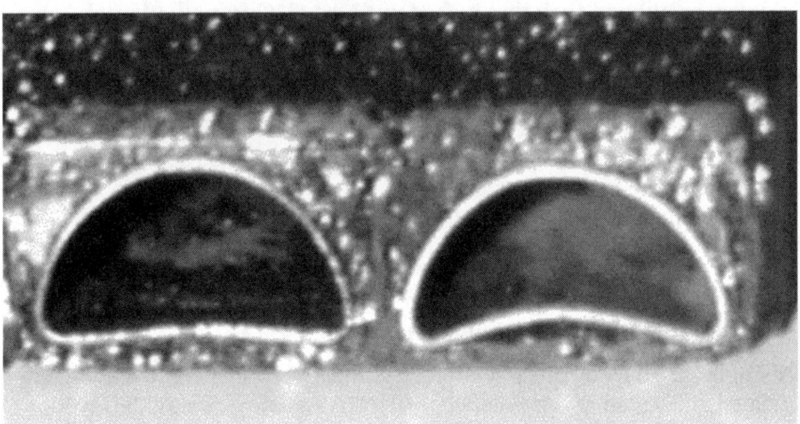

Fig. 4.51 Photograph of Mo–Re "D-shaped" tube embedded in C–C (different C–C design than used for tested subcomponent) that buckled inward to relieve thermal stresses.

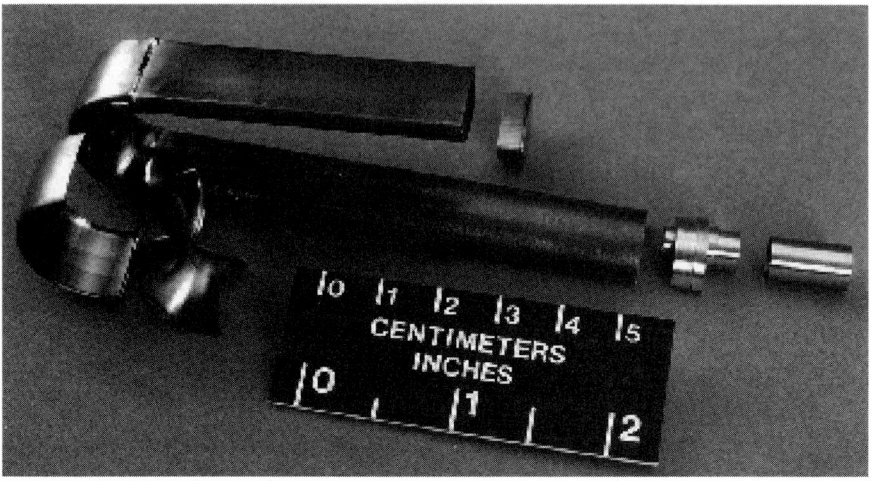

Fig. 4.52 Photograph of the individual machined nose components and the completed nose assembly with the end plugs and fill tube cover.

stress-relief buckling outward rather than inward. Inward buckling would have resulted in significant thermal contact resistance. An example of a "D-shaped" tube that buckled inward to relieve thermal stress is also shown in Fig. 4.51. Thus, if outward bowing is due to stress relief, the slightly convex surface was successful in directing the buckling in the "right" direction.

The goal of the test was a heat flux of 500 Btu/ft^2·s on a 1-in.-wide area of the C–C. This would simulate both the maximum heat flux (500 Btu/ft^2·s) and surface temperatures (~2400°F). As can be seen in Fig. 4.52, surface temperatures did not reach 2400°F. The calorimeter indicated that only one-third of the total anticipated heat was transferred through the heat pipes. It is believed that the water-cooled shield was positioned such that it shadowed the C–C from the Vortek heater. It is also believed that the maximum heat flux levels were obtained, but rather than a uniform heat flux over the 1-in.-wide area, the heat flux on the edges dropped off significantly.

Though the total anticipated heat load was not obtained, the test did demonstrate the successful startup and operation of heat pipes embedded in C–C. The heat pipes started up quickly and performed in a well-behaved manner. The C–C/heat-pipe subcomponent also operated successfully and survived as a hot, thermal structure. Characterization of the heat flux profile and further testing are planned.

J. Mo–Re Leading-Edge Shaped Heat Pipe

A single "D-shaped" cross section heat pipe in the shape of a leading edge was fabricated and tested at Los Alamos for NASA Langley. Several fabrication issues were resolved in the fabrication of the heat-pipe container and the wick. After fabrication, the heat pipe was tested at Los Alamos using induction heating in a vacuum chamber. Premature failure of the heat pipe occurred due to electrical discharge from the induction heating concentrator to the heat pipe.

Because of the small leading-edge radius of 0.5 in., fabrication of the stagnation region of the heat pipe was challenging. The "J-tube" heat pipe had a "D-shaped" cross section and was shaped like a sharp leading edge but had two unequal length condenser legs as described previously. Several techniques were evaluated to fabricate the stagnation region of the heat-pipe leading edge. Bending the "D-shaped" tube to the desired radius was not feasible, nor was bending sections and then welding the sections together. The selected technique was to machine the curved components out of a solid piece of Mo–41Re and weld them together to form the stagnation region. A photograph of the two curved pieces individually and welded together with the tubular condenser legs is shown in Fig. 4.52. The end plugs and a cover to be placed over the fill tube are also shown in the photograph. Thermo Electron Tecomet performed the machining, tube drawing, and welding.

The artery was fabricated by wrapping three layers of 400 × 400 mesh Mo–5Re screen around a 0.125-in. outer diameter steel mandrel, which was overwrapped with a 0.003-in.-diam steel wire spirally wound the length of the artery. The steel mandrel/screen assembly was inserted to a 0.25-in. outer diameter by 0.035-in. wall steel sheath. The steel sheath was drawn down to a 0.210-in. outer diameter. After the final draw, the assembly was immersed in a water/hydrochloric acid solution and the steel mandrel, sheath, and wire were dissolved leaving the formed screen artery.

The screen wick was fabricated by wrapping the first two layers of 400 × 400 mesh screen over a copper mandrel that had been machined to the wick's interior cross-sectional shape. The artery was placed over the first two layers of screen and the final two layers of screen were wrapped over the mandrel and artery and the edge of the final wrap was spot welded the length of the wick. The assembly was then removed from the copper mandrel.

Sections of the wick screen were removed in a portion of the wick as illustrated in Fig. 4.53. This allowed the screen wick to bend and still maintain a continuous artery through the curved section of the heat pipe. The angle of the screen sections that were cut out was 36°, with a 0.305-in. spacing on the outside surface. A photograph of the screen wick on a mandrel with a solid rod used to support the artery is shown in Fig. 4.54.

A photograph of the inside machined curved part of the heat pipe with the screen wick being wrapped around it is shown in Fig. 4.55. In the figure, the region of the wick where sections were cut out to allow its bending can be seen. The straight sections of the wick are also shown. The screen wick in its final configuration is

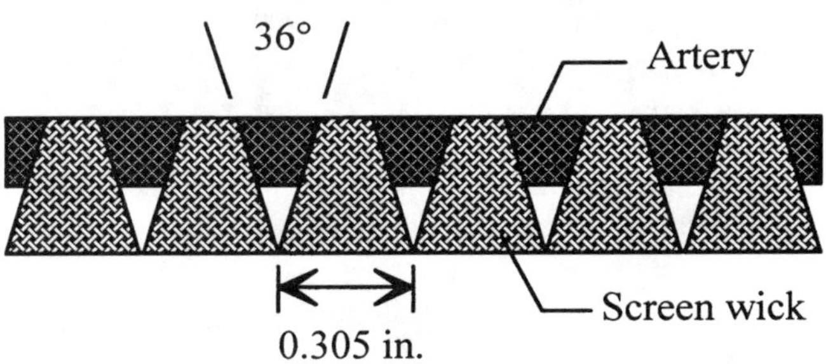

Fig. 4.53 Schematic diagram of the dimensions of the artery prior to bending for fabrication of the nose section.

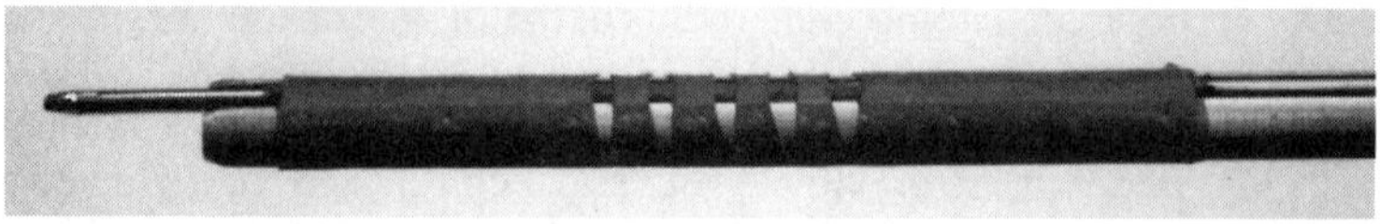

Fig. 4.54 Photograph of the nose portion of the wick on a mandrel.

shown in Fig. 4.56. The artery can be seen extending from the upper portion of the wick in the photograph.

After the heat pipe was assembled, 0.017 lb (7.5 g) of lithium were loaded into the heat pipe by distillation.[20] The heat pipe was wet in at 3170°F for 53 hours in a vacuum furnace. After wet in, the heat pipe was oriented with the short leg of the heat pipe on the top and the long leg on the bottom and placed in a vacuum test chamber. An rf-induction coil/concentrator was used as the heat source. The heating device was designed to heat the entire width of the flat face of the heat pipe for a running length of 1.6 in. in the curved region. Initial tests heated the pipe to 1700°F, with an estimated heat flux of approximately 362 Btu/ft^2·s. A photograph of the heat pipe during initial testing in the vacuum chamber is shown Fig. 4.57. The induction heating concentrator can be seen on the outside of the curved region. In the photograph, the curved region is hotter than the rest of the heat pipe because the photograph was taken during startup.

Four checkout tests were performed on the heat pipe before an accident that resulted in catastrophic failure of the heat pipe. In each case, the heat pipe was started up over the course of several hours. Though design temperatures were not obtained, the heat pipe was operational, where operational is defined as nearly isothermal over a significant length of the heat pipe. In some cases, the heat pipe

Fig. 4.55 Photograph of the screen wick in the nose region prior to bending around the machined nose part.

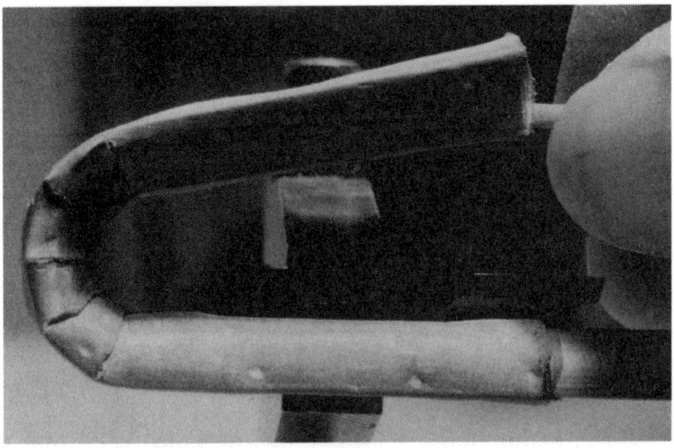

Fig. 4.56 Photograph of the screen wick in the nose region after fabrication.

did reach steady-state conditions. The insulation on the heat pipe had been baked out before testing but continued to outgas as temperatures rose. As a result of the outgassing and subsequent increase in chamber pressure, heat fluxes were kept low.

In the first two tests, the heat pipe was operational over most of the entire length, while in two later tests, it was operational only over a portion of the heat-pipe length. Data in Fig. 4.58 represent the highest temperatures obtained at a given time during the particular test, not steady-state values. The stagnation line is located at $s = 0$ in. Data for test 1 do not include a temperature at $s = 8.1$ in. because the thermocouple was not functioning properly at the time. From the

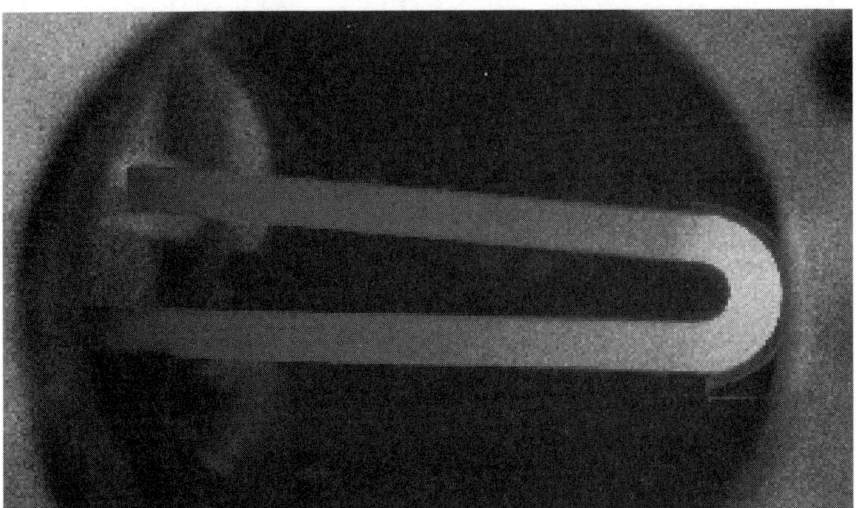

Fig. 4.57 Photograph of the heat pipe during initial testing with induction heating concentrator in the nose region.

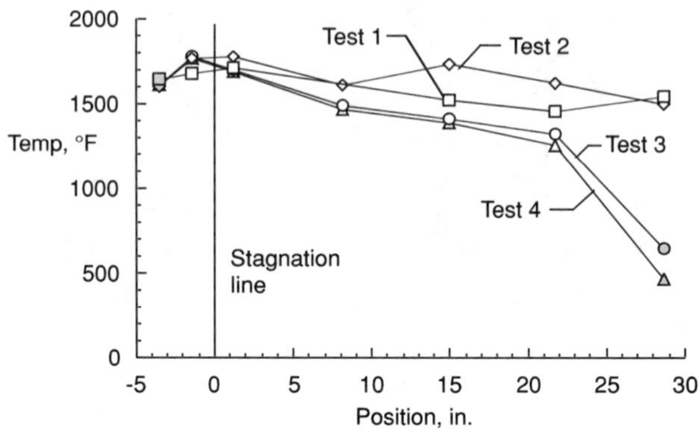

Fig. 4.58 Maximum heat-pipe temperatures during the four tests.

figure, it is apparent that the leading-edge-shaped heat pipe did operate as expected, though design temperatures and heat fluxes were not obtained.

K. Assessment of the State of Heat-Pipe-Cooled Leading Edges

Heat-pipe-cooled leading edges have been evaluated for hypersonic vehicles for three decades but have yet to be utilized on a flight vehicle. The technology has progressed from individual metallic heat pipes to metallic heat pipes embedded in a composite structure. The state-of-the-art refractory-metal/refractory-composite heat-pipe-cooled leading edge builds on prior work to offer an integrated thermal/structural design solution for hypersonic vehicles.

Though there seem to be no "show stoppers" to applying heat pipes to leading edge cooling, there are several areas in which more work would be beneficial. One is in the development of high-temperature barrier coatings on the external surface of the metal heat-pipe container. When a refractory metal heat pipe is embedded in C–C or C–SiC, reaction layers tend to form on the outer surface of the heat-pipe container. The growth of the reaction layer is a function of both time and temperature. A thin coating that would prevent the diffusion of C and/or Si into the heat-pipe container would extend the life of the heat pipe and allow a thinner (lighter weight) heat-pipe container. The fabrication of heat pipes with ultra sharp leading edges, i.e., 0.030-in. radius, is another area requiring further study. Once the ultra-sharp-leading-edge heat pipes are developed, optimally embedding them in a composite material is the next step.

Though there are areas for further development, much of the technology required to utilize heat-pipe-cooled leading edges on hypersonic vehicles is available for application. The weight, cost, and maintenance savings over an actively cooled leading edge would be significant. In any application where passive leading edges will not survive design conditions, heat-pipe-cooled leading edges are attractive design options.

VI. Conclusion

This chapter provides a small glimpse into the breadth of effort in extreme environment structures that has occurred over the years. To fully capture all the effort that has gone into this subject would require many more pages. For instance, the subject of cryogenic tankage, especially composite cryogenic tankage, is not even addressed and airframe actively cooled structure is just mentioned. These subjects alone could easily fill their own books.

Surprising levels of progress have been made in extreme environment structures. However, much remains to be done before this technology area will mature to a sufficient level to enable relatively routine and economical high-speed flight. In conventional flight regimes, the focus has to be on affordable design and manufacturing. One large barrier that has to be overcome is the lack of a national infrastructure to experimentally validate large to full-scale components in extreme environments.

References

Books

[1]Camarda, C. J., "Thermostructural Applications of Heat Pipes for High Speed Aerospace Vehicles," Third International Heat Pipe Symposium, Tsukuba, Japan, Sept. 1988.

[2]Faghri, A., *Heat Pipe Science and Technology*, Taylor and Francis, Washington, DC, 1995.

[3]Woloshun, K., A, Merrigan, M. A., and Best, E. D., "HTPIPE: A Steady-State Heat Pipe Analysis Program," LA-11324-M, Los Alamos National Laboratory, Nov. 1988.

[4]Dunn, P., and Reay, D. A., *Heat Pipes*, Pergamon, Oxford, England, U.K., 1976.

[5]Chi, S. W., *Heat Pipe Theory and Practice*, Hemisphere, New York, 1976.

[6]Silverstein, C. C., *Design and Technology of Heat Pipes for Cooling and Heat Exchange*, Hemisphere, Washington, DC, 1992.

[7]Glass, D. E., and Camarda, C. J., "Preliminary Thermal/Structural Analysis of a Carbon-Carbon/Refractory-Metal Heat-Pipe-Cooled Wing Leading Edge," *Thermal Structures and Materials for High Speed Flight*, edited by E. A. Thornton, Vol. 140, AIAA Progress Series, Washington, DC, 1992, pp. 301–322.

[8]Cao, Y., and Faghri, A., "Analysis of Transient and Steady State Performances of Nose-cap and Wing Leading Edge Heat Pipes," ASME HTD-Vol. 221, 1992, pp. 43–52.

[9]Wojcik, C. C., "Niobium Alloy Heat Pipes for Use in Oxidizing Environments," 8th Symposium on Space Nuclear Power Systems, Albuquerque, NM, January 1991, pp. 326–333.

Journals

[1]Cao, Y., and Faghri, A., "Transient Multidimensional Analysis of Nonconventional Heat Pipes With Uniform and Nonuniform Heat Distributions," *Journal of Heat Transfer*, Vol. 113, 1991, pp. 995–1002.

[2]Cao, Y., and Faghri, A., "Conjugate Modeling of High-Temperature Nosecap and Wing Leading Edge Heat Pipes," *Journal of Heat Transfer*, Vol. 115, 1993, pp. 995–1002.

[3]Glass, D. E., "Closed Form Equations for the Preliminary Design of a Heat-Pipe-Cooled Leading Edge," *Journal of Spacecraft and Rockets*, in press.

[4]Niblock, G. A., Reeder, J. C., and Huneidi, F., "Four Space Shuttle Wing Leading Edge Concepts," *Journal of Spacecraft and Rockets*, Vol. 11, No. 5, 1974, pp. 314–320.

[5]Camarda, C. J., and Masek, R. V., "Design, Analysis and Tests of a Shuttle-Type Heat-Pipe-Cooled Leading Edge," *Journal of Spacecraft and Rockets*, Vol. 18, No. 1, 1981, pp. 71–78.

[6]Glass, D. E., Camarda, C. J., Merrigan, M. A., and Sena, J. T., "Fabrication and Testing of Mo–Re Heat Pipes Embedded in Carbon/Carbon," *Journal of Spacecraft and Rockets*, Vol. 36, No. 1, 1999, pp. 79–86.

[7]Glass, D. E., Merrigan, M. A., Sena, J. T., and Reid, R. S., "Fabrication and Testing of a Leading-Edge-Shaped Heat Pipe," *Journal of Spacecraft and Rockets*, in press.

[8]Modlin, J. M., and Colwell, G. T., "Surface Cooling of Scramjet Engine Inlets Using Heat Pipe, Transpiration, and Film Cooling," *Journal of Thermophysics and Heat Transfer*, Vol. 6, No. 3, 1992, pp. 500–504.

[9]Colwell, G. T., and Modlin, J. M., "Heat Pipe and Surface Mass Transfer Cooling of Hypersonic Vehicle Structures," *Journal of Thermophysics and Heat Transfer*, Vol. 6, No. 3, 1992, pp. 492–499.

Proceedings

[1]Camarda, C. J., and Glass, D. E., "Thermostructural Applications of Heat Pipes for Cooling Leading Edges of High-Speed Aerospace Vehicles," *Proceedings of a Workshop on Current Technology for Thermal Protection Systems*, NASA Langley Research Center, NASA CP-3157, Feb. 1992, pp. 291–318.

[2]Merrigan, M. A., Sena, J. T., and Glass, D. E., "Evaluation of a Sodium/Hastelloy-X Heat Pipe Designed to Cool the Wing Leading Edge of an Advanced Space Transportation System," *Proceedings of the ASME Heat Transfer Conference*, Houston, TX, August 1996, pp. 333–341.

Reports, Thesis, and Papers

[1]Jang, J. H., "An Analysis of Startup From the Frozen State and Transient Performance of Heat Pipes," Thesis, Dep. of Mechanical Engineering, Georgia Inst. of Technology, Feb. 1988.

[2]Colwell, G. T., Jang, J. H., and Camarda, C. J., "Modeling of Startup from the Frozen State," Paper presented at the Sixth International Heat Pipe Conference, Grenoble, France, 25–29 May 1987.

[3]Anon., "Design, Fabrication, Testing, and Delivery of Shuttle Heat Pipe Leading Edge Test Modules," McDonnell Douglas Astronautics Company—East, Rept. MDC E0775, 20 April 1973, Contract NAS8-28656.

[4]Clark, L. T., and Glenn, G. S., "Design Analysis and Testing of Liquid Metal Heat Pipes for Application to Hypersonic Vehicles," AIAA Paper 88-2679, June 1988.

[5]Boman, B. L., and Elias, T., "Tests on a Sodium/Hastelloy X Wing Leading Edge Heat Pipe for Hypersonic Vehicles," AIAA Paper 90-1759, June 1990.

[6]Glass, D. E., Camarda, C. J., Sena, J. T., and Merrigan, M. A., "Fabrication and Testing of Heat Pipes for a Heat-Pipe-Cooled Leading Edge," AIAA 97-3876, August 1997.

[7]Silverstein, C. C., Gottschlich, J. M., and Meininger, M., "The Feasibility of Heat Pipe Turbine Vane Cooling," ASME 94-GT-306, Presented at the International Gas Turbine and Aeroengine Congress and Exposition, The Hague, Netherlands, 13–16 June 1994.

[8]Silverstein, C. C., "Heat Pipe Cooling for Scramjet Engines," NASA CR 4036, 1986.

[9]Silverstein, C. C., "Heat Pipe Cooling for Engine Cowl," Final Rep. Rockwell International Corporation Purchase Order No. R92CAA89251204, Jan. 1989.

[10]Henson, M., Stingley, R., and Haynes, C. C., "188 Heat Exchanger Panels X-30 Nozzle Actively Cooled Structures 1"x 4" Thermal Sub-Element Tests," NASP Contractor Rept. X30LK94013A, Dec. 1994.

[11]Rockoff, H., Bakalyar, A., Nguyen, S., and DeLangis, L., "Final Report for Thermal Tests of the Haynes 188 NASP Engine Heat Exchanger," NASP Contractor Rep. X30NA94-1749, Dec. 94.

[12]Silverstein, C. C., "A Feasibility Study of Heat-Pipe-Cooled Leading Edges for Hypersonic Cruise Aircraft," NASA CR 1857, Nov. 1971.

[13]Alario, J. P., and Prager, R. C., "Space Shuttle Orbiter Heat Pipe Application," NASA CR 128498, April 1972.

[14]Anon., "Study of Structural Active Cooling and Heat Sink Systems for Space Shuttle," NASA CR 123912, June 1972.

[15]Peeples, M. E., Reeder, J. C., and Sontag, K. E., "Thermostructural Applications of Heat Pipes," NASA CR 159096, June 1979.

[16]Clark, L. T., "Design of Wing Leading Edge Thermal Protection Using Refractory Metal Heat Pipes with Lithium Working Fluid for the National Aero-Space Plane," NASP CR 1139, Feb. 1993.

[17]Anon., "Design, Fabrication, Testing, and Delivery of Shuttle Heat Pipe Leading Edge Test Modules," NASA CR 124425, April 1973.

[18]Camarda, C. J., "Analysis and Radiant Heating Tests of a Heat-Pipe-Cooled Leading Edge," NASA TN D-8468, Aug. 1977.

[19]Camarda, C. J., "Aerothermal Tests of a Heat-Pipe-Cooled Leading Edge at Mach 7," NASA TP-1320, Nov. 1978.

[20]Clark, L. T., and Young, G. K., "Development of a Lithium Vapor Chamber Nosecap and a Transpiration Nosecap for the National Aerospace Plane," Boeing Rep. F33657-86-C-0061, Sept.1991.

[21]Boman, B. L., Citrin, E. C., Garner, E. C., and Stone, J. E., "Heat Pipes for Wing Leading Edges of Hypersonic Vehicles," NASA CR 181922, Jan. 1990.

[22]Glass, D. E., Merrigan, M. A., Sena, J. T., and Reid, R. S., "Fabrication and Testing of a Leading-Edge-Shaped Heat Pipe," NASA CR-1998-208720, Oct. 1998.

[23]Wojcik, C. C., and Clark, L. T., "Design, Analysis, and Testing of Refractory Metal Heat Pipes Using Lithium as the Working Fluid," AIAA Paper 91-1400, June 1991.

[24]Kuecker, B., Bejarano, T., Hackett, T., Dubose, W., and Lattimore, K., NASP Materials and Structures Augmentation Program Final Report, Refractory Composites, Volume IIA—Applications of Carbon/Carbon and Ceramic Composites to Hypersonic Structures, NASP Contractor Rep. 1148, March 1993.

Patents

[1]Glass, D. E., Camarda, C. J., and Merrigan, M. A, "Refractory-Composite/Heat-Pipe-Cooled Leading Edge," U.S. Patent No. 5,720,339, Feb. 1998.

[2]Camarda, C. J., and Ransone, P. O., "Reusable High-Temperature Heat Pipes and Heat Pipe Panels," U.S. Patent No. 4,838,346, June 13, 1989.

Chapter 5

Gossamer Structures: Space Membranes, Inflatables, and Other Expandables

A. B. Chmielewski
Jet Propulsion Laboratory, Pasadena, California

and

C. H. Jenkins
South Dakota School of Mines and Technology, Rapid City, South Dakota

I. Technology Background

A. Overview of Gossamer Structures Technology

RECENT technological advancements in space ultralight structures, especially in the areas of material rigidization and controlled deployment, have presented a new possibility to the space community with a low-cost, low-mass alternative to mechanically deployed space structures. This general category of space structures, because of the extensive use of membranes and ultralow mass components, has been referred to as *gossamer structures*, which include inflatables and many other forms of expandables. Gossamer structures have the potential to provide many benefits and advantages over current mechanical systems. They can have mass and be packaged into small volumes. This, in turn, can potentially reduce the overall program cost by reducing launch vehicle size requirements. Reduction in total system mass and deployment complexity can also increase system reliability.

To be clear, we use the term *gossamer structure* as a specific application of a *membrane* structure. We mean by *membrane* structures those structures (load-carrying artifacts or devices) composed of highly flexible (*compliant*) plate or shell-like elements. This usually implies thin, low-modulus materials, such as polymer films. Membrane structures have very little inherent bending stiffness and do not lend themselves well to carrying compressive loads. Thus, they are usually found either in tensioned–planar or inflated–curved configurations. In what follows, we will use the terms *membrane*, *inflatable*, and *gossamer* structures interchangeably.

There is great interest in gossamer structures for near-term space missions and a possibility of enabling several breakthrough missions in the more distant future. Membrane structures are envisioned for such applications as deep space antennas, earth radiometers, radars, concentrators, telescopes, sun shields, solar sails, solar arrays, and spacecraft booms. National Aeronautics and Space Administration

(NASA) and Department of Defense (DOD) sponsor technology development to advance the inflatable technology and make it available to 21st century space missions. The most important aspects of technology development are deployment techniques, membrane and rigidizable materials, and analytical tool development.

The analytical simulation of mechanical performance of membrane space structures is absolutely essential to support the development of technology and projection of performance for this new class of space structures. A number of elements of this technology must be understood analytically to develop applications. In addition to the modeling needed to characterize the structural dynamics of on-orbit configurations, highly specialized approaches are needed to simulate other specific mechanical events and structural relationships. Such specialized capability should include simulation of 1) specific control deployment techniques for structural members, 2) generic spatially organized and controlled deployment processes, 3) identification and application of scaling laws to enable the use of subscale test results to project full-scale mechanical performance, and 4) deflection characteristics of thin membranes under tension and loaded by differential pressure, among many others.

There are currently a number of different techniques for control of deployment of inflatable structural elements. The most promising techniques currently under development have been simulated and will be verified experimentally with scaled hardware. New and unique techniques for spatially controlling deployment through the management of a number of discrete masses have been developed. Potential benefits of this process have been identified. Scaling laws for this new class of space structures have been developed. They have been applied to test results of scaled hardware to show the relationship with full-scale structures. Basic equations have been developed to determine the deflection characteristics of a tensioned membrane under differential pressure loading. This characterization will be modified to include segmented membranes and stiffness effects of their seams. This will enable the development of an error budget for precision inflatable reflector structures.

And what is the future of membrane structures in the new millennium? It can be summarized with just two words—*gossamer spacecraft*. "Technology to find it and then shake its hand" is a long-term vision of the Gossamer Spacecraft Initiative. This somewhat humorous sounding motto appears even more outrageous when translated into scientific terms. NASA will develop technology to enable imaging of extrasolar planets to detect the ones that support life. Once we find such planets, we will not be satisfied with "just" creating somewhat blurred pictures of these worlds or finding spectroscopic "smoking guns" of life's existence—we will want to travel there and take a closer look at our cosmic friends. We do not know how long this search will take; we do not know how successful we will be. One thing we do know is that, to make this vision more than just science fiction, we will need to develop radically different observatories and spacecraft.

A gossamer telescope, which will produce the first detailed image of an extrasolar planet, will bear little resemblance to current space observatories. The gossamer telescope will be about 150 times lower mass per square meter of collecting area than the Hubble Space Telescope and its diameter will be 10 to 20 times greater for each single telescope. When such a telescope, or an interferometer consisting of an array of such telescopes, will finally produce an image of "the other world," there will be strong public summon to take an even closer look by means of interstellar travel.

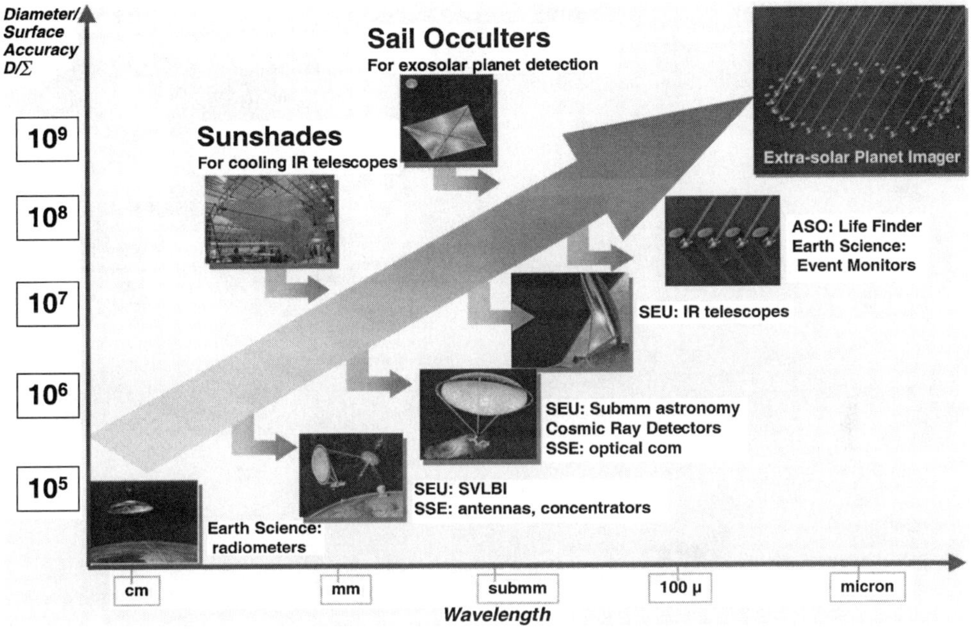

Fig. 5.1 Gossamer aperture technology development.

In this day and age we envision three major ways that offer the possibility of interstellar travel. They are matter–antimatter propulsion, nuclear propulsion, and solar sailing. The propulsion method that recently has been getting the most attention is solar sailing. The first generation interstellar sail missions are envisioned to use gravitational boost from the sun in addition to solar pressure. In the more distant future, lasers will propel the sails. Gossamer structures will eventually make interstellar sailing possible.

The "vision missions" for gossamer space structures are the interstellar probe and the extrasolar planet imager. Both of these missions can be considered the pinnacles of the technology development roadmap. Before the technology is ready to enable either of these vision missions, there will be a series of technology development products benefitting several types of space endeavors as shown on Figs. 5.1 and 5.2. Figure 5.1 shows a number of candidate missions that will be enabled on the way to developing 20 to 40-m-diam class optical apertures needed for a planet imager. Likewise, Fig. 5.2 shows what solar sail missions will benefit from progress in gossamer technology development.

In general, gossamer spacecraft technology will enable very large, ultralight weight systems for bold missions of discovery such as

1) Very large telescopes for imaging extrasolar planets, studying formation of large-scale structure in the early universe, and continuously monitoring the Earth from distant vantage points.

2) Large deployable and inflatable antennas for space-based radio astronomy, high-bandwidth communications from deep space, and Earth remote sensing with radar and radiometers.

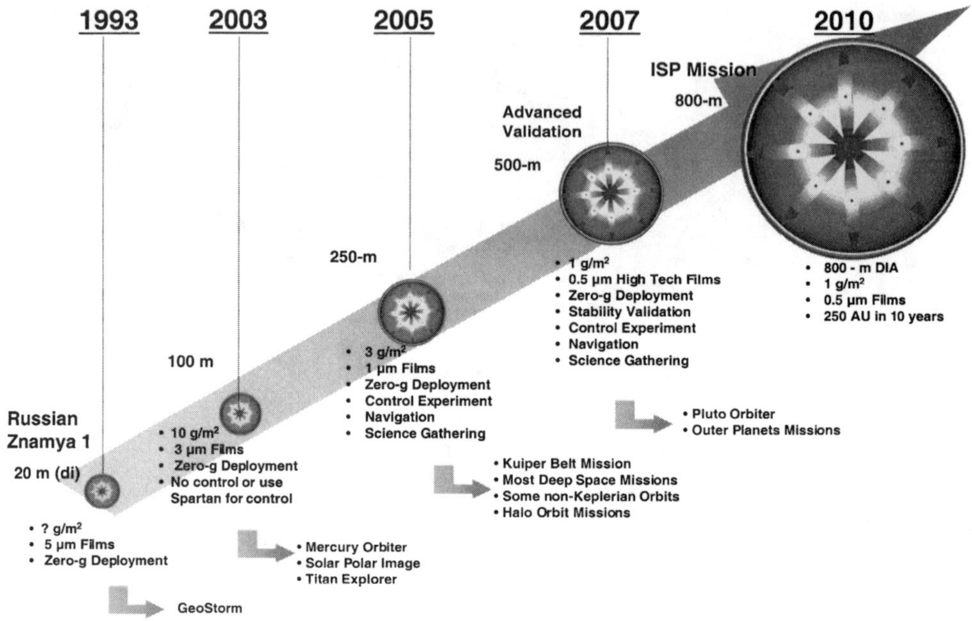

Fig. 5.2 Gossamer sails technology development.

3) Solar sails for low-cost propulsion, station keeping in unstable orbits, and precursor interstellar exploration missions.

4) Large solar power collection and transmission systems for human and robotic exploration missions and the commercial development of space.

The overarching goal of gossamer spacecraft technology development is to achieve breakthroughs in mission capability and cost, primarily through revolutionary advances in structures, materials, optics, and adaptive and multifunctional systems all described briefly in the following sections. For a more complete discussion of these topics, the reader should refer to the *Gossamer Space Structures* handbook in this same AIAA series.

B. History of Gossamer Structures

There has been interest in inflatable deployable space structures since the 1950s due to their potential for low-cost flight hardware, exceptionally high mechanical packaging efficiency, deployment reliability, and low mass.

A number of significant technology developments, focused on the demonstration of such potential, include the Goodyear antennas in the early 1960s; the Echo Balloon series from the late 1950s to the early 1960s; the Contraves antennas and sun shades in the late 1970s to the mid-1980s; the L'Garde, Inc., inflatable decoys in the 1970s and mid-1980s; and their space-shuttle-launched Inflatable Antenna Experiment (IAE) in May 1996. The IAE was the most recent and most significant demonstration of this technology.

NASA's interest in demonstrating the potential of this relatively new class of space structure resulted in their sponsoring the IN-STEP IAE, which flew on

Fig. 5.3 IAE during space flight.

STS-76 on 29 May 1996 (Fig. 5.3). The antenna concept used was developed by L'Garde, Inc., who has been designing, manufacturing, and ground and flight testing inflatable space structures for the past 25 years. The experiment objectives were to (1) verify that large inflatable space structures can be built at low cost, (2) demonstrate that large inflatable space structures have high mechanical packaging efficiency, (3) demonstrate that this new class of space structure has high deployment reliability, (4) demonstrate that large membrane reflectors can be manufactured with surface precision of a few millimeters, and (5) measure the reflector surface precision on orbit.

The IAE inflatable structure was composed of two basic elements, the inflatable reflector assembly and the torus/strut supporting structure (Fig. 5.3). The reflector assembly formed a 14-m offaxis parabolic aperture with a f/d of 0.5. The surface accuracy goal was 1.0 mm rms compared to a best fit parabola. The reflector film, 13-μm aluminized Mylar, was stressed to approximately 8.3 μPa by the inflation pressure of 2.1 Pa. This stress level was sufficiently high to ensure a good reflective surface for the accuracy measurement system. The canopy was constructed from 62 gores of 13-μm Mylar but was left transparent. The torus/strut structures are 0.61 and 0.46 m in diameter, respectively, and are made with 0.30 mm thick, neoprene-coated Kevlar. The function of the struts was to locate the reflector assembly at the effective center of curvature of the reflector parabola as required for operation of the surface accuracy measurement subsystem. The torus also provides the rim support for the reflector assembly without which the reflector assembly would take a spherical shape.

The experiment was flown on the recoverable Spartan spacecraft and demonstrated the main objectives. New, unique, and low-cost space structures technology was demonstrated on orbit by (1) building a large inflatable space antenna structure on the order of about $1,000,000, (2) demonstrating extremely efficient mechanical packaging by stowing a 14 by 28 m inflatable structure in a container the size of an office desk, (3) manufacturing an offset membrane reflector structure with a surface precision on the order of a few millimeters RMS, and (4) demonstrating the robustness of deployment for this new class of structure. The results of this experiment were used specifically to establish the technology database and were the basis of a technology roadmap for the continued development of this type of space structure. The results of this experiement have been thoroughly documented by Freeland and co-workers.

II. Applications

A. Solar Arrays

Future commercial and deep space satellites will require solar arrays with higher specific power densities than the current state of the art, which is of the order of ~40 W/kg, and lower costs to meet mission objectives. Planned communication, radar, DOD, and exploration satellites anticipate power requirements upwards of 30 kW. To meet these needs, revolutionary changes in solar array technology are required.

The advantages of using inflatable systems technology in designing a large solar array are reduced stowage volume and mass, increased specific power (greater than 100 W/kg), and reduced cost over current mechanically deployed solar arrays. The inflatable solar array is particularly attractive for missions that demand high power output with launch vehicle size restrictions.

L'Garde, Inc., [developed for DARPA and Phillips Laboratory (now Air Force Research Laboratory)] an inflatable solar array in 1993. The array supplied 274 W and weighed 2.94 kg (flight model). The array was 3.64 m long and 1.1 m wide. The support tubes were made from stressed aluminum laminate. Figure 5.4 shows the first controlled inflation of this large space system, deployed only by inflation in the supporting tubes. This system rigidized after inflation so that gas pressure need not be maintained. The array was successfully vibration tested at 12 *g* in all three axes and underwent five thermal cycles under this test program. Complete current/voltage (I–V) tests were conducted before and after the deployment in vacuum at −90°C.

Another good example of an inflatable solar array is one designed by a team consisting of ILC Dover, Inc.; L'Garde, Inc.; and AEC-ABLE under the auspices of the ST4 project in 1999 (Fig. 5.5). The configuration of the ST4 solar array was a modular split blanket style with the deployment tube located on the array centerline. When stowed, solar array modules were accordion-folded. Sharp's rigid, high-efficiency silicon photovoltaic assemblies were the basic building blocks for the modular blanket solar array. (Flexible thin film solar cells using amorphous silicon [Si], copper indium gallium diselenide (GAS), or other materials hold great promise to provide even lower cost and lighter weight photovoltaic modules in similar arrays in the future.) Structural components included the inflatable beam, stowage panels, plume-offset panels, launch ties, and launch tie release mechanism. The inflatable beam could have been fabricated from any of the rigidization systems. However, the thermal heating method was utilized for the flight

Fig. 5.4 ISAT array technology deployed in a vacuum tank at the Air Force Research Laboratory (L'Garde, Inc.).

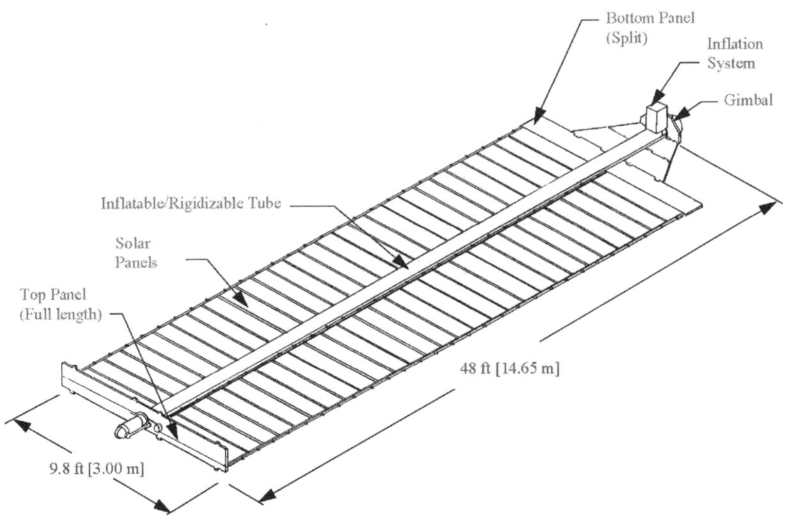

Fig. 5.5 Champollion (ST4) inflatable solar array (ILC Dover, L'Garde, AEC Able).

Table 5.1 Solar array performance projections (SP = specific power, HE = high efficiency, and TJ = triple junction)

	ST4, 5.6 kW	10-kW LEO		15-kW LEO		20-kW LEO		10-kW GEO		15-kW GEO		20-kW GEO	
Cell type	HE Si	HE Si	TJ GAS	HE Si	TJ GAS	HE Si	TJ GAS	HE Si	TJ GAS	HE Si	TJ GAS	HE Si	TJ GAS
Area, m^2	36	43.5	26.5	65.2	39.7	87.0	52.9	43.5	26.5	65.2	39.7	87.0	52.9
Width, m	3	6.0	6.0	6.0	6.0	6.0	6.0	6.0	6.0	6.0	6.0	6.0	6.0
Height, m	15	9.1	5.5	13.6	8.3	18.1	11.0	9.1	5.5	13.6	8.3	18.1	11.0
SP, W/kg	102	140	149	152	160	151	165	154	159	167	171	167	176

experiment design. The purpose of the inflatable beam was to provide a deployment mechanism and support structure for the solar array. It is similar in function to the mechanical deployment truss masts currently utilized on spacecraft. The beam was located on center of the split blanket and was stowed by rolling the tube.

The ST4 solar array design was reviewed for use in NASA's space solar power (SSP) program. The SSP program requires array sizes on the order of 18 kW and a specific power of 250 W/kg. The ST4 solar arrays were designed to be 5.6 kW, with a specific power of 102 W/kg. Table 5.1 shows a projection of the ST4 array design at larger power capacities.

The Teledesic program envisions a constellation of 288 satellites in low earth orbit to provide high data rate communication from anywhere on earth. Original concepts for the satellite included a pair of 6-kW solar array (see Fig. 5.6), with a cost target of $100/W in production. Inflatable technology was considered to be a leading candidate in meeting this goal and was, therefore, investigated by the Teledesic team.

ILC Dover, under contract to Boeing, designed an inflatable structure to support a 3-m by 10-m rectangular satellite solar array. A full-scale prototype demonstration unit was also fabricated and used in deployment trials (see Fig. 5.7).

NASA is developing several advanced rover vehicles for exploration of the Martian surface. Several concepts call for large deployable solar arrays to meet the power needs of the rover. One such concept (see Fig. 5.8) utilizes 1.5-m deployable wheels and an inflatable solar array to cover vast surface areas in rapid times.

The Mars rover solar array prototype (Fig. 5.8), is a working, full-scale inflatable solar array for rover application. This prototype is a system to be packaged in a small volume for launch and deployed in situ to collect solar energy. The array is parasol shaped and consists of four main components: 1) the canopy, which is the membrane that carries the solar modules; 2) the inflatable torus; 3) the inflatable column; and 4) 16 solar modules. The sixteen Kapton (polyimide) gores (the individual segments of the canopy) of the solar array are tensioned by a

Fig. 5.6 Teledesic satellite.

Fig. 5.7 Teledesic inflatable solar array (ILC Dover).

8.0-cm-diameter (tube) by 150-cm-diameter (major), 16-sided inflatable torus and supported by a 10.0-cm-diameter inflatable column.

Because of the requirement for multiple deployments, the torus of the prototype is constructed from a lightweight Mylar (aluminized polyester) assembled by pressure-sensitive adhesive tape of the same material. The prototype inflatable column is fabricated from a laminate of Kapton and aluminized Mylar. The inflatable torus and column of the future flight unit will be constructed from thin-walled aluminum/Kapton laminates or UV-cured laminates.

The Mars solar array prototype is capable of generating approximately 20 W in a terrestrial environment or 12 W in a martian environment. Further optimization in cell population pattern, cell efficiency, and mass will further increase specific power to weight ratio.

Fig. 5.8 Inflatable Mars rover solar array prototype (ILC Dover).

Fig. 5.9 Power sphere (ILC Dover/Aerospace Corp.).

ILC Dover, together with the Aerospace Corporation, developed concepts for a power sphere solar array, which would enable the use of micro- and nanosatellites. The design of the power sphere allows for constant power generation independent of the sphere's orientation. The sphere could be packaged in a very small stowage volume and would deploy and inflate using sublimating powder or compressed gas. The power sphere would generate approximately 5 W with a system mass of 600 g. ILC Dover fabricated a proof of concept unit for demonstration purposes.

B. Communication Systems

NASA and DOD have been studying innovative concepts for large space antennas for several years. One concept that has emerged and was demonstrated on the IAE flight experiment is a segmented inflatable reflector based on the premise of seaming together flat segmented gores of membrane material to form a highly accurate doubly curved surface (see Fig. 5.10). Inflatable space reflectors show tremendous promise for advancing all antenna parameters simultaneously. Their low mass, high deployment reliability, and small packaging volume reduce the mass, complexity, and cost of a host of missions and spacecraft. Studies estimate that, for very large antennas, inflatables are approximately 10 times less expensive, two to five times lighter, and have 15 to 100 times smaller stowage volume than their mechanical counterparts. The low fabrication costs, as well as reduced launch vehicle size, could result in order-of-magnitude cost savings for missions requiring large apertures.

Large antennas and similar structures have a multitude of applications. Large antennas for space radio telescopes have been extensively described by Chmielewski and co-workers for the advanced radiointerferometry between space and earth (ARISE) project (see Fig. 5.11). Earth-observing antennas, such as for soil moisture and ocean salinity measurements, are primary customers for inflatable antenna technology in the near term. These new antennas will be characterized by sizes

Fig. 5.10 Lenticular antennas (3 and 7 m) (L'Garde).

exceeding 25 m in diameter, fraction of a kilogram per square meter density, and operating frequencies between 1 and 100 GHz. NASA's structure and evolution of the universe (SEU) theme needs large apertures for space radio frequency, microwave, and submillimeter telescopes. The SEU is also considering a mission to image luminescence created by extremely high energy particles entering the Earth's atmosphere. Deep space applications will be developed for solar system exploration (SSE) and sun–earth connection (SEC), and high bandwidth communication systems for human exploration development of space.

A large calibration sphere was built by L'Garde and launched by the U.S. Air Force (USAF) in January 2000. The sphere provides concurrent RF and optical highly specular target for space testing. The sphere measures 4 m in diameter and mass of 5.7 kg. The sphere was made out of thin (30 μm) five-layer rigidizable laminate consisting of Kapton, aluminum, and optical coatings.

C. Human Habitats

Since the earliest of times, humans have desired shelter that was rapidly deployable. One of the earliest forms of constructed shelter was undoubtedly the tent, formed, for example, from animal hides draped over branches. The tent made a significant impact on culture because it now allowed people to be mobile, to leave their caves or other fixed shelters and migrate, following food and water. The tent exists throughout history, for example, playing an essential role in military

Fig. 5.11 Artist's concept of the ARISE observatory imaging the environments of black holes (JPL).

activities, even today. Modern architectural *tension fabric structures* are highly sophisticated versions of the ancient tent that are used to span large open spaces. The first *air-supported* roofs sprang up after the World War II, when dozens of radomes were constructed to cover radar installations during the intensification of the cold war. Since that time, air-supported roofs have found favor in applications where large clear spans are required, notably recreational facilities such as football stadiums and tennis courts.

Terrestrial aerospace applications of membrane/inflatable human habitats are, for the most part, associated with providing shelter for aerospace support personnel and equipment. Again, tents play a leading role. Recently, hybrid membrane/inflatables have been developed to provide large, clear span space for aero space vehicle maintenance, particularly for helicopter repair in the field. Curved beams of cylindrical cross section are fabricated from coated fabric and inflated on site. The beams span the space to be covered, supporting the membrane skin of the shelter.

A number of intriguing uses of membrane/inflatables are planned for human habitats in space. Nearest to earth, inflatables could provide efficient shelter for lunar habitation. Efficient for packaging and lightweight for launch, modules could

Fig. 5.12 Prototype of optical calibration sphere during the deployment test (L'Garde).

be inflated on the lunar surface then covered with regolith to provide radiation protection for occupants. Inflatable modules have been proposed for space travel as well, for example, as modules on planetary rovers or for the international space station (see Fig. 5.13).

D. Planetary Surface Exploration

All present planetary rovers are based on the standard Pathfinder Sojourner concept of a four-or six-wheeled vehicle that must generally travel *around* most obstacles. To give a specific example, during a month of operation after its July 1997 landing on Mars, the 35-cm tall Sojourner rover only traveled a total of about 100 m, and never more than 7 m/d. Other developmental rovers are somewhat larger but still are not intended to travel more than about 1 km from the landing point, and that would be over a period of weeks. Additionally, the 1-kg Nanorover presently in development, although very capable, would have a substantially shorter range.

Inflatable technology for surface exploration opens new parameter space. Inflatable rovers would use novel, large inflatable wheels to climb *over* rocks, instead of traveling *around* them, thus enabling the rover to traverse quickly over 99% of the martian surface. For example, an inflatable rover would travel tens of kilometers in a single day, with a range of hundreds of kilometers or more. The use of inflatables to increase speed and range is a critical enabling technology that will allow for robotic outpost development. Functions of inflatable rovers may include

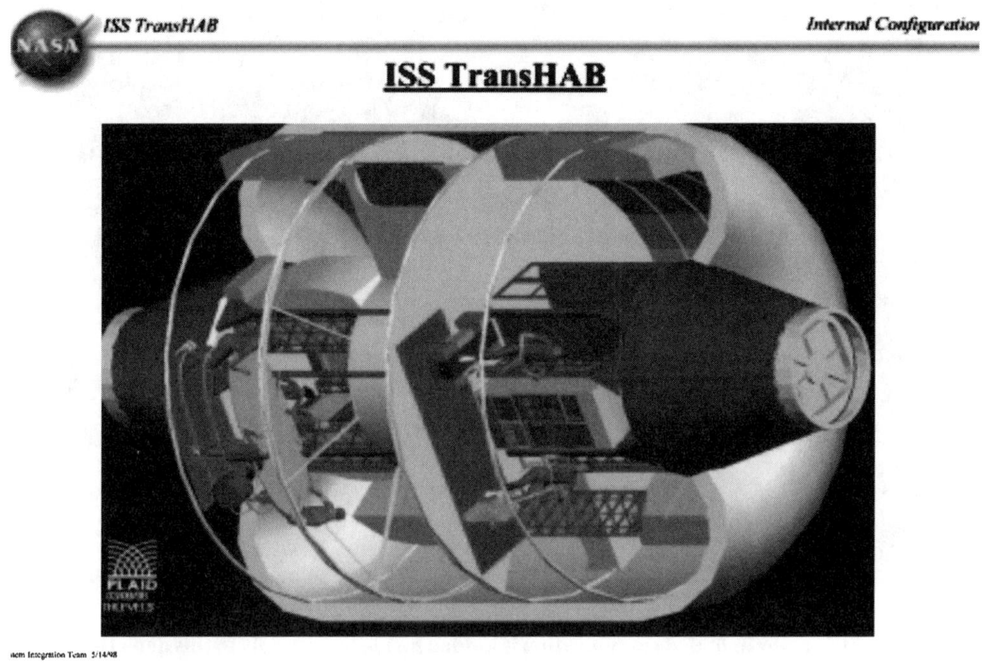

Fig. 5.13 Transhab module (NASA Johnson Space Center).

transporting other rovers to distant sites, transporting astronauts, and the long-distance transfer of heavy equipment or in situ resources, such as water ice from the Martian north pole.

The large, inflated spherical wheels allow the rover to basically “swallow” most smaller rocks, while providing a large contact surface to climb up steep hills or over larger rocks, and maintain excellent ground contact during wind storms. In fact, the present, simple 20-kg prototype, which uses commercial nylon balloons as tires (Fig. 5.14), has stood up well to 13-m/s (30-mph) gusts, which is equivalent in force to 130-m/s gusts in the extremely thin martian atmosphere (0.006-bar pressure). It requires very low power (18 W) to travel 2000 m/h and can climb steep hills.

Preliminary tests using commercial nylon balloons as tires, a rigid metal chassis, and a simple joy-stick control, have shown great promise. Rugged, lightweight inflatable wheels and an inflatable–extendable chassis provide the ability to pack the rover into a small planetary entry capsule. An autonomous control system allows the rover to follow an astronaut or uniquely climb over most obstacles to designated sites in a Mars-like terrain, while carrying smaller, conventional rovers in cooperative rover studies.

A recent, preliminary design of a Mars astronaut transporter system is shown in Fig. 5.15, which evolved from previous Lunar Apollo astronaut rovers. This design will likely be modified to include larger, more rugged inflatable wheels and may include a removable habitat. Control systems will also allow equipment rovers to follow walking astronauts or develop detached trains of equipment that follow in line.

Fig. 5.14 Inflatable rover and the artist's rendering of the rover deployment using a balloon (the rover in folded configuration is hidden inside the gondola in this depiction) (JPL).

Inflatable technology is also critical and enabling to explore Jupiter's moon Titan. An inflatable rover is the only one thus far capable of operating on the liquid methane lakes anticipated on Titan. Numerous materials, such as Kevlar, Mylar, and PBO, are known to be capable of functioning well in cryogenic conditions and the large tires can help insulate critical electronics from the frigid Titan surface. This insulating factor can also help prevent the rover from sublimating/melting into icy surfaces, such as on Europa, Callisto, and other moons, including the known polar ice crater on our own Earth's moon.

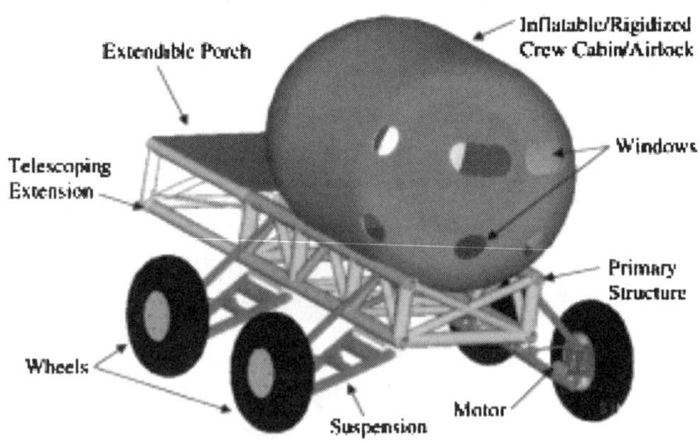

Fig. 5.15 Inflatable Mars rover trailer (JPL).

Fig. 5.16 Mars Pathfinder impact attenuation system with a size-representative lander, rendering of the landing event, and the view of deployed bags under the Pathfinder spacecraft and Sojourner rover on Mars (ILC Dover).

A non-NASA commercial area that is presently being pursued by Federal Fabrics and Fibers is the use of an inflatable rover for land mine traversing and detecting. With very low mass and a large contact area, the pressure on the ground is below the threshold for mine detonation. Outfitted with land mine detectors, this rover could potentially help rid the world of the 100 million active, buried land mines.

One of the most successful inflatable structures ever flown in space is the Mars Pathfinder impact attenuation system. This system of landing airbags enabled the Mars Pathfinder lander and rover to touch down softly on the surface of Mars in 1997.

In the more distant future, it is envisioned that there will be more low-mass, highly-packageable equipment operating on surfaces of planets. Such a vision is portrayed in Fig. 5.17, which shows inflatable rigidizable towers for cellular communications, inflatable solar power energy generation stations, in situ resource utilization tanks, rovers, and thin film solar sails delivering cargo to the planet.

E. Radar and Reflectarrays

There is a critical need for enabling technologies that will reduce the mass, physical size, and cost of antennas. The inflatable synthetic aperture radar (SAR) and reflectarray described in this section represent advanced antenna technology that has the potential for providing an enhanced communication capability with relatively lower cost (1/10), smaller size (1/100 launch volume), and smaller mass (1/2) than current technologies.

A program is currently under development to demonstrate the feasibility of a low-mass inflatable SAR. Figure 5.18 shows the prototype system, including a flat, multiple-layered, metalized, thin film microstrip array (a three-membrane assembly), tensioned and supported by an inflated frame that can be rigidized after deployment. The size of the full-scale array is a 10 m $\times$ 3.3 m. The system was stowed in a rolled up configuration on the spacecraft bus and was deployed

Fig. 5.17 Artist's rendering of inflatable equipment on Mars (JPL).

in a controlled manner via inflation gas. The objective of this program was to develop a functional subscale system that was less than 2 kg/m^2 of radiating area when projected to full scale. The depicted subscale prototype met the requirements and demonstrated the feasibility of the low-mass SAR concept. The deployment method and supporting frame concept used in the SAR could also be used for a flexible blanket-type solar array in future developments. That is, the membrane assembly of the SAR could be replaced with a flexible thin film solar cell blanket and the system would then act as an inflatable solar array.

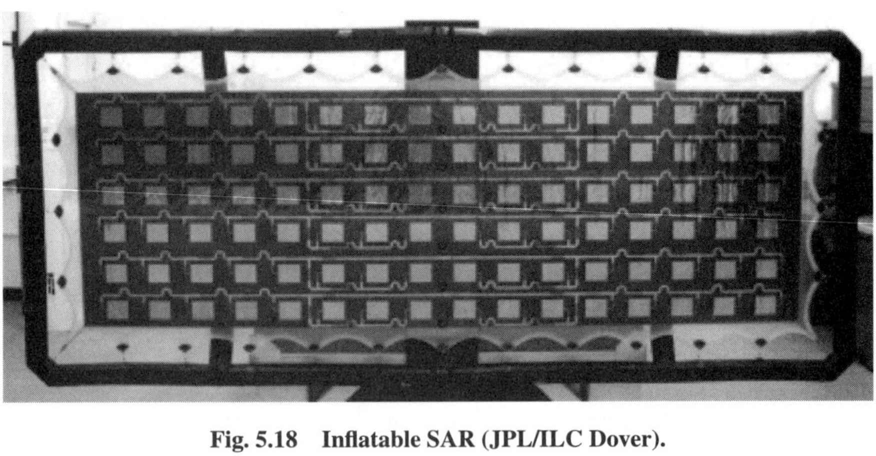

Fig. 5.18 Inflatable SAR (JPL/ILC Dover).

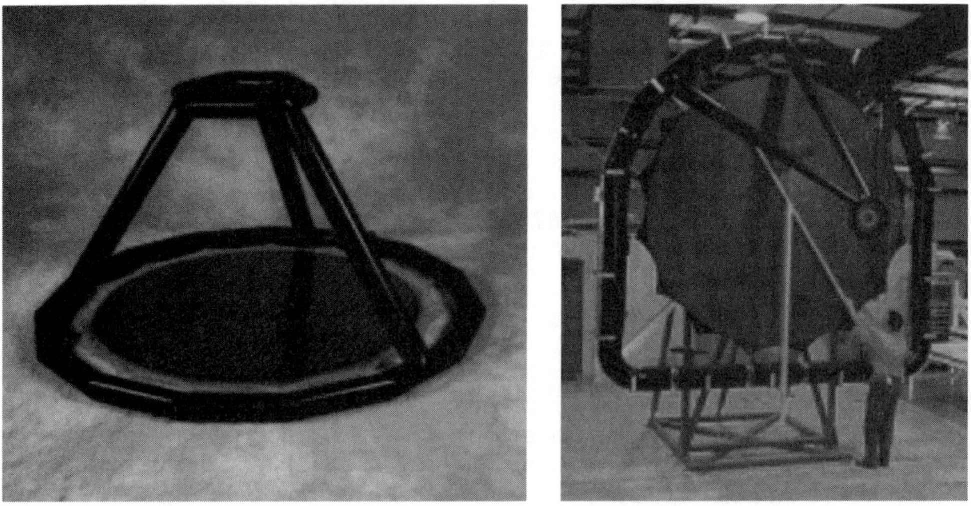

Fig. 5.19 Inflatable microstrip reflectarrays (1 and 3 m) (JPL/ILC Dover).

The 3-m Ka-band reflectarray program is another recent development effort to demonstrate the feasibility of super-low-mass telecommunication and space-borne SAR concepts. This 3-m prototype (see Fig. 5.19) consists of four major subassemblies: 1) the membrane assembly, 2) the rigid frame assembly, 3) the inflatable frame assembly, and 4) the suspension system.

The reflectarray membrane is suspended by a horseshoe-shaped structure assembled from a straight rigid frame assembly and a U-shaped inflatable frame assembly. The rigid frame assembly is made from graphite epoxy with aluminum end caps. The inflatable frame assembly is fabricated from urethane-coated Kevlar to simulate a rigidizable material in the final application. The membrane support frame is 25 cm in diameter and the feed horn support torus is 7.5 cm in diameter. The feed horn torus is supported by three vertical struts grouped within 90° of each other. The struts are tapered in diameter, from the larger diameter of the membrane support tube to the smaller diameter of the feed horn torus, to minimize material in the RF path as well as in the system mass.

One of the critical requirements for reflectarrays is the flatness of the membrane. For this reason, many packaging and deployment concepts are considered, with the current configuration being the best for this particular application. Again, similar to the rectangular frame used for the SAR array, the horseshoe frame concept could also be used for solar array deployment.

The main beam of the microstrip reflectarray, unlike that of a parabolic reflector, can be designed to point at a large fixed angle (up to 60°) from the broadside direction. With the reflectarray integrated with low-loss phase shifters, a very large spaceborne beam scanning array antenna is now feasible. By developing this novel phased reflectarray concept, the complicated beam forming network, high-cost phase shifters, and high-cost transmit/receive modules of a conventional phased array are not longer needed. A reflectarray antenna can even be combined with a solar array. Such system would allow launching very large high-power, large-aperture spacecraft on small launch vehicles.

The reliability of such an array is very high. Because each element in the reflectarray is isolated from others, the failure of one element will have insignificant impact on the performance of the complete antenna with thousands of elements. If 1/10 of the total number of elements fail, the drop in antenna gain is only 0.5 dB. The inflatable reflectarray lends itself to very large sizes. A 20-m X-band antenna is feasible. Because it is much simpler to manufacture an inflatable structure, such a structure will be much less expensive and reliable to deploy than conventional mechanically deployed systems. An inflatable "light SAR" would provide new capabilites in target detection and monitoring, foliage penetration, and topographic mapping.

F. Solar Concentrators

There has been interest in solar thermal propulsion since Kraft Ehricke first presented the concept in 1956. Solar thermal propulsion is an alternative to chemical, electrical, and nuclear propulsion for interorbital, lunar, and interplanetary transfer missions. Chemical rockets require both fuel and oxidizer that, while providing high thrust, are extremely heavy and inefficient. Conversely, electrical propulsion is extremely efficient but provides very low thrust values. Nuclear propulsion overcomes many of these shortfalls but has serious issues associated with safe ground handling and launch.

Only a single propellant is required for solar thermal propulsion, which can provide moderate thrusts and efficiencies, with very low safety concerns. Solar energy is collected and focused onto a thruster. The (hydrogen gas) working fluid is heated and expanded by solar heated material in or surrounding the cavity. The expanding gas provides thrust without ignition or combustion. Laug and co-workers

Fig. 5.20 Solar concentrator (5 m) (SRS, Inc.).

at the USAF Research Laboratory, Edwards site, have developed concepts to use membrane/inflatables as solar concentrators. Similar to a membrane/inflatable antenna, these large doubly curved concentrators could be packaged into a small volume for launch then inflated on orbit. Solar concentrators can be used directly to improve solar-electric conversion efficiency or as collectors for solar thermal propulsion systems. The shooting star Concentrator project was a joint NASA Marshall, Glenn, and Goddard effort to develop and demonstrate solar thermal propulsion technologies for application to space transfer vehicles.

Photovoltaic (PV) cells have a relatively low efficiency, and carrying more photocells is a trade with increased mass. Another approach is to concentrate more sunlight (photons) onto a smaller number of cells. The Naval Research Laboratory has investigated concepts for a deployable membrane PV concentrator. These planar sheets would be placed on either side of a photovoltaic panel, forming a trough for concentrating sunlight. Such an array was built by SRS, Inc., and flown on the Hughes Corp. Galaxy XI satellite.

G. Solar Sails

In the early part of this century when physicists first began to gain a better understanding of the nature of light, it became apparent that electromagnetic waves could, in certain circumstances, demonstrate properties typical of particles. Photons, or quantized "particles of light," each have an associated momentum equal to h/λ, where h is Planck's constant, and λ is the wavelength. Solar sails operate by using solar photon momentum exchange to accelerate a spacecraft; at 1 AU, the applied force is about 9 N/km^2. As such, a solar sail has "infinite" specific impulse because it requires no propellant.

Sunlight falling on a large lightweight surface is a highly efficient method of changing the orbital energy of spacecraft. There is no need to carry onboard propulsion or fuel, and thus the reduced mass allows greater payload capacity and much shorter trip times for high-energy missions. The constant low thrust counteracts gravity to enable unique non-Keplerian orbits and new vantage points for space physics measurements. Eventually, sails that use laser energy beamed from Earth may enable rapid access to all parts of the solar system and beyond.

Fig. 5.21 Solar sail quadrants (20 m) manufactured from a 3-μm thick polyester film with embedded ripstops (ILC Dover).

Solar sails represent one of the few propulsion technologies that can be developed rapidly within the next decade and lead to dramatic improvements in space exploration capability and affordability. Using solar photon pressure to accelerate a spacecraft, solar sails enable a large number of very propulsion-intensive missions, from non-Keplerian orbits, to solar polar missions at less that 1 AU, to fast trips to the edge of our solar system.

Advances in thin film materials, novel high-temperature sail substrates, and low-mass booms give sails the potential for high-performance, low-cost propulsion. This fact combined with the need for low total mission costs, make sails a high-payoff space transportation technology. Significant technology development that addresses issues of sail storage, deployment, and control is needed. Developing sails for propulsion applications is synergistic with other work on large gossamer thin film space structures, such as the next generation space telescope sunshade and large inflatable concentrators and antennas. Furthermore, solar sails represent the first step in beamed energy propulsion technology, where reflected laser light or microwave radiation provides the momentum to propel sailcraft on interstellar mission. Solar sails are a key to lowering the cost of accessing space beyond low Earth orbit as well as enabling new science missions.

In the near term, sails could be used for the NOAA geostorm warning mission and small-body rendezvous missions with microspacecraft. Intermediate applications could see the use of solar sails for fast missions to the outer solar system, heliopause, and perhaps the first mission into the interstellar medium. In the far term, a light sail, which uses light from sources other than the sun to accelerate, may provide the first propulsion technique to start a human-made object on a journey to another star. Sail use has the potential to reduce trip times dramatically because of the very high mission velocities achievable through close solar perihelion trajectories. Depending on sail materials, payload, and configuration, this could be well in excess of 100 km/s Δv for solar sails in the next decade. Similarly, propulsion system mass could see significant reductions as lower areal density materials and advanced support structures become available.

The primary performance parameter for solar sails is their areal density (g/m^2). This parameter is an important measure of sail performance because it determines the acceleration of the sail (i.e., solar pressure [N/km^2] divided by areal density [g/m^2] gives acceleration). Areal density is determined by the thickness and density of the sail material and the mass of the supporting structure. The term "loaded areal density" refers to the entire sailcraft mass, including payload, divided by sail area. Solar sails are large, gossamer structures. Dimensions range from tens of meters for small spacecraft with fairly modest mission Δv requirements, to hundreds of meters for more ambitious missions to the outer solar system. Solar sail areal density requirements range from around 20 g/m^2 to perform near-term demonstration missions to around 1 g/m^2 to accomplish fast missions to the heliopause.

Operating temperature is another figure of merit for solar sails, as it dictates minimum solar distance and, therefore, maximum achievable acceleration for a given sail areal density. Maximum operating temperature is governed by the sail material's reflectivity and emissivity.

Sailcraft survivablity also merits serious consideration by mission planners. Understanding sail performance degradation over time due to space environmental effects—micrometeroid impacts, radiation, and sail charging—will be critical for successfully completing sail missions.

Fig. 5.22 Sail occulter (Case Western Reserve University).

Sails can carry a spacecraft inbound toward the sun by using the solar photon pressure to slow down in the direction of orbital motion, the spacecraft then being accelerated inward by the sun's gravitational pull. Spacecraft can spiral outward from the sun by positioning the sail to increase heliocentric orbital velocity. Sails can also be used for inner planet rendezvous missions or for applications that would prohibit more conventional propulsion approaches such as planetary "pole sitters." At Earth, an initial sail deployment altitude of around 2000 km is needed so that atmospheric drag is significantly less than solar pressure. Nevertheless, sails can be used for planetary spiral-out and capture.

There are several solar sail design concepts that mission planners have considered; square and heliogyro sails represent the designs examined in the Halley comet rendezvous and subsequent mission studies of the 1970s. The inflatable sail concept is a recent innovation derived from inflatable antenna and structures technology. Deployment of round, spin-stabilized sails has been demonstrated in space with the Russion Znamya sails deployed by Progress vehicles.

The square sail consists of a thin sheet of metallized plastic stretched over a supporting low-mass boom. Small vanes are located at the corners of the sail; they are rotated to produce differential light pressure for use in maneuvering the sail. Alternatively, the sail can be maneuvered by shifting the center of mass relative to the center of (light) pressure. The World Space Foundation, a nonprofit organization which recently has been incorporated to the Planetary Society, built a prototype square sail 880 m^2 in area for demonstrating on-orbit deployment and maneuvering. However, the sail was never flown.

The second type of sail, the heliogyro, is spun like a helicopter blade; the thin-sheet sail material is unrolled and stabilized by centripetal acceleration. Maneuvering is accomplished by changing the pitch of the blades. The heliogyro sail is easier

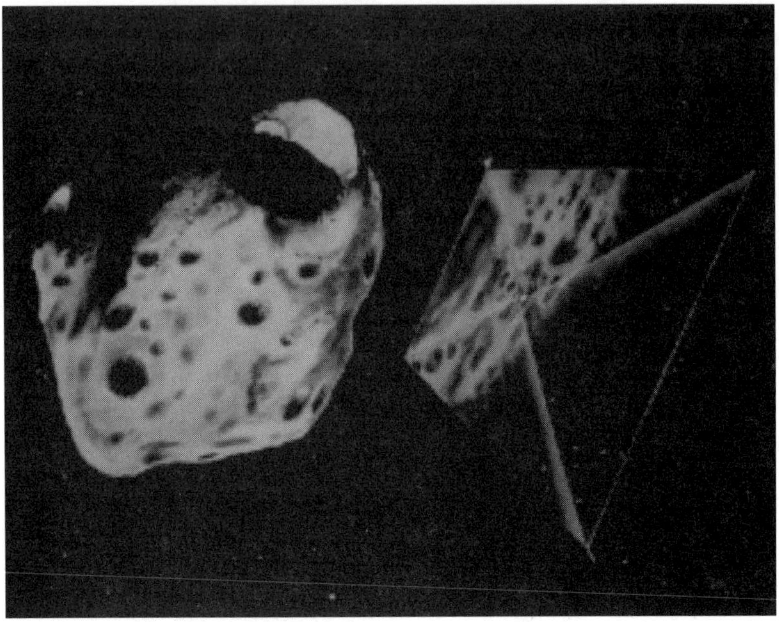

Fig. 5.23 Artist's rendering of a sail rendezvous with an asteroid (JPL).

to deploy than the square sail, has a greater stability from random disturbances (due to its rotational inertia), but has a slower maneuvering rate due to the rotational inertia. Thus, the two types of sails have different strengths and weaknesses, although the square sail, with its faster maneuvering response, might be favored for missions involving extensive planetary escape and capture spiral orbits.

More generally, light sails provide spacecraft propulsion by using photon pressure from any light source; laser sails and microwave sails are two concepts that would require a source for beaming energy to a spacecraft. The source might be either ground- or space-based, depending on mission requirements. Because of the infrastructure required for beamed-energy light sails, these concepts are further term and merit lower technology readiness level ratings. Beamed energy sail concepts are among the few propulsion techniques that hold promise for interstellar missions; such applications would require kilometer-scale sails with areal densities on the order of 0.1 g/m^2.

Solar sails are being considered for several missions planned in the NASA Sun-Earth Connection (SEC) theme, including a solar polar imager and the first-generation interstellar probe launched to explore the outer reaches of the solar system. Several missions of the NASA Solar System Exploration (SSE) theme would also be greatly enhanced by the use of solar sails, such as missions to comets and the outer planets. This technology could also enable large sunshields needed to stabilize and protect large gossamer apertures and other cryogenic telescopes.

H. Solar Shades

NASA is currently engaging in technology development for the next generation space telescope (NGST) mission scheduled to be flown in 2007. This mission,

Fig. 5.24 ISIS deployed configuration (NASA Goddard/JPL).

managed by NASA Goddard Space Flight Center (GSFC), envisions a near infrared, 8-m-aperture telescope to be positioned at sun–earth L2 orbit for 5 to 10 years. To passively cool the telescope to below 60 K for maximum science return, a very large (up to 32-m × 14-m) sunshield with multiple layers of thermal membranes is included in the baseline NGST architecture. Although both a mechanically deployed and inflatable sunshield are being considered at this time, the inflatable design offers several inherent advantages, including lighter weight and smaller launch volume, over its mechanically deployed counterpart. More importantly, an inflatable sunshield will be much less complex (e.g., at least one order of magnitude fewer number of parts) than a mechanically deployed sunshield. Less complexity leads to higher reliability and cheaper fabrication and assembly costs.

ILC Dover is working with JPL and GSFC on the NGST sunshield program. The goal of the program is to demonstrate a 1/3-scale model of the NGST sunshield on a shuttle flight experiment named inflatable sunshield in space (ISIS) (Fig. 5.24). A scale model has already been demonstrated in 1g and the current program calls for the 1/3-scale model to be demonstrated in 2001.

The 1/3-scale sunshield for the ISIS flight experiment will be a diamond-shaped membrane structure that measures approximately 5 m × 11 m × 1 m when deployed. The system consists of four layers of membranes that are deployed and supported by four inflatable beams, which later rigidize to provide a permanent structure. The membranes, when stowed, are accordion-folded or "Z" folded into a small packing volume around the bus structure. The roll-up method for controlled deployment and heat-cured thermoset composite laminates are baselined for this demonstration.

The system shown in Fig. 5.25 is a one-half-scale model deployed at ILC Dover with a gravity negation system attached to the sunshield. The booms are rolled and attached to the tips of the membranes and contain deployment control mechanisms. During deployment, the booms unroll and dispense the membrane from the packing container in a controlled fashion. At full deployment of the membranes, the booms

Fig. 5.25 NGST one-half-scale sunshield (ILC Dover).

(aluminum laminates in this case) are inflated to a prescribed pressure to "rigidize" them, then they are subsequently vented.

I. Structural Support Members

Membrane structures rarely exist of and by themselves. They usually require some support structure, especially to carry compression. Examples include the already mentioned hybrid membrane/inflatable tents and the IAE. The support tubes are clearly visible in Fig. 5.25.

Structures that respond to service loads primarily through membrane action do so typically because of a combination of small thickness and/or low modulus. For thin structures, bending resistance may be orders of magnitude smaller than membrane resistance, such that the former provides only a minor contribution to load resistance, if at all. An important byproduct of thinness is that membrane structures are typically *no-compression* structures. A (possibly localized) buckling or *wrinkling* will occur upon subjecting a membrane structure to compressive loads (also seen in Fig. 5.25), and this becomes a principal design driver.

The literature on membrane/inflatable structures members is significant and has been thoroughly reviewed by Jenkins and co-workers. Two works are of particular interest in the present context. An early interest in inflatable reentry vehicles motivated a study by Leonard and co-workers in 1960. They reported on an inflatable cantilever cylinder, made simple strength-of-materials projections of maximum tip load before collapse, and provided some brief experimental evidence. More recently, Main and co-workers conducted a similar investigation on inflated fabric cantilever cylinders.

Membrane/inflatable structural members may take the form of rods, beams, rings, and the like. These members may find applications individually, in combination with other structural members, or in conjunction with membrane structures. Research is underway to develop methods for rigidizing these members after inflation/deployment.

Inflatable rods of circular cross section are fitted with end caps that allow them to be used as struts or be linked to form trusses. The IAE launched in 1996 used three pneumatic rods as struts to connect the antenna reflector to the feed. Prototypes for inflatable trusses go back at least to the 1950s and work at Goodyear on inflatable radar systems. Overstressed metallic tubes are discussed in a later section.

Besides providing axial resistance as struts, inflated cylinders also have bending rigidity and thus act as beams. An inflated cantilever cylinder can act as a boom for extending the reach of an astronaut. In an inflatable airplane, they can act as wing girders. Curved pneumatic beams were discussed above as being used for structural support in inflatable habitats. Curved beams, closed onto themselves, form pneumatic rings or tori. The IAE used a pneumatic ring to provide reaction for the inflated membrane reflector and as the attachment point for the pneumatic struts. NASA/Marshall has recently investigated the static deflection and dynamic response of inflatable cylinders [see *Gossamer Spacecraft*].

Torsion resistance by an inflated cylinder or torus is also possible; however, this application has not been well studied.

J. Telescopes and Optics

Deployable space-based parabolic membrane optics is of current interest to NASA and DOD, and in particular to the Air Force Research Laboratory (AFRL). The interest is driven by the potential for large reduction in launch costs through reduced structure mass and stowed cargo volume. In addition to large space-based, optical-quality membrane telescopes, this technology can be used for radio and radar antenna as well as solar power and propulsion systems. Regardless of the use, one stringent requirement for the success of this project is the surface accuracy of the deployed optical membrane.

The surface precision required for optical quality mirrors is rigorous. A typical *uncompensated* "$\lambda/20$ specification" (i.e., $1/20^{\text{th}}$ the wavelength of interest) could require a mirror with a surface precision on the order of less than 1 μm rms. Precise grinding and polishing is essential in the production of conventional *rigid* optics to achieve such tolerances. For space-based applications, rigid optics are limited by both volume and mass due to launch cargo capacity and thrust capacity. Therefore, to produce large space-based optics, a lightweight, deployable but precise structure is necessary to minimize both mass and volume but maintain high performance.

However, rapid technology developments in wavefront reconstruction and adaptive optics provide an opportunity for the stringent precision requirements to be relaxed. In particular, holographic spatial light modulation (SLM) techniques show promise to work in conjunction with membrane optics to achieve desired performance levels. SLM will allow membranes with much relaxed surface figures, of perhaps 50λ, to act as optical primaries.

When AFRL began research into developing large, ultralightweight deployable optics in the late 1990s, the focus was on inflatable structures composed of a membrane mirror and highly transparent canopy (*lenticular* configuration), along with the surrounding inflatable/rigidizable support structure. The objective was to obtain a near-net parabolic shape by inflating either a plane monolithic circular membrane or cutting and assembling flat gores such that the final shape was parabolic. Through this research, it was determined that the requirements placed on both the transparent canopy and support structure were highly demanding. An alternative means of achieving and maintaining the desired figure in a membrane reflector needed to be found.

Spin-casting appears to be an ideal method for producing a near-net shape parabolic membrane that does not require an inflatable lenticular. Spinning mercury has proven effective in creating optically fast primary mirrors for telescopes. To produce the mold, this spinning method could be used with a hardening liquid that cures in a near-net parabolic shape. The solidifying substrate could then be produced atop the mold in the same spin-casting fashion. Finally, optical coatings can be applied such that the coating places tension on the substrate, thereby making it a "semirigid" structure capable of maintaining a near-net parabolic shape upon release from the mold.

Although the analysis of precision membranes is discussed in more detail below, the current status of activities in precision membrane mechanics is given in the following table.

The lenticular membrane described is generated by pressurizing an initially plane circular membrane (Hencky–Campbell–Croll models). Each of the items in Table 5.2 is briefly described here.

1) Initial shape: characterizing the deviation of the initial inflated shape of an unperturbed, isotropic, homogeneous membrane from a paraboloid.

2) Prestrain: the effects of prestrain of the inflated shape.

3) Boundary displacement: the effects of boundary displacement on the inflated shape.

4) Modulus variation: the effects of spatial variations in elastic modulus on the inflated shape.

5) Thickness variation: the effects of spatial variations in thickness on the inflated shape.

Table 5.2 Current status of theoretical mechanics of precision membranes

	Lenticular membrane (Hencky–Campbell–Croll)
Shape achievement	Initial shape (done)
	Prestrain (done)
	Boundary displacement (done)
	Modulus variation (done)
	Thickness variation (done)
	Seams (some)
	Rotation (current)
	Scaling (some)
	Physical explanation (done)
	Zernike relations (none)
Shape maintenance	Thermal deformation (some)
	Boundary compliance (current)

6) Seams: the effects of seams between gores on the inflated shape.

7) Rotation: the effects of centrifigual forces on the inflated shape.

8) Scaling: the development of scaling laws that predict how membrane/inflatable characteristics scale with aperture diameter.

9) Physical explanation: a physical explanation of the reason for the systemic occurrence of the so-called "W-curve" when comparing the membrane to a paraboloid reference.

10) Zernike relations: the mapping of the Zernike polynomial decomposition of optical error into membrane mechanics information.

11) Thermal deformation: the effects of thermal gradients on the inflated shape.

12) Boundary compliance: the effects of nonrigid boundaries on the inflated shape.

In most of the above areas, high-fidelity validation experiments still need to be performed.

NASA is studying future missions requiring very large space observatories. The long-range goal of the astronomical search for origins (ASO) and planetary systems theme in the space science enterprise is to detect and characterize planets in orbit around nearby stars. This grand challenge is a driver for gossamer technology development of large apertures.

Independent of any specific instrument concept, basic detection physics sets minimum requirements for optical apertures. Set against the benchmark of the largest ground-based telescopes and the 8-m NGST, terrestrial planet spectroscopic characterization requires a 10-fold increase in aperture area and low-resolution direct imaging requires an additional 25-fold increase in area. Such large collection area requirements most likely preclude implementing missions with a single telescope. Rather, such missions are currently envisioned to utilize constellations of large telescopes flying in formation and operating as interferometers.

The basic building block for these systems is likely to be diffraction-limited optical collectors of 20 to 40 m diameter. One of the critical metrics for such systems is the areal density of the fully loaded primary mirror (optical surface, reaction structure, actuators, and wiring). An areal density of 100 kg/m^2 is typical for conventional telescopes. The NGST is striving to achieve areal densities between 10 and 15 kg/m^2. For future ASO missions, areal densities of 1 kg/m^2 or less are required to enable affordable system architectures. Ultimately, to achieve higher resolution imagery and spectroscopy, even more aggressive architectures with collecting areas equivalent to much larger aperture diameters (hundreds to thousands of meters) and much lower areal densities (<0.1 kg/m^2) will be required.

Large aperture technologies also have direct applicability to the needs of other NASA missions. For example, the SEU theme needs large apertures for advanced x-ray, space radio frequency, microwave, and submillimeter telescopes. Large space radio telescopes and Earth-observing antennas, such as for soil moisture and ocean salinity measurements, will need great improvements in antenna technology in the near term. These new antennas will be characterized by sizes exceeding 25 m in diameter, fraction of a kilogram per square meter density, and operating frequencies between 1 and 100 GHz.

Other applications that could benefit from large aperture technology include high-resolution imagers for land and tropospheric wind studies, radar for Earth and space science, deep-space communications, and large-scale solar power generation.

The Hubble Space Telescope (HST) represents the current state of the art in space apertures. The aperture size of the HST is 2.4 m diameter and the areal density of the aperture system is approximately 150 kg/m^2. The Gossamer Initiative will advance these parameters to 20 to 40 m for aperture diameters and less than 1-kg/m^2 densities. The cost of these systems will also have to be drastically reduced. No firm metrics for cost exist yet, but it is expected that the current level of funding per mission will not increase in the future. Even the most capable telescopes will have to stay within a cost cap of the currently planned telescopes such as NGST or terrestrial planet finder. To quote Roger Angel and Neville Woolf from their 1996 *Scientific American* article "... we would need an enormous space telescope, close to 60 meters in diameter. With current [1996] technology, the cost of such a telescope would rival the national debt." Obviously, gossamer telescopes will have to break the cost paradigm along with its technology paradigm.

Helping to accomplish this paradigm shift are new architectures of space observatories. An example of such a revolutionary architecture is the design by Roger Angel and Neville Woolf from the University of Arizona (Fig. 5.26). Their design uses multiple flats as primary reflecting elements. This design breaks completely with conventional ground telescope designs where optical flats are not feasible due to their very long focal lengths. In space, large flat reflectors of high quality and extremely low mass could be simply made from thin stretched membranes. The requirements are that the membrane be of uniform thickness and that its perimeter

Fig. 5.26 Rendering of astronauts assembling a gossamer telescope using optical flats (U. Arizona).

be held in a plane. In practice, the surface would be defined by a discrete number of edge actuators, actively controlled from wavefront measurements to maintain flatness. The membranes might be formed on an optical flat of metalized plastic such as polyimide, or of metal alone.

Another revolutionary design has been presented by Mark Dragovan of JPL. Dragovan noted that a surface with zero Gaussian curvature is either flat or has the shape of a trough, so that one of the principal curvatures is always zero. Such a surface can be formed by bending along only one axis. If the shape of the surface in the curved direction is a parabola, then a line focus results for an incident plane wave. To produce a point focus, a system of two trough-shaped reflectors properly oriented with respect to each other must be used. A perspective view of such a system is presented in Fig. 5.27. This layout takes advantage of a two-mirror reflector system where individual reflectors are parabolic cylinders.

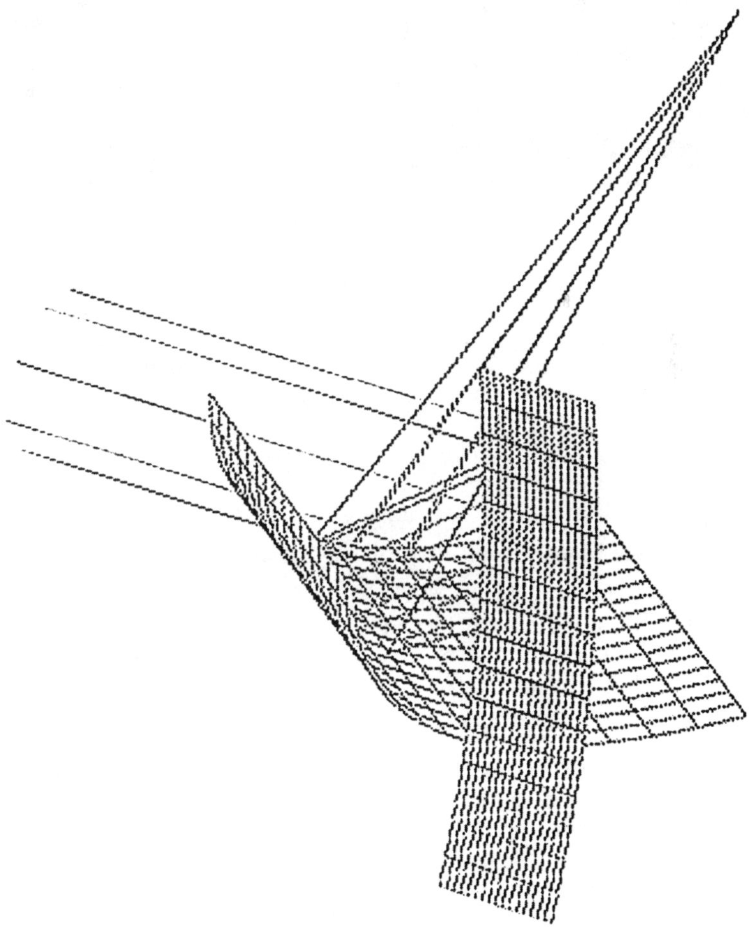

Fig. 5.27 Optical layout of the dual reflector telescope (DART) (JPL).

Fig. 5.28 Artist's rendering of DART telescope (JPL).

The orientation and curvatures of individual reflectors are chosen so that a point focus results for an incident plane wave.

The parabolic–cylindrical surfaces are formed by tensioning a reflective foil over a frame that has a parabolic contour along one axis and is rigid enough to support tensioning.

III. Membrane Materials

A. Materials for Membrane/Inflatables in Space

Two characteristics are common to all membrane/inflatables: flexibility and light weight. Both of these are achieved through some combination of low modulus and thinness of the material. Weight scales with mass density and geometric dimensions. Stiffness scales with modulus and geometric dimension.

As an elementary example, consider the axial stiffness of a rod under tension P displacing an amount x, or $P = k_r x$. Here, the stiffness k_r can be shown to be equal to EA/L, where E is the modulus, A is the cross-sectional area, and L is the length. Note that making the rod thinner decreases the stiffness. Similarly, the stiffness of a beam in bending is $k_b \sim EI/L^3$, where I is the area moment of inertia of the beam.

The stiffnesses of a plate and membrane can be shown to be $k_p \sim Eh^3$ and $k_m \sim Eh$, respectively, where h is the plate or membrane thickness. Hence, the thinner the membrane, the more flexible (compliant) it becomes, and consequently the lighter in weight.

Thus, current membrane/inflatable materials either typically come as thin polymer sheets (films) or thin-walled polymer tubes. A number of candidate polymer films exist today. One of the earliest space-based films in use was Mylar, a polyester film usually coated with a thin layer of metal such as aluminum. Mylar was used to construct the Echo series of satellites, and is still used in high-altitude balloons and other space applications today. Kapton, a polyimide, has been a commonly used membrane material for recent space applications. More recently, several polyimide derivatives (e.g., TOR, COR, CP1, and CP2) have been developed with various improved properties.

A number of mechanical and other properties are of special interest in space membrane/inflatables, including

1) Coefficient of thermal expansion (CTE),
2) Modulus,
3) Low-temperature toughness,
4) Low-temperature ductility,
5) Dimensional stability,
6) Environmental durability,
7) Aging,
8) Surface smoothness,
9) Thickness uniformity,
10) Bend radius (especially after coating),
11) Potential to be "active" or integrated with active materials,
12) Potential to be "rigidized," and
13) Strength.

(Although strength is included in the list, only a small amount of strength is typically required. Membrane/inflatables will, in general, be lightly loaded, with stress and strain levels at much less than their ultimate values. Low modulus values will also help to keep stress levels low, even though strain levels may be high. Failure by fracture or collapse is much more likely than tensile failure.) A sample of properties for common polymer films is provided in Table 5.3.

Other properties are also important for different applications. For example, optical properties are important in imaging applications. Electrical and thermal conductivity and electromagnetic properties may be of interest across a number of applications. Environmental compatibility and survivability are important in nearly every application, and these issues are discussed in a separate section of this chapter.

Table 5.3 Selected properties of representative polymer films

Flim	Polymer	Density, g/cm^3	Mod, GPa	Strength, MPa	CTE, ppm/C	Ult. Elongation, %
CP1&2	Polyimide	1.4	2.6	124	47 to 51	
Kapton	Polyimide	1.42	3.0	172	20	72
Mylar	Polyester	1.38	3.8	172		
TOR	Polyimide	1.4	3.4	138	42	6.5
Upilex–R	Polyimide	1.39	3.7	248	28	
Upilex–S	Polyimide	1.47	8.8	393	12	42

B. Rigidization Technology

For any long-term space application, an inflatable structure needs to be rigidized. This is because leaks will be developed through small holes in the walls of the inflatable, created by impacting micrometeoroids and space debris. If the structure is rigidized upon completion of its inflation deployment, the need to carry a large amount of makeup gas to compensate for leaks can be eliminated.

Since the early 1960s, many researchers have engaged in finding the best rigidization methods for space inflatable structures. These include the development of a number of polymers that can be cured by space environments, such as vacuum, UV light, infrared (IR) energy, and low temperature. There are advantages of simplicity, high reliability, and reversibility of several pioneering methods involving solvent loss systems. In a solvent loss system, a volatile plasticizer leaves the impregnated fabric that forms the wall of tubular inflatable structure as soon as the structure is deployed in vacuum. One of the solvent loss systems uses a water–gelatin solution.

The desired features for space rigidization include

1) Easily adaptive to high-efficiency packaging,
2) Compatible with controlled deployment scheme,
3) No or low space power required,
4) No or low in-orbit outgassing or contamination,
5) Predictable postrigidization configuration accuracy, and
6) Adaptive to ground handling and testing.

The space rigidization methods currently under favorable considerations and/or active development include

1) Stretched aluminum laminate,
2) Hydro-gel rigidization,
3) Open-cell foam,
4) Heat-cured thermoset composite laminates (thermal heating),
5) Thermoplastic composite laminates (passive cooling),
6) UV-curable composite laminates,
7) Foam injection,
8) Inflation gas reaction laminates, and
9) Cold hibernated elastic memory (CHEM) materials.

A self-rigidization process of stretched aluminum laminates if possible. These sandwich laminates are made by bonding thin aluminum foils to polyester films, such as Kapton. While the polyester films provide tear resistance and a gas seal, aluminum foils are stretched by pressure just above the yield point to provide rigidity of the inflatable structure. Figure 5.29 illustrates the stretched aluminum laminate method.

In the thin aluminum wall approach of rigidization, a laminate is fabricated from Kapton film and ductile aluminum, where the Kapton film is positioned on both sides of the aluminum as shown in Fig. 5.29.

To rigidize the cross section, the structure is inflated to eliminate wrinkles in the laminate to a point of just yielding the aluminum. After yielding the aluminum, the Kapton/aluminum laminate is 'rigidized' and will maintain structural integrity. The inflation gas is then vented to space. A typical aluminum laminate cross section consists of two multiple-layered components: 1) the multiple layer insulation (MLI) blanket and 2) the support tube laminate. The MLI blanket is similar to that used for thermal cured laminates, and it is tailored to mission environment and

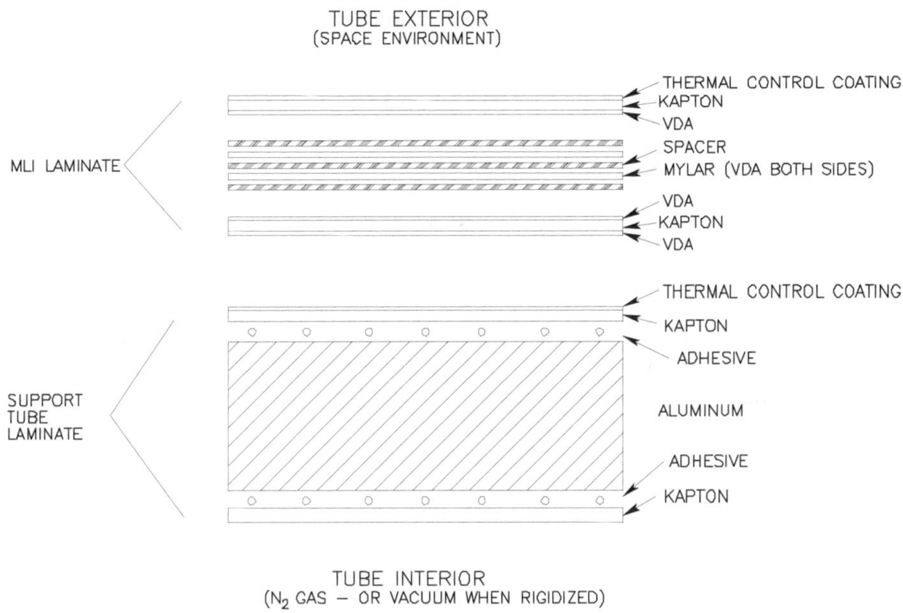

Fig. 5.29 Aluminum laminate rigidization.

requirements. The support tube laminate is fabricated from layers of Kapton–adhesive–aluminum–adhesive–Kapton laminate. The thickness of each layer and the number of layers used can be tailored and designed to mission specific requirements.

In addition to taking advantage of the inflation system that is already needed for deployment, this method also offers the following distinct advantages: 1) it

Fig. 5.30 Inflatable space frame manufactured with shape memory composite materials along a graphite/epoxy tubular column (ILC Dover).

can accommodate inflatable tubes that are either rolled up or Z-folded; 2) it does not require power; and 3) it has a negligible level of outgassing. More importantly, both of the component materials, aluminum and Kapton, of the stretched aluminum laminate have a long heritage of space applications. On the other hand, there remains a major shortcoming of the stretched aluminum laminate rigidization method that needs to be overcome. Because, due mainly to packaging constraints, only a very thin (no more than 0.13 mm) aluminum layer can be incorprated to the laminate, the inflatable/rigidizable tubes made of stretched aluminum laminates can only be used to carry relatively low axial loads. For the cases that involve high compresion and/or significant bending loads, the thin-walled aluminum laminate tubes tend to fail by local crimping. It is believed that, before the stretched aluminum laminate method can be considered as the preferred rigidization method for general inflatable structures, major improvements in materials selection, laminate layup, and fabrication processes are needed.

Hydro-gel rigidization is a rigidization method that has potential applications to certain classes of space inflatable structures. This method has been successfully applied by L'Garde to develop an inflatable/rigidizable space truss . The truss (see Fig. 5.31) is formed by tubular elements that are assembled together with metal joints. Each tubular element is basically a cylindrical tube that is made of woven graphite fabric and impregnated with a water-soluble resin (hydro-gel). Evaporation of the water content in the impregnated hydro-gel will occur after inflation deployment in vacuum and the dehydrated hydro-gel rigidizes to give structural stiffness to individual tubular elements. Like other solvent loss systems, one major advantage of the hydro-gel method is that the rigidization process is completely reversible. That is, the rigidized inflatable structure can absorb water, get softened, and be rerigidized repeatedly to facilitate ground testing and measurements of the flight inflatable structure.

The thermal heating method is currently under further development at ILC Dover for flight experiment in the year 2001. The composite laminate system, which consists of a thermoset matrix resin and a fiber reinforcement such as graphite, is cured or rigidized by heating. The thermoset resin hardens after being heated to a specified temperature and cure time. This rigidization method can be designed

Fig. 5.31 Truss assembly (L'Garde).

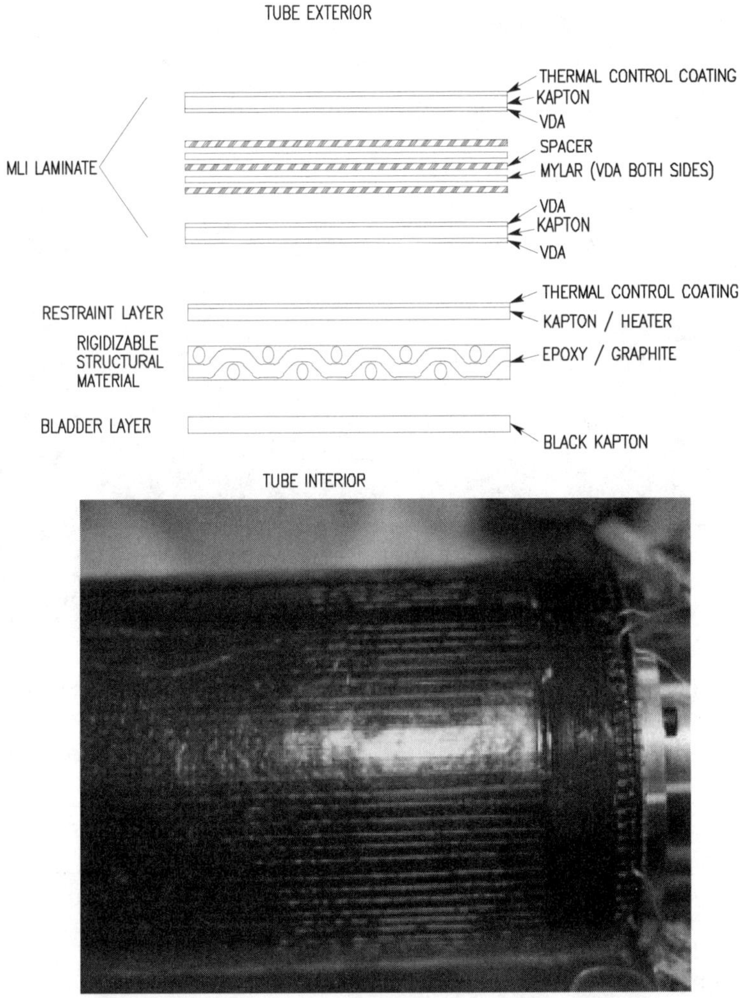

Fig. 5.32 Composite laminate cross section (top) and a picture of the thermoset graphite/epoxy boom segment with heater elements (for use on ISIS program) (ILC Dover).

to cure from solar energy, from the spacecraft power, or a combination of both. Properties of the composite material are consistent with those used in today's spacecraft design. A typical composite laminate cross section, as shown in Fig. 5.32, consists of two multiple-layered components: 1) the MLI blanket and 2) the support tube laminate.

The MLI heating blanket is designed and fabricated from various layers of vapor-deposited aluminum (VDA) Kapton, VDA Mylar, and spacers. The purpose of the MLI blanket is to keep out the harsh space environment and, at the same time, maintain the required curing temperature inside. The design of the MLI blanket is tailored to the specific mission environment and requirements. The support

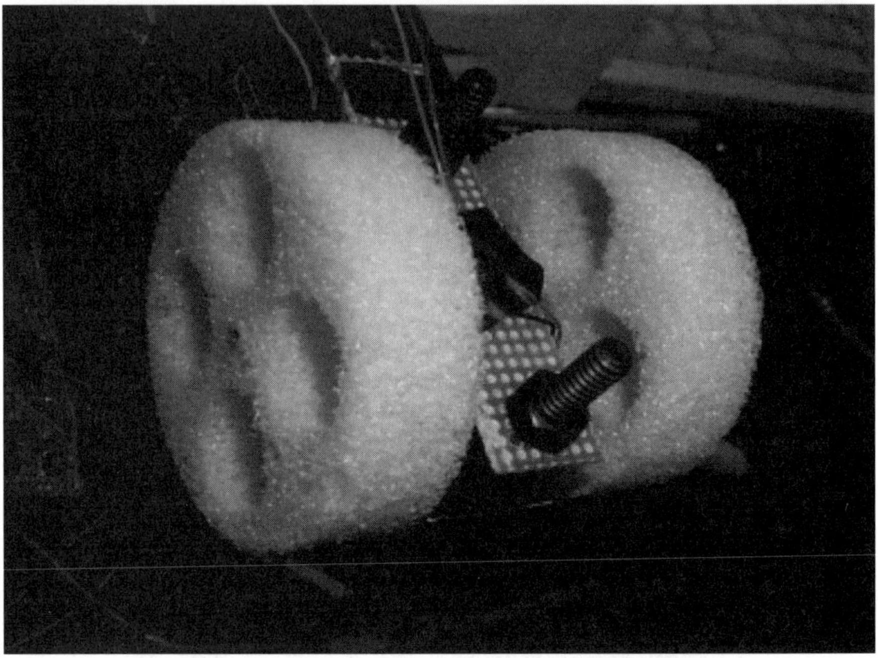

Fig. 5.33 Subscale nanorover wheels for manufactured using the CHEM material (JPL).

tube laminate typically consists of four separate layers: 1) the *restraint layer* that maintains the shape of the inflatable structure; 2) the *heater assembly layer*, which provides the proper temperature for deployment and curing (in some designs the restraint layer and the heater layer are combined to become one assembly); 3) the *rigidizable composite laminate layer*, fabricated from prepreg materials such as epoxy/graphite, which is the support structure when cured; and 4) *the bladder layer*, manufactured from black Kapton, which keeps and maintains inflation pressure during deployment. Thermoplastic composite laminates, which use passive cooling as the curing method, have a similar construction.

Witold Sokolowski at JPL, working with Mitsubishi Heavy Industries of Japan, is also developing a new alternative to space rigidizable material, called CHEM (see Fig. 5.33). CHEM is formulated by incorporating shape memory polymers to open-cellular foam. This material has a maximum deployed/stowed volume ratio of 30 and is self-expanded when heated above its glass transition temperature. This means that a space-deployable structure made of CHEM does not need an inflation system to deploy. Other on-going space rigidization development efforts include heat-cured composites (thermoplastic and thermoset) at ILC-Dover, cold rigidization at L'Garde, and UV-cured composites at Adherent Technologies.

L'Garde has developed a new rigidization technique for space applications, which is based on low-temperature hardening elastomer impregnated composites, which become rigid below a tailorable glass transition temperature. Inflatable

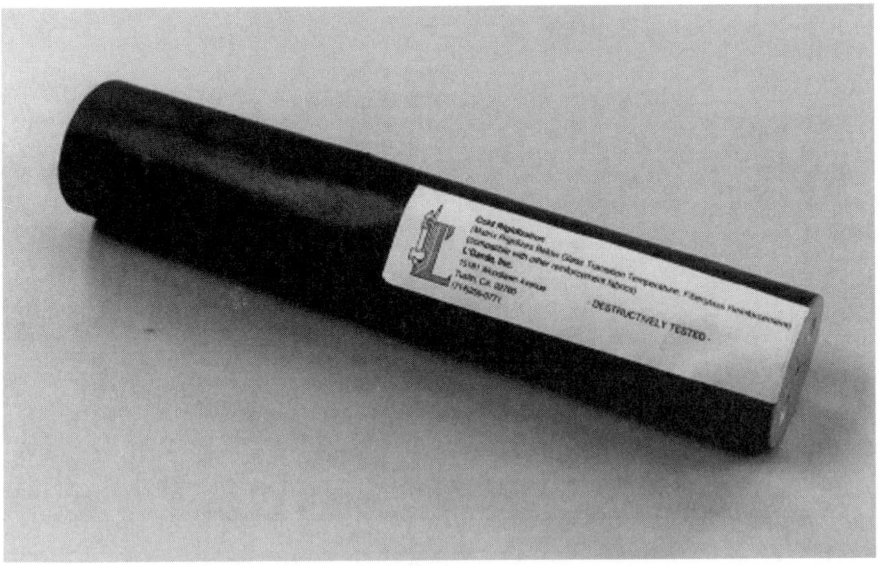

Fig. 5.34 Sub-Tg rigidizable article (L'Garde).

systems will be produced from these composites and then inflated in space for utilization in a variety of load-bearing structures. These structures will be high strength, low mass, and exhibit low stowage volumes.

L'Garde has already tested cylinders made of such composites, with impressive results. The tests were performed below the glass transition temperature for evaluation of compression, bending, and stiffness.

C. Micrometeoroid and Space Debris Effects on Membrane Materials

Micrometeoroid and space debris are major space environmental effects that must be considered in the design of space missions utilizing membrane components. Because of these environmental effects, membrane structures are not considered for Earth-orbiting missions with a perigee lower than 750 km. Many sails and telescopes are being designed to be stationed at different L points where the environment is somewhat uncertain.

Micrometeoroids with diameters comparable to the film thickness or larger are expected to penetrate at least the first film layer of any membrane structure (Fig. 5.35). If not sufficiently broken up or vaporized by its initial impact, then the meteoroid will also penetrate subsequent layers. Particle debris generated from an impact with one layer can also produce multiple holes in subsequent layers. Hole sizes are typically not more than three times the particle diameter, with this ratio becoming smaller for larger particles.

Meteoroids that impact at shallow angles may produce highly elongated holes that, under the influence of mechanical stress, could lead to tears, which greatly increase hole area. Shallow-angle impacts may also generate more vapor and

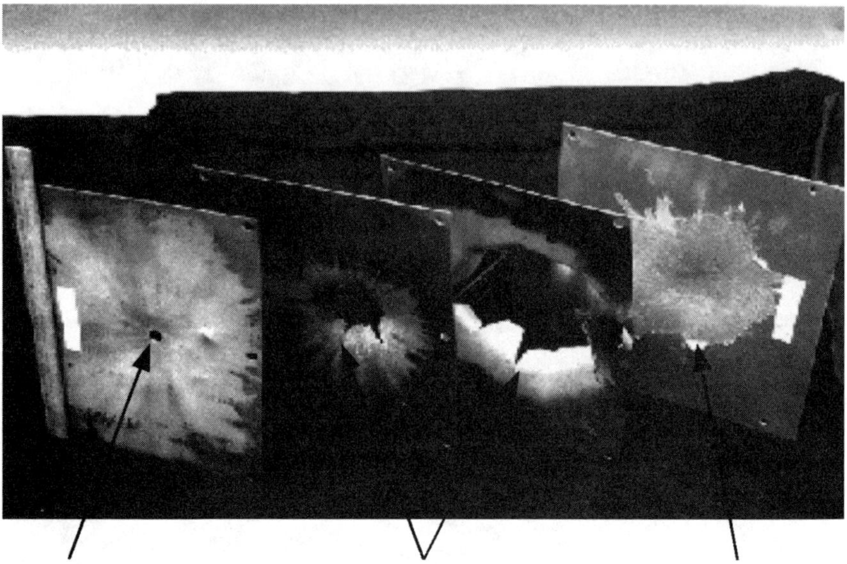

Fig. 5.35 Illustration of meteoroid impact effects—holes, gross failure, and surface damage.

debris than normal-angle impacts and may cause greater damage to subsequent layers.

Tearing or petaling of film can occur in inner layers of multilayer structures, such as lenticulars, radars, or sunshields, due to the impact of high-velocity vapor clouds generated by the impact on the previous layer. This type of failure is referred to as *gross failure*. Gross failure is a major concern because the area of the resulting hole can be several orders of magnitude larger than the meteoroid.

Shock effects, debris particle impacts, and condensed gases can increase film emittance without causing penetrations in the film layer. Such effects may degrade an area considerably larger than the typical hole size. Because the emittance of aluminized film is very low (2~3%), a small increase in emittance over a relatively large area can be a significant performance degradation factor.

At the L2 point, which determines the design parameters for the NGST inflatable sunshield as well as for many future telescopes, the mass density for meteoroids spans a wide range. Meteoroids from comets are composed of low-density mixtures of ice and dust, while meteoroids from asteroids are mostly mineral with some fraction being metallic. Density ranges reported in the literature range from low values of 0.2 g/cm^3 for comet debris to an upper limit of 8 g/cm^3 for solid iron particles. The recommended mean values for L2 are

1) 2.0 g/cm^3 for meteoroids smaller than 1E-6 g;
2) 1.0 g/cm^3 for meteoroids between 1E-6 and 0.01 g; and
3) 0.5 g/cm^3 for meteoroids larger than 0.01 g.

The use of an effective density of 2.5 g/cm^3 has been suggested. The uncertainty in meteoroid density is not too serious because the flux models are, in part, derived from crater and impact data with a similarly assumed particle density. For meteoroids less than 1E-6 g, the mass is uncertain to within a factor of about 0.2 to 5. This means that the particle flux is uncertain to within a factor of 0.3 to 3 at a given mass. For meteoroids above this size, the particle flux is well known, but the associated mass is even more uncertain. Consequently, the uncertainty in particle flux at a given mass is a factor of 0.1 to 10.

Meteoroid flux is generally stated in terms of cumulative flux, which is the number of particles per unit surface area per unit time of mass M or greater that hit a randomly tumbling surface. In interplanetary space (e.g., at 1 AU from the sun), the cumulative meteoroid particle flux $F(M)$ is given in hits per square meter per second.

For LEO, corrections are applied to this flux to correct for both Earth shielding and focusing effects of the Earth's gravity. Both of these effects are inversely proportional to the orbit altitude and so are insignificant at L2. In interplanetary space, the meteoroid flux decreases with increasing distance from the sun and the following correction factor is applied: $F = F_{1\mathrm{AU}} R - 1.8$, where $F_{1\mathrm{AU}}$ is the flux at 1 AU from the sun and R is the heliocentric orbit radius in AU.

In a multilayer structure like the NGST sunshield, which has been studied most extensively, the debris cloud travels until it impacts the next layer. The angular spread of the cloud is usually not more than 45° (one-half-cone angle). The momentum impulse of the debris cloud on the subsequent shield layer sets that layer in motion. The film layer stretches, converting kinetic energy to potential energy. Depending on the magnitude of the momentum impulse, the area over which it is applied, and the strength of the film, the film layer may either be stressed to a level below its yield limit, into the plastic deformation range, or to failure. If stressed below the yield limit, then the film layer will remain essentially undamaged structurally, although coatings may be damaged. Particle and droplet debris impact may locally crater, spall, or penetrate the surface. If stressed into the plastic deformation range, then the second layer will exhibit a permanent deformation. Coatings will very likely be damaged at this point independent of the additional effects of impacting particles and droplets. If stressed beyond the ultimate tensile or shear range, the second film layer will fail over a relatively large area and typically exhibit petalling or tearing.

Figure 5.36 presents a set of photographs of surface-damage effects. Reverse flow debris clouds can degrade surface coatings far beyond the diameter of the impact crater or penetration. Forward flow debris clouds can damage succeeding surfaces over a diameter comparable to the separation between layers. Even without penetration, the impact shock can damage coatings through spallation or delamination over an area far beyond the diameter of the crater. Because surface degradation effects can extend over a much larger area than the area of a crater or perforation, small reductions in surface properties due to impacts may be more detrimental to membrane performance that the penetrations themselves. Surface effects extending to 10 times the diameter of a penetration cover 100 times the area. With few exceptions, surface degradation models do not exist and those that do are not standardized. The effect of surface degradation from debris impact must thus be addressed on a case-by-case basis.

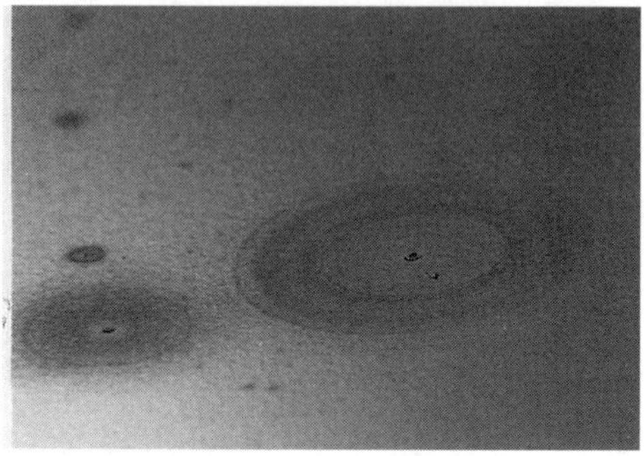

a) Impact into a painted aluminum surface (LDEF)

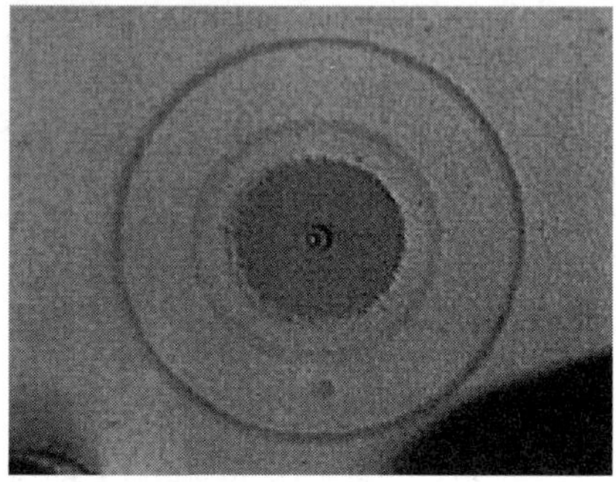

b) Impact into a painted aluminum surface (LDEF)

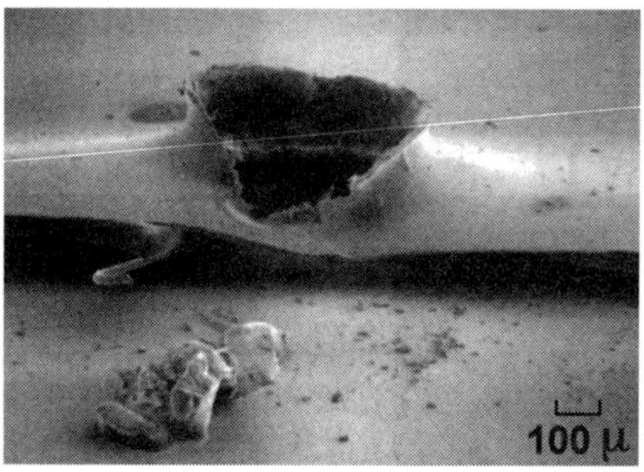

c) Impact into Teflon tape (Solar Max)

d) Penetration in aluminized Mylar film (LDEF)

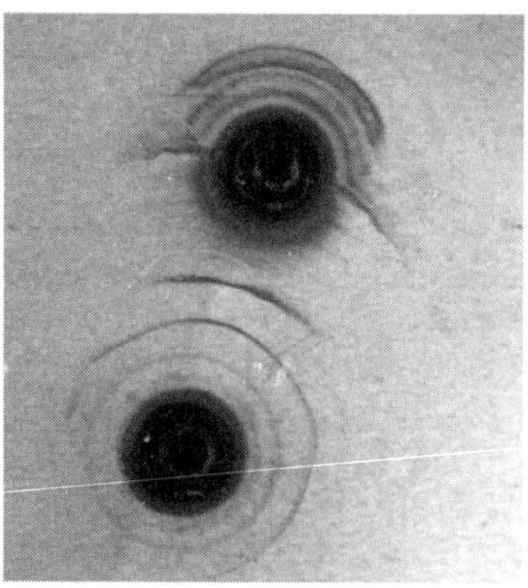

e) Penetration in Ag-Teflon MLI (LDEF)

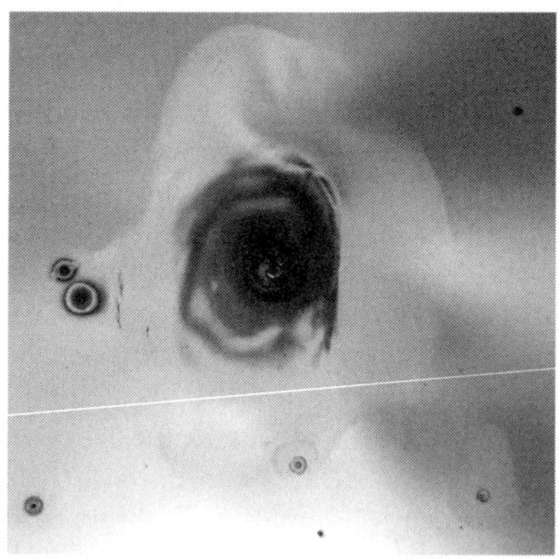

f) Penetration in Ag-Teflon MLI (LDEF)

Fig. 5.36 Impact damage surfaces showing that surface damage area can be many times larger in diameter than penetration hole diameter.

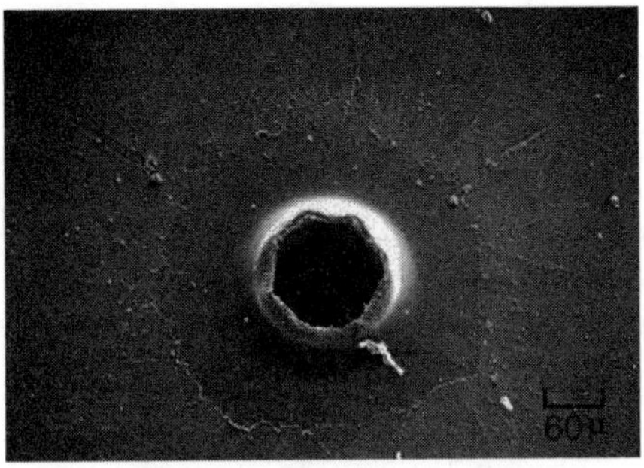

g) Penetration through thin film

h) Oblique-angle impact on MLI blanket (LDEF)

Fig. 5.36 (Continued)

IV. Analysis of Gossamer Structures

A. Static and Dynamic Analysis

The modeling of membrane response may be carried out by one or more of the three basic categories of analysis: classical mathematical, numerical, or experimental. Only for very special classes of membrane problems will classical analysis lead to closed-form solutions; usually numerical analysis must be relied upon. For example, for the case of a square membrane with initial in-plane preload and uniformly distributed transverse load, stress and transverse deflection response may be found in closed-form; however, for the more general case of a rectangular membrane, numerical methods must be applied. Experimental analysis may be used to generate results for specific cases or validate and "tune" more general mathematical or numerical models.

The equation formulation for the combined initial-boundary value problem that models the response of membrane structures has roots in classical elasticity. Membrane response analysis may take a variety of different formulations for the initial-boundary value problem, including Föppl–von Kármán theory, large deformation continuum mechanics, and energy/variational principles. One must be careful to distinguish between the membrane response of an idealized structure and the response of a true membrane. Thus, membrane field equations are applicable in two distinct categories. First, a plate or shell structure may be *idealized* to respond as a membrane away from its boundaries, in its 'interior.' That the stress couples may be neglected in this interior region is an approximation about which the analyst need be aware.

Second, structure models that cannot support stress couples *anywhere* are by definition 'true membranes,' examples of which are discussed throughout the present work. True membranes (hereafter simply called 'membranes') are inherently nonlinear, with the degree of nonlinearity incorporated to field equations dependent on the formulation philosophy. Thus, theories may be developed ranging from linear, to small strain–finite rotation (geometric nonlinearity only), to finite strain formulations that are fully nonlinear both in geometry and materials (geometric and physical nonlinearity). The linear theory will not be discussed here.

Below, we briefly present the four basic parts of the large deformation continuum theory: kinematics (strain, deformation), stress, balance laws, and constitution. Space constraints prevent providing here a self-contained treatment of membrane analysis [see *Gossamer Spacecraft* for a more complete discussion]. We try then only to give the reader a *taste* of the language of nonlinear continuum mechanics to give some sense of the challenges inherent in membrane analysis. The following conventions are used: Latin indices take the values 1,2,3; capital and lower case Latin letters refer to the undeformed and deformed state, respectively; bold type indicates vector or tensor quantities; other conventions are given where required.

1. Kinematics

Geometric nonlinearity requires formulating the equilibrium equations in the *deformed* configuration of the body, which may be substantially different from the *undeformed* configuration (and which is not known in advance). We also, however, consider an elastic body to have a *natural, unstressed, reference state* and we expect the elasticity to be derivable from a thermodynamic potential written in this reference state. Hence, we have a need for coordinate systems in the reference and current configurations. Also, for analysis of membranes (as in shells), it will prove convenient to have coordinate systems aligned with the membrane *midsurface*.

Consider the rectangular Cartesian (RC) coordinates *X-Y-Z* of a point X on the reference membrane midsurface that becomes point x on the current midsurface with RC coordinates *x-y-z*. (Note that *rectangular* refers to *orthogonal* coordinates, while *Cartesian* refers to *straight* coordinate lines.) We also define *curvilinear midsurface* coordinates X^I and x^i in the reference and current configurations, respectively. These coordinate lines will be in general neither straight nor orthogonal, but they remain tangential and normal to the midsurface at every point. It is obvious, then, that a single coordinate system will not suffice for midsurface coordinates. Also, reference and current configurations are typically taken as the undeformed and deformed states, respectively.

Vectors (*basis vectors*) are chosen that characterize the coordinates. In the RC coordinates, the vectors $\boldsymbol{E}_I$ and $\boldsymbol{e}_i$ are orthogonal and of unit magnitude (*orthonormal*). In the curvilinear coordinate system, *covariant* basis vectors $\boldsymbol{G}_I$ and $\boldsymbol{g}_i$ lie *tangent* to the respective coordinate curves X^I and x^i. In general, they are neither orthogonal nor of unit magnitude. (A *reciprocal basis*, the *contravariant* vectors $\boldsymbol{G}^I$ and $\boldsymbol{g}^i$ can also be defined, such that $\boldsymbol{G}_I \cdot \boldsymbol{G}^J = \delta_I^J$ and $\boldsymbol{g}_i \cdot \boldsymbol{g}^J = \delta_i^j$ but these coordinates will not be discussed further here.)

A *metric tensor* **g** in the deformed state is defined by *metric coefficients* $g_{kl} = \boldsymbol{g}_k \cdot \boldsymbol{g}_l$, where $\boldsymbol{g}_i = \partial \boldsymbol{r}/\partial x^i$ is the basis vector in x^i and $\boldsymbol{r}$ is the position vector from o to x; similarly, in the undeformed state, $G_{KL} = \boldsymbol{G}_K \cdot \boldsymbol{G}_L$, where $\boldsymbol{G}_I = \partial \boldsymbol{R}/\partial X^I$ is the basis vector in X^I and **R** is the position vector from 0 to X. The significance of the *metric tensor* is that it contains all of the information about how *length* (specifically *squares of length*) is measured in different coordinate systems (hence the name *metric*). As a simple example, consider that the (square of the) length of the infinitesimal element dl in rectangular Cartesian coordinates (x, y, z) is

$$(\mathrm{d}l)^2 = (\mathrm{d}x)^2 + (\mathrm{d}y)^2 + (\mathrm{d}z)^2 \tag{5.1}$$

But in cylindrical polar coordinates (r, θ, z), the same value is given by

$$(\mathrm{d}l)^2 = (\mathrm{d}r)^2 + (r\mathrm{d}\theta)^2 + (\mathrm{d}z)^2 \tag{5.2}$$

because the angular coordinate θ is not of and by itself a measure of length. So to measure length in the θ direction, one uses the metric coefficient $(r)^2$ in this case. It should also be noted that in the rectangular Cartesian basis the metric tensor is the *identity tensor* **1**.

The use of general curvilinear coordinates (i.e., nonrectangular, non-Cartesian) provides for an elegant and fundamental formalism when developing the general theory of nonlinear membrane response. Noting, however, that the general curvilinear basis is always referred to a rectangular Cartesian basis, some simplifications can be made in the formulation, particularly for computations and in reporting of engineering quantities. To this end, we can establish a *local* rectangular Cartesian coordinate system at each point on the deformed midsurface.

It is convenient to formulate deformation measures that have *zero* value when *undeformed*. These measures are called *strain* and are nonunique. It is intuitive, then, to refer strain from the undeformed or reference configuration, i.e., a Lagrangian strain definition. A very fundamental approach to developing a strain definition is to simply measure the difference in gauge lengths before and after deformation (stretching and/or rotation). This accounts for the fact that rigid body rotation should not contribute to strain, i.e., $(\mathrm{d}r)^2 = (\mathrm{d}R)^2$. Then it can be shown that

$$(\mathrm{d}r)^2 - (\mathrm{d}R)^2 = g_{ij}dx^i dx^j - G_{IJ}\mathrm{d}X^I\mathrm{d}X^J \tag{5.3a}$$

$$\equiv 2E_{IJ}(\boldsymbol{X}, t)\mathrm{d}X^I\mathrm{d}X^J \tag{5.3b}$$

Now defining a *deformation gradient* tensor $\boldsymbol{F}$ such that $\mathrm{d}x^i = F^i_J\mathrm{d}X^J$, where $F^i_J = \partial x^i/\partial X^J$, and comparing Eqs. (5.3a) and (5.3b) above, it must be that

$$E_{IJ}(\boldsymbol{X}, t) = \tfrac{1}{2}\left(g_{ij}F^i_I F^j_J - G_{IJ}\right) \tag{5.4}$$

which are the components of the *Green–Lagrange strain tensor*, and $g_{ij}F^i_I F^j_J$ are components of the so-called *Green's deformation tensor*. It is readily seen that this stress measure is referred to the reference configuration. If, as we have assumed, the midsurface coordinates are RC coordinates, $\boldsymbol{g}$ and $\boldsymbol{G}$ become the identity tensor **1**, and the Green–Lagrange strain tensor may be written as

$$\boldsymbol{E} = \tfrac{1}{2}(\boldsymbol{F}^T\boldsymbol{F} - \mathbf{1}) \tag{5.5}$$

(A similar development leads to an *Eulerian strain* tensor, but this will not be discussed further here.)

2. *Stress*

The idea of stress in a body is a way of characterizing the internal force reaction to loads. Physically, the loads create deformations that lead to changes in interatomic distances relative to the unloaded equilibrium state. Resistance to the deformation (e.g., due to van der Waals forces), averaged over a large group of atoms (the continuum *point*), is what we call *stress*. The goal then is to relate the loads to the (internal) stress.

A remarkable hypothesis attributed to Cauchy allows us to do this. It is an extension to deformable media of Newton's law of action/reaction. The first step is to define a *stress vector* or *traction* $\boldsymbol{t}$, which is the limit of an increment of force Δp per unit increment of current area Δa:

$$\boldsymbol{t} = \lim_{\Delta a \to 0} \frac{\Delta p}{\Delta a} \tag{5.6}$$

Without additional development, we then simply state *Cauchy's stress hypothesis*:

The surface traction $\boldsymbol{t}$ (and body load $\boldsymbol{b}$, which is neglected here) acting on a body, or portion of a body, is related to the *stress* in the neighborhood of the traction by

$$\boldsymbol{t} = \boldsymbol{\sigma} \cdot \boldsymbol{n} \tag{5.7}$$

where $\boldsymbol{n}$ is the outward unit normal to the surface and $\boldsymbol{\sigma}$ is the *Cauchy* (or true) stress tensor.

(Newton's law of action/reaction follows since $\boldsymbol{t}(-\boldsymbol{n}) = -\boldsymbol{t}(\boldsymbol{n})$.) The components of $\boldsymbol{t}$ and likewise $\boldsymbol{\sigma}$ depend on the coordinate system chosen as a basis.

The Cauchy stress is the most accurate measure of stress at a point. However, for constitutive relation development, other forms of the stress tensor are desirable. Several options are available:

1) *Kirchhoff stress*: $\boldsymbol{\tau} = j\boldsymbol{\sigma}$ (simply a "weighted" Cauchy stress);

2) *1st Piola–Kirchhoff* stress: $\boldsymbol{P} = j\boldsymbol{F}^{-1}\boldsymbol{\sigma}$ (which is not a symmetric tensor); and

3) *2nd Piola–Kirchhoff* stress: $\boldsymbol{S} = j\boldsymbol{F}^{-1}\boldsymbol{\sigma}\boldsymbol{F}^{-T}$ (for small strain, the 2nd *P–K* stress can be shown to be merely the Cauchy stress rotated as if acting on the originally oriented surface).

In membrane analysis, *stress resultants* are often defined, which are simply the stresses integrated over the membrane thickness.

3. *Balance Laws*

A number of *balance* or *conservation* laws can be defined. Laws of mass conservation and energy conservation are familiar. Equilibrium equations are laws of balance of forces and moments. For global translational equilibrium, we sum all of the surface tractions $\boldsymbol{t}$ and body forces $\boldsymbol{b}$ over any arbitrary portion of the body of volume v and surface area a in the current configuration:

$$\int_a \boldsymbol{t}\,da + \int_v \boldsymbol{b}\,dv = \boldsymbol{0} \tag{5.8}$$

The internal stress within the volume can be related to the surface tractions through the Cauchy stress hypothesis:

$$\int_a \boldsymbol{\sigma} \cdot \boldsymbol{n}\,da + \int_v \boldsymbol{b}\,dv = \boldsymbol{0} \tag{5.9}$$

The surface integral on the left can be converted to a volume integral through use of Gauss's theorem, giving (after some manipulation):

$$\int_v (\nabla \cdot \boldsymbol{\sigma} + \boldsymbol{b})\,dv = \boldsymbol{0} \tag{5.10}$$

Because the original volume v was arbitrary, it must follow that

$$\nabla \cdot \boldsymbol{\sigma} + \boldsymbol{b} = \boldsymbol{0} \tag{5.11}$$

This one vector equation contains the familiar three scalar equations of translational equilibrium.

Rotational or moment equilibrium follows in a similar fashion, namely

$$\int_a (\boldsymbol{t} \times \boldsymbol{x}) da + \int_v (\boldsymbol{b} \times \boldsymbol{x}) dv = \boldsymbol{0} \tag{5.12}$$

Again, use is made of Gauss's theorem to convert the surface integral to a volume integral. The conclusion drawn from the result is that the stress tensor σ must be symmetric. Conversely, the symmetry of the stress tensor σ automatically satisfies rotational equilibrium. (This result assumes, however, that there are no local "stress couples," which is satisfied by the membrane hypothesis.)

4. *Constitutive Relations*

Three general categories of *elastic* constitutive relations can be listed: linear elastic, hyperelastic, and hypoelastic. Linear elastic relations have a general form

$$\sigma = \boldsymbol{C}{:}\varepsilon \tag{5.13}$$

where $\boldsymbol{C}$ is a 4th order tensor that contains information about material properties and ε is the small strain or linear strain tensor.

The general nonlinear hyperelastic stress–strain relation has the form

$$\boldsymbol{S} = \frac{\partial W}{\partial \boldsymbol{E}} \tag{5.14}$$

where W is the strain energy function to be defined for specific materials or classes of materials. (Hypoelastic relations are rate dependent and will not be discussed here.)

To represent the behavior of real membrane/inflatables structures and materials, the constitutive relation should take into account no-compression behavior (discussed later under "wrinkling") and may need to account for viscoelasticity or viscoplasticity, among other features. For example, Jenkins and others have developed "viscohyperelastic" constitutive relations.

B. Analysis of Precision Membranes

Three key factors are paramount for the success and user acceptance of gossamer technology: deployment, longevity, and performance. Performance hinges critically on the precision of the membrane surface. The amount of precision required is highly mission dependent and may entail one or more of the following issues: surface smoothness, deviation from desired surface profile, and slope error. Parabolic profiles are most commonly desired. The most prominent systemic error found in circular inflated or vacuumed membranes is a so-called "spherical aberration," often referred to as a "W-profile error," which is a measure of the deviation of the actual surface from that of the desired configuration (that happens to have the characteristic shape of a "W"). Other measures are also in use, including "rms" and "wavefront" error metrics.

For example, consider the task of creating a large optical-quality membrane telescope. The membranes on these telescopes will range in thickness from ten to a few tens of micrometers. The maximum acceptable peak-to-peak figure error over the entire surface will range from 10 to 20 μm. (This value assumes that a certain amount of adaptive optics will be used to correct image errors.) It has been common thus far to use the membrane form of the von Kármán plate equations for the analysis in this application and we review that theory briefly here.

Consider a geometry that represents a plane elastic sheet, with circular boundary of radius a. The geometry is rotationally symmetric (axisymmetric); we also assume the loading to be a uniform (normal) pressure p. The sheet has modulus E and Poisson's ratio ν. The problem of the inflation of an initially plane membrane with circular boundary begins with Hencky. Earlier, Föppl had arrived at equilibrium equations for a membrane plate. These equations were essentially modified von Kármán plate equations with the bending rigidity set to zero. The von Kármán plate equations are based on strains that have nonlinearity introduced through rotations (squares of slope):

$$\varepsilon_r = \frac{\mathrm{d}u}{\mathrm{d}r} + \frac{1}{2}\left(\frac{\mathrm{d}w}{\mathrm{d}r}\right)^2 \tag{5.15}$$

$$\varepsilon_\theta = \frac{u}{r} \tag{5.16}$$

where u and w are displacements in the r and z directions, respectively.

The von Kármán plate equations in axisymmetric form can be written as

$$D\left(\frac{\mathrm{d}^2}{\mathrm{d}r^2} + \frac{1}{r}\right)^2 w - \frac{h}{r}\frac{\mathrm{d}}{\mathrm{d}r}\left(\frac{\mathrm{d}\Phi}{\mathrm{d}r}\frac{\mathrm{d}w}{\mathrm{d}r}\right) = p \tag{5.17}$$

$$r\frac{\mathrm{d}}{\mathrm{d}r}\left(\frac{\mathrm{d}^2\phi}{\mathrm{d}r^2} + \frac{1}{r}\frac{\mathrm{d}\Phi}{\mathrm{d}r}\right) + \frac{E}{2}\left(\frac{\mathrm{d}w}{\mathrm{d}r}\right)^2 = 0 \tag{5.18}$$

where $D = Eh^3/12(1 - \upsilon^2)$ is the bending rigidity and Φ is a stress function to which the stress resultants in the radial and circumferential direction are related, respectively, by

$$\sigma_r = \frac{1}{r}\frac{\mathrm{d}\Phi}{\mathrm{d}r}, \qquad \sigma_\theta = \frac{\mathrm{d}^2\Phi}{\mathrm{d}r^2} \tag{5.19}$$

Linear elastic constitutive relations are assumed. It is important to note that the above equations are not *geometrically exact* in that small angle assumptions have been made.

Hencky provided a solution of the above for the case of zero boundary displacements ($u = 0 = w$ at $r = a$) and $D = 0$ by assuming power series for stresses and transverse displacement. Cambell allowed an arbitrary initial tension in the membrane. Wilkes has given a thorough review of the Hencky–Cambell problem and applied it to determine the shape of a membrane mirror. Jenkins and Faisal have applied the Hencky–Cambell theory to the problem of thermal deformation of inflatable precision membrane/inflatables and compared this to finite element method (FEM) analysis.

As a tractable example, it is convenient to take as the solution to the governing equations the three-term polynomial series:

$$w = w_0 \left(1 + \alpha_2 \frac{r^2}{a^2} + \alpha_4 \frac{r^4}{a^4}\right) \tag{5.20}$$

where w_0 is the deflection at $r = 0$, and α_2 and α_4 are constants that satisfy boundary conditions. Using the membrane assumption and $\nu = 0.4$, the central deflection w_0 can be shown to be

$$w_0 = 0.626a \left(\frac{pa}{Eh}\right)^{\frac{1}{3}} \tag{5.21}$$

In the present case, boundary conditions (at $r = a$) are given by the displacement $w = 0$, and either one of the following two equivalent conditions:

$$M_r = D \left(\frac{\mathrm{d}^2 w}{\mathrm{d}r^2} + \frac{\nu}{r} \frac{\mathrm{d}w}{\mathrm{d}r}\right) \tag{5.22}$$

or

$$\epsilon_\theta = 0 \Rightarrow \sigma_\theta - \nu \sigma_r = 0 \Rightarrow \frac{\mathrm{d}^2 \Phi}{\mathrm{d}r^2} - \frac{\nu}{r} \frac{\mathrm{d}\Phi}{\mathrm{d}r} = 0 \tag{5.23}$$

Taking the assumed deflected shape and substituting into the above boundary conditions leads to the following expressions for the αs:

$$\alpha_2 = -\frac{6 + 2\nu}{5 + \nu} \tag{5.24}$$

$$\alpha_4 = \frac{1 + \nu}{5 + \nu} \tag{5.25}$$

For the Poisson's ratio of 0.4, the deflected shape becomes [from (5.20)]

$$w_b = w_0 \left(1 - 1.259 \frac{r^2}{a^2} + 0.245 \frac{r^2}{a^4}\right) \tag{5.26}$$

The corresponding Hencky solution would reveal (from truncation *after* solution):

$$w_m = w_0 \left(1 - 0.899 \frac{r^2}{a^2} + 0.101 \frac{r^4}{a^4}\right) \tag{5.27}$$

The equivalent parabola would be given by

$$w_p = w_0 \left(1 - \frac{r^2}{a^2}\right) \tag{5.28}$$

It is obvious from comparing the last three equations that an initially plane circular membrane never inflates to a parabola (at least not without some other corrective action).

C. Computational Technology for Membrane Analysis

Computational methods for structural analysis of membrane/inflatables are required in the following areas:

1) Static structural analysis (static thermoelastic, buckling),
2) Dynamic transient analysis (launch, deployment, shock, slewing), and
3) Dynamic steady-state analysis (vibration).

Other computational needs exist in areas such as antenna and optical performance and environmental simulation, but those will not be discussed here.

Classically, rate effect problems of continuum mechanics are couched in terms of partial differential equations. To solve these equations numerically, the domain is discretized. The most popular discretization techniques are associated with the finite difference method and with the FEM. FEMs play the most significant role in computational structural analysis. Other methods, such as boundary element and finite difference methods, play a smaller part. One of the great advantages of the FEM is the ease with which a variety of substructures, composed of different materials, may be linked together for system level analysis. NASTRAN has been used extensively, due in part to its wide acceptance and long history within NASA. Other commercial FEM codes have been used, notably ABAQUS, due to its advanced capabilities in nonlinear analysis and its extensive element and material libraries.

Space and time domains are discretized. The most general approach for discretization is through a finite difference scheme and then marching forward through time by computing the state at the current instant based on the recent history of the system. It is noted that nonlinear static problems are treated in a similar manner through advancing from some baseline state, by taking small advances through nonlinearly varying conditions, to the equilibrium state. Similarly to the initial value problem, information from recent advances in the analysis process is used to compute the current advance, i.e., the step advance is treated as a "pseudotime increment." Hence, the mathematical formulation of the nonlinear static problem bares a certain resemblance to that of the dynamic problem and often the numerical algorithms used are identical.

The governing field equations are most easily analyzed in static loading configurations, thus it is not surprising that the bulk of work in the membrane problem has been of this type. This is not necessarily as restrictive as it may seem since many loading regimes may be considered as quasi static, i.e., when the inherent time scale (e.g., the period of oscillation) of the loading is such that inertial effects (e.g., added mass for a surrounding fluid) may be satisfactorily ignored. The slow inflation of a plane membrane would be of this type.

Furthermore, because membranes are inherently thin, light structures, their weight and hence inertial effects may be ignored even for more rapid loading schemes.

Two general types of solution schemes are available, each of which have their advantages and drawbacks. Explicit integration schemes compute the current state directly from some forward difference scheme. These schemes are only conditionally stable (numerically), thus they require small time steps. The maximum size of the time step depends on the spatial discretization: a finer spatial mesh requires smaller time steps. Full advantage of explicit schemes is taken when the inertial matrix is diagonal; in that case, no matrix inversion is performed.

Implicit integration schemes are inherently stable and much larger time steps can be taken. However, they require matrix inversion, which is computationally expensive and requires considerably more storage than is needed for explicit schemes. A typical finite element discretization for a combined incremental (implicit)/iterative (modified Newton–Raphson) method uses an isoparametric formulation. With the notation that the increment of time is Δt, Ψ_I = isoparametric shape function for node I, X_i^I = ith initial coordinate of node I, U_i^I = ith displacement of node $I, I = 1, \ldots,$ are the number of nodes (per element), and $k = 1, \ldots,$ are the number of iterations per time step, the membrane equation of motion is

$$[\tilde{M}_{ijIJ}]\left\{\frac{\mathrm{d}}{\mathrm{d}t}U_j^J(t+\Delta t;k+1)\right\} + [\tilde{K}_{ijIJ}(t+\Delta t)]\{\Delta U_j^J(t+\Delta t)\}$$
$$= \{\tilde{P}_{iI}(t+\Delta t;k+1)\} - \{\tilde{F}_{iI}(t+\Delta t;k)\} \tag{5.29}$$

where $[\tilde{M}_{ijIJ}]$ = consistent mass matrix, $[\tilde{K}_{ijIJ}(t+\Delta t)]$ = tangent stiffness matrix, $\{\tilde{P}_{iI}(t+\Delta t;k+1)\}$ = external force vector, $\{\tilde{F}_{iI}(t+\Delta t;k)\}$ = internal force vector, and $\{\Delta U_j^J(t+\Delta t)\}$ = incremental displacement vector between iterations k and $k+1$. Advantage is taken of the fact that, by use of the modified Newton–Raphson method, $[\tilde{K}_{ijIJ}(t+\Delta t)]$ is constant during a time step.

Two important requirements for computational methods in analyzing membrane/inflatables are the ability to adequately deal with inherent nonlinearities and low structural stiffness. Nonlinearities may be either geometric or material but typically include both. Of course, geometric nonlinearity is related to low structural stiffness, and material nonlinearity may influence structural stiffness as well. Thin membranes are inherently no-compression structures. Potential compressive stress and/or loss of prestress are handled via changes in membrane geometry, i.e., large out-of-plane deformations. These 'wrinkles' are a localized buckling phenomenon. Analysis of wrinkling is important to prediction of structural response.

Another issue related to low stiffness arises at system level studies of membrane/inflatables. It is usual for membrane/low stiffness members to be tied somewhere to structures of higher stiffness. If the stiffness mismatch is too great, then the structural stiffness matrix may become ill conditioned.

D. Analysis of Membrane Wrinkling

One of the major advantages of inflatables is the ability to deploy very large area structures with extremely low launch masses and small launch volumes. Thin-film membranes that are pretensioned to maintain their desired configurations and shapes form many of these large area structures. Two classes of wrinkles are describable: elastic wrinkles (called "wrinkles" here), which are completely reversible, and plastic wrinkles, which are permanent creases.

Wrinkles significantly influence structural and dynamic behaviors of the space system containing the pretensioned structure. Examples of space systems containing large wrinkled membranes include solar sails, sunshield, solar arrays, and synthetic aperture radar antennas. Currently, research is being conducted to develop and test/verify computationally effective analysis methods to predict in-space behavior of wrinkled membranes with various geometry and boundary and loading conditions.

The presence of wrinkles in large membrane apertures significantly influences dynamic behaviors of large space structural systems. Although several investigators have studied the subject of wrinkled membranes, how wrinkles are formed by in-plane tension is not yet well understood. Currently, a semiempirical approach, in which the wrinkled patterns are assumed on the basis of laboratory observation on scaled membrane models, is being used to model and analyze modal characteristics of the inflatable sunshield for the NGST (see Fig. 5.25).

Jenkins and Leonard, and Jenkins, have provided thorough reviews of the literature on membrane wrinkling. Jenkins and co-workers have developed finite element algorithms to predict wrinkling parameters (location, extent, orientation, stress/strain state) in large elastic deformations and have also measured wrinkle parameters experimentally. Recently, they developed methods to predict specific details of the wrinkles themselves (number, wavelength, and wave amplitude).

Mikulas and co-workers have investigated the effects of creases on thin polymer film response. It was shown that creases reduce the stiffness of the film by an order of magnitude and the film will be shorter than expected. To obtain high-performance and/or accurate systems, researchers must account for these effects.

V. Membrane/Inflatable Systems

System level architecture and control is vital to optimizing membrane/inflatable structures, which in turn is critical to meeting weight, performance, and cost targets for mission applications. Table 5.4 summarizes the drivers for control issues in membrane/inflatables.

Table 5.4 Membrane/inflatables control drivers

Global			
	1) Areal density: <1 kg/m^2, <1 g/m^2, ...		
	2) Sensor/actuator *access*: discrete or distributed		
	3) Sensor/actuator *influence*: space, time		
	4) Other conventional issues: power limitations, redundancy, ...		
	Deployment	Shape achievement	Shape maintenance
Local			
	Highly compliant structures	Precision requirements	Short-term perturbations
	No-compression materials (global buckling)	Plane vs curved configuration	1) Thermal loads 2) Shock/vibration
	High packaging ratios	Inflatable vs nonlenticular configuration	Long-term perturbations
	Time- and temperature-dependent material properties	No-compression material (local wrinkling)	1) Material aging 2) Environmental attack (UV, AO, micrometeorites, ...)
	Creases and crinkles	Rigidization	3) Makeup gas vs rigidization
	Residual gas		

A. Adaptive Compensation of Membrane Reflectors

The ever-increasing need for higher antenna gains and efficiencies at ever-higher frequencies has led to the consideration of various types of inflatable precision apertures. Recent advances in reflector materials, design, analysis, predictive modeling, and manufacturing are expected to yield accuracies almost an order of magnitude higher than present practice for surfaces of <1 mm RMS. However, one of the problems related to high-frequency performance of these inflatable reflectors is the surface distortion caused by thermal effects in addition to the initial distortion caused by imperfections in the manufacture process, and long-term changes in material properties; there is also a problem with accurate pointing of the structure. Several solutions to this problem are available.

1. Adaptive Secondary Reflectors

Combining inflatable apertures with mechanically shaped secondary reflectors will further improve aperture efficiency, as recently demonstrated on a 3.3-m secondary reflector by Composite Optics, Inc., in a project for the National Radio Astronomy Observatory. The final wave front correction can be accomplished using phased array electronic compensation at the aperture focal region. This combined effect of improved primary, adaptive secondary, and phased array compensation will improve the effective aperture surface accuracy by two orders of magnitude and is expected to push the inflatable aperture performance into high radio frequencies and submillimeter applications.

2. Holographic Correction

One of the most promising methods for active compensation is a passive holographic technique utilized to allow existing reflector technology to be used in the optical spectrum, yielding apparent surface precision in the 1.0 μm range. Thin membrane inflatable reflectors utilizing this technology could be used for space projects requiring large apertures for optics.

3. Boundary Manipulation

It has been well proven analytically that manipulating the boundary of certain classes of inflatable reflectors can make significant improvements to their surface accuracy. Initial experimental studies have also verified this effect.

4. Selective Heating

It has been shown analytically that temperature gradients over the surface of a membrane/inflatable cut both ways: they may either distort or improve surface accuracy. Selective heating will be considered as potential future shape control strategy.

5. Deformable Flat Plate

There have been many studies and experiments that use either array feeds or a deformable flat plate (DFP) to compensate for reflector distortions recently investigated by William Imbriale of JPL. In many cases, especially for the DFP, it is assumed that the distortions are both stable and known; while, for the array feed, generally an adaptive system has been proposed. The disadvantage of the array

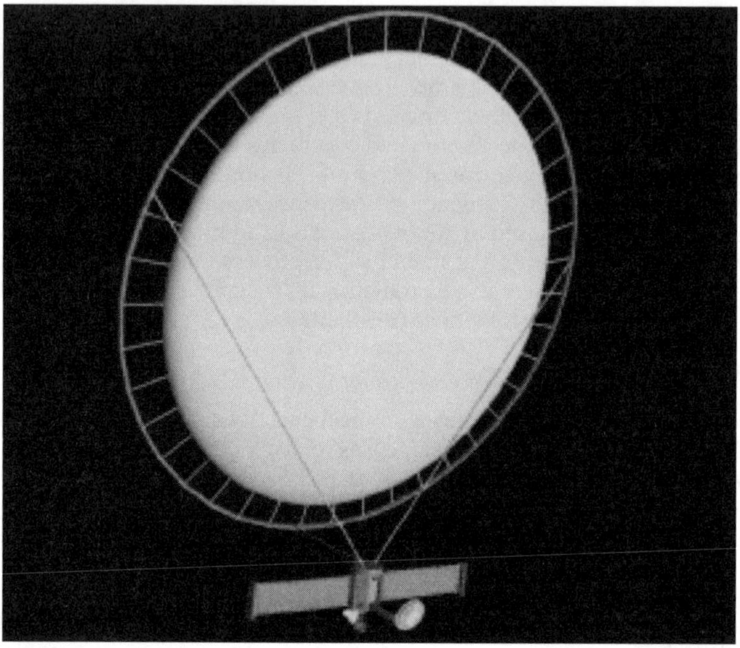

Fig. 5.37 Elements of an aperture system designed for maximum antenna performance: the main reflector with torus, smart support structure, active secondary, and electronic feed (JPL).

approach is that the spread of the energy in the focal plane is quite extensive for modest distortions requiring a large number of feeds. The advantage of the combined approach is the reduction of the number of feeds required for compensation. The DFP can focus all the energy into a small area in the focal plane. Also, by including a monopulse center feed, accurate pointing information can be obtained. Thus, such a system can provide almost compete restoration of the gain and also provide accurate pointing information to the active control system.

B. Multifunctional Membranes

Miniaturization, multifunctionality, and full integration are the major drivers for achieving low mass and low cost in current and future space structures. Low-cost, low launch volume, high-performance space structures may be developed by fully integrating technologies such as "system-on-chip" to the body of the deployable space membrane. In particular, polyimide films offer an outstanding opportunity for integration/packaging of multifunctional devices/chips on light-weight membranes. Today's thick silicon chips are realtively large multilayer polyimide systems. Mechanical integration and functional interfacing is of significant importance and is the foundation of many future space applications. The establishment of the related know-how and systematic study forms the basis of future research in this area. Currently, the technology has been established for reliable attachment of thin (25 to 35-μm) silicon chips as well as thin chip carrying polymer patches, to polyimide membranes.

Fig. 5.38 COI's 3.3-m adaptive reflector for 90–900-GHz applications.

Technologies of specific interest in the upcoming years in the area of multifunctional membranes will be

1) Advanced materials and processes for integrating membranes, electronics, sensors, actuators, power sources, and their associated interconnects to a unified and adaptable structure.

2) Methodology and modeling techniques needed to optimize materials and process selection for integrating MEMS devices with CMOS electronics.

3) Lightweight, distributed power systems and active or passive thermal control systems integrated with membrane structures.

4) Models that optimize power storage and thermal management designs.

Fig. 5.39 Thin electronics on a Kapton substrate (U. Arkansas/Irvine Sensors).

5) Materials with controllable surface properties that, when combined with integral control electronics, could adapt to changing environmental conditions or mission needs (e.g., changing optical properties to steer a solar sail).

6) Concepts and components for active, adaptive wavefront control systems, including shape control of membrane mirrors, with correction to < 1 wavelength.

7) Adaptive systems for measuring and correcting surface figure and wavefront errors and for controlling structural geometry and dynamics.

8) Thinning processes for substrates and bonding processes for attachment to thin films and membranes.

C. Smart Structures

One of the most important technology drivers for demonstration of the feasibility of ultralightweight inflatable structures is the requirement for reliably maintaining surface precision during the mission's life. It is impractical, if not impossible, to qualify such large lightweight systems on the ground before committing to flight. The basic approach is to develop the required methodology to ensure confidence for maintaining the desired surface accuracy on orbit by integrating actuators and sensors to the precision surface, and support structure of the inflatable system. Sensors will then be used to evaluate surface accuracy on orbit and make required adjustments using the actuators during the mission to achieve and maintain required precision surface. This methodology will provide project managers confidence that requirements for the precision of the surface can be achieved before committing to launching an inflatable structure as part of a space mission.

The performance of inflatable optics and antennas critically depends on surface precision, which is difficult to preassure and is affected by service conditions. Emulating biological muscles allows control and reconfigurability of inflatable structures, while benefiting from packaging ability, low mass, and power. Embedding actuators in membranes addresses the critical need for shape control of inflatable structures.

Recent advances in materials technology have resulted in several possible new actuator and sensors that can be placed on any thin membrane material with an actuation range of up to 5%. These new actuator/sensor materials are compatible with polymer membrane materials being considered in current inflatable structures and packaging approaches. Two of the potential materials are the new long stroke piezo-polymer and a thin film shape memory alloy studied by Yoseph Bar-Cohen and colleagues at JPL.

NASA Langley, in collaboration with JPL, is also developing smart materials to actively control the geometry and structural dynamics of large apertures. The main goals of the smart structures research program are to develop thin piezoelectric actuators that can be embedded in rigidizable composite structures and demonstrate active damping and position control for inflatable columns. Langley has developed an MFC piezoelectric actuator that consists of a thin PZT ceramic wafer encapsulated with polyimide films. The PZT wafer is cut into narrow longitudinal strips that function as strain-producing fibers when high voltage is applied.

MFC actuators have been embedded in thermoset and thermoplastic composite columns. The columns have been Z-folded, inflated, and rigidized. Active damping of rigidized columns has been demonstrated by using velocity feedback signals

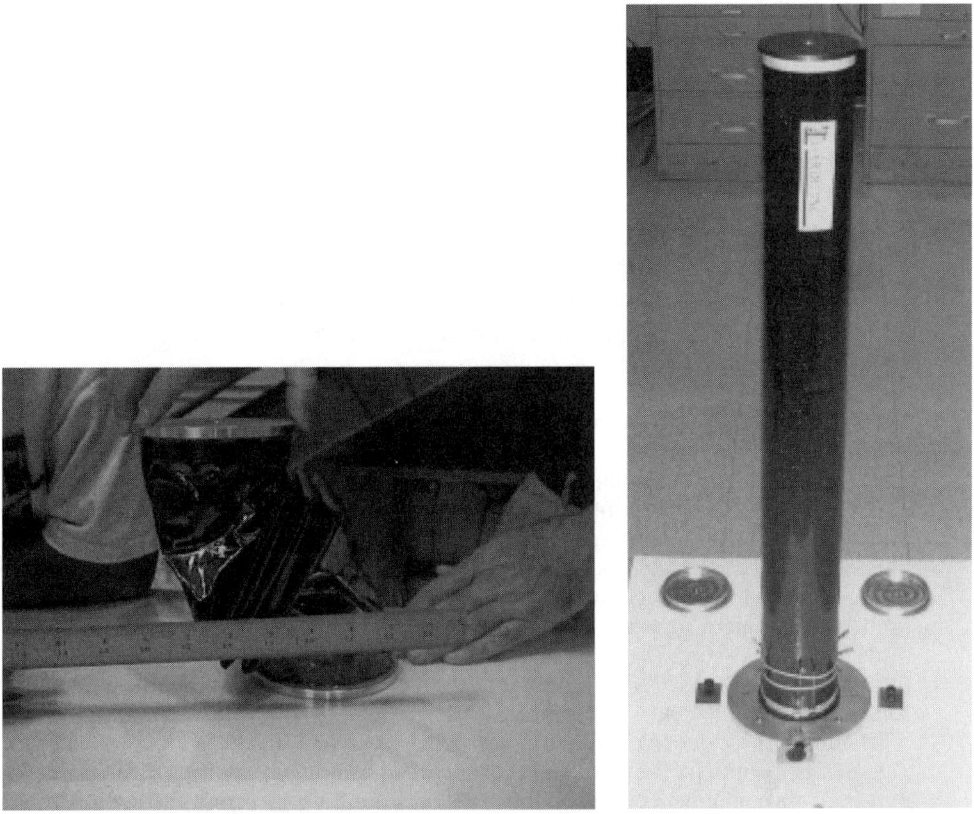

a) b)

Fig. 5.40 Macrofiber composite (MFC) actuators were embedded in 1 m long and 0.1 m diameter inflatable/rigidizable composite active strut test article: a) Z-folded boom and b) boom after deployment and curing (NASA Langley/L'Garde).

from an external laser vibrometer and a tip-mounted accelerometer to drive the MFC actuators. Damping was increased by a factor of 18x for the first bending mode of the column.

New configurations for the MFC actuators are under development to adjust the length of inflatable columns during the rigidization process. These position control actuators will be integrated with an inflatable truss support structure and used to correct postdeployment errors and compensate for manufacturing tolerances.

In addition to piezoelectric ceramic actuators, Langley is developing electrostrictive polymer actuators. These new electrostrictive polymers have 20x the strain capability of conventional PVDF polymer films. Electrostrictive films can be laminated with a reflective membrane to form an ultra-lightweight telescope mirror. Experiments are investigating the feasibility of a concept to adjust the local curvature of a membrane telescope mirror by charging patches of the electrostrictive layer with a non-contacting electron beam.

VI. Testing and Deployment

A. Ground Testing

Ground testing of any structural system may be required for the following reasons:

1) Physical insight for mathematical model development. Physical testing and experience are often necessary to develop a sense of what parameters should be included in a model. For example, an impact model might require high-speed video of an impact to determine whether rebound of the target or projectile were significant parameters.

2) Determination of model data. Models that attempt to predict the response of real structures will require physical data about the structural system, such as material properties, loads, temperature, etc.

3) Validation of mathematical models. *Every* mathematical model needs a reality check. Model predictions must be compared where possible to actual physical response.

4) Prediction beyond model capability. Many physical phenomenon will exist beyond the current capabilities of mathematical models due to complexity, size, time, etc., and physical testing of actual or prototype systems is required.

Specific to membrane/inflatables, the following tests are of interest:

1) Static loading. Static load tests range from determining mechanical material properties (strength, modulus, etc.) to structural component or system response to static loads (stress, strain, deflection). These tests may be time independent (elastic) or "quasistatic" time dependent (viscoelastic). Loads may be mechanical forces or thermal loads.

2) Vibration or modal. Vibration or modal tests seek to determine "modal" parameters: natural frequencies, mode shapes, and damping properties. These tests may be at the component or system level. Due to the ultra-low weight and flexibility of membrane/inflatables, vibration tests often require use of a vacuum chamber to simulate the response in space.

3) Surface precision. Because a primary application of membrane/inflatables is for antennas or mirrors, the characterization of their surface accuracy is important. Test methods include stereoscopic video topography for long wavelength antennas, with results given in millimeters rms. Optical reflector characterization consists of Shack–Hartmann local slope measurement, with results given in waves of error.

4) Deployment. At the time of writing, deployment tests have been limited to slow deployment of structural members such as tube and struts and planar membranes such as sunshields. Full ground deployment of a packaged inflatable reflector has not been accomplished.

5) Space environmental. This testing is discussed further in the next section.

Especially challenging with membrane/inflatables is the unique need to make test measurements via noncontact methods. Because of their extreme low weight and flexibility, contact with the test article may intolerably disturb what is to be measured. For typical membrane material,

1) A strain gauge will be much stiffer,

2) The smallest accelerometers will be orders of magnitude more massive, and

3) Thermocouple leads will be more massive and stiffer.

Alternatives to strain gauges are clip on or optical extensometers for material testing, or optical methods such as moire or electronic speckle interferometry for general strain measurements. Alternatives to accelerometers are laser doppler velocimeters (laser vibrometers). Alternatives to thermocouples include IR thermography. Jenkins and co-workers have demonstrated the use of both laser vibrometry and IR thermography on membrane/inflatables.

B. Micrometeoroid and Debris Testing

This testing is intended to provide data on micrometeoroid impact effects on spaced thin film layers representative of long life space missions. The NGST sunshield team led by Linda Pacini of GSFC and Michael Lou (JPL) has recently proposed a comprehensive test plan for micrometeoroid and debris testing. A key issue with respect to the feasibility of the inflatable sunshield concept is the thermal performance degradation caused by micrometeoroid impact effects. Micrometeoroid impacts can reduce sunshield thermal performance through the damage mechanisms listed below. These damage mechanisms led to the questions listed in Table 5.5. Table 5.5 is a good example of methodology that should be followed in micrometeoroid testing.

Three basic types of tests should be conducted on any membrane structures in particular multilayer structures designed for long life in space:

1) Gross-failure test: these tests are for the purpose of 1) providing data on the susceptibility of spaced thin films to gross failure and 2) on the number and size of holes generated throughout the multiple film layers in the sunshield from a single particle impact. Layer-to-layer spacing and impact particle size should be varied. Multiple shots should be made at a given layer-to-layer spacing and impact particle size to provide data on various impact velocities and increase the size of the statistical database. Tests should be conducted at a normal-impact angle and utilize multiple square film layers per test shot.

2) Shallow-angle test: shots should provide a check for the development of highly elongated holes that might tear under subsequent stressing. These shots should also measure any increase in the number of particles that impact the second layer.

3) Thick-coating test: shots should be made at normal angle at single film layers with a coatings to check for cracking, delamination, or other impact effects that could degrade performance.

C. Deployment/Inflation Methods

One of the great advantages of membrane/inflatables is their ability to be compactly packaged. That compactness is a double-edged sword, however, because is leads to additional challenges in going from the stowed to the deployed state.

Deployment can either proceed as sequential or parallel, in open or closed-loop control. In sequential deployment, one event is *planned* to occur at the conclusion of another. Parallel deployment consists of one or more events *planned* to occur simultaneously. Whether the events occur as planned depends on a number of factors, primarily on the adequacy of factors such as engineering, design, fabrication, etc. Here, however, we concentrate on the additional factor of deployment control.

Open-loop deployment control is nonfeedback control. For example, in open-loop sequential control, a timer may be set to stage one event after the conclusion (with a possible delay) of another event. However, there is no feedback as to

Table 5.5 Sunshield performance degradation questions to be addressed by micrometeoroid impact simulation testing

Question	Relevant data to be obtained
1. Are film layers susceptible to gross failure? If so, what minimum separation distance is required to prevent gross failure?	Visual inspection of film layers by the HYPER facility will determine presence of gross failure.
2. As a function of impact particle size, what number of layers are penetrated, and what are the numbers and sizes of holes in each layer?	Number and sizes of holes in each layer to be measured by the HYPER facility.
3. Are thick, multilayer coatings viable for the outer layer of the NGST sunshield?	Visual inspection for coating loss, cracking, delamination, or other failure modes at GSFC.
4. What is the change in film layer surface emittance?	Thermal emittance measurements at GSFC as appropriate.
5. How does hole size differ for shallow angle impacts and is there the potential for initiating tears?	Hole size measurement by the HYPER facility. Subsequent mechanical stressing at GSFC as appropriate.
6. Are there synergistic degradation effects due to the combination micrometeoroid and solar-UV exposure?	GSFC to perform solar-UV exposure testing on selected film layers as appropriate after micrometeoroid testing at the HYPER facility.
7. How will micrometeoroid impact debris contribute to the contamination of optical surfaces?	Total mass of debris generated equals mass of particle and mass loss from film layers, which is a function of the number and size of holes generated. Number and sizes of holes to be measured by the HYPER facility. Witness plate will collect impact-generated contaminates downstream from last layer. GSFC inspection of witness plate and film layer surfaces will provide data on debris particle size.

HYPER = hypervelocity impact facility at the Space Power Institute at Auburn University.

whether the previous event actually concluded or in what state it left the system if it did conclude. The deployment of the IAE was an example of sequential open-loop control.

Closed-loop control seeks to minimize an error function, given some input, some reporting of the current state of the system (the feedback), and some desired system state. Continuing the example above, under closed-loop control, a new event would not begin until it was determined that the previous event had concluded and had left the system in the desired state to begin execution of the new activity.

Techniques for deployment of membrane/inflatable structural members, such as tubes and struts, is under development. Three techniques currently being investigated are roll-out method, mandrel method, and fan-fold method.

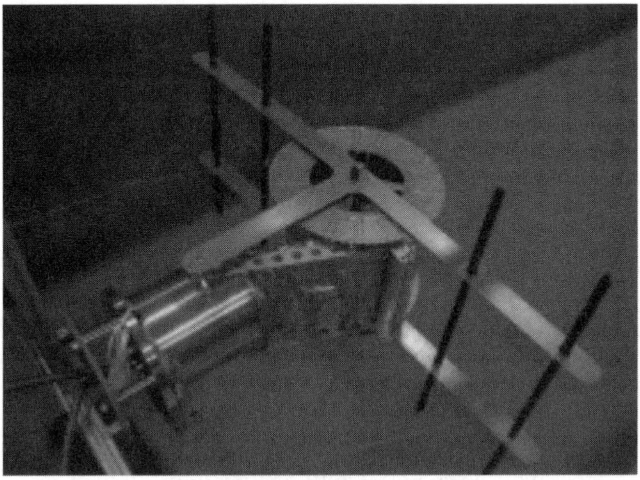

Fig. 5.41 ISIS wire brake controlled deployment mechanism in test (ILC Dover).

In the roll-out method, a tube is flattened and rolled out under pressure, analogous to the children's party favor. Judiciously placed Velcro tabs provide sequential open-loop control by supplying resistance for the tube to inflate against. In the mandrel method, a Z-folded tube is extruded under pressure over a tapered mandrel. The mandrel provides passive directional control for the tube as well as friction resistance to control deployment rate. The fan-fold method is similar to the mandrel, utilizing Z-folded tubes, but the internal mandrel is replaced with an external framework.

Modeling of membrane/inflatable deployment is in its infancy. Modeling is hampered by highly nonlinear deformations, displacements, and material properties encountered with membrane/inflatables. Moreover, although initial and final deployment states may be well defined, it is likely that the intermediate states will be nonunique and nonrepeatable.

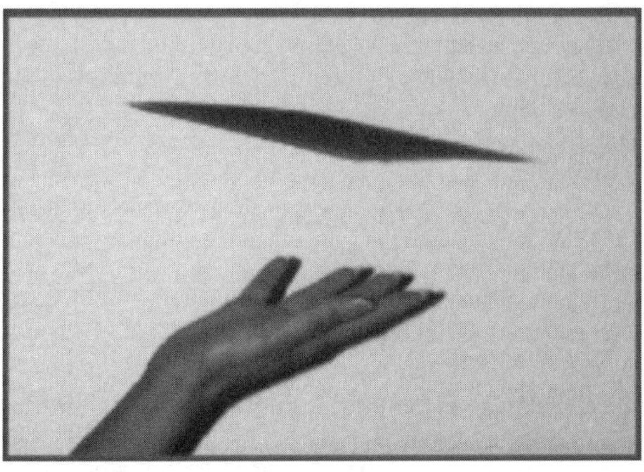

Fig. 5.42 Innovative carbon microtruss fabric (Energy Science Laboratories, Inc.).

VII. Future Directions

A. Future Requirements

It is expected that the focus of future work in the area of membrane and inflatable structures will be on technology development for solar sails and large apertures. For large apertures the areas of specific interest are

1) Very large, ultralightweight (areal density <1 kg/m^2) optical systems for astronomical telescopes and Earth imagers observing in the submillimeter/far IR, IR, Visible, UV, and x-ray spectral regions.

2) Large deployable and/or inflatable antennas for use in space-based radio astronomy, microwave radiometry, radar, and communications.

Desirable performance characteristics for these systems and, in particular, for near-term proof-of-concept technology and hardware development are

1) High packaging efficiency for small launch volume,
2) Rapid, low-cost manufacturing,
3) Design traceable to space-qualified materials,
4) Robust system response to repointing of aperture, and
5) High surface reflectivity after repeated deployment.

The visionary program goal is an areal density of ≤ 1.0 kg/m^2. However, it is expected that initial proof-of-concept hardware will exhibit areal density >1 kg/m^2.

In the solar sail area, the main objective will be to rapidly advance the readiness of key technologies so that a solar sail space flight experiment could be conducted within the next 5 to10 years.

It is expected that both large apertures and solar sails will rely heavily on multifunctional and adaptive systems. The development of multifunctional and adaptive systems applicable to large, ultralightweight deployable or inflatable structures will be very important. To achieve breakthroughs in lowering the cost, launch volume, and mass of future missions, it will be necessary to develop highly integrated multifunctional membranes with embedded thin-film electronics, sensors, actuators, and power sources.

B. Material Requirements

1) A comprehensive membrane/inflatable material property database containing what we now know and how well we know it (test methods).

2) Better interaction of material scientists with applied mechanicians, system designers, mission planners, etc.

3) Multifunctional materials that integrate load carrying, electronics, thermal management, shielding, etc.

4) Materials with reduced densities and improved mechanical, thermal, optical, and electrical properties.

5) Self-healing materials.

C. Analysis Requirements

1) The greatest challenge is to bridge length and time scales, from the nano- and microlevels to structural level response.

2) Better interaction of applied mechanicians with materials scientists, system designers, mission planners, etc.

3) Environmental effects such as aging and degradation must be better linked to mechanical property changes and all incorporated to predictive models.

4) Thorough coupling of numerical and experimental analysis needs to occur.

D. Testing and Deployment Requirements

1) Deployment control. Closed-loop control is essential, whether sequential or parallel. Passive techniques, such as Velcro, shape memory materials, and friction resistance need further development. Sensors are required for state measurement. Active actuators and materials need development.

2) Shape achievement. The disadvantages with lenticular systems include the need for makeup inflatant when tears and micrometeroid punctures occur and the loss of information caused by the double-pass through the canopy. Nonlenticular methods for achieving service shape need to be developed, such as use of stress coatings on near-net shape membranes and the use of precision rigidizable materials.

3) Shape maintenance. Sensors and systems to determine the current shape of large membrane/inflatables need to be developed. Considerable effort needs to be made in modeling deployment dynamics, including employment of stochastic methods. Non-model-based algorithms, such as fuzzy-logic approaches, should be investigated. Integrated material/sensor/actuator systems are required.

4) Damping control. Because membrane/inflatables tend to have high inherent damping, we have not included a discussion of damping control here. However, configurations may exist for certain applications where damping is a significant driver, such as precision instrumentation positioned on the end of a long strut. Hence, some efforts need to be made to explore passive and active damping of such systems.

Acknowledgment

The following personnel contributed to the respective sections of chapter 5:

Overview of Gossamer Structure Technology—Michael Lou, JPL.

Historical Perspective of Inflatable Aerospace Structures—Robert Freeland, JPL, Gordon Veal, L'Garde

Solar Arrays—Mark Grahne and David Cadogan, ILC Dover, Leo Lichodziejewski, L'Garde

Planetary Surface Exploration—Jack Jones, JPL

Radar and Radiometers—Michael Lou, Alfonso Feria, JPL, Mark Grahne, David Cadogan, ILC Dover

Solar Sails—Charles Garner, JPL,

Solar Shades—Michael Lou, JPL, Linda Pacini, GSFC

Telescopes and Optics—Mark Dragovan, JPL, Roger Angel, University of Arizona

Rigidization Technology—Michael Lou, JPL, Mark Grahne and David Cadogan, ILC Dover

Micrometeoroid and Space Debris Effects on Membrane Materials, C. Perrygo, Swales Aerospace Inc., K. Parrish, E. Wooldridge, GSFC

Wrinkling Analysis—Michael Lou, JPL

Deployment Analysis—Moktar Salama, JPL

Adaptive Compensation—Yahya Rahmat-Samii, UCLA, William Imbriale, JPL

Multifunctional Membranes—Saverio D'Augostino, JPL; Ajay Molshe, University of Arkansas
Smart Structures—Yoseph Bar-Cohen, JPL, Chris Moore, LaRC
Micrometeoroid and Debris Testing, C. Perrygo, Swales Aerospace Inc.

Bibliography

Angel, J. R. P., and Woolf, N. J., *ApJ.* 475, 1997, pp. 373–377.

Ash, J. A., Jenkins, C. H., and Marker, D. K., "Deployment of a Membrane Mirror with a Center Plunger," 41st *AIAA/SDM Space Inflatables Forum*, Atlanta, GA, 2000.

Basart, J. P., Mandayam, S. A., and Burns, J. O., "An Inflatable Antenna for Space-Based Low-Frequency Radio Astronomy," *Proc Space '94: Engineering, Construction, and Operations in Space IV*—Vol. 2, Albuquerque, NM, 1994.

Broderick, L., and Jenkins, C. H., "Experimental Investigation of a Fluid-Filled Membrane Breakwater," *J. Waterway, Port, Coastal, Ocean Eng*, 119, 1993, pp. 639–656.

Burge, J. H., Angel, J. R. P., Cuerden, B., Martin, H. M., Miller, S. M., and Sandler, D. G., *Proc SPIE* 3356, 1998, pp. 690–701.

Cadogan, D., and Grahne, M., "Development Control Mechanisms for Inflatable Space Structures," 33rd Aerospace Mechanisms Conference, 1999.

Cadogan, D., Grahne, M., and Mikulas, M., "Inflatable Space Structures: A New Paradigm for Space Structure Design," 49th International Astronautical Congress, 1998.

Cadogan, D., Lin, J., and Grahne, M., "The Development of Inflatable Space Radar Reflectarrays," AIAA 99-1517, 40th AIAA/ASME/ASCE/AHS/ASC SDM Conference, 1999.

Cadogan, D., and Lin, J., "Inflatable Solar Array Technology," 37th AIAA Aerospace Sciences Meeting and Exhibit, 1999.

Cadogan, D., Lin, J., Huang, J., and Feria, A., "An Inflatable Microstrip Reflectarray Concept for Ka-Band Applications," 41st AIAA/ASME/ASCE/AHS/ASC SDM Conference, 2000.

Cadogan, D., Sandy, C., and Grahne, M., "Development and Evaluation of the Mars Pathfinder Inflatable Airbag Landing System," 49th International Astronautical Congress, 1998.

Cadogan, D., Stein, J., and Grahne, M., "Inflatable Composite Habitat Structures for Lunar and Mars Exploration," 49th International Astronautical Congress, 1998.

Cambell, J. D., "On the Theory of Initially Tensioned Circular Membranes Subjected to Uniform Pressure," *Q J Mech Appl Math*, Vol. 9, 1956, pp. 84–93.

Cassapakis, C., and Thomas, M., "Inflatable Structures technology Development Overview," *AIAA* 95-3738, 1995.

Chmielewski, A. B., (ed.), *Arise Mission and Spacecraft*, 2nd Ed., 1999 JPL 99-14, 1999.

Chmielewski, A. B., Moore, C., and Howard, R., "The Gossamer Initiative," IEEE, 1999.

Chmielewski, A. B., Noca, M., and Ulvestad, J., "The Arise Antenna," IEEE, 1999.

Chow, P. Y., "Construction of Pressurized, Self-Supporting Membrane Structure on the Moon," *J Aerospace Engineering*, Vol. 5, 1992, pp. 274–281.

Dragovan, M., "Dual Reflector Telescope (DART)," submitted to *Astrophysical Journal.*

Feria, V., Huang, J., and Cadogan, D., "3-Meter Ka-Band Inflatable Microstrip Reflectarray," Antennas & Propagation Conference, 2000.

Fichter, W. B., "Some Solutions for the Large Deflections of Uniformly Loaded Circular Membranes," *NASA Technical Paper 3658*, 1997.

Föppl, A., "Vorlesungen Uber Technische Mechanik (Leipzig, Germany)," *B.G. Teubner*, Bd. 5., 1907, p. 132.

Freeland, R. E., Bilyeu, G. D., Veal, G. R., Steiner, M. D., and Carson, D. E., "Large Inflatable Deployable Antenna Flight Experiment Results," IAF Paper 97-1.3.01, 48th

Congress of the International Astronautical Federation, Turin, Italy, 6–10 October 1997.

Freeland, R. E., and Veal, G. R., "Significance of the Inflatable Antenna Experiment Technology" 39th AIAA/ASME/ASCE/AHS/ASC Structures, Structural Dynamics and Materials Conference, AIAA/ASME/AHS Adaptive Structures Forum, Long Beach, CA, 20–23 April 1998.

Grahne, M., and Cadogan, D., "Revolutionary Design Concepts for Inflatable Space Structures," 1999-01-5520, World Aviation Congress, 1999.

Grahne, M., Cadogan, D., Sandy, C., and Moriyama, K., "Inflatable Space Structures: A New Paradigm for Space Structure Design," 21st International Symposium on Space Technology and Science, 1998.

Grahne, M. S., Cadogan, D. P., and Lin, J. K., "Inflatable Solar Arrays—Concept to Reality," *IAF*-99, 1999.

Greschik, G., Mikulas, M. M., and Palisoc, A., "Approximations and Errors in Pressurized Asisymmetric Membrane Shape Predictions," 39th *AIAA SDM Conference Proceedings*, AIAA-98-2101-CP Long Beach, CA, 20–23 April 1998.

Greschik, G., Palisoc, A., Cassapakis, C., Veal, G., and Mikulas, M. M., "Approximating Paraboloids with Axisymmetric Pressurized Membranes," 39th *AIAA SDM Conference Proceedings*, AIAA-98-2102-CP Long Beach, CA, 20–23 April 1998.

Grossman, G., and Williams, G., "Inflatable Concentrators for Solar Propulsion and Dynamic Space Power," *J Solar Energy Engineering*, Vol. 112, 1990, pp. 229–236.

Hart–Smith, L. J., and Crisp, J. D. C., "Large Elastic Deformations of Thin Rubber Membranes," *Int J Eng Sci*, Vol. 5, 1967, pp. 1–24.

Hedgepeth, J. M., "Accuracy Potentials for Large Space Antenna Reflectors with Passive Structures," *J Spacecraft*, Vol. 19, No. 3, 1982, pp. 211–217.

Hencky, H., "Uber den Spannungszustand in kreisrunden Platten," *Z. Math. Phys.*, Vol. 63, 1915, pp. 311–317.

Jenkins, C. H., "Membrane Wrinkling: Theoretical Predictions and Experimental Verifications," (invited) *Symposium on Advances in Mechanics of Elastic and Bioelastic Membranes at the Int Mech Engr Conf Expo, ASME*, San Francisco, CA, 1995.

Jenkins, C. H., "Nonlinear Dynamic Response of Membranes: State of the Art—Update," *Appl Mech Rev*, Vol. 49 No. 10, 1996, pp. S41–S48.

Jenkins, C. H., "Membrane Vibrations: a Review and New Experimental Results," *1999 ASME Joint Applied Mechanics and Materials Summer Conference*, Blacksburg, VA, 1999.

Jenkins, C. H., and Faisal, S. M., "Thermal Load Effects on Precision Membranes," 1999 *AIAA Adaptive Structures Forum*, St. Louis, MO, 1999.

Jenkins, C. H., and Kalanovic, V. D., "Issues in Control of Space Membrane/Inflatable Structures," *IEEE Aerospace Conference*, Big Sky, MT, 2000.

Jenkins, C. H., and Khanna, S. K., "Determination of Membrane Wrinkling Parameters using Shadow Moire," *1999 Spring Conference, Society for Experimental Mechanics*, Cincinnati, OH, 1999.

Jenkins, C. H., and Kondareddy, S., "Dynamics of a Seamed Membrane," *1999 AIAA Adaptive Structures Forum*, St. Louis, MO, 1999.

Jenkins, C. H., and Leonard, J. W., "Nonlinear Dynamic Response of Tension Fabric Structures—A Review," *Proc Eighth Struct Conf*, Baltimore, MD, 1990.

Jenkins, C. H., and Leonard, J. W., "State of the Art: Nonlinear Dynamic Response of Membranes," *Proc Forth Rail Bridge Conf*, Edinburgh, 1990.

Jenkins, C. H., and Leonard, J. W., "Transient Deformation of Viscoelastic Membranes," *Proc First Int Offshore Polar Eng Conf*, Edinburgh, 1991.

Jenkins, C. H., and Leonard, J. W., "Nonlinear Transient Deformation of Viscoelastic Membrane Structures," *Struct Eng Rev*, Vol. 3, 1991a, pp. 197–204.

Jenkins, C. H., and Leonard, J. W., "Transient Response of Wrinkled Viscoelastic Membranes," *Dev Theoretical Appl Mech* (*Proc S E Conf Theoretical Appl Mech*) XVI, Nashville, TN, 1992.

Jenkins, C. H., and Leonard, J. W., "Dynamic Wrinkling of Viscoelastic Membranes," *J Appl Mech*, Vol. 60, 1993, pp. 575–582.

Jenkins, C. H., and Liu, X., "Computational Issues in the Modeling of Wrinkling During Parachute Inflation," *Symposium on Deployable Structures*, Cambridge, UK, 1998.

Jenkins, C. H., and Marker, D. K., "Surface Precision of Inflatable Membrane Reflectors," *J Solar Energy Eng*, Vol. 120 No. 4, 1998, pp. 298–305.

Jenkins, C. H., and Najdawi, H. F., "Experimental Investigation of Wrinkling in a Bi-Thickness Membrane," *1998 Spring Conference, Society for Experimental Mechanics*, Houston, TX, 1998.

Jenkins, C. H., and Tampi, M., "Local Membrane Vibrations and Inflatable Space Structures," *Space 2000*, Albuquerque, NM, 2000.

Jenkins, C. H., and Vinogradov, A., "Active Polymers for Space Inflatables: Properties and Applications," *IEEE Aerospace Conference*, Big Sky, MT, 2000.

Jenkins, C. H., and Leonard, J. W., "Nonlinear Dynamic Response of Membranes: State of the Art," *Appl Mech Rev*, Vol. 44, 1991, pp. 319–328.

Jenkins, C. H., (ed.) (to appear), Gossamer Spacecraft: *Inflatable Structure Technology for Space Applications*, AIAA Progress in Astronautics and Aeronautics Series, Reston, VA.

Jenkins, C. H., Fitzgerald, D., and Liu, X., "Wrinkling of an Inflated Membrane with Thermo-Elastic Boundary Restraint," *41st AIAA/SDM Space Inflatables Forum*, Atlanta, GA, 2000.

Jenkins, C. H., Freeland, R. E., Bishop, J. A., and Sadeh, W. Z., "An Up-to-date Review of Inflatable Structures Technology for Space-based Applications," *Space 98: 6th Int Conf & Expo on Engineering, Construction, and Operations in Space*, Albuquerque, NM, 1998.

Jenkins, C. H., Haugen, F., and Spicher, W. H., "Experimental Measurement of Wrinkling in Membranes Undergoing Planar Deformation," *Exp Mech*, Vol. 38, 1998, pp. 147–152.

Jenkins, C. H., Haugen, F., Kalanovic, V. D., and Najdawi, H. F., "Experimental Measurement for Control of Membrane Wrinkling," *1997 Spring Conference, Society for Experimental Mechanics*, Bellevue, WA, 1997.

Jenkins, C. H., Kalanovic, V. D., Faisal, S. M., Padmanabhan, K., and Tampi, M., "Adaptive Shape Control of Precision Membrane Antennae and Reflectors," *Adaptive Structures and Material Systems Symposium, Int Mech Engr Conf Expo, ASME*, Anaheim, CA, 1998.

Jenkins, C. H., Kalanovic, V. D., Padmanabhan, K., and Faisal, S. M., "Intelligent Shape Control for Precision Membrane Antennae and Reflectors in Space," *Smart Matls. Struct*, Vol. 8, 1999, pp. 1–11.

Jenkins, C. H., Marker, D. K., and Wilkes, J. M., "Improved Surface Accuracy of Precision Membrane Reflectors Through Adaptive Rim Control," *AIAA Adaptive Structures Forum*, Long Beach, CA, 1998.

Jenkins, C. H., Marker, D. K., and Wilkes, J. M., "Improved Surface Accuracy of Precision Membrane Reflectors Through Adaptive Rim Control," *1998 AIAA Adaptive Structures Forum*, Long Beach, CA, 1998.

Jenkins, C. H., Najdawi, H. F., and Haugen, F., "Wrinkling of Plane Elastic Sheets," *1997 Joint ASME, ASCE, SES Summer Meeting*, Northwestern University, Evanston, IL, 1997.

Jenkins, C. H., Spicher, W., and Vedoy, A., "Noncontact Measurement of Membrane Wrinkling," *24th Midwestern Mechanics Conf*, Ames, IA, 1995.

Jenkins, C.H., Spicher, W. H., and Vedoy, A., "Experimental Measurement of Wrinkling in Plane Elastic Sheets," *1996 Spring Conference, Society for Experimental Mechanics*, Nashville, TN, 1996.

Jenkins, C. H., Tampi, M., Kalanovic, V. D., and Padmanabhan, K., "Practical Aspects of Precision Membrane Antennae Shape Control," *IEEE Int. Conf. on Systems, Man, and Cybernetics*, La Jolla, CA, 1998.

Jenkins, C. H., Wilkes, J. M., and Marker, D. K., "Surface Accuracy of Precision Membrane Reflectors," *Space 98*, Albuquerque, NM, 1998.

Jenkins, C. H., Wilkes, J. M., and Marker, D. K., "Surface Accuracy of Precision Membrane Reflectors," *Space 98: 6th Int Conf & Expo on Engineering, Construction, and Operations in Space*, Albuquerque, NM, 1998.

Kalanovic, V. D., Jenkins, C. H., and Haugen, F., "Control of Membrane Wrinkling via Intelligent Control Strategies," *6th IEEE Conference on Control Strategies*," Hartford, CT, 1997.

Kalanovic, V. D., Jenkins, C. H., and Haugen, F., "Fuzzy Control of Membrane Wrinkling," *Intell Automation Soft Comput.*, Vol. 5, 1999, pp. 139–148.

Kalanovic, V. D., Padmanabhan, K., and Jenkins, C. H., "A Discrete Cell Model for Shape Control of Precision Membrane Antennae and Reflectors," *Adaptive Structures and Material Systems Symposium, Int Mech Engr Conf Expo, ASME*, Nashville, TN, 1999.

Laug, K., personal communication.

Leonard, R. W., Brooks, G. W., and McComb, H. G., "Structural Considerations of Inflatable Reentry Vehicles, *NASA Technical Note* D-457, 1960, pp. 1–23.

Lin, J., Cadogan D., Sandy, C., and Folke, J., "Inflatable Boom Controlled Deployment Mechanism for Inflatable Sunshield," 34th Aerospace Mechanisms Conference, 2000.

Liu, X., and Jenkins, C. H., "Issues in Modeling of Parachute Wrinkling," *25th Midwestern Mechanics Conf.*, Rapid City, SD, 1997.

Liu, X., and Jenkins, C. H., "Issues in Modeling of Parachute Wrinkling," *25th Midwestern Mechanics Conf.*, Rapid City, SD, 1997.

Liu, X., Jenkins, C. H., and Schur, W. W., "Fine Analysis of Degenerate Membrane States using a Penalty Parameter Modified Constitutive Relation," *Fifth U.S. National Congress on Computational Mechanics*, Boulder, CO, 1999.

Liu, X., Jenkins, C. H., and Schur, W. W., "Large Deflection Analysis of Membranes by a User-Supplied Penalty Parameter Modified Material Model," *Fifth U.S. National Congress on Computational Mechanics*, Boulder, CO, 1999.

Lou, M., and Feria, V. A., *Development of Space Inflatable/Rigidizable Strucutres Technology*, JPL.

Main, J. A., Peterson, S. W., and Strauss, A. M., "Load-Deflection Behavior of Space-Based Inflatable Fabric Beams," *J. Aerospace Engineering*, Vol. 7, No. 2, 1994, pp. 225–238.

Malone, P. K., and Williams, G. T., "A Lightweight Inflatable Solar Array," *Proc 9th Annual AIAA/USU Conference on Small Satellites*, Logan, UT, 1995.

Marker, D. K., and Jenkins, C. H., "Surface Precision of Optical Membranes with Curvature," *Optics Express*, Vol. 1, 1997, pp. 324–331.

Marker, D. K., Carreras, R. A., and Jenkins, C. H., "Autonomous Control of Large Membrane Optical Space Sensors," *World Automation Congress*, Anchorage, AK, 1998.

Marker, D. K., Carreras, R. A., Wilkes, J. M., Jenkins, C. H., Duneman, D., Rotge, J. R., and Hogge, C. B., "Optical Evaluation of Membranes Mirrors with Curvature," *9th Conf on Lasers and Optics 98*, St. Petersburg, Russia, 1998.

Marker, D. K., Jenkins, C. H., and Schoof, L., "On the Systematic "W" Profile Error in Uncompensated Isotropic Membrane Reflectors," *25th Midwestern Mechanics Conf.*, Rapid City, SD, 1997.

Murphy, L. M., "Stretched-Membrane Heliostat Technology," *J Solar Energy Eng*, Vol. 108, 1986, pp. 230–238.

Murphy, L. M., "Moderate Axisymmetric Deformations of Optical Membrane Surfaces," *J Solar Energy Eng*, Vol. 109, 1987, pp. 111–120.

Murphey, T. W., and Mikulas, M. M., "Nonlinear Effects of Material Winkles on the Stiffness of Thin Polymer Films," AIAA Paper 99-1341, *Proc. 40th AIAA SDM*, St. Louis, MO, 1999.

Najdawi, H. F., and Jenkins, C. H., "Passive Control of Membrane Wrinkling," *25th Midwestern Mechanics Conf.*, Rapid City, SD, 1997.

Natori, M., Shibayama, Y., and Sekine, K., "Active Accuracy Adjustment of Reflectors Through the Change of Element Boundary," AIAA 89-1332, 1989.

Nowak, P. S., Sadeh, W. Z., and Janakus, J., "Feasibility Study of Inflatable Structures for Lunar Base," *J Spacecraft Rockets*, Vol. 31, 1994, pp. 453–457.

Palisoc, A. L., and Huang, Y., "Design Tool for Inflatable Space Structures," AIAA paper 97-1378, 1997.

Palisoc, A., and Thomas, M., "A Comparison of the Performance of Seamed and Unseamed Inflatable Concentrators," *Solar Engineering, Proc. ASME/JSME/JSES Int. Solar Energy Conf.*, Vol. 2, 1995, pp. 855–864.

Perrygo, C., Parrish, K., and Wooldridge, E., "Sunshield Meteoroid Impact Test and Analysis Requirements Definition," NGST Memorandum, 1999.

Rapp, D., "Prospects and Limitations of Technical Approaches for Ultra Lightweight Space Telescopes," *JPL Rep. D-13975*, 1996.

Rogers, C. A., Stultzman, W. L., Campbell, T. G., and Hedgepeth, J. M., "Technology Assessment and Development of Large Deployment Antennas," *J Aerospace Engineering*, Vol. 6, No.1, 1993, pp. 34–54.

Ruggiero, T. J., and Mikulus, M. M., "A One-Dimensional Constitutive Model for Wrinkled Thin Polymer Films," NASA CR-1999-209129, 1999.

Sadeh, W. Z., and Criswell, M. E., "A Generic Inflatable Structure for a Lunar/Martian Base," *Space IV, Proc Space '94*, Albuquerque, ASCE, 1994, pp. 1146–1156.

Sandy, C., Next Generation Space Telescope Inflatable Sunshield Development, IEEE, 2000.

Stevens, H. H., "Behavior of Circular Membranes Stretched Above the Elastic Limit by Air Pressure," Experimental Stress Analysis, Vol. 2, 1944, pp. 139–146.

Thomas, M., and Veal, G., "Highly Accurate Inflatable Reflectors," AFRPL TR-84-021, 1984.

Vaughn, H., "Pressurizing a Prestretched Membrane to Form a Paraboloid," *Int J Eng Sci*, Vol. 18, 1980, pp. 99–107.

von Kármán, T., "Festigkeitsproblem im Maschinenbau," *Encyk D Math Wiss* IV, 1910, pp. 311–385.

Wilkes, J. M., Jenkins, C. H., Marker, D. K., Carreras, R. A., Duneman, D. C., and Rotge, J. R., "Concave Membrane Mirrors from Aspheric to Near-Parabolic," *Int. Symp. Optical Science, Engineering, and Instrumentation (SPIE)*, Denver, CO, 1999.

Woolf, N. J., and Angel, J. R. P., Planets and Beyond the Solar System, edited by D. R. Soderblom , A.S.P. Conference Series Vol. 119, 1997, pp. 285–293.

Chapter 6

Smart Air and Space Structures

Janet M. Sater
Institute for Defense Analyses, Alexandria, Virginia

C. Robert Crowe
Virginia Polytechnic Institute and State University, Alexandria, Virginia

Richard Antcliff
NASA Langley Research Center, Hampton, Virginia

and
Alok Das
Air Force Research Laboratory, Kirtland AFB, New Mexico

I. Introduction

DURING the past decade, the multidisciplinary field of smart materials and structures has experienced rapid growth in terms of individual technologies and applications. The integration of sensors, actuators, and controllers with structures that enable adaptation to environmental and operational conditions has progressed to such a point that numerous systems applications are being demonstrated. This chapter reviews the current status, results to date, and issues associated with a number of these projects. This review is not comprehensive to the entire body of literature but instead focuses on realistic sub- or full-scale system demonstrations and relevant characterization and testing. The status of individual technologies important to achieving the ultimate objective of a "smart" system is also addressed in some detail.

The idea of synthesizing smart materials and structures dates at least as far back as 1968[1] when Henry Clauser published this idea in a broad, conceptual form. By 1975, Clauser had fully developed the concept of engineered materials as dynamic systems that could replace mechanical and electrical components and respond to service conditions.[2] By 1978, the idea had received international attention. R. L. Forward, among the first researchers to investigate the possibility of using piezoceramic devices as passive dampers in mechanical systems, patented several innovative concepts in the late 1970s.[3–5] In 1983, Forward, Swigert, and Obal[6] completed one of the first successful vibration control demonstrations using surface-mounted piezoceramic sensors and actuators. A substantial amount of jitter reduction was achieved on a cavity resonator mirror by combining a passive tuned mass damper with an active rate feedback vibration control network that included piezoceramic devices. Following these early efforts, much of the work into the early 1990s focused on vibration suppression applications in spacecraft, work funded primarily by the Ballistic Missile Defense Organization (BMDO, then Strategic Defense Initiative Organization, or SDIO) and the U.S. Air Force (USAF).[7–11] It is important to note that the bulk of work in this field has been supported by the Department of Defense for reasons that will become obvious.

In 1993, the Defense Advanced Research Projects Agency (DARPA) recognized that smart materials and structures technology provided a specific opportunity for many technological breakthroughs. An 8-year program was initiated to develop new, affordable smart materials and structures and demonstrate performance gains achievable in system applications. The focus of the DARPA program is to use smart materials to achieve aerodynamic and hydrodynamic flow control and reduce noise and vibration in a variety of structures. A portion of this effort is also directed toward improving the authority of actuation materials and their use in actuators to expand potential applications of the technology.

Since the early 1990s, the Army has supported more basic research in smart structures and materials. Current efforts of the Army Research Office in the smart structures area include vibration suppression, noise reduction, antenna shape control, health monitoring, and nonlinear control systems. Most of these programs are focused on helicopter performance issues. Other groups in the Army are addressing advanced rotary wing concepts, including active on-blade control and active blade twist as well as vibration issues for large guns. Since the early 1990s, the USAF, the National Aeronautics and Space Administration (NASA), and the Navy have also initiated several programs to demonstrate the application of smart structures in a variety of systems—most specifically, fixed-wing aircraft, helicopters, spacecraft, missiles, and submarines. For example, NASA has worked closely with the Services (USAF and Army) and DARPA to address specific problems on helicopters and aircraft. They have also initiated programs to investigate the application of these and other advanced technologies to new aircraft and spacecraft concepts. These and related USAF efforts are focused on aerodynamic flow control, vibration and noise suppression, and optimization of lifting surfaces. They also include modifying structural dynamics and aeroelastics, providing for flight path controls, and integrating electronics in structures.[12] In addition, the Navy has funded significant work in the development of actuator materials, especially ceramics and, most recently, single crystal piezoceramics. Other focus areas for the Navy include structural health monitoring via fiber optics; control of large, precision spacecraft; hydrodynamic flow control; vibration and noise suppression; and optimization of lifting surfaces for underwater vehicles.

The idea of using adaptive structures to improve the performance of military aircraft is not new. A joint NASA/Wright Laboratory demonstration program—the mission adaptive wing, flight-tested during the 1980s on an F-111A test aircraft—investigated active control of chordwise camber, spanwise camber, and wing sweep while maintaining a smooth, continuous airfoil.[13–16] Variable leading and trailing edge shapes were of particular interest. Shapes with large camber, large leading-edge radii, and large depths are desired for low-speed flight because they provide high maximum lift coefficients. In high-speed flight, on the other hand, drag becomes the dominant parameter; therefore, low camber, small leading-edge radii, and shallow depths are highly desirable. Observed benefits for the F-111A wing with variable leading and trailing edges were improved performance and terrain following, control of maneuver loads, and reduced radar cross section (RCS). While stealth and aerodynamic benefits were achieved, the devices and linkages needed to obtain the shape changes were so complicated as to be impractical for implementation.

At present, there are three approaches to the development of smart materials and structures. The first approach attempts to synthesize new materials at the atomic and molecular level to produce new materials with smart functionality. The success of this approach will depend on new scientific discoveries. As a result, the technologies derived from this approach are very immature. For the second approach, actuators and sensors are attached to a conventional structure. This approach is the most mature, but the active components are parasitic to the structure, a feature that may make it less attractive for implementation and that does not take full advantage of potential "smart" systems capabilities. The third approach attempts to develop new materials by synthesizing composite systems from known materials. These composites contain active constituents (sensors and actuators) and are used to fabricate the structure. The conceptual difference between the latter two approaches is as follows: by adding parasitic sensors and actuators (second approach), the control system makes the structure respond; by integrating sensors and actuators into the material itself to form a composite (third approach), the *material* responds to the control system.

Sensors include conventional strain gauges, fiber optics, and piezoelectric ceramics and polymers, among others. Embedded sensors can be used in discrete or distributed locations to provide built-in structural quality assessment during composite processing and system operation. For system performance, it is important that the right features be measured by the sensors (e.g., in a shape-adaptive airfoil, strain associated with curvature or deflection is measured to reflect changes in airfoil shape) and that the signal be interpreted with respect to the desired performance outcome (e.g., a change in lift). Typical smart structure actuators are shape memory alloys (SMAs), piezoelectric and electrostrictive ceramics, and magnetostrictive materials. When combined with a sensor/signal processing network and an appropriate control system, actuators allow structural performance to be changed or adapted to meet various operational performance criteria (e.g., to change lift). Actuator devices can be used either dynamically, such as for vibration suppression, or quasi statically, such as for shape control. Electro- and magneto-rheological fluids and elastomers, although not strictly sensor or actuator materials, find use in actively changing attributes such as structural stiffness. The challenge, then, is one of designing and synthesizing the material and fabricating the structure to realize anticipated performance gains.

II. Vision

Structural integrity monitoring represents a more near-term opportunity for implementation of smart materials and structures technologies. Recent advances in sensors, data acquisition capabilities, electronics miniaturization, and sensor system integration offer unprecedented opportunities for an integrity monitoring system.[17] Such integrated systems could provide accurate, detailed load histories for the air or space vehicle. These histories can be used to determine the location of damage (possibly including the amount of damage), thus simplifying inspection and life-cycle monitoring. This, in turn, could reduce staff time expended on current manual methods and provide significant cost savings.

Vibration problems occur in all types of air and space vehicles. When applied to structural dynamics problems, the use of smart materials and structures technology is expected to stabilize dynamic instabilities and significantly alleviate vibrations and, hence, enhance fatigue life. Acoustic signature affects military operations by increasing the detectability of air vehicles. It also affects commercial operations: high noise levels in fly-overs or landing result in limited community acceptance. Vibrations impact passenger/crew comfort and weapons accuracy and the fatigue lives of most structural and electronic components are adversely affected. As a result, system reliability is reduced and maintenance activity increases. Vibrations, acoustics, and shock stemming from large launch loads dominate designs for spacecraft and launch vehicles. These design limitations, in turn, lead to significant, excess mass and high launch costs. Billions of dollars in lost satellites or degraded performance of precision payloads are attributable to failures arising from launch load vibrations. Several specific vibration problems have been identified and targeted for systems demonstrations using smart materials and structures technologies. These demonstrations include launch isolation, precision pointing, isolation from shock caused by pyrotechnic release devices, isolation of electronic components from forced vibrations, interior noise suppression, tail buffet damping, wing flutter control, helicopter blade–vortex interaction (BVI), and helicopter blade tracking adjustment.

The concept of shape-adaptive structures and aerodynamic flow control figures predominantly in current thinking about applications for smart materials and structures. Ultimately, the need for these concepts is driven by the fact that fixed geometry structures like aircraft wings and engine inlets exhibit nonoptimal performance over a range of flight conditions, although they may exhibit exceptional performance at a specific flight condition. Design concepts of interest include, for example, wing warping, camber shaping, inlet lip blunting, and inlet wall shaping. Among expected performance benefits are enhanced maneuverability, improved aeroelastic effects, reduced signature and drag, increased takeoff gross weight, and increased range capabilities.

Other smart materials and structures concepts use multifunctional structures [i.e., structures that are designed to include multiple functional components, such as radio frequency (RF) antennas, signal processors, various types of sensors, wiring, and cabling into a conformal structure] in a smart skin. Nonconformal, externally mounted antennas degrade vehicle aerodynamic performance, require substantial maintenance, and increase vehicle signature. Potential benefits of these integrated systems include reduced weight and volume; low observability (especially for conformal antennas); reduced energy consumption; improved system performance, including flexible capabilities to enable new missions; and lower costs because of reduced duplication and potentially easier repair and maintenance. Significant weight and volume reductions are particular advantages for spacecraft containing integrated power distribution and data-transmission cabling. The need for large weight and volume reductions will become increasingly important with the advent of nanosatellites (about 10 kg) and microsatellites (about 100 kg).

Future commercial and military air and space systems will benefit substantially from smart materials and structures technology. Expected benefits include, among others (ranked, more or less, from near term to far term):

1) Allowing for "maintenance on demand" (or condition-based maintenance) by monitoring system health to include damage detection and, eventually, mitigation and repair.

2) Increasing passenger/crew comfort by reducing interior cabin/vehicle noise.

3) Increasing system/component structural life by reducing structural and acoustic vibrations caused by panel flutter, buffet, BVI, and spacecraft launch environments, including those vibrations caused by other mounted components like motors or gear boxes.

4) Improving precision pointing and/or sensing capabilities by reducing structural, acoustic, and shock vibrations in on-orbit spacecraft.

5) Enhancing aircraft and rotorcraft handling by manipulating lift or reducing drag, changing control surface shape, or affecting flow conditions over the lifting surface.

6) Improving aerodynamics and possibly enabling new flight profiles by producing twist in aircraft wings or helicopter rotor blades.

7) Improving aerodynamics and low observable characteristics and reducing manufacturing and assembly costs by integrating power and electronic systems to the structure.

III. Significant Demonstration Projects

This section highlights major demonstration programs addressing structural health monitoring, vibration suppression, shape control, and multifunctional structural concepts for spacecraft and launch vehicles, aircraft, and rotorcraft. These demonstrations are focused on showing potential system-level performance improvements using smart technologies in realistic structures.

A. Structural Integrity Monitoring

Aircraft structural maintenance has evolved considerably since the late 1950s when fatigue problems in aircraft were first observed. Continuous integrity monitoring of aircraft structures has received particular, significant attention.[18–20] Recent advances in sensors, data-acquisition capabilities, electronics miniaturization, and sensor system integration offer unprecedented opportunities for a structural integrity monitoring system (SIMS) and, eventually, for "maintenance on demand."[17,21] In general, an integrity monitoring system consists of

1) Sensors for acquisition of structural properties,

2) Signal acquisition and analysis electronics to process the sensor output, and

3) A mathematical algorithm to extract information about damage in the structure.

Such integrated systems could provide accurate, detailed load histories for the aircraft and locations of damage (including the amount of damage). Large area damage, such as that which would result from gunfire, is much easier to assess. Finding small, damaged areas that may later lead to failure is a much more difficult and challenging problem. Recent concepts look to combine local preprocessors with sensors capable of detecting loads and environments. The central processor interrogates each of the local preprocessors to obtain sensor data that are then analyzed and stored as required. This type of system offers flexibility because it can be enlarged to handle growth. Having a network of sensors also offers fault tolerance and redundancy. An issue with this approach, however, is information management. Synthesis and analysis of signals from large numbers of sensors to derive appropriate diagnostic information and prognoses are still evolving. Efforts are currently underway to reduce the total number of sensors required by using available aircraft usage data (altitude, Mach number, weight, fuel consumption, total flight hours, etc.) to infer stresses and damage using suitably validated parametric and analytical models.

Under a jointly sponsored USAF/Navy program, Northrop Grumman developed and demonstrated a prototype SIMS on a simulated wing spar and a wing carry-through bulkhead (a multibay structure), as shown in Fig. 6.1.[17] Results indicated that the acoustic emission (AE) sensors could detect cracks as far away as 18 in. in simple geometries (e.g., wing spar web). Cracks were much harder to detect in complicated geometries (e.g., stiffeners between the bays in the bulkhead). For complex structures, the AE sensors had to be located within a few inches of the flaw to obtain positive identification. Fiber-optic sensors were more sensitive, detecting a small torsion in the wing spar that was not detected using conventional strain gauges. These approaches were evaluated during a full-scale fatigue test of an F/A-18 wing attach bulkhead.[22] The bulkhead failed at 9,000 equivalent flight spectrum hours of loading at a somewhat unexpected location away from the sensors: the collected broadband AE data revealed growth of the failure crack at about 7,000 hours; however, because processing and pattern recognition were not automated, the crack detection was performed posttest by an AE expert. Northrop Grumman is currently developing automation routines. Expected benefits of this technology include improved safety, reduced and simpler maintenance, and reduced life-cycle costs. Cost savings could be quite substantial: >$35 million is predicted for the F-18 (assuming 33 flight-hours/aircraft/month), 1,000 in fleet);[23] >$9 million is predicted for the T-38 (assuming 420 flight-hours/aircraft, 720 in fleet).[24] More significant savings are expected from the elimination of logistics personnel: an estimated $100,000 per year in staff and equipment with the automation of just one logistics function.[25]

Researchers at NASA–Langley have also investigated the application of fiber optics for health monitoring.[26] In this project, fiber-optic distributed strain sensors were used to measure strains on a dual cantilever beam. This technique is appropriate for monitoring the cure processing of composite structural components as well as their in-flight operations.

Another approach that has been used is to excite the structure with piezoelectric actuators and use piezoelectric sensors to monitor the response at selected points.[19] Changes of the structural response to excitation are measured over time and are then correlated with damage through analysis of measured frequency response functions of the structure. This technique has been tested on a rib of the vertical fin of the ATR 42 aircraft with good results. In this test, rivets were removed to simulate damage.

McDonnell Douglas (now Boeing) developed a similar approach, referred to as active damage interrogation (ADI). It uses an array of piezoelectric transducers—either embedded or attached to a structure—that act as both sensors and actuators.[27] This system, which is model independent, actively interrogates the structure via broadband excitation of multiple transducers across the desired frequency range. The severity of the damage and its location are determined via statistical analysis of changes in transfer functions. ADI methodology was successfully tested on a composite flexbeam on the MD-900 Explorer rotor system.

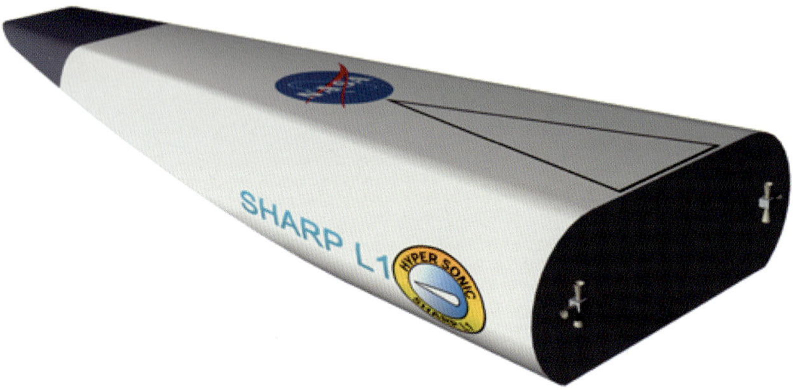

Fig. 1.2

Fig. 1.3a

Fig. 1.3b

Fig. 1.4a

Fig. 1.4b

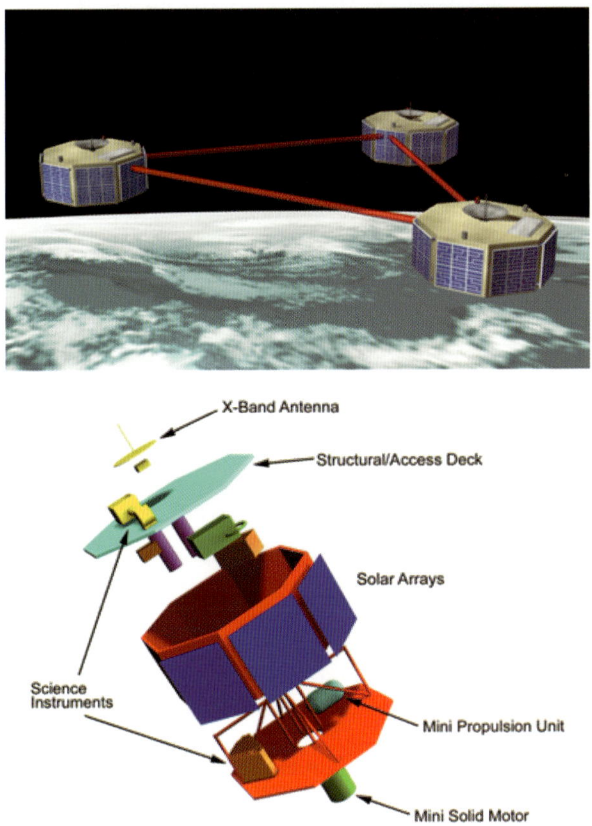

Fig. 1.5

Fig. 1.6

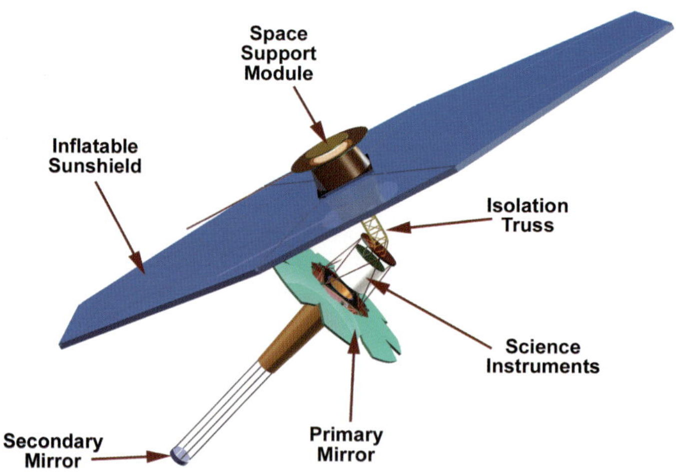
Space
Support
Module
Inflatable
Sunshield
Isolation
Truss
Science
Instruments
Secondary
Mirror
Primary
Mirror

Fig. 1.7

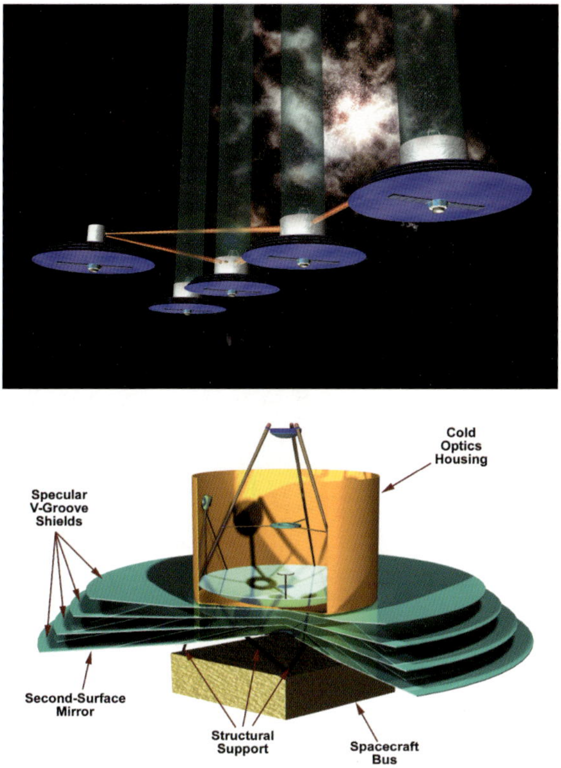
Cold
Optics
Housing
Specular
V-Groove
Shields
Second-Surface
Mirror
Structural
Support
Spacecraft
Bus

Fig. 1.8

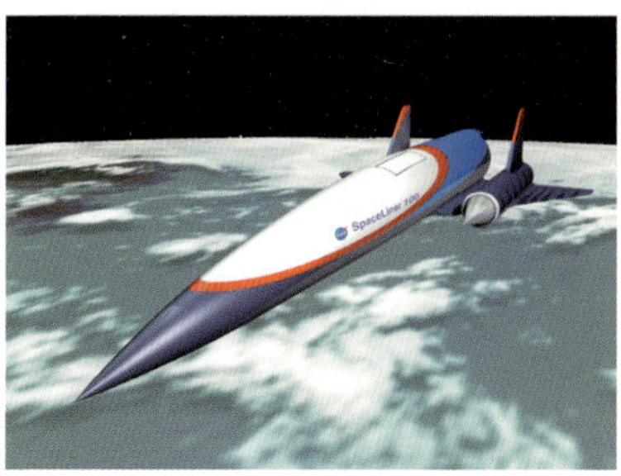

Fig. 1.9

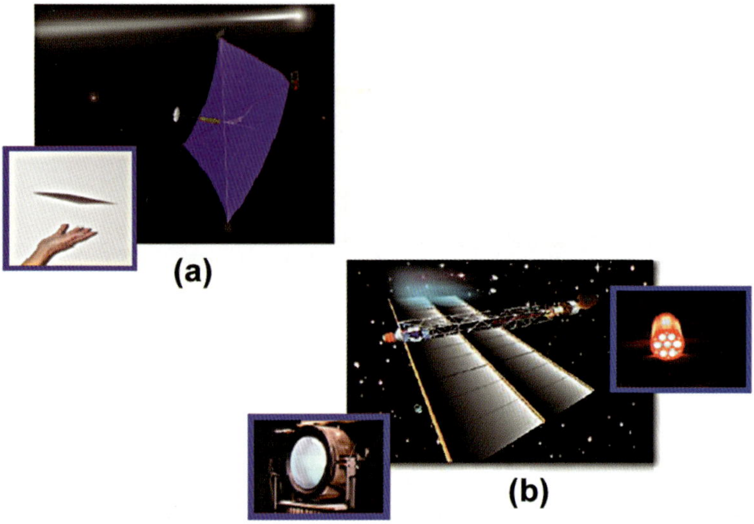

Fig. 1.10

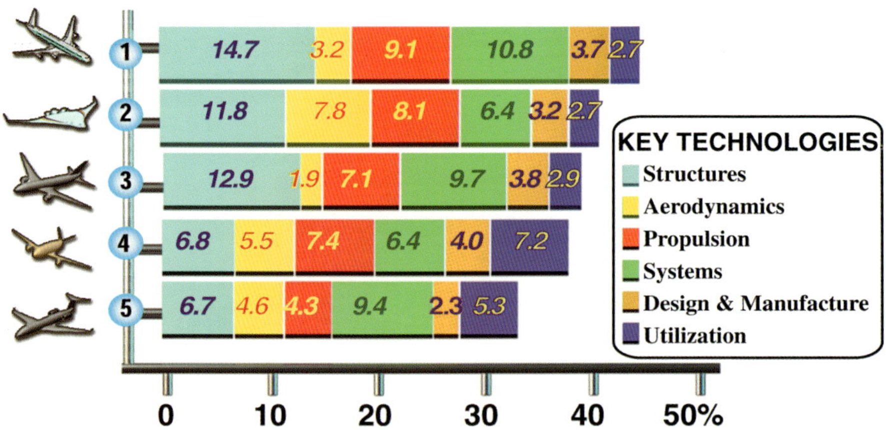

Fig. 1.11

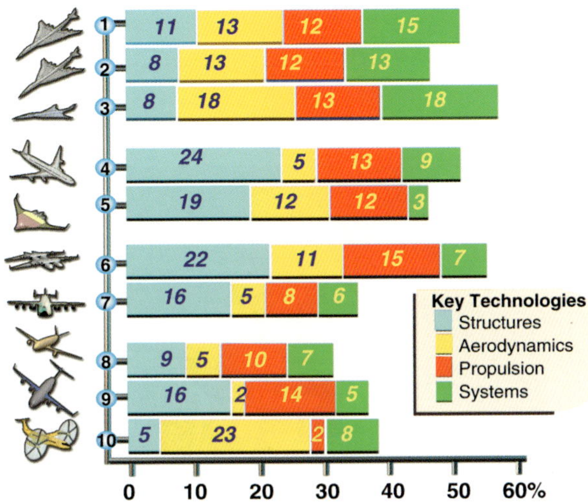

Fig. 1.12

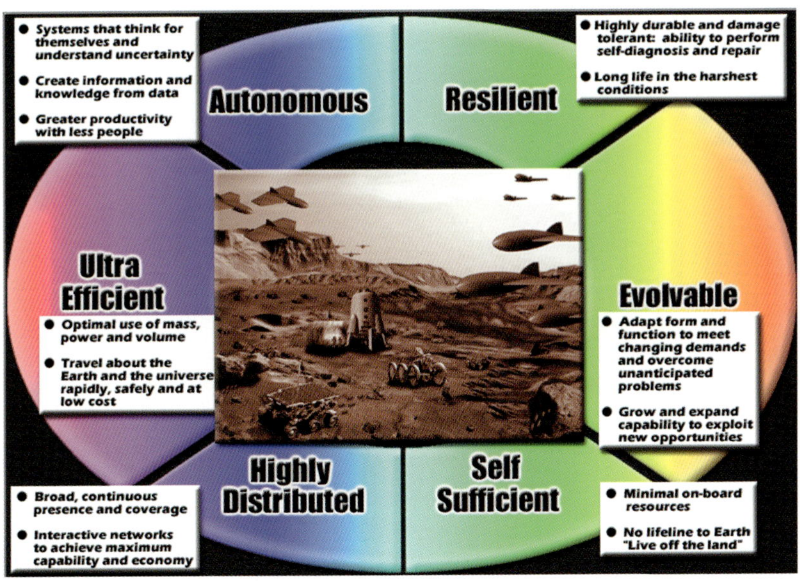

Fig. 1.13

Fig. 1.14

Fig. 1.15

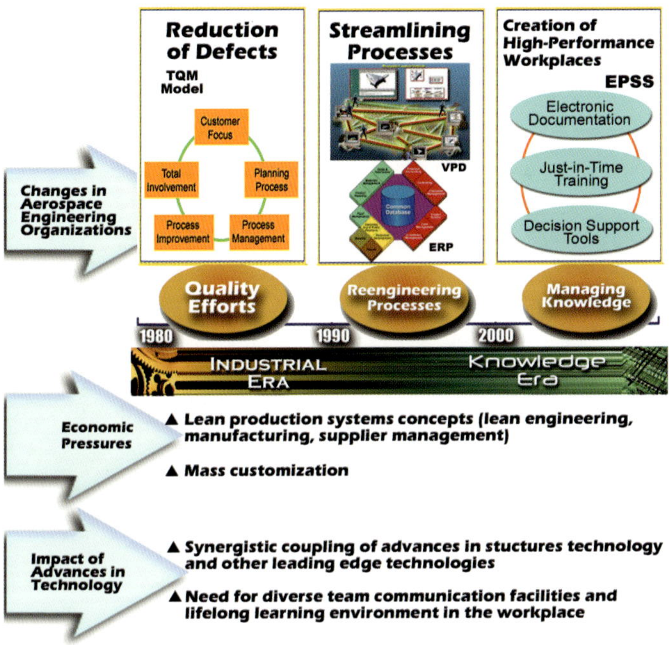

Fig. 1.17

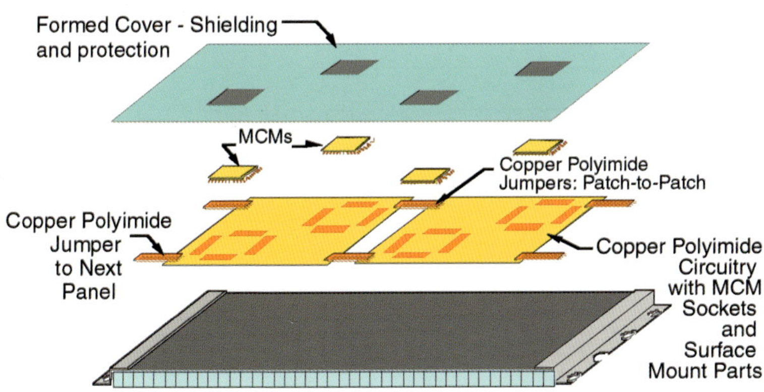

Fig. 1.19

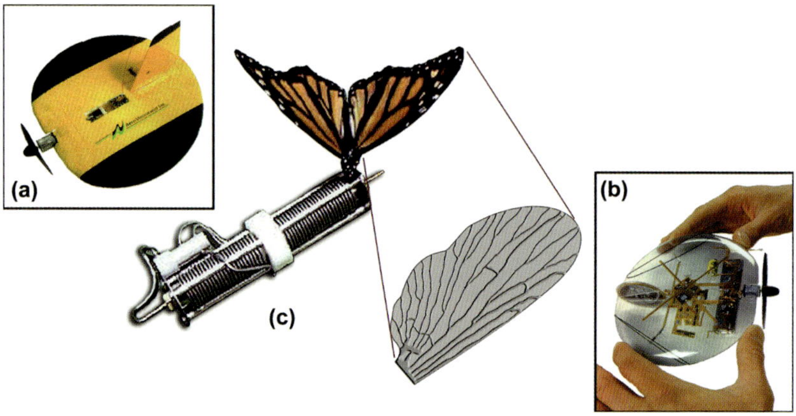

Fig. 1.20

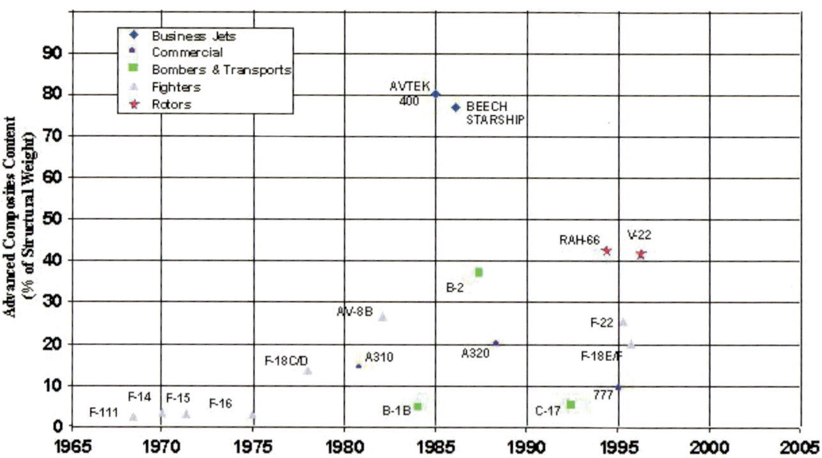

Fig. 2.3

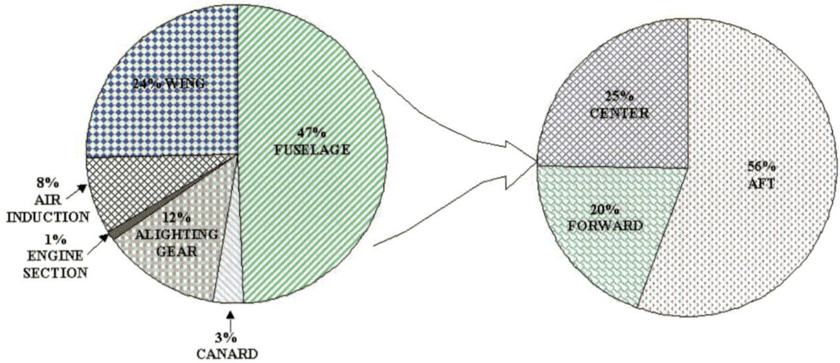

Fig. 2.5

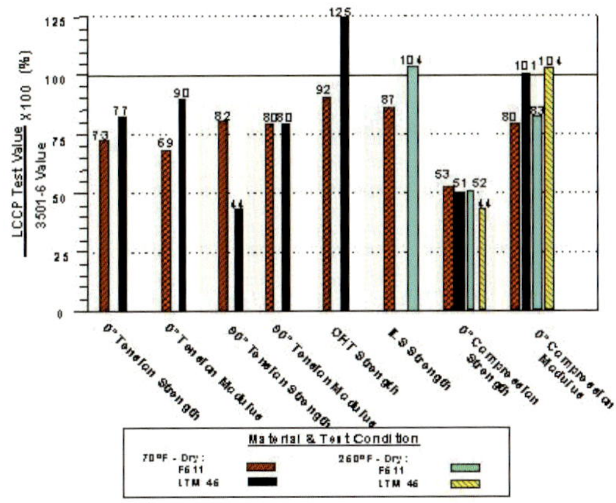

Fig. 2.11

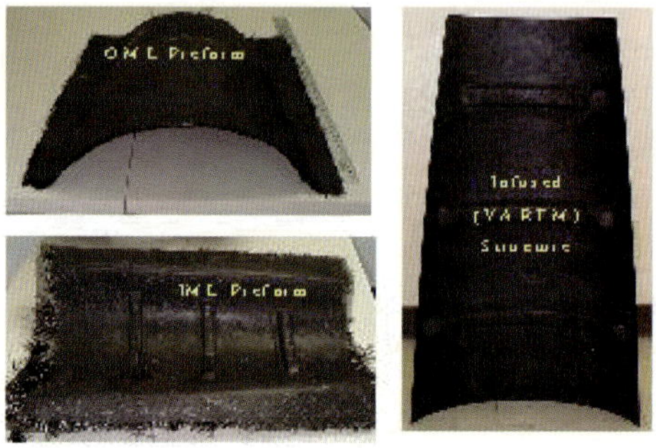

Fig. 2.20

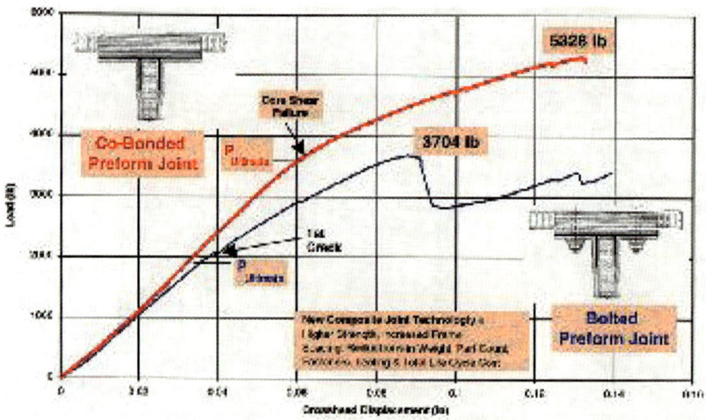

Fig. 2.30

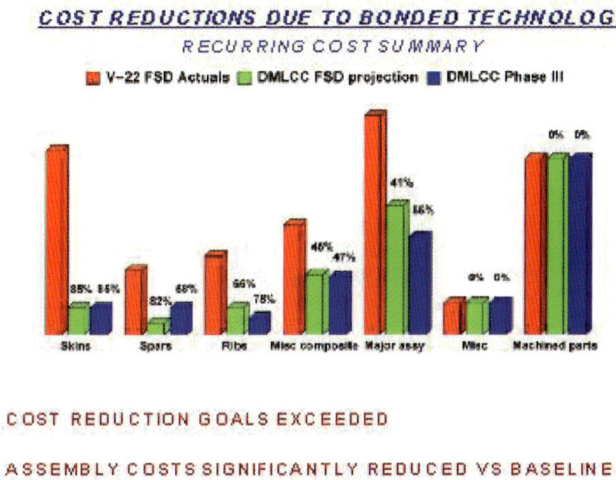

Fig. 2.34

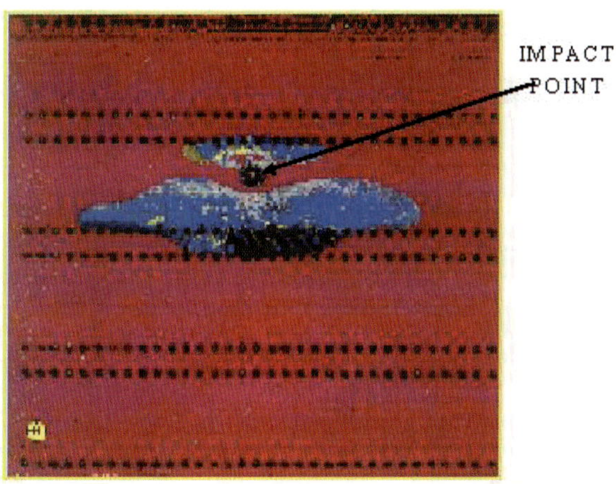

Fig. 2.47

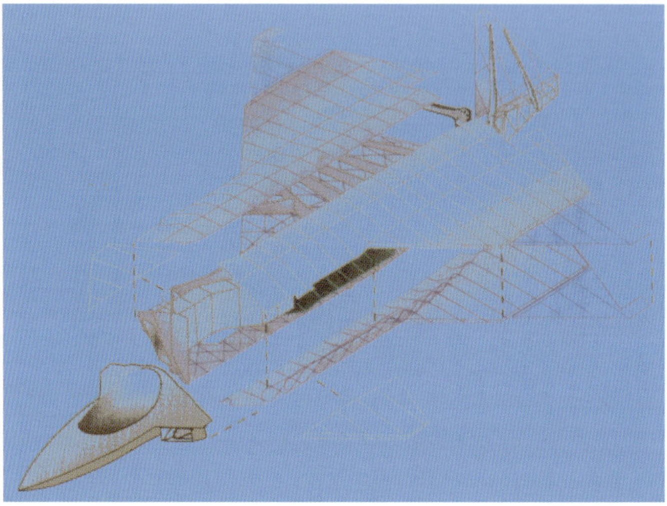

Fig. 2.57

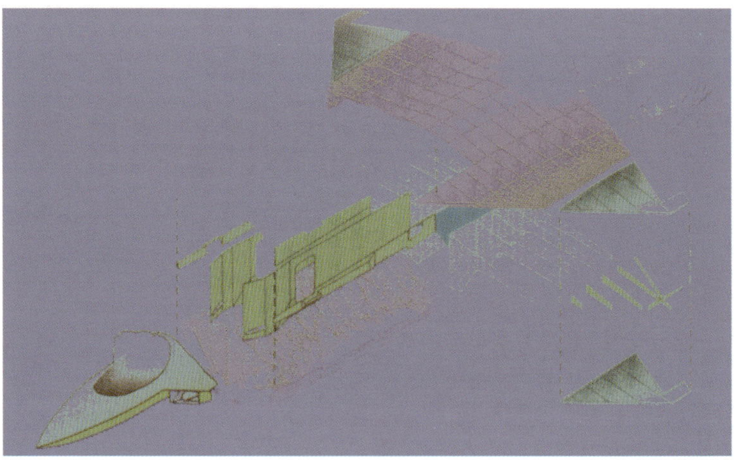

Fig. 2.58

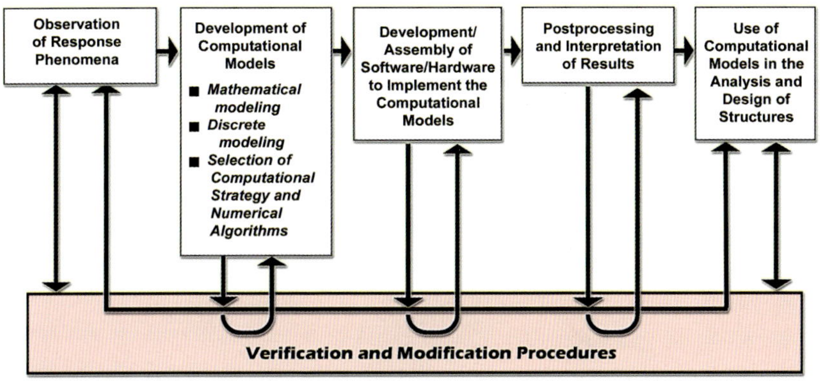

Observation of Response Phenomena
Development of Computational Models
Mathematical modeling
Discrete modeling
Selection of Computational Strategy and Numerical Algorithms
Development/ Assembly of Software/Hardware to Implement the Computational Models
Postprocessing and Interpretation of Results
Use of Computational Models in the Analysis and Design of Structures
Verification and Modification Procedures

Fig. 7.1

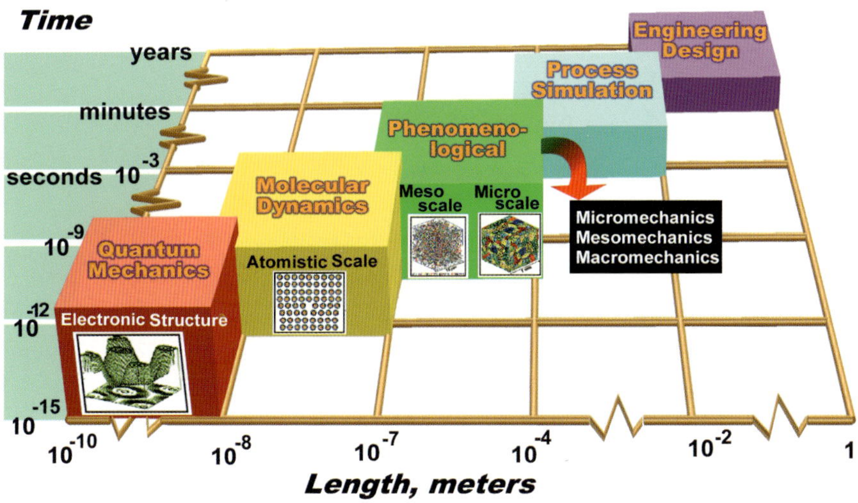

Time
years
minutes
seconds 10^-3
10^-9
10^-12
10^-15
Engineering Design
Process Simulation
Phenomenological
Meso scale
Micro scale
Molecular Dynamics
Atomistic Scale
Quantum Mechanics
Electronic Structure
Micromechanics
Mesomechanics
Macromechanics
10^-10
10^-8
10^-7
10^-4
10^-2
1
Length, meters

Fig. 7.3

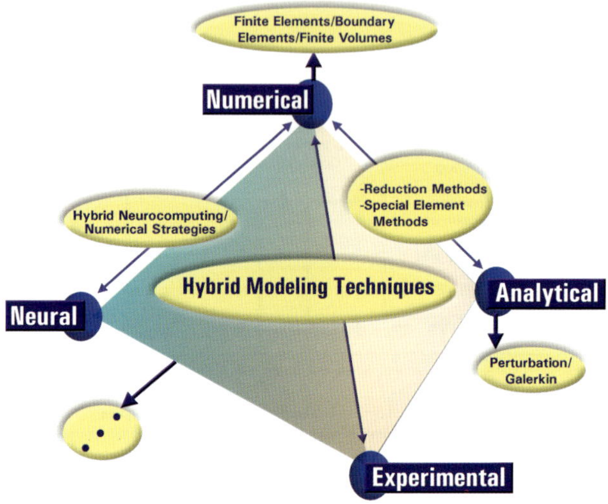

Fig. 7.4a

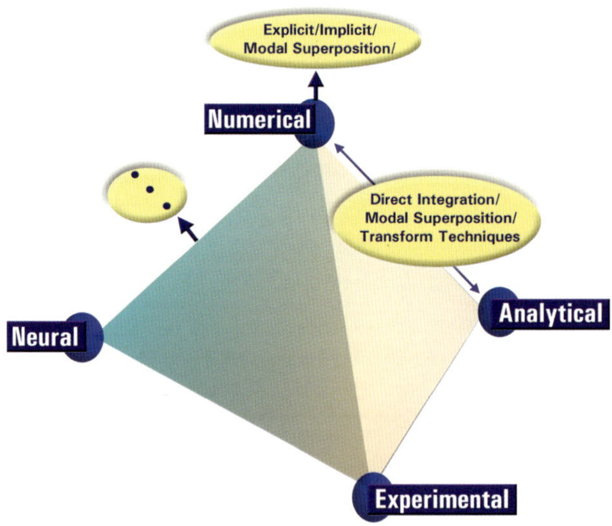

Fig. 7.4b

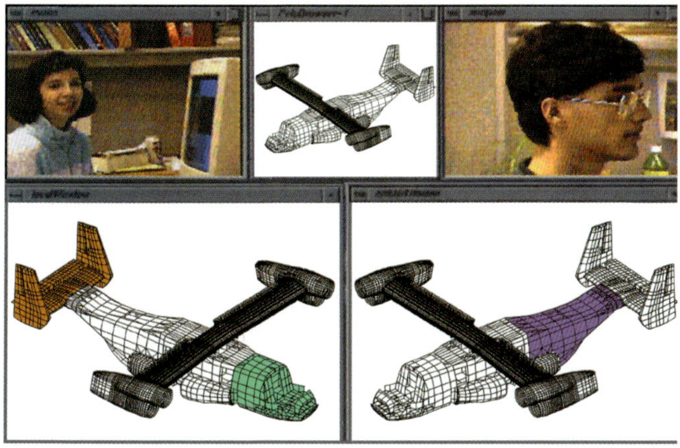

Fig. 7.5

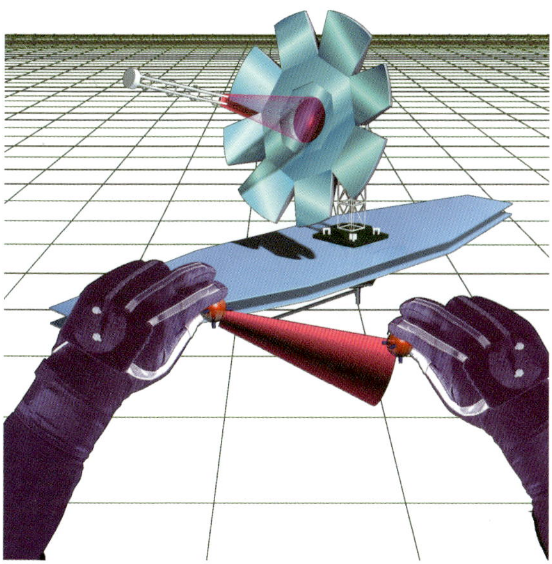

Fig. 7.6

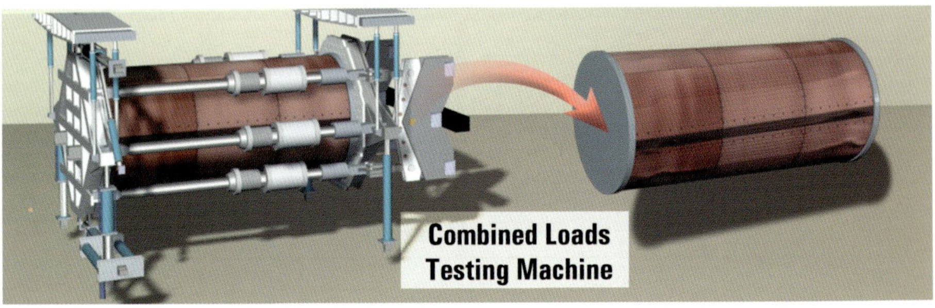
Combined Loads
Testing Machine

Fig. 7.7

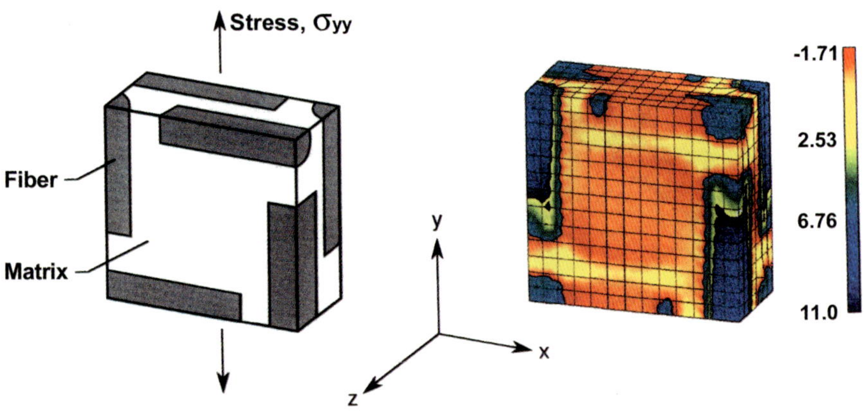
Stress, σyy
Fiber
Matrix
y
x
z
-1.71
2.53
6.76
11.0

Fig. 7.8

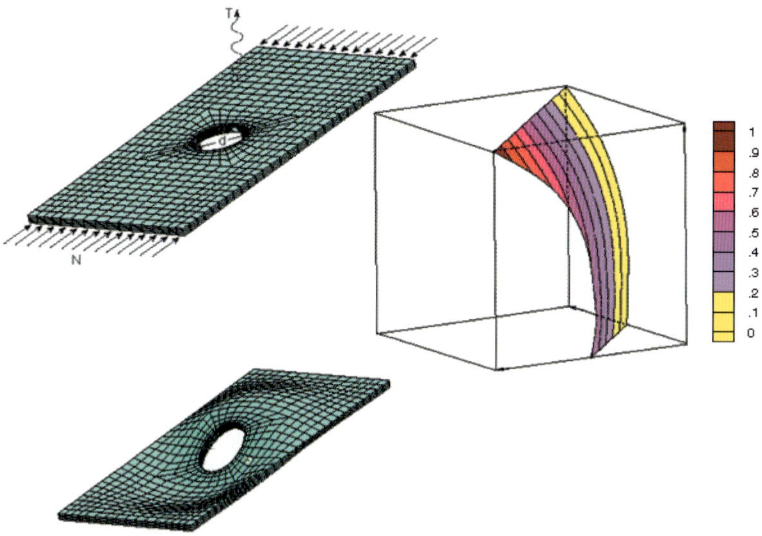

Fig. 7.9

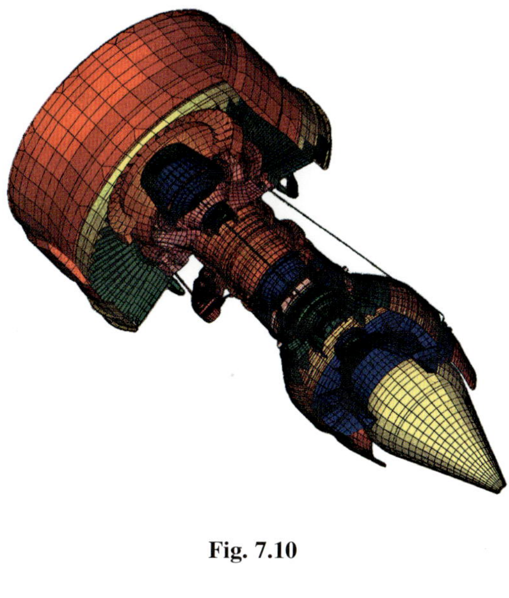

Fig. 7.10

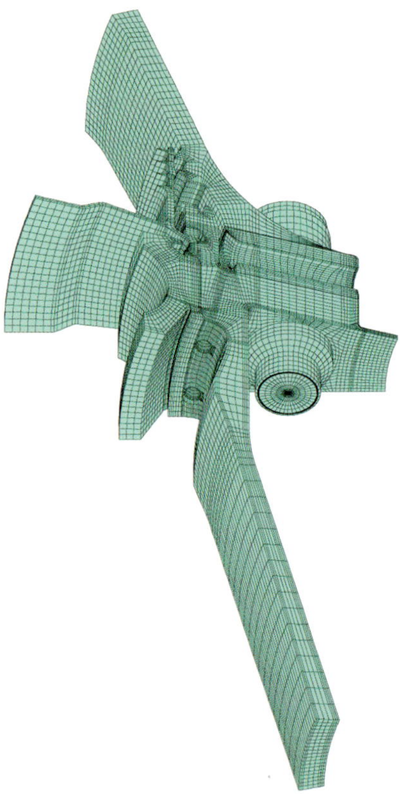

Fig. 7.11

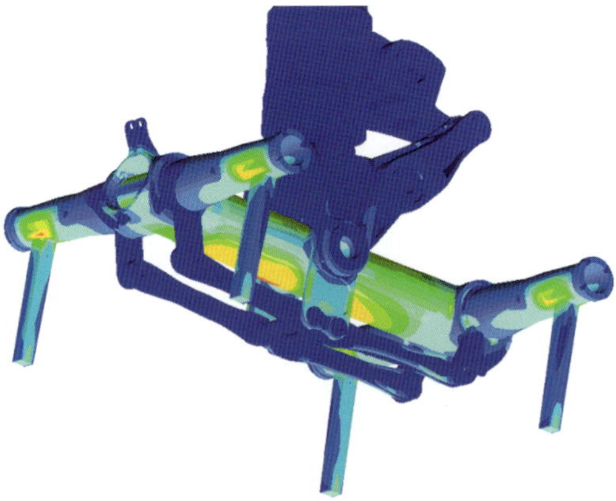

Fig. 7.14b

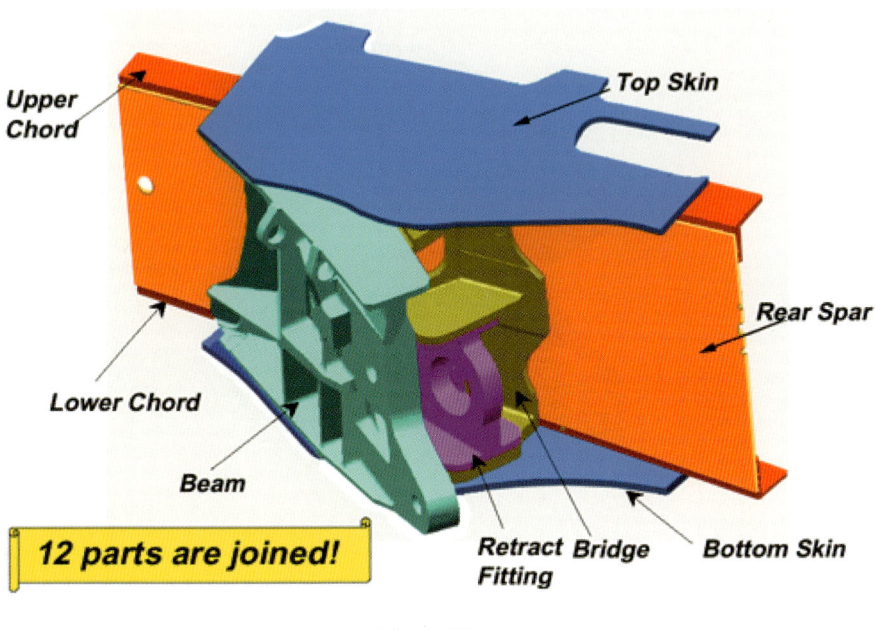

Fig. 7.15a

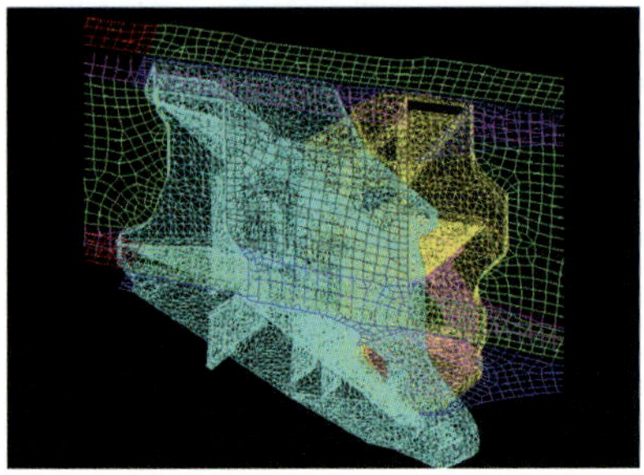

Fig. 7.15b

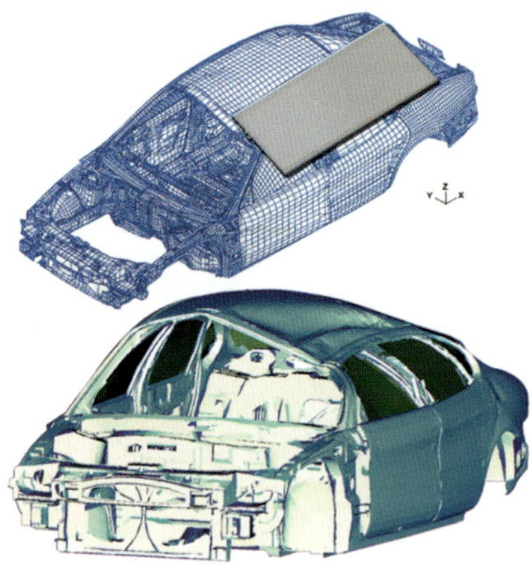

Fig. 7.17

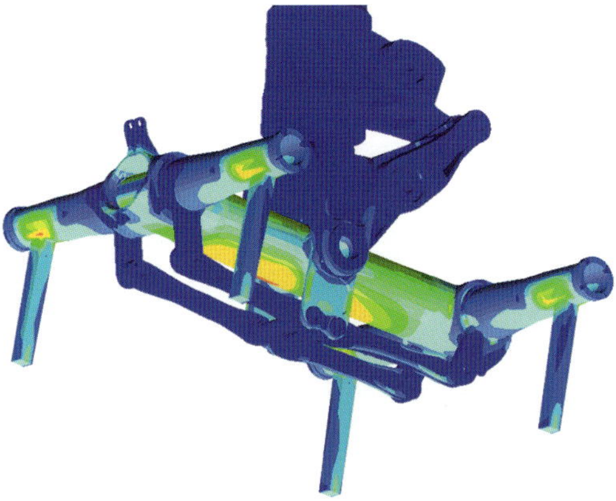

Fig. 7.14b

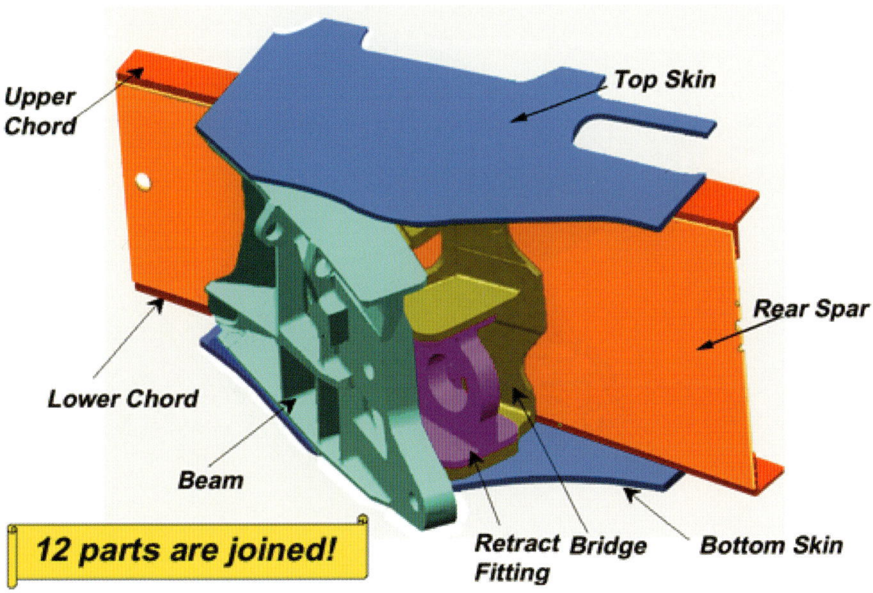

Fig. 7.15a

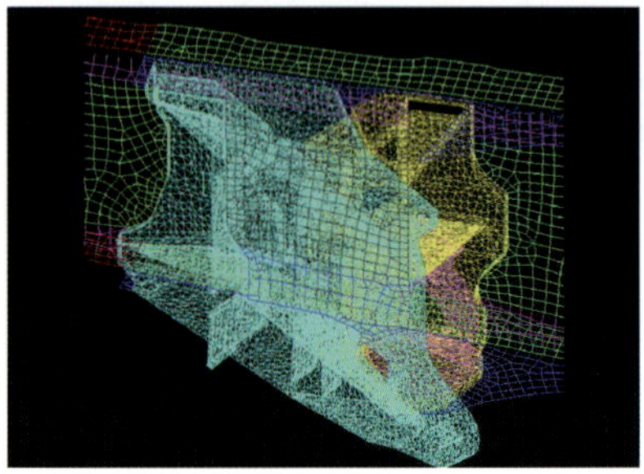

Fig. 7.15b

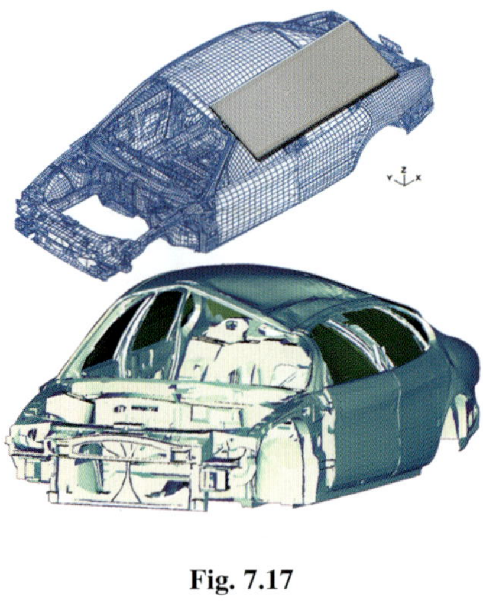

Fig. 7.17

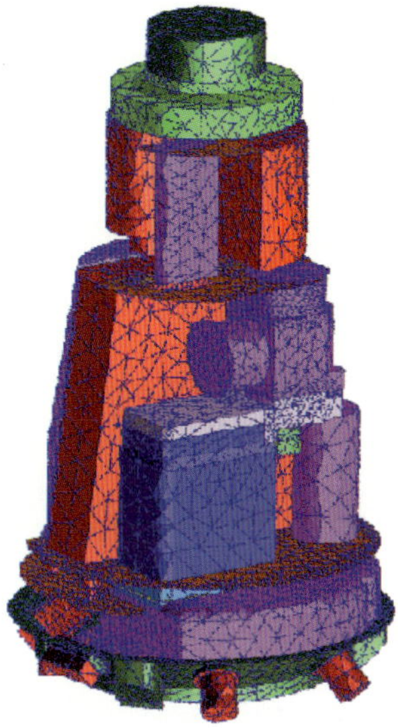

Fig. 7.18a

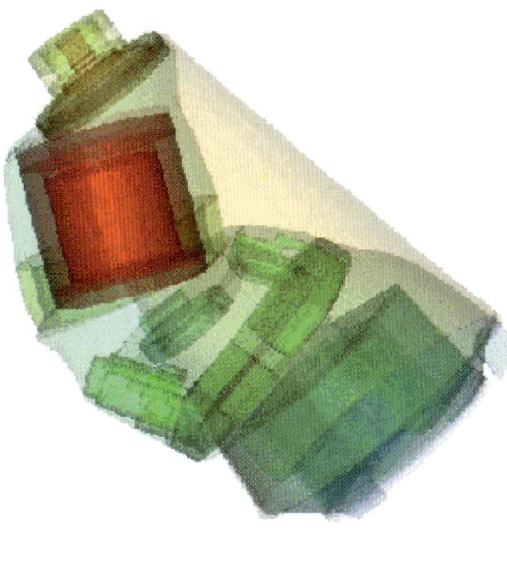

Fig. 7.18b

a)

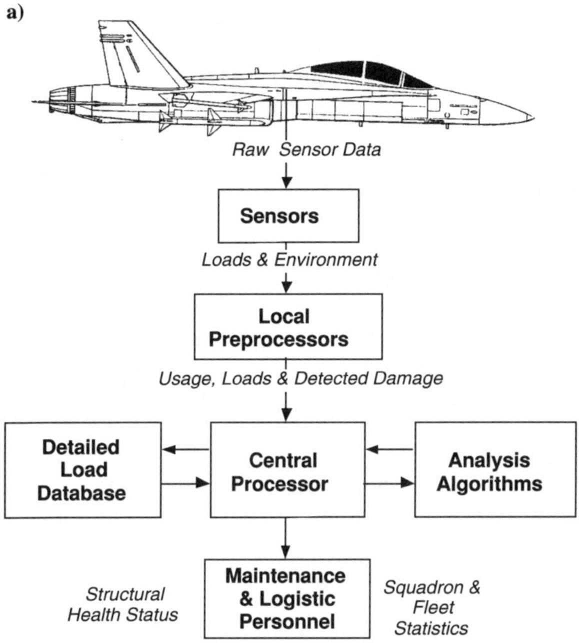

b)

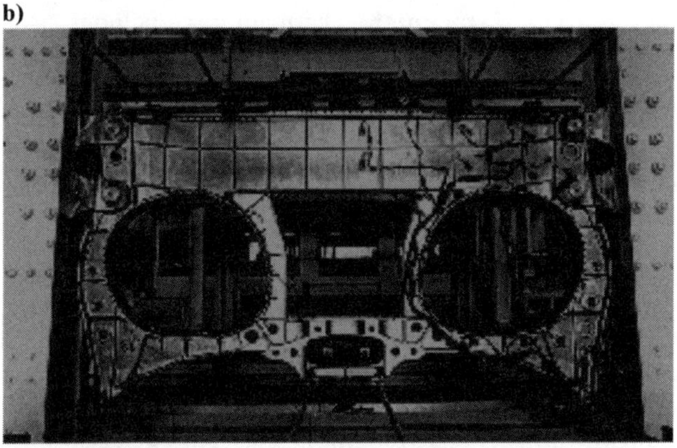

Fig. 6.1 Prototype structural health-monitoring system developed and tested by Northrop Grumann: a) overview of the components b) the wing attach bulkhead mounted in the test frame (Courtesy of Northrop Grumman).

The SIMS for the Eurofighter 2000, being designed by British Aerospace, is an integral part of the avionics system. It consists of a set of strain gauges attached to the aircraft structure to monitor cyclic strain and significant structural events. This developmental system is flexible and can be configured either as a parametric-based or strain-gauge-based, fatigue-monitoring system. Accumulated and real-time load cycle data are analyzed to determine fatigue life consumed and monitor significant structural overload events.[28,29] The system is integrated with the logistics support functions of the aircraft to minimize maintenance costs.

A prognostics and health management (PHM) system is being developed for the joint strike fighter (JSF). The JSF program is examining prognostic methods for analyzing sensor data based on an artificial intelligence scheme that may use rule-, model-, and/or case-based reasoning; artificial neural networks; fuzzy logic; and genetic algorithms. This system will also be integrated to logistics support for the aircraft.[18] The JSF is currently sponsoring demonstrations of appropriate sensors, signal processors, and reasoners on aircraft parts. Seeded fault experiments in jet engines are being conducted and data will be used to formulate and validate models for the system. The JSF program office envisions that this PHM system will become part of a distributed information network that integrates diagnostic support, aviation maintenance information, and USAF integrated information systems to a joint distributed information system encompassing both on- and off-aircraft information.

JSF, NASA, and the Navy are supporting the development of wireless PHM sensors. Wireless data-acquisition accelerometers and strain sensors for monitoring airframes have been developed and tested in a Navy flight test. Algorithms have been developed to extract relevant "features" from the data for classification. By using artificial neural networks, the ability to classify maneuvers in real time has been demonstrated.[30]

The Navy has supported efforts to develop remotely queried microsensors capable of being embedded into structural composites.[31] The remotely queried embedded microsensor (RQEM) program is funded by the Office of Naval Research (ONR) and led by MTS Corporation. The team also includes two laboratories, three universities, and two companies. Several prototype devices have been fabricated and tested. RQEM packages are being field tested on an AV-8B aircraft, on the experimental composite mast structure of the *USS Radford*, and in a one-half scale hull section fabricated using the SCRIMP™ process.

The rationale for applying structural integrity-monitoring methodologies to aircraft applies equally as well to space systems like reusable launch vehicles. A SIM system was developed as part of the joint Delta Clipper–experimental advanced (DC-XA) reusable rocket program between NASA and McDonnell Douglas (now Boeing).[32,33] This system was required to monitor and validate the performance of several key advanced structural components during ground and flight tests. It was also used to assess readiness of these same structural components to support rapid, safe turnaround flight tests. Both fiber-optic and conventional sensors were successfully demonstrated in four flight tests, although issues associated with installation and alignment of fiber-optic sensors still need to be resolved.

B. Vibration Suppression

When applied to structural dynamics problems, the use of smart materials and structures technology is expected to reduce dynamic instabilities and vibrations significantly (and, hence, fatigue damage caused by vibrations). Acoustic signature affects military operations by increasing the detectability of air vehicles, especially rotorcraft. It also affects commercial operations: high noise levels in flyovers or landing result in limited community acceptance. Vibrations impact passenger/crew comfort and weapons accuracy and the fatigue lives of most structural and electronic components are adversely affected. As a result, system reliability is reduced, and maintenance activity increases. A number of vibration problems have been identified and targeted for systems demonstrations using these smart technologies: various spacecraft vibrations, including launch, on-orbit, and shock vibrations; interior cabin noise; tail buffet; wing flutter; isolation of electronic components from forced vibrations; and helicopter BVIs and blade tracking.

Both the USAF and NASA and, to a lesser extent, DARPA, have supported extensive work in the area of vibration and noise suppression for a variety of applications. Specific vibration problems have been addressed to date with scaled models in various ground tests and wind tunnel tests. Full-scale component and flight tests have also been conducted on several systems.

1. Spacecraft and Launch Vibration Suppression

Much of the early work on vibration suppression of space structures was and continues to be focused on directed energy weapons (DEW) and precision sensor platforms. For the large, flexible DEW platforms, weapon effectiveness depends on sustained submicroradian pointing accuracy against accelerating targets. The sensors are typically viewing targets that are hundreds of miles away and required tolerances are often fractions of the wavelength the sensor is built to detect, with precision on the order of millionths of a degree or tens of nanometers. These large DEW systems typically exhibit very low stiffness and light damping, factors that result in open-loop errors of the order of a football field. Structural flexibility cannot be reduced without adding mass, which is undesirable from a launch perspective. Multiple disturbance sources on spacecraft (e.g., solar panel arrays, attitude control devices, cryocoolers) can affect the performance of space-based sensor systems, and it is difficult to eliminate these vibrations completely. In addition, the increased use of lightweight, high-stiffness composite materials in new satellites gives rise to other concerns about the severity of the acoustic environment: fewer mechanical connections translate to less damping, which, in turn, implies increased vibration levels and acoustic transmission. The response to these challenging, complex problems has been a hierarchical one, incorporating passive and active vibration isolation concepts with structural control and active optics.

One of the first major demonstrations of active vibration suppression was the Advanced Composites with Embedded Sensors and Actuators (ACESA) program,[8,9] sponsored by the Air Force Astronautics Laboratory [now Air Force Research Laboratory (AFRL) at Kirtland Air Force Base] and SDIO. This project focused on designing, fabricating, and testing graphite–epoxy composite components containing embedded piezoelectric lead–zirconate–titanate (PZT) sensors and actuators and microprocessors in logical steps—first, demonstrating feasibility, and then fabricating and testing sub- and full-scale components. TRW demonstrated that precise attitude and alignment stability of a flexible structure could be enabled by an active damping/shape control system, which operates on the flexible body modes and does not interact adversely with the rigid body attitude control system.

The ACESA program culminated with the installation of three, 16-ft-long, 5-in.-diameter "smart struts" and custom control electronics in the Phillips Laboratory's Advanced Space Structures Technology Research Experiment (ASTREX) facility (Fig. 6.2). ASTREX was a ground testing facility for large-angle retargetting, precision pointing, and vibration suppression testing on a dynamically scaled,

Fig. 6.2 ASTREX facility at AFRL.

three-dimensional structure. The damping control system was able to eliminate first mode vibrations within two cycles after being turned on: 20% damping levels achieved represent an increase of 100X over damping levels inherent in the structure. The project also demonstrated the ability to dampen vibrations over a frequency band from 10 to 80 Hz, with a very simple control approach. A stroke of several micrometers was achieved in the "smart" struts, with virtually no hysteresis or creep. Each active member had a high stiffness-to-weight ratio, very high levels of vibration damping through local feedback from sensors to actuators, and the capability for a limited amount of shape control by commanding actuators from a central control computer. In addition to some fabrication difficulties, one issue that became obvious with these early programs was the size and weight of ancillary support electronics (e.g., amplifiers, processors, cabling). These would make application of this technology difficult, if not impossible, on a real system. Miniaturization of electronic devices is key to addressing this issue.

A more important result, perhaps, was the demonstration of the scalability of the active damping concept to realistic space structures. A mission study[8] showed that active members similar to the ones employed on ASTREX could be used to settle slew-induced vibrations of a space-based radar spacecraft in two cycles, allowing mission pointing requirements to be met. Validation testing confirmed that 100 fatigue cycles of 2,000 μ-strains and thermal cycling over the range $\pm 100°C$ had virtually no effect on the dynamic performance of active members that had a layup identical to full-scale struts. Even for very large space-based structures, such as spaced-based radar, results have shown that damping forces demanded of the active members remain in a reasonable and achievable region, in the active device linear range.

The ASTREX facility was also used to evaluate a variety of different control approaches and algorithms. Control is a particularly important issue for spacecraft due to the difficulty of predicting on-orbit behavior (e.g., disturbances, 0-*g* dynamics, nonlinearities) and reliability (e.g., device failures) as shown, for example, by the March 1995 test of the NASA/Massachusetts Institute of Technology (MIT) Middeck Active Control Experiment (MACE I) on STS-67.[34] It is, therefore, important to have autonomous, self-adapting systems that can learn and recover from failure. One particularly successful approach tested on the ASTREX facility involved adaptive neural control, an approach developed by Harris Corporation in conjuction with AFRL. This is believed to have been the first large-scale "intelligent structure" demonstration. In this test, a multitone disturbance was controlled with no prior system knowledge: the system converged in about 7 minutes with a root mean square (RMS) attenuation of 27 dB across the bandwidth of interest. In another test, the structure recovered from failure of 33% of the actuators: the system reconverged in about 7 minutes with similar levels of RMS attenuation. In these tests, a Kalman filter learning algorithm was used.[35] The MACE II precision pointing experiment—involving a large team that includes the USAF and NASA, with three universities and four industry participants—will be a shuttle flight test of these adaptive learning algorithms.[36] However, full flight validation is needed to mature this technology completely.

The ACESA program was followed by a space flight test of a tripod structure (1 ft × 1 ft × 2 ft volume) built by TRW and shown in Fig. 6.3; it is believed to be the first active structural experiment to be flown in space. Shape memory alloys (SMAs) were used to change the stiffness properties of one strut. Piezoelectric

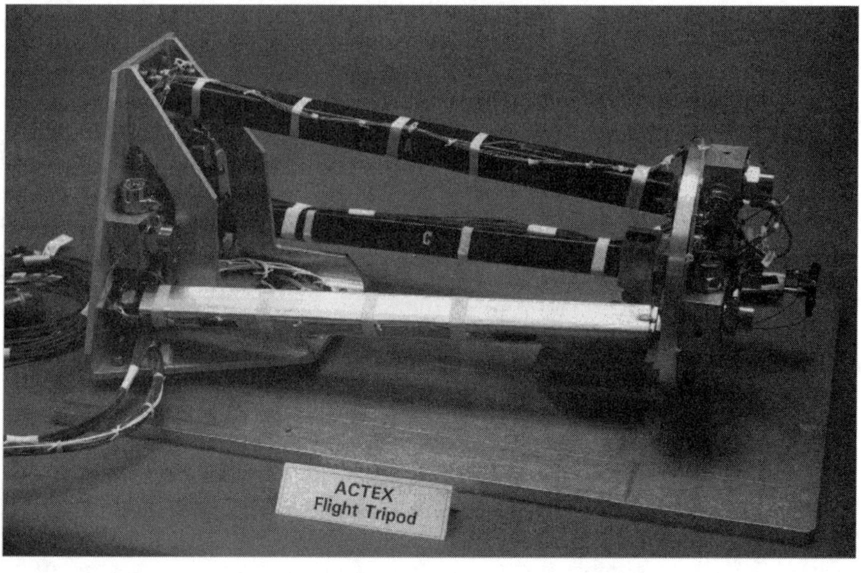

Fig. 6.3 Advanced Controls Technology Experiment (ACTEX-1) flight hardware.

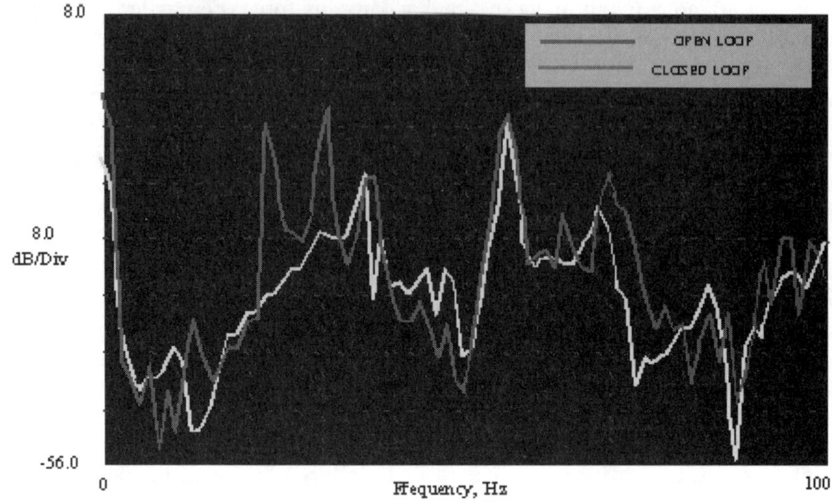

Fig. 6.4 ACTEX-1 flight test data showing vibration suppression.

sensors and actuators embedded in the other graphite–epoxy composite struts were used to demonstrate active structural control. The Active Controls Technology Experiment (ACTEX) has been in orbit for more than 2 years and has provided system identification and closed-loop control data over a broad range of temperatures: up to 29 dB of vibration suppression has been demonstrated (Fig. 6.4).[37] The SDIO (now BMDO) and the Air Force Phillips Laboratory (now AFRL) funded this flight experiment. The structure is currently being used for on-orbit testing of control algorithms as part of a guest investigator program sponsored by the AFRL.

The objective of the follow-on ACTEX II experiment,[38] also funded by BMDO and the USAF, was to demonstrate improved jitter control of an active solar array yoke on an experimental space test platform. It was to be a completely autonomous experiment with flexible, programmable operation. The active yoke—containing sensors, actuators, and power and signal conductors embedded in a space-qualified, load-bearing graphite–polycyanate composite—had adjustable structural dynamics, ultimately to provide dynamic control of the solar array over the long term while in orbit. This experimental system also had a passively damped solar array drive assembly. The key technology advancement in this experiment was the development of a modular control patch (described later), a small electronic patch integrated with the active struts in the yoke. The ACTEX II structure was successfully ground tested but never flown. A flight was attempted on the first Pegasus XL launch, but the launch was aborted. This program pioneered the idea of multifunctional structural concepts.

As part of a joint program with the United Kingdom to build a small experimental satellite [Space Test Research Vehicle-2 (STRV-2)], AFRL, and BMDO funded a project with Honeywell, Trisys, Inc., and the Jet Propulsion Laboratory. The team designed, fabricated, and tested the Vibration Isolation and Suppression System (VISS). Its purpose was to isolate an optical system from spacecraft bus disturbances by a minimum of 20 dB at >5 Hz; reduce sensor motion induced by

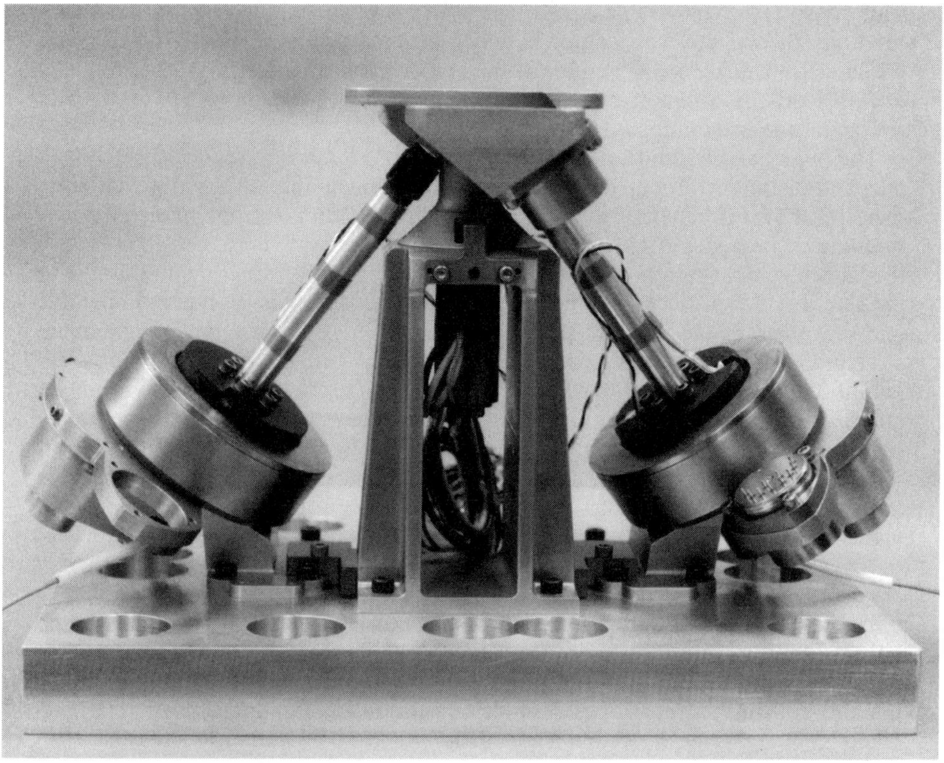

Fig. 6.5 VISS to be launched on STRV-2 (courtesy of Honeywell).

cryocooler operation by 20 dB at the first three harmonics of the cryocooler vibration; and provide a limited amount of precision, fast steering ($\pm 0.30°$). VISS, using six hybrid passive/active isolation struts in a hexapod configuration (Fig. 6.5), was successfully space qualified via acoustic and random vibration testing and thermal-vacuum testing, both as a stand-alone system and as an integrated payload on the STRV-2.[39] It is scheduled to fly on a Pegasus launch in the near future. A future project, jointly funded by AFRL and DARPA, involves the design, fabrication, and test of a miniature-vibration isolation system. The goal is a $\sim$1-in.^{3} unit containing a piezoelectric actuator and a tunable passive element, with each unit providing three-axis isolation.

More recent projects supported by the USAF are focused on incorporating isolation systems at the satellite/launch vehicle interface to reduce effects of launch vibrations on spacecraft components. This is analogous to automobile suspension systems but avoids costly and heavy ad hoc component isolation systems. Approaches being investigated for launch isolation include linear passive systems, such as viscous dampers and viscoelastic materials; nonlinear passive systems that can achieve greater isolation with lower stroke; and hybrid systems[40] that combine active and passive components (like the VISS). As an example, the world's first whole spacecraft isolation system was demonstrated by the AFRL for the Navy geosat follow-on satellite, launched on a Taurus launch vehicle in February 1998.

This passive system was designed and built by CSA Engineering under AFRL funding. The system was located between the space vehicle separation ring and the launch vehicle forward cone. It reduced critical vibrations by 86%, reduced overall loads by a factor of two, and saved 6 months and $8 to $10 million in redesign time and component cost.[41]

The objective of acoustic satellite protection systems is to reduce the acoustic energy transmitted through the payload fairing to the satellite, a reduction that saves mass for the launch vehicle and payload. It is a challenging problem: acoustic vibrations are broad band and high amplitude. Dynamic behavior that interferes with the launch vehicle system cannot be introduced, nor can the solution interfere with the satellite. Several new AFRL projects are addressing this problem via passive and active techniques. Passive approaches involve innovative, simple-to-fabricate structural concepts, such as grid-stiffened structures with integrated functional capabilities. Active approaches include noise cancellation, dissipative cavity mode control, and structural control concepts. Because each of these has drawbacks, a hybrid approach may ultimately be most effective. In one project, AFRL is working with several industrial and academic partners to demonstrate a full-scale active acoustic suppression system for the orbital/suborbital program Minotaur launch vehicle payload fairing. This project culminates in a qualified noise reduction test of the active system. Modeling tools (e.g., integrated structural–acoustic models) are also being developed as part of this program.

Another serious vibration problem for spacecraft is attributed to shock-related failures: 83 such failures were noted in over 600 launches (up to 1984), and more than 50% of these 83 failures led to a catastrophic loss of mission while the remaining systems suffered from degraded performance. Examples of such failures could include broken wires and leads, dislodging of contaminants, and others, ultimately leading to component and/or subsystem failure and possibly loss of the spacecraft. Pyrotechnic release devices contribute to shock failures: these explosive devices cause large forces and accelerations, exceeding the 500-*g* limit at some frequencies. Solar array deployment hinges can also cause shock problems. Lockheed Martin, with funding from AFRL, has investigated the use of SMA-based release mechanisms to reduce the shock to acceptable levels, as illustrated in Fig. 6.6. Example concepts include a two-stage nut (TSN) and a low force

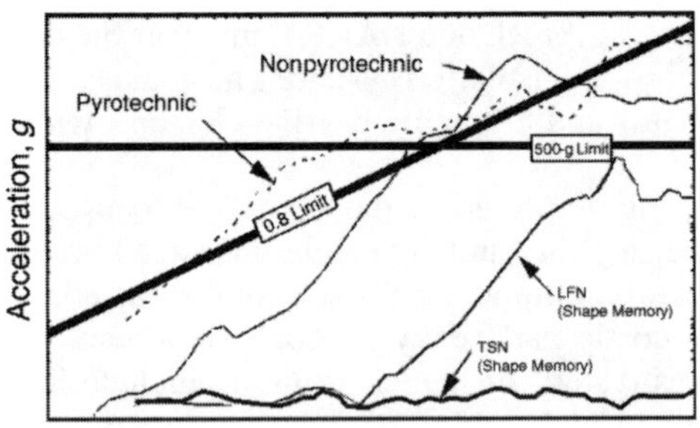

Fig. 6.6 Shock vibration design limits for spececraft compared with the shock vibration performance of shape memory alloy release mechanisms and conventional pyrotechnic devices.

nut (LFN).[42] Such concepts have been flight tested: the Shape Memory Actuation Release Devices (SMARD) experiment on the MightySat 1, a USAF experimental satellite. MightySat was released from the shuttle STS-88 in late spring 1999. Results from the SMARD experiment showed performance levels consistent with predicted behavior for the two concepts: the TSN produced less than 200 *g*'s of shock; the LFN produced less than 500 g's of shock; and the pyrotechnic device produced shock levels in excess of 6,000 *g*'s.[43] AFRL has also been working with Starsys Research, Inc., to develop second-generation devices. These SMA devices will be tested in the FalconSat low-shock release experiment on OSP-1. They expect to reduce separation shock by a factor of 10 or more, reduce the number of shock-related failures, and demonstrate improved reliability of release mechanisms by using these devices. Because these devices are nonpyrotechnic, increased range safety and contaminant avoidance are additional benefits. A low-shock, reliable SMA mechanism for solar array deployment was tested in the lightweight flexible solar array hinge experiment on STS-93 (summer 1999), sponsored by DARPA and AFRL. Hinges are the primary mechanism used to deploy spacecraft solar arrays that are folded together for launch. Six SMA hinges were tested to verify mechanical design data and evaluate dynamic properties of hinges in a realistic environment. These SMA hinges offer key advantages over other hinges: low-shock controlled deployment, fewer parts, lighter weight, higher reliability, and ease of production and assembly. Preliminary reports from the shuttle crew indicated that hinge operations were nominal: the hinges performed as expected although no data have been analyzed yet.[43]

2. *Air Vehicle Noise and Vibration Suppression*

Active acoustic control, or the use of one acoustic source (or secondary source) to cancel another (or primary source), has a long history, particularly with respect to air vehicles. In a recent survey paper, Fuller and Von Flotow[44] described the earliest practical demonstrations of the technique and the earliest known U.S. patent. Active structural acoustic control (ASAC) for air vehicle interior noise reduction is an area of particular interest to all the Services and NASA and commercial air vehicle companies. The most obvious difference between ASAC systems and early acoustic control systems is that ASAC uses structural actuators like shakers or PZT patches attached to the aircraft fuselage rather than acoustic actuators such as loudspeakers inside the fuselage. This concept is attractive because structural actuators are more effective by weight and consume less interior volume than competing active or passive noise control options.[45]

One important area of ASAC research involves the determination of optimal locations for actuators and sensors. Early theoretical investigations[44–48] established the importance of actuator and sensor architecture and suggested optimization strategies and goals. Much of the work focused on the development of mathematical models, with some supporting experimental work to validate them. Some more recent activities have been focused on developing concepts for new actuators[49,50] and demonstrations on real aircraft.[51–54]

Many researchers recommend a modal method for ASAC: actuators are placed to excite a selected structural mode and sensors are placed to observe each important acoustic mode.[44] Lyle and Silcox[45] tested this modal method on a simulated aircraft fuselage with mixed results. Although significant global interior noise reduction was obtained at a frequency at which primary and secondary sources excited the same dominant acoustic mode, the same actuator and sensor configuration was not effective at a second frequency because a global increase in interior noise was observed. This behavior, attributed to amplification of other modes (e.g., control spill-over) could be reduced via an alternate set of actuators and sensors. While the benefits appear obvious, no practical solutions using smart technologies have been implemented on real air vehicles.

The Australian Defence Science and Technology Organization has conducted laboratory tests of out-of-plane vibration suppression on the tailplane of a CT-4 aircraft previously used by the Royal Australian Air Force as a basic trainer.[55] These tests used four PZT patches—two as strain sensors and two as strain actuators—to actively suppress out-of-plane vibrations. A method to find a suitable location for sensors and actuators was developed, and a digital controller system was designed and implemented. The system significantly reduced out-of-plane vibrations and demonstrated robust characteristics.

An active aft fuselage skin panel subjected to engine noise and unsteady flow-induced vibrations has been flight tested on the B-1B.[56,57] This is the first known flight test of such a system on primary aircraft structure. The system used a piezo-ceramic lead–lanthanate–zirconate–titanate (PLZT) patch vibration suppression system. The primary objective was to suppress forced responses from separated flow, but the project also demonstrated that piezoelectric actuators could withstand the operational environment and loads of a real aircraft. Additional objectives of the flight test were to demonstrate fundamental and higher order mode suppression of the panel. The fuselage panel was relatively thick (0.80 in.), with a radius of curvature of approximately 38 in. PLZT patches were attached to the inside of the skin panel, although they may have been more effective attached to the outside. The system was designed to withstand temperatures ranging from $-60°$ to $185°F$ as well as a large number of vibration cycles. A greater than 8-KHz digital processor system was designed and used to suppress the critical modes over the 400-to 800-Hz range. The active control system maximized the force generated by PLZT patches. The system was successful in reducing fundamental panel modes by as much as 79% for the takeoff condition and by about 46% for transonic flight conditions, with a 25% response reduction for a higher order mode.

Under another DARPA-sponsored program, Lucent Technologies applied active structural control to reducing structural vibrations in gas turbine engines.[58] They worked with Pratt & Whitney to develop solutions for particular vibration-induced problems in engine fan blades and vanes, engine cases, and external engine components. Expected benefits included improved engine durability, improved engine performance, reduced operating and support costs, and shorter engine development cycles.

3. *Buffeting Suppression*

Buffeting is an aeroelastic phenomenon that plagues high-performance aircraft, especially those with twin vertical tails. Twin tail buffet arises at high angles of attack (AoAs) when unsteady vortices generated near the wing leading edge impinge on the tails, as shown in Fig. 6.7, and induce severe structural vibrations. This can lead to premature fatigue failure and more frequent inspections.[59,60] The buffet problem is particularly severe for the F/A-18 and F-15, as evidenced by high costs associated with special 200 flight-hour inspections, repair and replacement of damaged tails, and redesign.

Fig. 6.7 An F-18 climbing at a high AoA, with leading edge vortices impinging on the vertical tails.

To reduce fatigue and thus increase the life of a vertical tail, stresses caused by buffeting must be reduced. This reduction may be accomplished by modifying the load-carrying structure within the tail, reducing buffet loads by altering the flowfield around the vertical tail, or reducing the buffeting response through active control of effectors on the tail. The success of a proposed fix to the buffeting problem has generally been measured by percentage reduction in the RMS of the strain at the root of the vertical tail.

To understand the buffeting problem, several USAF and NASA programs have focused on quantifying buffet loads by acquiring response measurements and surface pressures on the vertical tails of scaled models in a wind tunnel and on an actual aircraft during flight.[59–66] In general, the results of these studies are published as spectra and pressure coefficients for the inboard and outboard surface of the tail. Based on some of these studies, Boeing (formerly McDonnell Douglas) implemented an interim solution on the F-18, relying entirely on structural modifications to the vertical tail to reduce dynamic stresses in critical areas.[62] Even with the structural enhancement, dynamic stresses were still too severe. This led to the investigation of an alternative approach that offered some improvement but also compromised the aircraft's high-alpha performance by reducing the unsteady lift and pitching moment of the aircraft.[63] Therefore, other options are being explored.

There are two approaches being considered to address the buffeting problem. The first of these uses an actively controlled rudder. The second approach uses a smart materials and structures solution to the problem. In 1992, the concept of an actively controlled rudder was proposed to alleviate vertical tail buffeting on the F-18 and F-15 aircraft.[67] This analysis showed that an actively controlled rudder of an F-18 might be effective in adding damping to the vertical tail, resulting in reductions in the RMS of the root bending moment.

The active vertical tail (AVT) project to reduce buffet response was a joint venture between McDonnell Douglas (now Boeing) and Parks College of St. Louis

University.[68] The AVT was a 5% scale aeroelastically tailored structure having similar vibration response to that of a full-scale aircraft. In 1995, vertical tail buffeting alleviation was achieved using piezoelectric actuators on this double delta wing wind-tunnel model with twin vertical tails that were not canted.[69] Piezoelectric actuators were attached to the spar to control the first two bending and torsion modes. This structure was tested in a low-speed wind tunnel over AoAs ranging from 20 to 55° and dynamic pressures ranging from 0.5 to 7 psf. The peak response of the vertical tail was reduced by as much as 65% over the uncontrolled response. These results were achieved using simple control algorithms employing collocated strain gauges.

In 1995, the use of actively controlled piezoelectric actuators on an F/A-18 vertical tail was analyzed.[70] This analysis showed that actively controlled piezoelectric actuators might increase damping greater than 60% in the first bending mode for a less than an 8% increase in the weight of the vertical tail.

In a NASA–Langley-sponsored program, with participation from Wright Laboratory (now AFRL) and Daimler Benz Aerospace (now DaimlerChrysler), a 1/6-scale F-18 wind-tunnel model was tested in the NASA–Langley transonic dynamics tunnel (TDT) in 1995–1996, as shown in Fig. 6.8, which was part of the Actively Controlled Response of Buffet-Affected Tails (ACROBAT) program to assess the use of active controls in reducing vertical tail buffeting.[59,61] The research objectives of ACROBAT program were twofold: to determine spatial relationships of differential pressures during open- and closed-loop conditions at various AoAs and to apply active controls technology using a variety of force producers to perform buffeting alleviation on twin vertical tails of the wind-tunnel model. The starboard vertical tail was equipped with an active rudder and other aerodynamic devices, while the port vertical tail was equipped with piezoelectric actuators. By using single-input/single-output control laws, the power spectral density of the root strains at the frequency of the first bending mode of the vertical tail was reduced by as much as 60% for AoAs up to 37°. RMS values of root strain were reduced by as much as 19%.

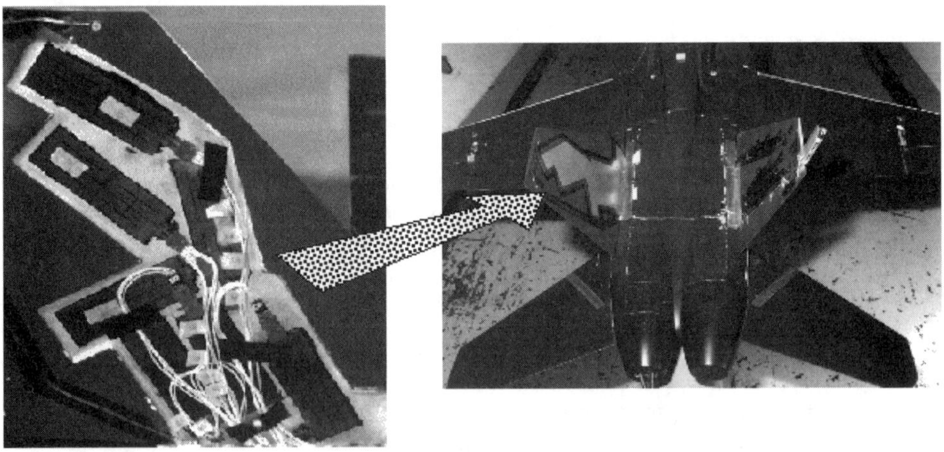

Fig. 6.8 A 1/6-scale F-18 model mounted in the NASA–Langley TDT during the ACROBAT program.

Fig. 6.9 Full-scale ground test of an active buffet suppression system (courtesy of Active Control Experts, Inc., and AMRL, Australia).

ACROBAT program results demonstrated that buffeting alleviation of vertical tails on an F-18 could be achieved using active piezoelectric actuators or rudder articulation.[61] This investigation was the first known experimental demonstration of active buffeting alleviation on a scaled F-18 wind-tunnel model using an active rudder and piezoelectric actuators.

Ground tests of an active buffet suppression system were conducted on a full-scale F/A-18 in Australia under the auspices of The Technical Cooperation Program.[60,71,72] These tests were completed in February 1998 at the Aeronautics and Maritime Research Laboratory (AMRL) in Melbourne, Australia, using the international follow-on structural testing project rig (Fig. 6.9). The test facility used air bags and shakers to simulate buffet loads on the aircraft. Patch piezoelectric actuators produced by Active Controls eXperts (ACX), Inc., were surface-bonded to the starboard vertical fin. ACX also designed and built the control system. Four different simulated flight conditions, as shown in Fig. 6.10, were evaluated in the tests. To maximize control authority for each simulated flight condition, a separate optimal controller was designed and tested by ACX for each of the four flight conditions. Figure 6.10 also shows the results in percent reduction of RMS bending moment. Buffet alleviation targets were not quite achieved in these experiments: nonlinear effects of the shaker load on the modal response had not been considered in the mathematical model of the plant. Also contributing to the results is the fact that, during the worst-case flight condition, only one control law functioned over the entire duration of the load cycle without overdriving the amplifiers. Overdriving the amplifiers caused shutdown of the

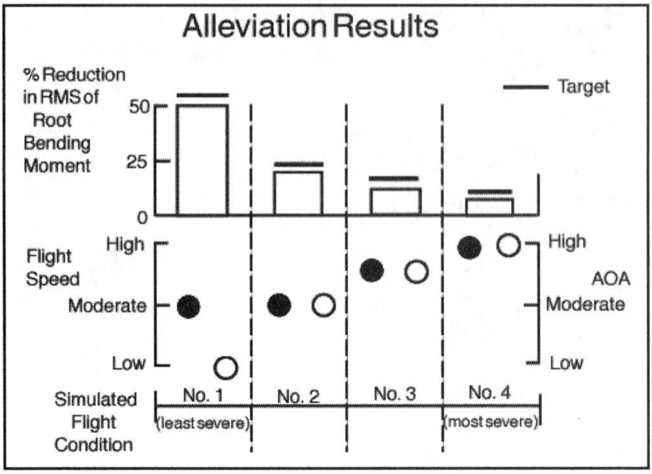

Fig. 6.10 Flight conditions and simulated buffet alleviation results in percent reduction in RMS bending moment achieved using piezoelectric actuators during the full-scale ground test.

affected piezo-system so that it was no longer active in reducing root bending moment.

The USAF has sponsored other projects addressing buffet load alleviation. Rohini International examined stacked piezoelectric actuators to solve this problem. Because stacked piezoceramic actuators provide only uniaxial motion, an assembly was designed to transform the longitudinal motion into moments that would provide control actuation. The Rohini system was demonstrated in ground and wind-tunnel tests. A 1/16th-scale F-15 model was tested in the Subsonic Aerodynamic Research Laboratory facility at AFRL and in the Research Institute Model Test facility at Georgia Tech.[73] Some of the experiments performed during this study included four different AoAs (14°, 17°, 20°, and 23°) at free stream dynamic pressures that varied from 5 to 13 psf. Results showed that the system was effective over the entire buffet domain. As the disturbance increased, however, the effectiveness of the control decreased.

A 2-year, interagency study currently underway will down-select the most effective buffet load alleviation approach for a planned follow-on effort. A flight demonstration program is scheduled for years 2001 to 2003.

Similar, ongoing efforts in Europe are also focused on the application of smart structures technologies to the buffet problem.[74–80]

4. *Flutter Suppression*

The interaction of structural dynamics with aerodynamic characteristics of an air vehicle at particular flight conditions causes flutter, a series of divergent and destructive oscillating motions. It is a safety of flight concern. In fighter aircraft, flutter is aggravated by the presence of under-wing weapons.[81] Solutions generally involve increasing structural stiffness, mass balancing, or modifying geometry, all of which typically increase weight and cost while decreasing system performance.

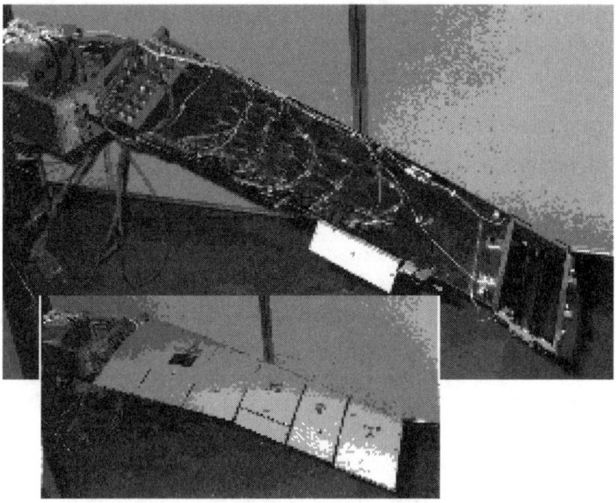

Fig. 6.11 The PARTI wind-tunnel model.

A cooperative project between NASA–Langley and MIT—the Piezoelectric Aeroelastic Response Tailoring Investigation (PARTI)—focused on the wing flutter problem.[82–84] The PARTI experiments, which took place from 1991–1996, were analytical and experimental studies based on a relatively large, multi-degree-of-freedom aeroelastic testbed. Program objectives were to demonstrate the ability of strain-actuated adaptive wings to control aeroelastic phenomena, including wing flutter suppression and gust load alleviation, and to develop experimental and analytical techniques. To accomplish these objectives, a wind-tunnel model was designed and fabricated, aeroservoelastic analyses were performed, and the model was ground and wind-tunnel tested in NASA Langley's TDT.[85]

The existing PARTI wind-tunnel model, shown in Fig. 6.11, consists of a composite plate in a sandwich construction (graphite fiber-reinforced epoxy facesheets with an aluminum honeycomb core) with piezoelectric patches surface-bonded to each side of the plate. Conventional strain gauges and accelerometers are also included. The piezoelectric patches are arranged into 15 groups to be used either as 15 actuators or sensors or a combination of both. During this program, active flutter suppression and reduced gust loads using piezoelectric actuation were successfully demonstrated in wind-tunnel testing of the 4-ft-long semispan wing model: flutter dynamic pressure was increased 12% and wing root bending moment caused by gust was reduced by 75%. Although the model tested during the PARTI program was a platelike model (vs a conventional monocoque structure), experimental results provided evidence that piezoelectric technology may offer a viable alternative to conventional aeroelastic control techniques.

As a result of the PARTI program, an extensive database of experimental information was gathered and is being used to help understand the many issues associated with applying strain actuation technology to dynamic problems. Three key issues identified during the PARTI program have provided guidance to NASA in establishing research focus areas and related follow-on activities. These activities include

1) Development of detailed structural and aeroelastic models to promote a better understanding of local and global effects of piezoelectric actuation. Finite element and aeroservoelastic modeling and validation research are both underway to address this issue.

2) Investigation of piezoelectric power consumption characteristics during active control to enable a realistic evaluation of the economic viability of using piezoelectric actuators on full-scale vehicles. Ground tests are being conducted to investigate the efficiency of active and passive control schemes. These tests use a method for predicting the power consumption of piezoelectric actuators that was developed in a follow-on study to the PARTI program.

3) Development of improved control and optimization techniques to use the piezoelectric capabilities to their fullest extent. To date, two studies have used data from the PARTI wind-tunnel tests to examine optimal control using the piezoelectric actuators, and further research is planned in this area using the PARTI model for open- and closed-loop ground tests.

5. *Rotorcraft Vibration Suppression*

The aerodynamics of rotating blades is quite complex. The aerodynamic environment varies with blade position around the azimuth and leads to suboptimal performance throughout much of the flight envelope. BVI noise is caused by the wake from the previous helicopter blade meeting the leading edge of the next blade.

Blade-tracking adjustments, required to account for slight physical differences between helicopter blades, are done infrequently because of the high cost, significant setup time, and subsequent maintenance. Vibrations increase when blades are out of track. Active, real-time, blade-tracking adjustments can produce large savings in maintenance costs and have an additional feature important in military operations: downtime for tracking adjustments can be significantly reduced.

Boeing Defense and Space Group formed a DARPA-supported consortium—Smart Structures for Rotorcraft Control—involving the Boeing Helicopter Division (Philadelphia), MIT, Pennsylvania State University (PSU), and Analytic Engineering. This team addressed control of trailing edge flaps and trim tabs as well as active twist control of helicopter rotor blades in the Phase I program. Expected performance improvements included an 80 to 90% reduction in vibrations, 5- to 10-dB reduction in BVI noise, blade twist of $\pm 2^\circ$, and trailing edge flap motion of $\pm 3^\circ$, all to be achieved without compromising required safety levels. Design and performance benefits were based on an existing CH-47 Chinook blade design. Actuation schemes under consideration included a discrete piezoelectric actuator for flap motion, SMA torsional actuators for trim tab motion, and a distributed, interdigitated electrode–piezoelectric fiber composite (IDE-PFC) or active fiber composite (AFC) actuator for blade twist. The SMA actuator was bench-tested, and the two piezoelectric concepts were spin-tested on model CH-47 blades. Preliminary experiments to evaluate integrity of the embedded piezoceramics and characterize their performance under loading conditions (especially fatigue) were completed. One 1/16 Froude-scale CH-47 blade was fabricated using the IDE-PFC and bench-tested to evaluate twist actuation performance: a maximum twist of 1.4° was measured (at 2,000 V) under static conditions.[86] This concept will be addressed in more detail later. Issues associated with any integrated devices included actuator ability to handle the large centrifugal loads, drag associated with external actuator elements, and harsh operational environments.

Fig. 6.12 The MD 900 helicopter (courtesy of Boeing–Mesa).

The McDonnell Douglas Helicopter (now Boeing Helicopter in Mesa) Phase I demonstration—Smart Materials Actuated Rotor Technology program (also supported by DARPA)—considered active control of a rotor blade trailing edge flap and trailing edge trim tab.[87] Expected performance improvements for the flap included a 10-dB reduction in BVI noise while landing, an 80% reduction in airframe vibrations, a 10% gain in rotor performance (lift/drag), and improved maneuverability from stall alleviation. For the trim tab, the goal was to eliminate manual tracking requirements, relax blade manufacturing tolerances, and reduce vibrations. Design and performance benefits were based on an existing MD 900 Explorer system. The MD 900 (Fig. 6.12) is a twin engine, 6,000-lb helicopter, with a 34-ft, all-composite main rotor. Phase I actuator concepts for the flap included a multilayer, cofired electrostrictive stack with a preload mechanism and stroke amplification flexure. Two torsional SMA tubes with a locking mechanism (to hold the tab in place so that power was not continually required) were considered for the tab actuator device. Integration issues included actuator ability to handle centrifugal force loading, mass balance in the chordwise direction, actuator reliability and durability, actuator size with respect to blade geometry constraints and aerodynamic profile considerations, and blade structural integrity with integrated devices. Risk reduction tests included characterization of actuator materials, actuator bench and spin tests, and integrated system tests.

Both of these helicopter projects were combined in the DARPA Phase II effort and are managed by Boeing in Philadelphia. Boeing is currently developing an MD 900 full-scale, active flap rotor using smart materials to achieve reduced vibrations, noise, and improved performance as noted for the Phase I program. Trailing edge flaps, extending from 74 to 92% blade span, will be actuated by piezolectric actuators installed inside the blade spar (Fig. 6.13). This active flap will provide the modification of aerodynamic forces required to affect helicopter

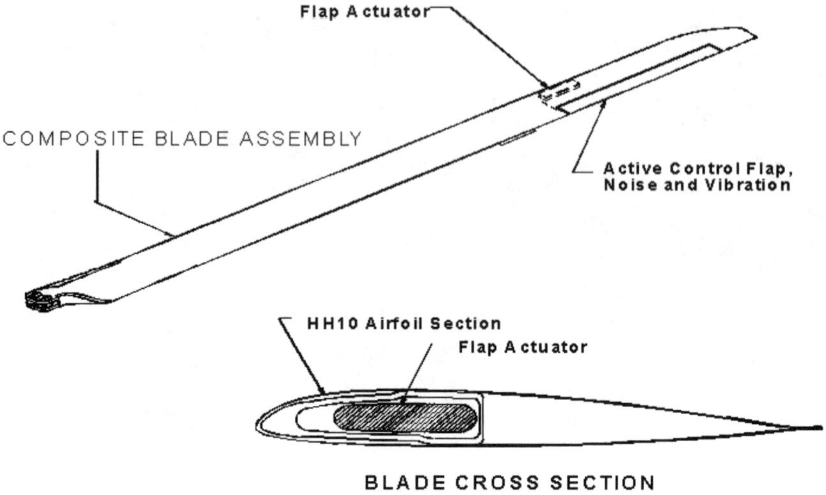

Fig. 6.13 The layout of the MD 900 smart helicopter blade (courtesy of Boeing–Mesa).

performance. Several actuator concepts have been bench tested in Boeing–Mesa's laboratory to evaluate basic actuator performance. Important actuator characteristics included stroke, force, bandwidth, power consumption, and thermal behavior. Stiffness, damping, and inertia were adjusted to simulate a range of mechanical loading conditions on actuators. For additional testing, the bench test rig and actuator were mounted on a shaker table or in a centrifuge to evaluate actuator performance under simulated rotor blade vibratory motions and centrifugal force loading. A unique X-frame actuator concept,[88–90] developed by MIT, has been selected for the final demonstration articles, as shown in Fig. 6.14. Table 6.1 lists

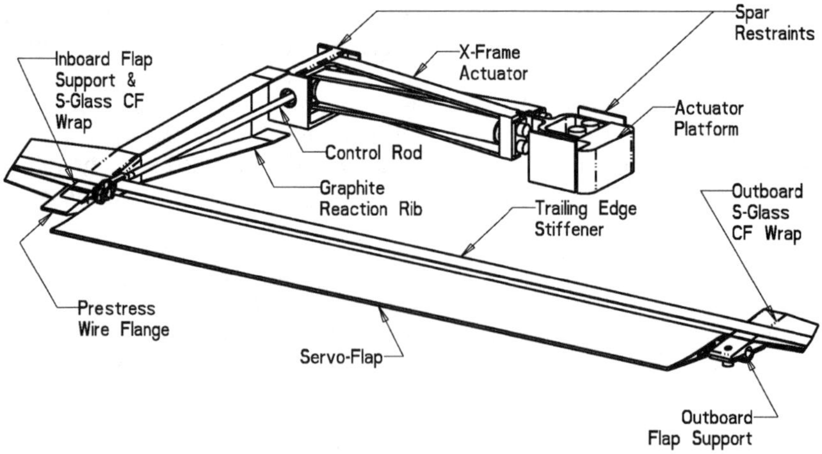

Fig. 6.14 Schematic of the X-frame actuator integrated with the trailing edge flap (courtesy of MIT).

Table 6.1 Model-scale properties of the MIT X-frame actuator

Property	Value
Peak-to-peak free stroke	0.056 in.
Peak-to-peak blocked force	20.7 lbf
Tip stiffness	370 lbf/in.
Bandwidth*	660 Hz
Weight	0.086 lbm

*Driving a nearly impedance matched load.

some of the model-scale properties of X-frame actuator properties. The MD 900 rotor will be whirl tower tested, wind-tunnel tested in the NASA–Ames 40-ft × 80-ft wind tunnel, and flight tested in the 2000–2001 timeframe.

The AFC work at MIT is being worked on a parallel path to this Phase II flight demonstration task. Before proceeding further, defining what is meant by the term, "active fiber composite," is important. AFCs consist of piezoelectric fibers embedded in an epoxy matrix with other inactive reinforcements to improve the durability characteristics of the actuator.[91,92] An interdigitated electrode poling method[93] is used to generate large directional actuation strains in the actuator plane. Combining both of these technologies results in a piezoelectric actuator laminate with induced stress, endurance, and conformability characteristics superior to typical monolithic piezoceramic actuators. Of no small significance is the fact that ultimate realization of this actuator concept will require the development of manufacturing technology and an industrial base for the production of piezoelectric fibers and composite laminates. DARPA has sponsored a project with Mide Corporation, Continuum Control Corporation, and CERANOVA to address these needs. A major focus of this parallel task in the DARPA helicopter program is to demonstrate the benefits of an active twist rotor system via a test of a Mach-scale rotor, complete with predictions and extrapolation to a full-scale rotor. Additional focus areas include prequalifying the AFC design for full-scale demonstration (via detailed design and analyses) and studying system integration issues. Shortly after the Phase II task was initiated, a Mach-scale blade, shown in Fig. 6.15, was spin-tested. Although the test was successful, it highlighted a number of issues, all of which affected the measured performance: voids, delaminations, AFC pack failures, and electrical connection failures, caused in large part, by the AFC and blade-manufacturing processes. On-going tasks have addressed these issues. The final demonstration in this task is expected to be a hover test of a Mach-scale rotor in 2000.

In another DARPA-sponsored program, Lucent Technologies and Sikorsky Aircraft have addressed two helicopter problems: reducing rotor blade noise via active rotor control and cutting vibration and noise levels in helicopter cabins via active control of noise and vibrations carried through transmission mounts.[58] The project was concerned with both low (e.g., from the rotor blades) and high (1 kHz from the transmission system) frequencies. Approaches considered in tradeoff studies included blade root control, blade twist, blade flaps, and other active airfoil concepts. Technical issues identified to date are similar to those identified in the Boeing programs. Plans include designing and fabricating a model active rotor

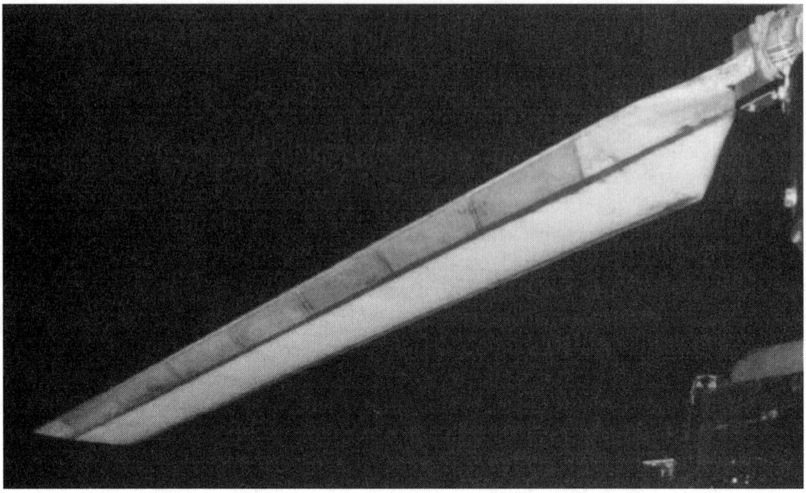

Fig. 6.15 Mach-scale active fiber composite blade spin tested at MIT (courtesy of MIT).

control system. The model will be tested at Sikorsky rotor test facilities and in an acoustic wind tunnel.

Concurrent with these DARPA programs are a number of closely related Army- and NASA-sponsored efforts. These programs are focused on similar activities but generally at a more fundamental level. Examples include active circulation and trailing edge flap or tab control,[94–96] active suspension for adaptive vibration isolation, active blade twist control, active fuselage walls for vibration damping and noise reduction, and active vibration control in a 20-mm gun.

Active twist concepts are of special interest. One program, in particular, a cooperative effort between NASA, the Army Research Laboratory (ARL), and MIT, focuses on the active fiber composites also being investigated in the DARPA Phase II effort. Recent analytical and experimental investigations[86,97–101] suggest that helicopter rotor blades containing embedded active fiber composites may be capable of meeting performance requirements necessary for a practical individual blade control (IBC) system.[102] An IBC system would allow for elimination of the swashplate, swashplate actuators, rotor blade torque tubes, and lag dampers. This would be a very significant step forward for improving helicopter performance (e.g., an 85% range increase is predicted by combining advanced rotor technologies with on-blade flight control for a UH-60 baseline system.[103])

In the AFC rotor blade concept, piezoelectric AFC plies are oriented at ±45° within the primary structure of the blade to generate dynamic blade twisting. Analytical studies indicate that twist amplitudes of 1 to 2° over a relatively wide-frequency bandwidth are possible using strain actuation capabilities of AFC plies. System studies also indicate that this magnitude of twist actuation authority should be possible at full scale, with only modest increases in blade weight and, possibly, relatively low levels of power consumption.

As part of this joint effort, an aeroelastically scaled, actively twisted, rotor research model is being fabricated for testing in the heavy gas environment of

Fig. 6.16 The ARES 9-ft-diam rotor testbed in the NASA–Langley TDT.

the NASA–Langley TDT. The test section of the TDT in Fig. 6.16 shows the Aeroelastic Rotor Experimental System (ARES),[104] which will be used to operate the active twist rotor model. These wind-tunnel tests will serve as an important demonstration of the active twist rotor concept and provide valuable experimental data for validation of active twist rotor analytical tools.

These activities are expected to form the foundation of future advanced active twist rotor research efforts for the Army and NASA. This future technology rotor will incorporate advanced airfoils, plan form geometry, and active twist capability in an optimized, integrated, intelligent rotor blade structure. By considering active twist capabilities from the beginning of the rotor design process, it should be possible to create an advanced rotor with aerodynamic performance, vibratory loads, and acoustic characteristics much superior to those obtainable with purely passive rotor blade structure designs.

C. Shape-Adaptive Structures

The concept of shape-adaptive structures and aerodynamic flow control figures predominantly in current thinking. Design concepts of interest include wing warping, camber shaping/control surface deformation, and variable stiffness structures (Fig. 6.17), among others. Specific objectives in the shape-adaptive structures area are developing innovative design processes, eliminating discrete control surfaces, and enhancing maneuver performance. Among expected performance benefits are reduced signature and drag and increased takeoff gross weight and range capabilities.

Several small systems have been built and tested since the early 1990s to demonstrate that shape-adaptive structures are possible and that flight control through the use of smart materials and structures is feasible and can offer significant performance advantages for a variety of systems.[105] The very earliest demonstrator used directionally attached piezoelectric actuators to achieve twist in a subsonic National Advisory Committee on Aeronautics (NACA) 0012 aircraft wing. Other systems included a supersonic variation of the subsonic active twist wing[106] and an

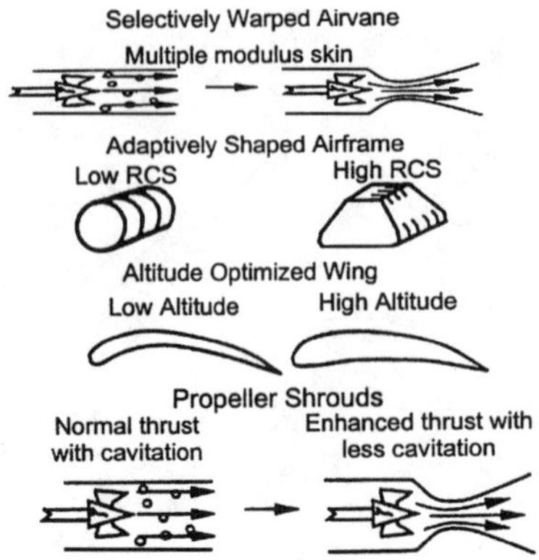

Fig. 6.17 Concepts for shape-adaptive structures.

active torque-plate concept for a NACA 0010 missile fin.[107] Later demonstrations along these lines (in the mid-to-late 1990s) focused on flight control for small vehicles, such as fixed-wing aircraft,[108] helicopters,[109] barrel-launched munitions,[110] and microair vehicles.[105] These demonstrations are the first known applications of smart technologies for air vehicle control surfaces. Other programs concurrent with these have pursued active shape control concepts for larger systems.

Two DARPA programs, the Smart Wing program and the Smart Aircraft and Marine Propulsion System Demonstration (SAMPSON) program, are particularly concerned with shape-adaptive structures for aircraft. Other groups, including the Deutsche Forschungsanstalt Fur Luft-Und Raumfhart (DLR), which is Germany's national aerospace research center (Germany),[111] USAF, and NASA, are also working on such structures for aircraft.

1. *Aircraft Wings*

Benefits of active control of wing shape have been demonstrated and well documented.[14–16] As an illustration, consider the use of conventional vs. continuous control surfaces (Fig. 6.18). Deployment of conventional control surfaces can, in effect, change the overall wing camber. But these same rigid surfaces give rise to discontinuous boundaries, which, in turn, result in early air flow separation and, ultimately, reduced lift and increased drag. Use of smoothly contoured control surfaces, however, delays flow separation and improves lift and stall angle characteristics.

Techniques for wing twist and camber control using smart structures approaches are being developed under the DARPA Smart Wing program.[112,113] Northrop Grumman was awarded contracts for Phase I (September 1994) and Phase II (August 1998). Other members of the large team of researchers involved in the program include Lockheed Martin Astronautics, Lockheed Martin Control Systems, Naval Research Laboratory (NRL), Mission Research Corporation, Rockwell Science Center, Fiber & Sensor Technologies, Etrema Products, SRI International, University of California at Los Angeles, Georgia Institute of Technology, and the University of Texas at Arlington. Wind-tunnel testing for the program is provided by NASA–Langley's TDT.[85] AFRL currently manages the program.

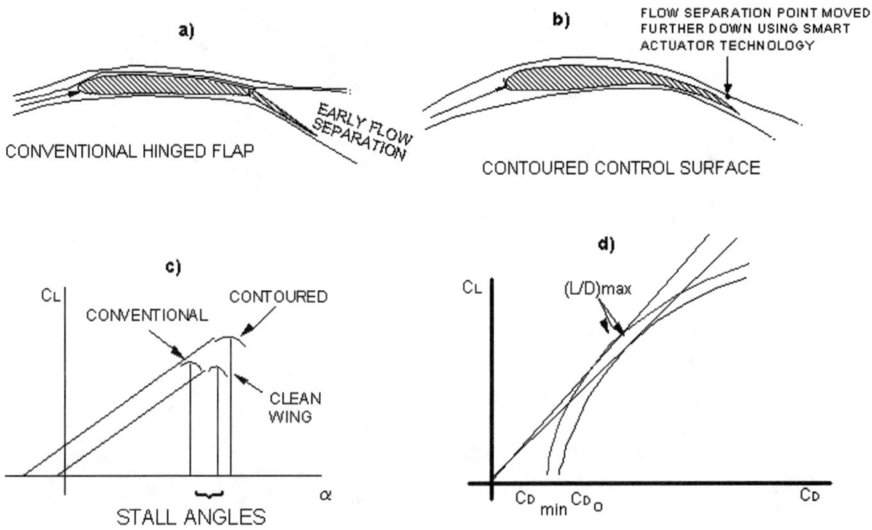

Fig. 6.18 Schematic illustrating the benefits of continuous vs conventional control surfaces.

The overall objective of this program is to design, develop, and demonstrate the use of smart materials and structures to improve the aerodynamic performance of military aircraft, including improvements in lift-to-drag ratio (L/D), maneuver capabilities, and aeroelastic effects.[113,114] Estimated performance improvements include an 8% increase in allowable takeoff gross weight, 30% increase in weapons payload, and a 10 to 15% increase in maneuver rates. The approach includes designing, fabricating, and testing scaled semispan and full-span wind-tunnel models; addressing power, reliability, packaging, and system integration issues; and laying the groundwork for technology transition in a potential follow-on program.

During Phase I of the program, a 16% scaled semispan model of the F/A-18 aircraft was designed and fabricated, incorporating three key features: hingeless, smoothly contoured, trailing edge control surfaces; variable spanwise wing twist; and fiber-optic strain and pressure transducers.[113] On this model, the hingeless aileron and flap were actuated using SMA tendons as shown in Fig. 6.19.[113,115] For the first tunnel entry on the smart model, wing twist was accomplished by using two SMA-actuated torque tubes. Only one torque tube was used in the second wind-tunnel test (see Table 6.2 for a more detailed comparison).[113,115,116] Figure 6.20 shows a layout of the smart wing. Fiber-optic pressure and strain sensors were included in the first wind-tunnel test: the strain sensors were used as part of the feedback and the specially developed pressure sensors were shown to be highly accurate. Another identically scaled model of conventional construction—hinged control surfaces and no wing twist—was fabricated and used as a baseline for comparison; this model was used in both wind-tunnel tests.

The first wind-tunnel test in Phase I took place in May 1996. Figure 6.21 shows a photograph of the smart wing model in the TDT. During this test, 1.25° of twist was achieved using SMA torque tubes, resulting in approximately an 8% improvement in rolling moment. The hingeless control surfaces deployed up to

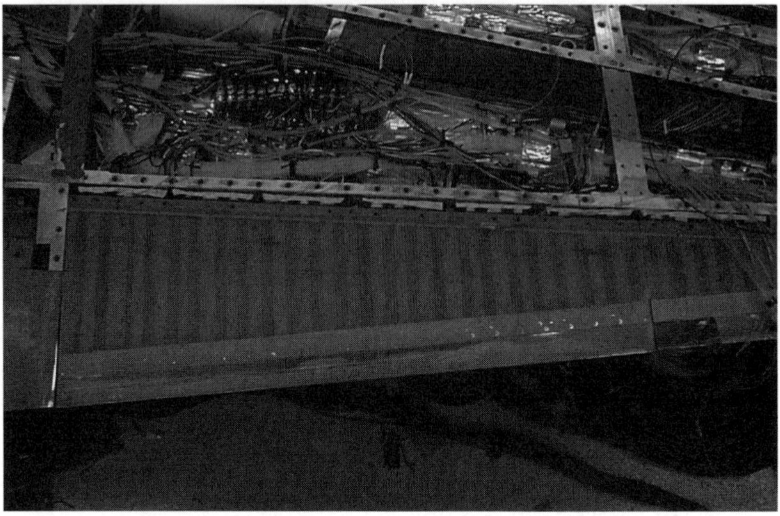

Fig. 6.19 Control surface hardware for the first smart wing wind-tunnel entry (courtesy of Northrop Grumman).

10°, provided between an 8 and 18% increase in rolling moment and approximately an 8% increase in lift.[117]

The second wind-tunnel test in Phase I took place in June–July 1998. In this test, the smart wing contained a single, redesigned torque tube and hingeless control surfaces similar to those used in the first test. During this test, 5° of twist was achieved resulting in a 15% increase in rolling moment. In addition, deflections of up to 10° on hingeless control surfaces were obtained with improved controllability and repeatability. Table 6.3 summarizes important results from both tests.[113,118]

The actuator scale-up results shown in Table 6.4 were an important conclusion from the Phase I tests. Analyses indicate that torque requirements for a full-size

Table 6.2 Comparison of smart wing shape memory alloy torque tube actuators

Wind-tunnel test 1	Wind-tunnel test 2
Two SMA torque tubes: Inboard 1-in. diameter, 1200 in. lb torque Outboard 0.5-in. diameter 600 in. lb torque	One SMA torque tube: 1.125-in. diameter 0.060 in wall thickness 3600 in. lb torque
Nichrome wire heater 1.25° spanwise twist (measured)	Nichrome wire heater 4.5° spanwise twist (measured)

Fig. 6.20 Layout of the smart wing (courtesy of Northrop Grumman).

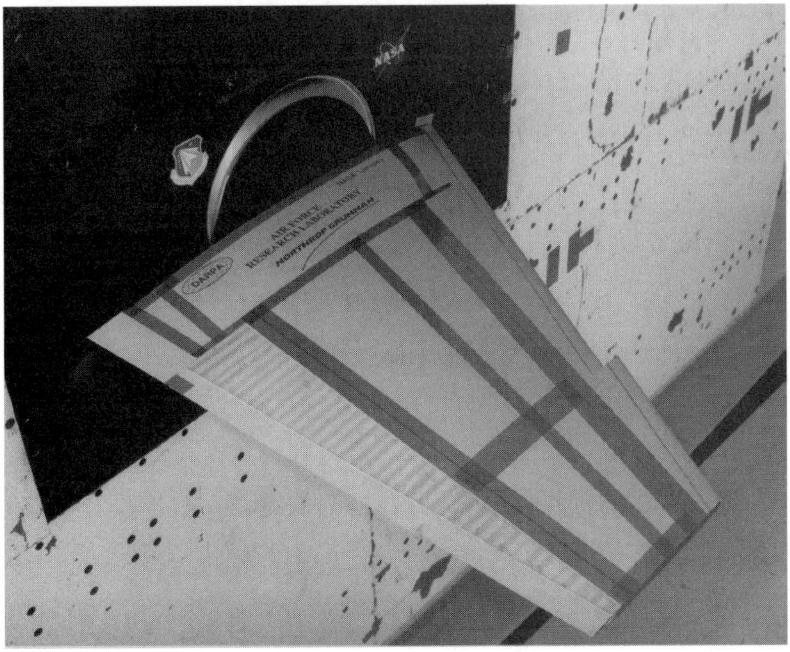

Fig. 6.21 The smart wing model in the NASA–Langley TDT (courtesy of Northrop Grumman).

Table 6.3 Summary of performance improvements for smart wing AoA = 8°

Improvements configuration	Deflection or wing twist,°	Lift, ΔC_L	Roll, ΔC_L	Lift and Roll, % Lift	Lift and Roll, % Roll
Flap only[1]	7.5	0.058	0.019	9.7	10.2
Combined flap aileron[1]	7.5	0.092	0.039	17.6	17.1
Aileron only[1]	5		0.015		8.0
Aileron only[2]	10		0.019		10.5
Wing twist[2]	3	0.034	0.019	8.0	10.0
	5	0.05	0.03	11.5	15.6
Wing twist[1]	1.4	0.041	0.022	10.0	12.8
Combined aileron and wing twist	+10° aileron, +4.5° wing twist	0.057	0.031	15.3	17.3

[1]Wind-tunnel test 1.
[2]Wind-tunnel test 2.

fighter aircraft wing are so large that they are beyond the capability of present-day actuation materials, even with innovative actuator designs.

Another Phase I Smart Wing task examined approaches to allow subtle changes in wing cross section to reduce transonic drag. At the speeds where most transport aircraft fly, just under Mach 1, air flowing over the wing can be supersonic in some locations and subsonic in others. When the airflow changes abruptly between those regions, a shock wave forms over the wing. The objective of this task was to provide a capability for shockless transonic cruise by subtly changing the airfoil shape at various stages of the flight profile, thus allowing up to 75% reduced drag as well as reduced fuel consumption. The concept involved a trusslike system of magnetostrictive actuators, designed to fit a full-scale Gulfstream III wing.[119] This system was not tested because analysis indicated that the added weight of the actuation system would effectively cancel the benefits expected from reshaping the

Table 6.4 Derived torque and rotation requirements for a shape-adaptive wing

Component	Full-scale aircraft (torque, in./lb)	16% model (torque, in./lb)	20% model (torque, in./lb)	Rotation,°
Wing box twist				
at 50% span	2.000×10^6	1.5×10^3	3.0×10^3	2
at wing tip	0.160×10^6	3.0×10^2	6.0×10^2	5
Leading-edge flap				
Inboard	0.185×10^6	1.3×10^3	2.5×10^3	10
Outboard	0.250×10^6	0.5×10^3	9.3×10^2	10
Trailing-edge flap				
Inboard	0.160×10^6	8.8×10^2	1.7×10^3	10
Outboard	0.032×10^6	3.1×10^2	6.0×10^2	10

airfoil. This analysis also led to the conclusion that improved lightweight actuators are required to realize the benefits of real-time, active airfoil shape optimization. It was determined that the same benefits could be realized by suitably deploying two flaps with built-in actuators (e.g., like Fowler flaps) rather than by changing the entire wing cross section.

A key limitation in the Phase 1 effort was the low bandwidth of the control surfaces. The SMA actuation system performed at a fraction of a hertz. Even with improvements in materials technology and active cooling concepts, the system would not be able to provide the tens of hertz response required for operational aircraft. Phase II of the Smart Wing program includes plans to mature further the technologies developed in Phase I and investigate new actuation concepts, (e.g., hybrid piezoelectric devices and piezoelectric motors) to address bandwidth limitations. The structure of choice is an unmanned vehicle. Two wind-tunnel tests in the NASA–Langley TDT are also currently planned for Phase II.

USAF efforts in shape-adaptive structures are focused on enhancing vehicle performance by eliminating discrete control surfaces and structural dynamic problems on current and future aircraft. Some of the USAF demonstration activities have been conducted in conjunction with NASA. The Active Aeroelastic Wing (AAW) (also called Active Flexible Wing) is a more complex project involving vehicle structure, controls, and aerodynamics to maximize performance.[120–122] This multidisciplinary effort takes advantage of inherent aeroelastic flexibility in the wing, with the leading- and trailing-edge control surfaces being used as aerodynamic tabs. The airstream twists the wing with minimal deflections of control surfaces, and surfaces can be optimized as a set for most efficient control under all flight conditions. This technique is expected to be flight tested. Expected benefits for this AAW technology generally include substantially increased control power, reduced drag and aircraft structural weight, and increased design latitude for wing span, sweep, and depth. Combining variable stiffness mechanisms with this concept may reduce the size and power of control surfaces and actuation systems presently being used in the AAW, as well as the weight, by as much as 30%, depending on the aircraft configuration and mission profile. Variable stiffness devices may also result in improved stealth characteristics and cruise efficiency and reduced drag, attractive in both manned and unmanned vehicles.

2. *Engine Inlets*

The inlet system of jet-powered aircraft preconditions the air entering the engine. Jet engines require air to enter the engine at approximately Mach 0.5 or less. Because of the wide range of Mach speeds, altitude, AoA, angle of slip, and engine airflow conditions, a fixed geometry inlet cannot provide ideal performance under all conditions. At low speeds, having large inlets with very blunt lips is desirable. This allows the high airflow associated with takeoff conditions to be drawn into the inlet without flow separation. At subsonic cruise, sharp inlet lips are desirable because they produce less drag. Sharp inlets also reduce RCS. At supersonic conditions, losses caused by rapidly decelerating the flow from supersonic to subsonic result in substantial losses in pressure and thrust. To overcome these limitations, variable geometry inlets have been used. The variable geometry inlets used on the F-15, for example, improve performance over a range of conditions, but their mechanical complexity adds weight and cost to the aircraft. Compliant mechanisms using smart materials and structures technology for variable leading and trailing edges are thought to be a simpler alternative to implement. These are flexible structures that change shape by deformation rather than by conventional rigid body motions.

Smart materials and structures technologies have advanced to the point where it is now feasible to demonstrate such physical shape control. The DARPA-sponsored SAMPSON program involves Boeing–St. Louis (formerly McDonnell Douglas) with the following team members: Lockheed Martin Astronautics, Georgia Institute of Technology, NRL, Electric Boat Corporation, BBN (now GTE), and PSU's ARL and Center for Acoustics and Vibration. This program is managed jointly by NASA–Langley and ONR and is focused on both aircraft and submarine applications, with a specific interest in shape and flow control approaches for inlets. Wind-tunnel testing of the full-scale aircraft inlet will be performed at NASA–Langley facilities.

Objectives for the aircraft portion of SAMPSON include designing, fabricating, and demonstrating a full-scale adaptive inlet for a tactical F-15 Eagle aircraft (Fig. 6.22), with a particular focus on validating control of inlet geometry and internal flows.[123] Potential benefits include improved range (20% for tactical aircraft) and maneuverability, flutter and buffet control, and reduced signature. Important core technology efforts are addressing needs for quasi-static shape control for inlet shaping, leading-edge lip deflection, and leading-edge blunting via compliant surfaces actuated with SMAs. Piezoelectric motor concepts are being evaluated for high-rate shape control of inlet walls. As part of this wind-tunnel test program, an actuator system based on SMAs will be used to rotate the cowl about a pivot point to vary capture area, similar to the way the current inlet operates. The forces to achieve the necessary stroke (about 6 in.) are quite large, on the order of 20,000 lb. The advantage of using SMAs in this way is their integration to the structure, which eliminates the need for a separate subsystem (e.g., hydraulics) to run it. Bench tests of some of the actuator devices are underway. These devices were integrated to subcomponents and tested in 1999. A wind-tunnel test of the full-scale section is planned for February 2000. This first test is expected to demonstrate cowl actuation, flow separation control with quasi-static shape changes, and high-rate

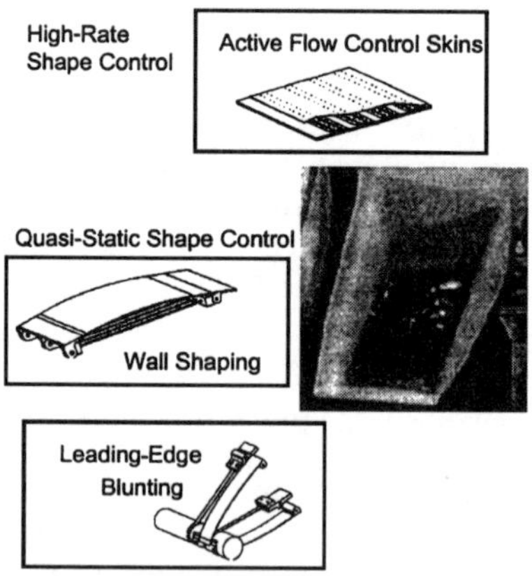

Fig. 6.22 Concepts for the shape-adaptive inlet demonstration in the SAMPSON program (courtesy of Boeing Phantom Works).

flow control of inlet walls. A second-wind-tunnel test is planned for later in the year.

Active flow control skins using synthetic jets to alter the boundary layer (for compression ramp generation) were also considered in the early SAMPSON studies. Preliminary wind-tunnel tests showed promising results, but this concept will not be part of the final demonstration article. Boeing has a related program with NASA–Langley and Georgia Tech for an integrated MEMS flight maneuvering system. Objectives of this effort are to develop and test an integrated synthetic jet flight control module using micromachined fluidic drivers and multifunctional composite structures design and manufacturing. NASA–Langley also has a cooperative effort with Lockheed Martin on innovative control effector design issues using distributed fluidic drivers and their effect on aircraft control.[124] All of these programs leverage the others.

3. *Missile Fins*

Modern, fast-maneuvering, guided missiles require highly dynamic fin actuation with excellent feedback control. Strong aerodynamic loads act on the control surfaces of a missile in flight: up to 2,000 lbf may be applied to the fin. Therefore, it requires a great deal of power to actuate the fin to produce rapid maneuvers in the missile. The power required can be up to 1 kW per fin shaft. LFK GmbH is conducting research in the use of smart materials to develop new actuators for application to missiles.[125] This work is examining the use of magnetostrictive alloys, piezoelectric ceramics, and SMAs in various actuator configurations. Barrett used directionally attached piezoelectric torque-plates to pitch a missile fin statically to ($\pm 4.5^\circ$ and dynamically at rates in excess of 30 Hz.[126] Barrett has also shown that adaptive materials could be used to generate fin pitch deflections useful for missile flight control with proper force and bandwidth capability.[127]

D. Integrated Electronics

Military air vehicles typically contain large numbers of antennas that protrude from the surface. For example, the F-18 has 66 antenna apertures located at 37 sites. These apertures cover a broad range of frequency bands from megahertz to gigahertz levels for radar and communications functions. These externally mounted antennas require local reinforcement of the structure to accommodate electromagnetic windows, a feature that adds weight and cost. Nonconformal antennas degrade vehicle aerodynamic performance, require substantial maintenance, and increase vehicle signature. The military services desire to minimize the number of antenna sites by combining functional capabilities of different devices at fewer sites while still maintaining equivalent or better coverage. Some believe that up to 50% of the vehicle's surface could be used to exploit this capability.[128] Controllable, reconfigurable antennas or conformal antennas are probably required to achieve the desired mission flexibility. Potential benefits include reduced weight and volume; low observability (especially for conformal antennas); reduced energy consumption; improved system performance, including flexible capabilities to enable new missions; and lower costs because of reduced duplication and potentially easier repair and maintenance.

All of the military services are interested in technologies that result in low observability. Specific approaches that allow a smooth surface shape to be maintained are, therefore, desirable. As in all of the smart structure applications identified previously, the miniaturization of processing and control electronics will be one step toward achieving this goal. The USAF supported a demonstration program at Northrop Grumman—Smart Skin Structures Technology Demonstration—to address some of these problems. A key feature of this program is the development of conformal load-bearing antenna structures, wherein the antenna element and associated structure carry structural loads, obviating the need for heavy local reinforcements and eliminating structural cutouts. Analyses indicate that many requirements can be met with a single broad band element, but selected narrow band elements are necessary to achieve the right gain levels as well as for special functions (e.g., global positioning system). Expected performance benefits include

1) An entirely new electronic warfare (EW) capability for tactical fighter aircraft at very low frequencies, e.g., for threat identification, threat angle of arrival, and situational assessment.

2) The potential for higher EW frequency bands using the same materials and processing techniques.

3) Significant low observable performance at relatively low cost.

4) Reduced drag and improved range.

5) Enhanced low frequency performance for voice communications.

Secondary benefits include reduced weight and ease of manufacturing due to eliminated fasteners and doublers, reduced structural cutouts (up to 75% reduction[129]), and reduced costs. The USAF predicts a cost savings of about $250,000 per airframe with a weight savings of about 70 lb for the F-22. More optimistic projections estimate cost savings ranging from $0.5 million to $3.3 million per aircraft and weight savings ranging from 260 to 1,000 lb per aircraft.[129]

Future smart skin efforts will look at improved avionics and, eventually, completely integrated antennas and avionics (load-bearing) combined with adaptive structures, vibration suppression, and structural integrity monitoring capabilities.

Northrop Grumman, TRW, and PSU have designed and fabricated a full-scale, load-bearing antenna to be embedded in typical fighter fuselage skin structure (a hypothetical F-18 for a next-generation mission).[130] Several subscale panels were evaluated for their structural and electrical performance and manufacturability. Figure 6.23a shows antenna components for the final demonstration article. The final demonstration article in this program was a full-scale, curved panel (about 3 ft $\times$ 3 ft) that was tested under combined axial and shear loads for both static and fatigue loading conditions (Fig. 6.23b). The panel withstood 4,000 lb/in. of running load (equivalent to 148,000 lb total load),[129] and a 4700 μstrain was achieved. The panel survived a fatigue lifetime and ultimately failed at 150% of its design limit load. Experimental data agreed well with predicted responses.

Northrop Grumman also investigated conformal antenna installation in the vertical tail of a military aircraft.[131] Their analyses indicated that communication link ranges—VHF-FM (30 to 88 MHz), VHF-AM (108 to 156 MHz), and UHF (225 to 400 MHz)—could be significantly improved by such installation. Gain improvements of up to 25 dB were demonstrated in flight tests on the NASA–Dryden systems research aircraft F-18 flight test bed. Structural integration considerations were significant because of the large number of items already located there. Tail buffet and flutter were also considered in the design. Material concerns, which included moisture absorption and acoustic fatigue of the candidate structural foam core, were addressed by environmental and fatigue testing. Material and

a)

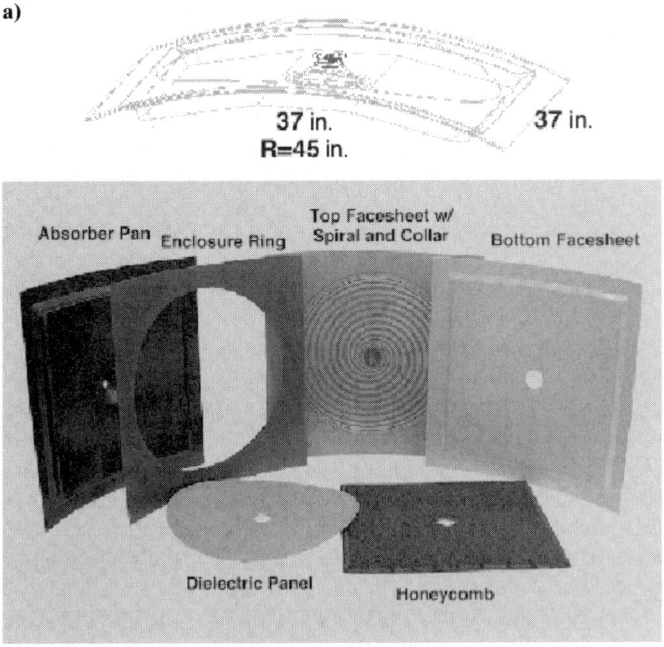

b)

Fig. 6.23 Integrated antenna/structure: a) antenna components and b) full-scale panel test article in loading rig (courtesy of Northrop Grumman).

device susceptibility to lightning damage remains an issue. The resin transfer molding process (RTM) was used for fabrication (primarily for reasons of cost) to produce the vertical tail end cap. Both of these Northrop Grumman projects were constrained, to some degree, by the requirement to use existing avionics/electronics. With advanced avionics, even greater benefits may be realized.

As another example, the Joint Attack Strike Technology (now the JSF) program funded design studies and limited demonstration work at Northrop Grumman (formerly Westinghouse) and Raytheon (formerly Hughes) for multifunction RF systems,[132] the goal being lighter weight and lower life-cycle cost relative to current systems. These systems included multifunction nose array antennas, support electronics, appropriate application software, and sensor/resource management control. These nose arrays may be used for air-to-ground and air-to-air radar functions, traditional radar electronic warfare, and communications, navigation, and identification.

Other programs on integrated electronics focus on spacecraft systems. A jointly sponsored (USAF, BMDO, and DARPA) program—the Spacecraft Integrated Electronic Structure (SIES)—addressed the weight issue by embedding power distribution and data-transmission lines into structural skins.[38] General program objectives included reducing weight and volume by eliminating brackets and connectors and incorporating bulky cables to the composite skin, as shown in Fig. 6.24a. Electronics enclosure and harness weight reductions of 70% were predicted. These items typically represent about 68% of the total weight of a small spacecraft. Lockheed Martin–Denver has addressed issues associated with panel-to-panel connections and multichip module (MCM) attachment to (and removal from) the skin. A multifunctional structural panel was flight tested on the Deep Space-1 (DS-1) comet/asteroid fly-by mission (launched October 1998), part of NASA's New Millenium Program (NMP).[133,134] The panel, shown in Fig. 6.24b, included the following components: a microcontroller printed circuit board for the spacecraft data-collection interface, a high/low power distribution MCM, a thermal simulator MCM, and a radiation-hardened composite cover. A number of key technology elements were demonstrated: embedded electronics interconnect system using flexible circuitry, MCMs and MCM socketing, flexible circuit jumpers, anisotropic electrical bonding, temperature sensors, and thermal doublers—all on a spacecraft structural panel. In this experiment, electrical-interconnect continuity tests were successfully performed. In addition, measures of thermal performance were consistent with predicted levels. Data are still being downloaded at regular intervals. This structural concept will also be evaluated in other space flight tests, including NMP DS-2, STRV-1d, MightySat II.1 (Sindri), and the Advanced Technology Demonstration Spacecraft.

ITN Energy Systems is developing several other integrated multifunctional system concepts in which structural, power supply, and electronic functions are combined.[135] The lithium battery core (LiBaCore) honeycomb structure in Fig. 6.25 uses large area, roll-to-roll lithium batteries as the honeycomb material sandwiched between a face sheet containing a photovoltaic material on the one side and a face sheet with flex-circuit electronics on the other side. This multifunctional material is envisioned as the structural material for fabrication of high-altitude, long-duration unmanned air vehicles (UAVs) and microspacecraft. The ITN multifunctional concepts extend to integrated structure/electronics/power packs for use in habitable spacecraft and space-based radar systems.

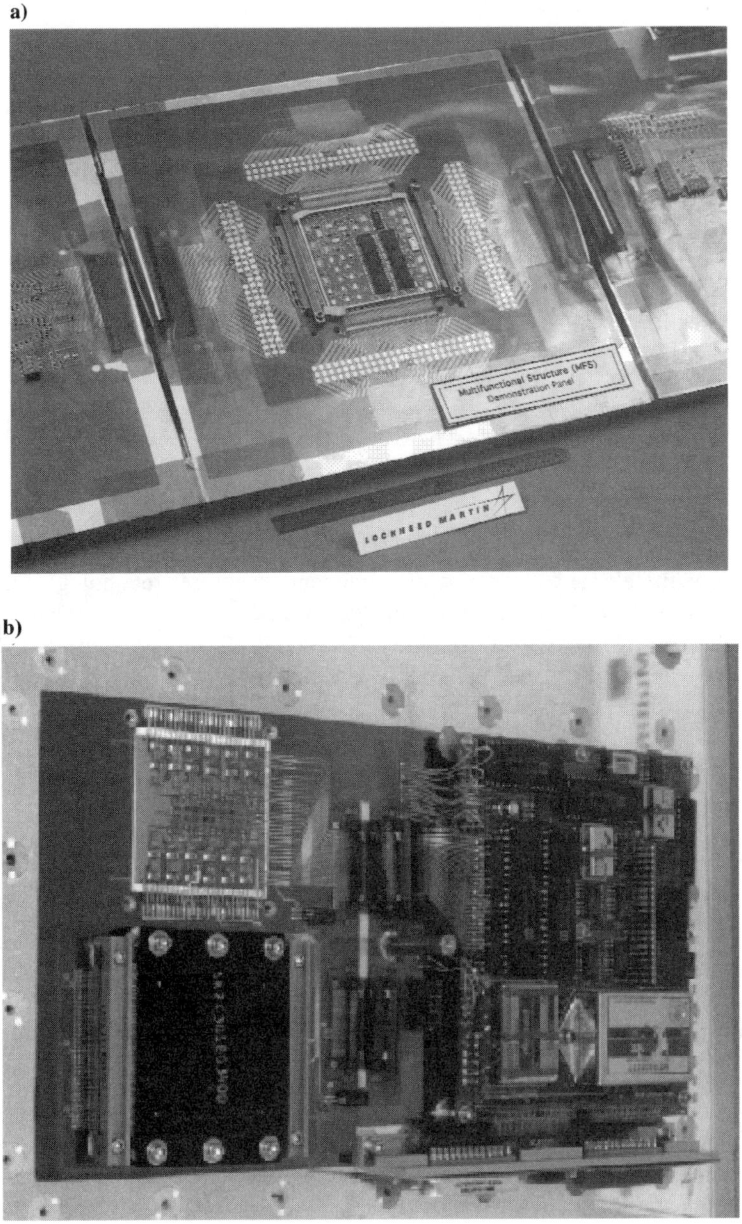

Fig. 6.24 SIES panels: a) development panel and b) flight hardware tested on the Deep Space−1 mission in 1998.

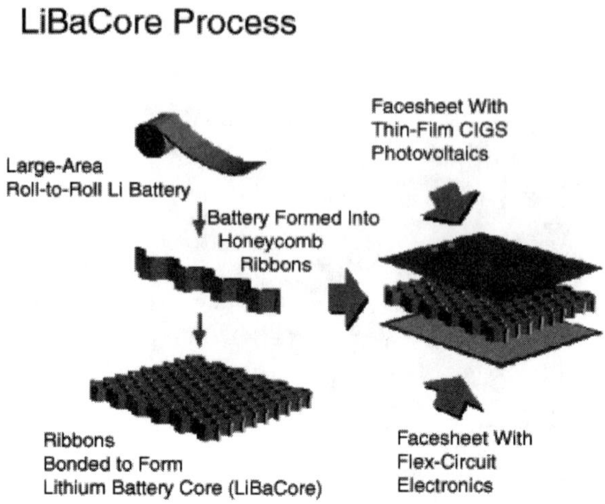

Fig. 6.25 Schematic of the lithium battery core honeycomb structure being developed by ITN (courtesy of ITN).

IV. Technology Status and Issues

While some elements of smart materials and structures technology are being demonstrated in current programs, many technical issues still have to be addressed. Technical issues fall into the broad areas of materials, devices, electronics, control, modeling and analyses, and integration. Universities, government and national laboratories, and industry are addressing many of the specific concerns in ongoing, mostly government-supported research programs.

A. Materials

The term "smart materials" has been applied to a broad range of materials that have one or more physical properties that can be varied with some input. By using these materials, a device that once consisted of separate structural, sensing, and actuation components can now exist as a single component. For some applications, a smart material can be substituted for existing components, reducing overall size and complexity. Smart materials have also opened the door for the development of many novel sensors, actuators, and smart structural components never before possible.

Recent progress in developing improved actuation materials, including electroactive ceramics, magnetostrictive alloys, and SMAs, is notable. Figure 6.26 summarizes some of the characteristics of various actuation materials.[136] Improvements in conventional actuator materials, discovery of new actuation materials, and their commercialization are described below.

1. Electroactive Ceramics

The best-known group of smart materials is piezoelectric ceramics, which produce a small shape change from an applied voltage. Piezoelectricity is a linear coupling between electric and mechanical variables. When an external force is

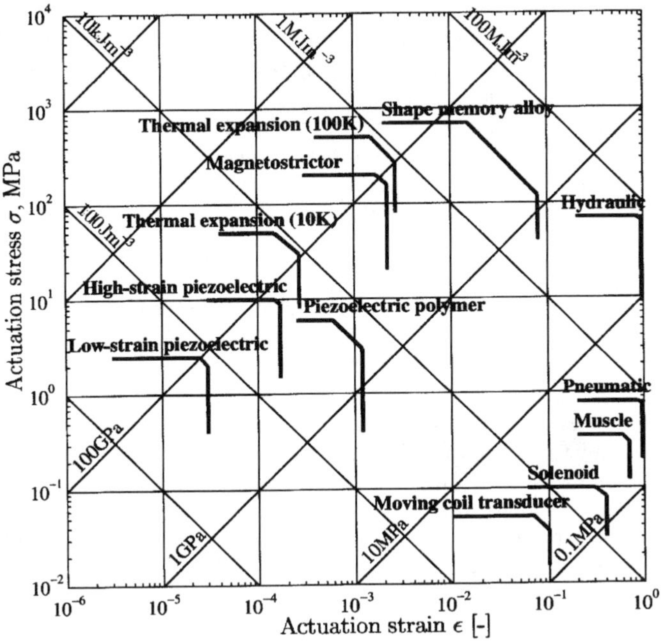

Fig. 6.26 Performance comparison of various actuator materials: actuator stress as a function of actuation strain with heavy lines bounding the upper limits of performance (courtesy of the Royal Society of London).

applied, the charge centers of the crystal structure separate, creating electric charges on the surface of the crystal and a change in polarization in the material. Electric polarization is proportional to mechanical stress in the direct piezoelectric effect. The thermodynamically related converse effect relates mechanical strain to the applied electric field. Electric charges on the crystal will cause a mechanical deformation. Piezoelectric materials can, therefore, be used for both sensing and actuation. Only solids lacking a center of symmetry show piezoelectricity, a third rank tensor property. Piezoelectric ceramics have limited strain-to-failure capability but a wide-frequency response range and fast response time as well as a fairly large force output. Shape changes can happen very quickly, so piezoelectric components have been utilized in high-frequency applications like vibration control, audio speakers, and ultra-sound generators. PZT is the most widely used piezoelectric ceramic for smart structure applications. Piezoelectric polymers such as polyvinylidene difluoride can also be used as actuators in some specific applications, but they are more often fabricated into ultra-thin film sensors that can be bonded onto a variety of materials, including metallic or composite sheet. Other polymers with unique properties are being developed at NASA–Langley.[26,137,138]

Relaxor-ferroelectrics are similar to piezoelectrics except that strain is produced by the second-order electrostrictive effect as opposed to the first-order effect. Electrostriction is a quadratic relationship between mechanical strain and the square of the electric polarization. Electrostriction is a fourth-rank tensor property that is observed in all insulating solids regardless of symmetry. It is normally a small effect, but electrostrictive strain can be surprisingly large near the Curie temperature of a relaxor-ferroelectric material. Lead magnesium niobate–lead titanate (PMN-PT), an example of such a material, can produce 0.1% strain at 1 kV/mm. Advantages of these materials in actuators include improved stroke, low hysteresis, and a return to zero displacement when voltage is suddenly removed. These materials are not susceptible to stress depoling, contrary to conventional piezoelectric materials, although they are sensitive to temperature and exhibit nonlinear performance that requires application of a bias voltage.

Fine-Grained Ceramics

Fine-grained piezoceramics have been shown to provide improved mechanical and dielectric strength and improved machinability over conventional, coarse-grained materials. Currently, fine-grained ceramics have a 30% higher bending strength than conventional ceramics. This improved strength translates to higher manufacturing yields during the slicing or lapping of thin plates, important in multilayer stack actuator fabrication. The use of thinner layers in multilayer actuators, in turn, enables the use of lower driving voltages. In addition, the higher dielectric breakdown strength of fine-grained ceramics means that actuators made from them can be reliably driven to higher electric fields, a feature that results in a corresponding two- to threefold increase in strain capability. Figures 6.27a and 6.27b illustrate achievable performance improvements in fine-grained materials.[139]

Ferroelectric to Antiferroelectric Phase Switching Ceramics

Ferroelectric/Antiferroelectric (FE/AFE) switching materials undergo a phase change (with accompanying unit cell dimensional changes) upon application of an electric field. Electric field-induced AFE-to-FE phase changes are accompanied by

1) High strains arising from the AFE tetragonal to FE rhombohedral phase change,
2) Large hystersis, and
3) Shape memory effects (for some compositions).

Much of the recent work in developing AFE–FE phase switching ceramics has been focused on the lead–lanthanate–stannate–zirconate–titanate (PLSnZT) system shown in Fig. 6.28.[140] In this system, strains of 0.5% can be achieved in single crystals and 0.2% in polycrystalline materials. These materials can be tailored to some degree to fit specific applications.

Single-Crystal Piezoelectrics

Recent breakthroughs in crystal growth methods have resulted in the development of larger FE single crystals with extremely large piezoelectric effects. Current growth methods result in large numbers of crystals approximately 1 cm^3 in size (Fig. 6.29), which are ideal for fabricating plates for high-displacement, high-force actuator stacks. The new crystals are also stronger than other piezoelectric crystals such as quartz and lithium niobate ($LiNbO_3$). The crystals have very high electromechanical coupling factors (greater than 90%), making them attractive candidates for broad bandwidth transducer applications. Field-induced strains are greater than 1%, which is an order of magnitude greater than the strains produced in conventional piezoceramics. These high strains result from a combination of high d_{33}'s (greater than 2,000 pC/N) and breakdown fields greater than 150 kV/cm.[139]

a)

b)

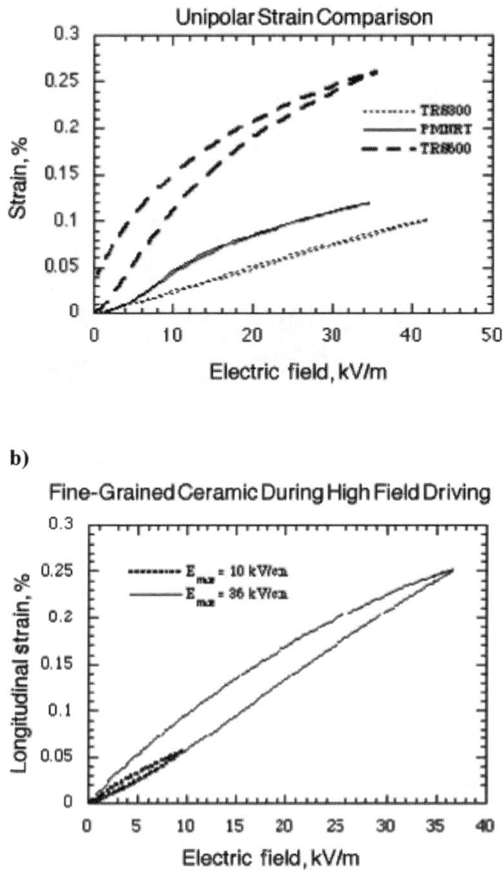

Fig. 6.27 Benefits of fine-grained PZT actuator materials: a) comparison of fine-grained PZT with conventional PZT and PMN and b) behavior of fine-grained PZT during high field driving (courtesy of TRS Ceramics).

2. *Shape Memory Alloys*

SMAs make up another important group of smart materials. The term "shape memory" refers to a thermomechanical phenomenon in which a solid possesses the ability to undergo shape change at low temperature and retain this deformation until it is heated, at which point it returns to its original shape. This "memory" is reversible. In returning to its original shape, the SMA generates large forces. Large shape memory effects occur in certain intermetallic compounds with martensitic phase transformations. At least 18 known alloy systems exhibit the shape memory effect, but, at present, only three are of commercial importance: nickel–titanium alloys (NiTiNOL: Nickel Titanium Naval Ordinance Laboratory), copper–zinc–aluminum, and copper–aluminum–nickel. Such metals are specially deformed in the low-temperature martensitic state but return to their original

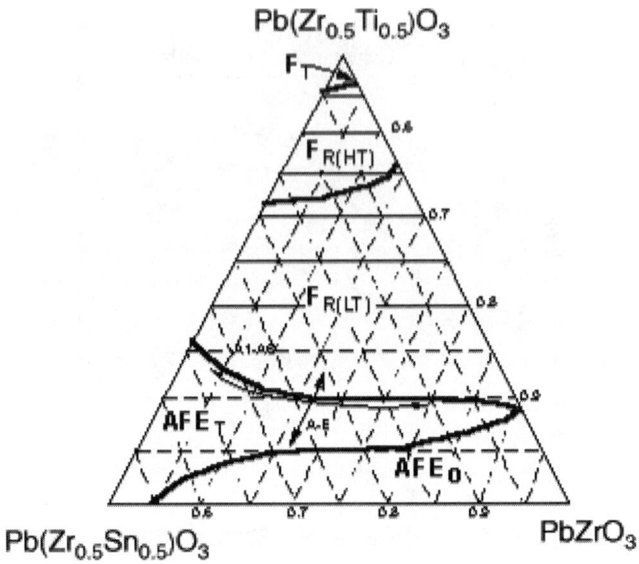

Fig. 6.28 Phase diagram of the PLSnT system showing the composition region of FE to AFE phase switching materials under development (courtesy of PSU).

shape when transformed to the high-temperature austenitic state. For example, a 0.01-in.-diameter NiTiNOL (nickel and titanium) wire can generate 10 N of force (26,000 psi) while straining 5% of its original length. SMAs have been used in many temperature-sensitive applications such as fuel control valves, surgical equipment, and aerospace equipment. Much research has been done on making shape-changing structural components. When embedded in composites, NiTiNOL wires can dampen vibrations and actively reduce stress concentrations by shifting the natural frequency of the surrounding material. Thus, NiTiNOL has potential in powertrain mounts and suspension bushings and in tendon devices that replicate muscle action. For dynamic applications, the cycle times are severely limited to the range of 0 to 5 Hz because of thermal time constants.

New classes of SMAs that can be actuated with a magnetic field are under development.[141–143] To overcome the bandwidth limitations of conventional SMAs, nickel–manganese–gallium, iron–palladium, and iron–nickel–cobalt(titanium) alloys are some of the targeted compositions being examined, as are shape memory

Fig. 6.29 PZN-PT single crystal grown by flux method at TRS Ceramics, Inc. Courtesy of TRS Ceramics).

ceramics materials such as lanthanum–doped PZT. Strains of 1.4% have been achieved, but the blocking stress is low at 8 MPa.[141]

3. *Magnetostrictive Materials*

Certain ferromagnetic materials undergo elastic strains when subjected to an external magnetic field. Induced strains and maximum stresses are on the same order of magnitude as those for piezoelectrics. One common magnetostrictive material—TERFENOL-D [TER (terbium) FE (Iron) Naval Ordinance Laboratory]—produces a 0.2% strain in a 100-kA-turns/m magnetic field. Although such strains are larger than those achievable in piezoceramics, optimum performance requires that a stress be applied to the magnetostrictor before actuation. One major disadvantage of magnetostrictive actuators is the need for a magnetic field to produce actuation. This is typically accomplished using a coil wrapped around the material, which, in turn, makes the device bulky. Losses in the coils can also be high.

4. *Other Materials*

Many other types of smart materials exist, including electro- and magnetorheological fluids, which change their viscosity with electric or magnetic field. Shape memory thermoplastics that display behavior similar to that of SMAs are also being developed, as are photostrictive and chemostrictive materials. A sizable change in shape is observed in these latter two classes of materials when they are exposed to light or chemical environments. These effects are caused by changes in electronic structure or in chemical bonding and are usually not connected with domain-wall motion. These latter materials are not covered in this chapter.

B. Devices

Performance of individual elements in a smart structure (e.g., sensors, actuators, signal processing electronics, power conditioning, and control electronics) will be critical to the successful demonstration of completely integrated systems.

1. *Sensors*

The most important performance parameters for fiber-optic and piezoelectric sensors include:[144,145]

1) Reliability/durability;
2) Consistent, predictable response;
3) Low cost (important if many devices are required);
4) Mechanical tolerances (device fit and finish for improved manufacturability);
5) Ability to handle an acceptable range of temperatures (composite processing temperatures for embedded devices as well as operational temperatures);
6) Aging and environmental effects (stable performance over long periods of time);
7) Sensitivity (figure of merit, varies with sensor type); and
8) Electrical bias (for electrostrictive ceramics, complicates use).

Various types of fiber-optic sensors (e.g., extrinsic Fabry–Perot interferometers, short- and long-period Bragg grating sensors) are frequently used because of their low weight, small size, high sensitivity to strain, multiplexing capability, and no electro-magnetic interaction (EMI) problem.[145] Figure 6.30 shows schematics of several types of these sensors. While these sensors have a relatively large strain

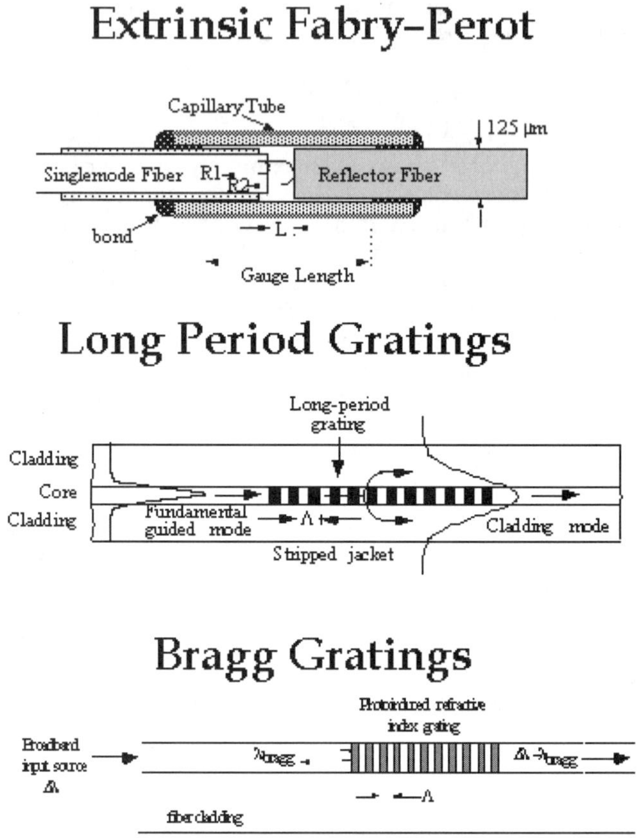

Fig. 6.30 Schematic diagram illustrating various fiber-optic sensors (courtesy of Luna Innovations).

to failure relative to other devices, they are fragile and require special handling procedures in manufacturing, particularly when they are being embedded. Fiber optics require an external power supply to generate the light source and analyze the signal. Signal interpretation is also an issue since the output is not electrical, but they have a demonstrated ability to measure strain[145,146] and detect vibration.[147] The issue of separating thermal effects from strain caused by loads—a particular signal interpretation issue—has been addressed using a number of approaches. A particularly promising approach is a special variation of long-period Bragg grating sensors.[145] A significant amount of work is being done to develop fiber-optic sensors for real-time composite cure monitoring, integrity monitoring, and configuration monitoring of aerospace structures.[148] A fiber-optic sensor capable of measuring chemical composition, strain, and temperature is currently under development. Both single and multimode optical fibers with and without Bragg gratings have been investigated: chemical spectra of a high-performance epoxy resin were obtained using both types of fibers.

Piezoelectric ceramics and polymers are also highly sensitive to strain and have easily measured electrical outputs, with the added advantage of no requirement

Fig. 6.31 Remotely queried strain gauge rosette designed to be embedded in graphite-reinforced polymer matrix composites (courtesy of NRL and MTS Corporation).

for an external power source. However, they suffer from EMI constraints. If embedded in composites, electrical shielding is necessary and there are temperature limitations. The ceramics are also limited by a relatively low strain-to-failure behavior.

Some manufacturing techniques for embedding sensor packages with minimal disturbance to the structure have also been developed. Figure 6.31 shows a strain gauge rosette designed to be embedded 0.25 in. below the surface of a graphite fiber-reinforced epoxy composite. The strain rosette is remotely powered and interrogated using an RF link that powers the device and reads the strain results.

2. *Actuators*

Desirable features for actuators include[7,8,9,136,144]

1) Consistent device-to-device and cycle-to-cycle response;

2) Extreme mechanical tolerances (especially surface parallelness and flatness in electroactive ceramics);

3) Low (or at least predictable) hysteresis;

4) Low creep (important when actuators must hold a position for a "long" time);

5) Large force/load-carrying capability (generally a tradeoff with displacement capability) and large dynamic range;

6) High-frequency response (application-dependent need);

7) Linearity (displacement with respect to applied field); and

8) Capacitance (for electroactive ceramics, important in high-frequency applications for which drive current is a concern).

Piezoelectric ceramic actuators. The ceramic actuators are found in a variety of forms: thin plates,[149] multilayer stacks,[150] injection-molded shapes,[151] and fibers.[152] Thin plate-type actuators are more or less a standard product and have been used in many vibration suppression applications.[9,73,105]

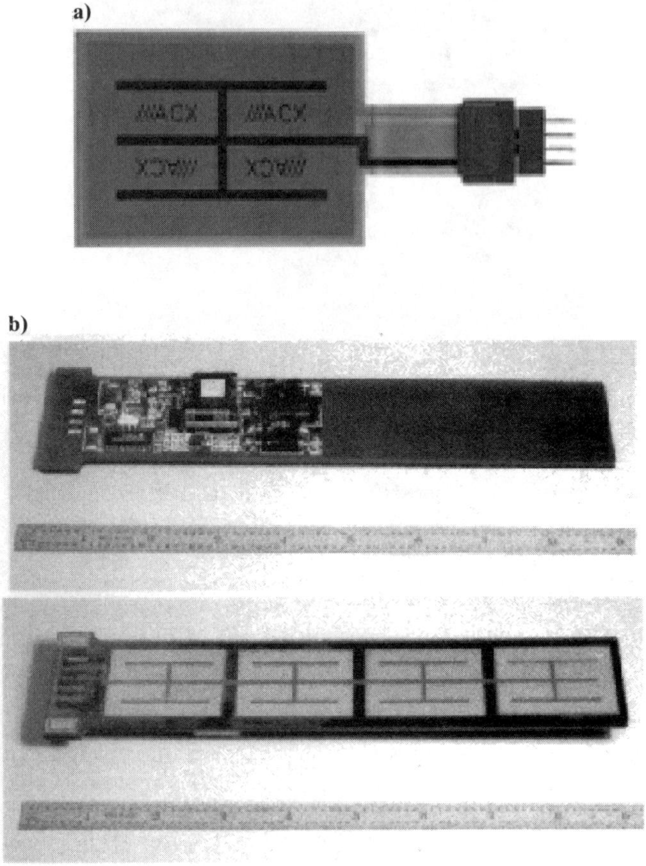

Fig. 6.32 ACX™ actuators: a) the QuickPack™ actuator and b) the SmartPack™ actuator containing integrated electronics (courtesy of ACX).

The ACX QuickPack® actuator, used in several of the major demonstration projects reviewed previously, is a thin, flat piezoceramic device that creates linear motion (Fig. 6.32a). Used as a strain actuator, it extends with applied voltage while bonded to the surface of a structure. Used as a bimorph actuator, it bends with applied voltage while clamped firmly at one end. ACX has developed an adaptive SmartPack™ containing sensors, actuators, and power and control circuits into an integrated package that automatically detects and adapts to uncertain structural mode frequencies (Fig. 6.32b).

NASA–Langley has also sponsored and developed two unique configurations of piezoelectric actuators: Reduced and Internally Biased Oxide Wafers (RAINBOW)[153–156] and Thin-Layer Unimorph Driver (THUNDER).[157,158] In contrast to piezoelectric patch actuators, which are flat and actuate primarily in plane, these actuators are dome shaped and actuate out of plane. Both devices exhibit displacements that are several orders of magnitude greater than conventional piezoceramic actuators but at the expense of actuator authority.

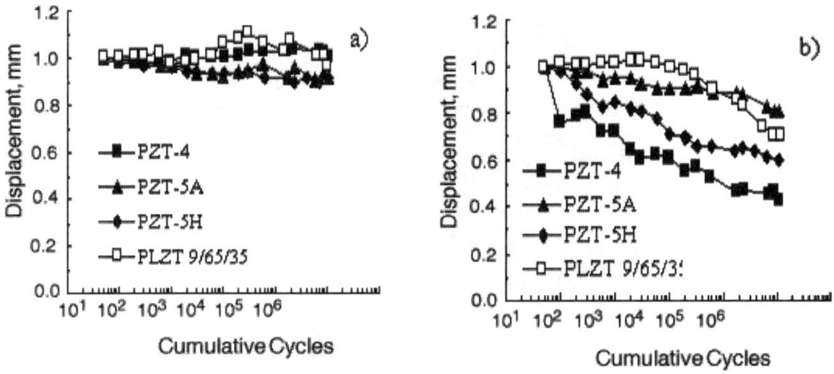

Fig. 6.33 The fatigue performance of RAINBOW devices produced from different starting piezoelectric compositions: a) no-load and b) 300 g static load.

RAINBOW actuators are produced by reducing one side of a piezoelectric wafer during firing by using an additional heat treatment step. In the RAINBOW process, developed at Clemson University,[153] typical PZT wafers are lapped, placed on a graphite block, and heated in a furnace at 975°C for 1 hour. The heating process causes one side of the wafer to become chemically reduced. This reduced layer, approximately one-third of the wafer thickness, develops internal strains in the wafer that shape it into a dome. These internal strains cause the material to have higher displacements and higher mechanical strength than a typical PZT wafer.[159] RAINBOWs with displacements of 3 mm and point loads of 10 kg have been reported.[154]

Because many applications will require continuous, long-term operation of high-displacement piezoelectric actuators, fatigue characteristics of these materials and devices are under investigation. Figures 6.33a and 6.33b illustrate performance properties for RAINBOW devices produced from different starting piezoelectric compositions.[160] In these measurements, displacement and/or polarization properties of the actuators have been evaluated under continuous application of electric fields for more than 10 million cycles. To improve these results, researchers are currently evaluating techniques to increase the load-bearing capability of actuators through enhanced processing techniques and alternative mounting configurations.

THUNDER[157,158] is a fairly new actuator that was developed at NASA–Langley. These devices are unimorph-type actuators that consist of a piezoelectric ceramic layer bonded to one or more nonpiezoelectric secondary layers. THUNDER actually refers to the process for packaging the ceramics: a metal sheet is laminated to the ceramic under heat and pressure to place the ceramic in a prestressed state, giving it a domed shape as shown in Fig. 6.34. Such packaging offers increased environmental resistance and improved durability while maintaining the basic electrical properties of the piezoceramic. THUNDER actuators operate from the DC into the megahertz region and have displacement-to-weight ratios that are orders of magnitude greater than those of other actuators in the market today. The efficiency of this design has made this type of actuator attractive for applications such as air pumps, speakers, and motors, among others. THUNDER has also been used in aerodynamic, vibration, and noise-control applications.[161]

Fig. 6.34 THUNDER actuator showing its domed shape.

Currently, the processing and characterization of THUNDER high-displacement actuators are under investigation. The major features that determine the operating parameters of a THUNDER actuator are the type and thickness of piezoceramic, the curvature, number of layers, thickness and placement of the foil stressing member, and the thickness of the adhesive. One recent characterization effort[162] studied the effects of electric field, load, and frequency on the displacement properties of rectangular THUNDER devices. Results showed that individual actuators were capable of free displacements in excess of 3 mm when tested at ±9 kV/cm. Increasing device stiffness through metal selection and thickness resulted in improved load-bearing performance at the expense of displacement, allowing devices to be designed with a range of performance capabilities. In this test, the actuator demonstrated a maximum of 1.7-mm total stroke with no load decreasing to approximately 1-mm total stroke with a 250-g point load. The resonance frequency (Fig. 6.35) was found to vary significantly with clamping configuration. As a result, improved clamping methods are currently being developed for application specific mounting configurations.

The NASA–Langley Airfoil THUNDER Testing to Ascertain Characteristics (ATTACH) project[161,162] was a concept validation study that focused on identifying material characteristics (e.g., creep, hysteresis, and fatigue) and airfoil shaping

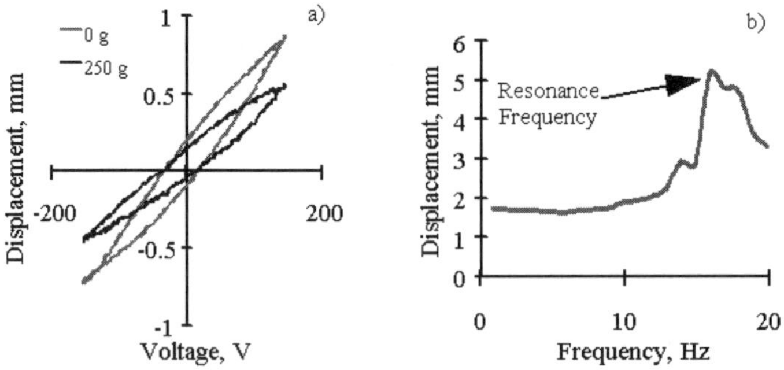

Fig. 6.35 Displacement performance of a THUNDER actuator as a function of a) voltage and b) frequency.

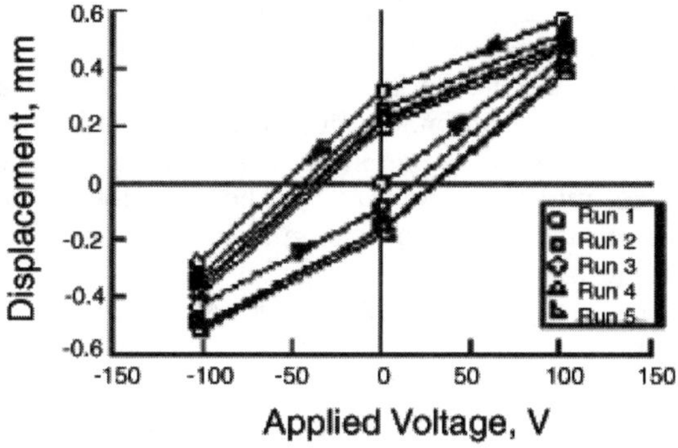

Fig. 6.36 Typical THUNDER wafer response to five cycles of applied voltage (±102 V).

effectiveness of the new THUNDER piezoelectric technology under aerodynamic loading. Characterization of the material behavior of a THUNDER actuator under aerodynamic loading was the objective for Phase I, while Phase II examined its ability to reduce drag over an airfoil.

To assess the ability of the THUNDER wafer to reduce drag over the airfoil, tests were conducted for wind-on conditions. Comparisons of wake velocity for different test conditions provided qualitative indications of the drag, reducing the potential of this piezoelectric actuator for this subscale model. Results indicate that the new THUNDER actuator technology is a promising candidate for future airfoil shaping investigations. Despite decreases in displacements of the wafer because of aerodynamic loading, noticeable drag reductions were obtained. ATTACH experiments also conclusively identified the presence of creep and hysteresis of the wafer under wind-on (loaded) and wind-off (unloaded) conditions.[161] Figure 6.36 shows an example of wafer displacement in response to five cycles of applied voltage. In the figure, creep is characterized by increasing positive (up) and negative (down) displacements exhibited by the wafer. Hysteresis appears as the repeatable symmetric "loops," which, in this case, are offset slightly due to creep. Phase I testing also revealed that wafer displacements are smaller at lower tunnel velocities and become higher with increasing AoAs. Both of these results were most likely caused by increased suction on the upper surface of the airfoil under those conditions. Further, the condition of the wafer was frequently checked to gain a preliminary understanding of material fatigue. After 2 weeks of testing, the performance of the wafer began to degrade noticeably. During subsequent examination, no visible flaws were found, but a 33% drop in capacitance was discovered, although repoling returned the wafer to its full capabilities. Thus, as with other piezoelectric adaptive materials, the performance of THUNDER appears to be a function of capacitance.

NRL is developing a piezoelectric torsional actuator.[163] This device uses segmented piezoelectric materials assembled to form a cylinder. Application of an electric field causes the cylinder to twist in torsion. The device is being configured to operate as a solid-state motor so that large rotation angles can be achieved.

Shape memory alloy actuators. SMA actuators, which work via a temperature-induced microstructural phase change in the material, exhibit relatively large actuation force (recovery force) and high strain (recovery strain) output and damping capabilities but may also exhibit large hysteresis. The temperature at which the phase transition occurs and the amount of observed hysteresis can be adjusted by changing the material composition.[164] Typical forms include wires, rolled foils or sheets, and torque tubes. Some researchers are examining sputtering techniques to deposit thin films for microscale devices.[165] The actuation mechanisms and control approaches themselves can be fairly simple in design, an advantage over other types of actuators. However, because of their slow response time—attributed partly to the slow cooling capacity of the material—they are best suited for low-frequency applications, on the order of a few hertz or less. Operational parameters, such as temperature, time at temperature, stress levels, required transformation strains, and number of transformation cycles, will affect the long-term durability of these devices. SMAs are quite suitable for slow motion of control surfaces, such as flaps in helicopters[87] or trailing edges in aircraft.[113]

Both Northrop Grumman and Boeing have evaluated nickel–titanium–copper SMA torque tubes for application in helicopter rotor blade twist, tilt-rotor blade twist, and shape-adaptive wings. Smaller torque tubes sized for 100 in.-lb and operating at ~0.2 Hz have been fabricated for use in helicopter blades. More robust SMA torque tubes have been developed for use in smart wing programs. For example, the Smart Wing Phase I program actuator (Fig. 6.37) has a blocking torque of 3,500 in.-lb and operates at ~0.03 Hz. This work established design guidelines for sizing SMA torque tubes.[87,116,165]

Other SMA devices include shockless release mechanisms[42] for spacecraft application. These devices provide considerable improvement over pyrotechnic release devices, as shown in Fig. 6.38. Shock separation reductions of 10X or more are achievable using these SMA devices. These devices have been flight tested in

Fig. 6.37 The smart wing phase I (wind-tunnel entry 2) SMA torque tube (courtesy of Northrop Grumman).

Fig. 6.38 SMA shockless release devices (courtesy of Lockheed Martin Astronautics).

the SMARD MightySat-1 experiment, as described earlier. SMA rings for turbine engine blade tip clearance control[166] are also being examined.

Researchers at NASA LaRC are studying the use of embedded SMA wires for noise suppression and panel flutter suppression in aircraft.[167,168] SMA hybrid composite panels provide active methods of controlling the structural acoustic response. These composites also have the potential of improving thermal buckling and postbuckling panel behavior, dynamic response, fatigue life, and structural acoustic response.

Magnetostrictive actuators. Rare-earth magnetostrictive alloys have high force and strain capabilities but are extremely heavy, and magnetic field shielding is an issue. These actuators are often found as rods or multisegmented stacks.[169] A group at Northrop Grumman and Etrema Products, Inc., has examined use of a magnetostrictive linear motor for controlling air foil shape of transport aircraft to optimize performance (e.g., improved L/D drag, increased range, reduced fuel consumption) over a range of flight conditions.[119] The motor has achieved ± 20.1 lbf with a stroke of ± 0.75 in.[170] As mentioned previously, performance benefits were not sufficient to pursue this concept without significant reduction in the actuator mass.

NASA Goddard is supporting efforts at Etrema Products, Inc., to develop a magnetostrictive actuator that operates at cryogenic temperatures for use on the Next-Generation Space Telescope and on other cryogenic devices such as switches and valves. These devices use modified TERFENOL alloy chemistry to adjust magnetostrictive properties for low-temperature use.[171]

3. *Other Concepts and Issues*

Because performance capabilities of currently "available" actuators are still orders of magnitude inadequate to address many application needs for large structures like aircraft wings, some researchers are investigating hybrid devices combining hydraulics and electroactive ceramics or magnetostrictives.[172] Others are investigating the use of small-scale devices (MEMS) to alter airflow over a surface.

For example, the location of local flow separation over a wing could be changed by the placement of microactuators at the leading edge.[173] The problem with this concept is that MEMS devices are fragile, which may preclude their use in many situations. As another example, aerodynamic shapes could be modified or turbulence, pressure gradients, and flow noise could be controlled by the placement of synthetic jets or "puffers" across the skin of the wing or in an engine air inlet.[174]

To help meet expected future requirements for environmentally robust, durable, low-power, high-authority, high-strain, fast-response actuators, DARPA initiated a number of new programs to develop higher authority actuator materials, including, among others, SMAs, single crystal electroactive ceramics, piezoelectric fibers, and injection-molded piezoelectric ceramics. Another new DARPA program will focus on compact hybrid actuation schemes that address integrated electronics and power issues.

Considering issues not only associated with the materials and devices themselves but also how they will be used in real applications is important. For example, one research issue identified in the PARTI program concerned piezoelectric power consumption during active control. Research in this area at NASA LaRC is focused toward the long-term goals of developing techniques to control larger, more realistic structures using piezoelectric actuators and assessing the economic viability of using piezoelectric actuators for control by examining power consumption. One of the techniques being investigated in the aircraft morphing program is examining passive or active damping augmentation using shunted piezoelectric actuators. This method, which uses a parallel inductor and resistor to shunt the piezoelectric actuator electrically, may provide a simple, low-power alternative to active control using conventional (unshunted) piezoelectric actuators. A recent NASA study[175] demonstrated that shunted piezoelectrics can significantly reduce aeroelastic response: reductions of up to 70% in plunging response were realized. The study also showed that the effectiveness of shunted piezoelectrics was a strong function of the inherent structural and aerodynamic damping. While this method may not be as effective for highly damped structures, shunted piezoelectrics can provide a simple, low-power, fail-safe vibration suppression method for lightly damped structures. Future research activities include conducting open- and closed-loop experimental ground tests on large-scale structures to assess tradeoffs in piezoelectric control effectiveness. Power consumption and optimal control strategies using active and passive control techniques with shunted and conventional, unshunted piezoelectric actuators will also be investigated. Shunted piezoelectric actuator studies will be accomplished in collaboration with Boeing (Huntington Beach).[176] During ground tests, comparisons and tradeoff between control effectiveness and power consumption of the active control vs active/passive damping control will be performed.

Two key benefits of using shunted piezoelectric actuators for active or passive damping augmentation are the very low power required for operation of the circuit and simplicity of mechanical operation when the shunted actuators are used passively for control. This is in contrast to several issues arising when piezoelectric actuators are used for active vibration control: the potentially high amount of power needed and the corresponding weight and volume of the actuators, power supplies, and amplifiers required. Some applications, however, may require active control to achieve the desired system performance.

Previous analytical and experimental tests showed that the maximum power required to control a structure can be estimated without modeling the complex dynamics between the piezoelectric actuator and the host structure.[177,178] Experimental results demonstrated that power consumption of piezoelectric actuators is a function of the material and geometric properties of the piezoelectric actuators for a perfectly controlled structure (all motion is stopped). Furthermore, as control effectiveness decreases, power consumption of the actuators decreases. Hence, maximum power consumption occurs when the structure is perfectly controlled. Experimental tests have further demonstrated that the structural dynamics of the host structure and piezoelectric actuators have a minimal effect on the power consumed by the actuators during active control. Consequently, for typical surface-bonded piezoelectric actuators, the power required for active vibration control can be conservatively determined within 90% accuracy (generally within 97% accuracy) without modeling the complex structural dynamics of the actuator and host structure.

C. Electronics

Of primary concern in the area of electronics are the large size, weight, and high power requirements, particularly for ancillary support equipment, e.g., processors, amplifiers, cabling. Miniaturization of electronics devices will be a key advantage in addressing this problem. The USAF and BMDO supported work at TRW on a Modular Control Patch (MCP).[38] The MCP, a 1-in. $\times$ 2-in. multichip module patch, provided retrofittable, miniaturized diagnostic and control electronics for vibration sensors and actuators. This patch integrated all the necessary features to a single, small, lightweight, low-power but reasonably capable package that could be embedded or attached to a structure. An individual patch can be used in a local manner at a discrete location or in concert with a network of other patches in a global manner. A device like this was used to control embedded piezoceramic actuators in the yoke of a solar array structure to demonstrate vibration suppression capabilities.[179] Researchers noted a thermal dissipation issue with the electronics, particularly since composites may act as thermal insulators.

DARPA supported the development of high-frequency switching amplifiers for use with electrostrictive "chiplet" actuator arrays.[180] The ultimate goal of this Virginia Tech/Virginia Power Technologies effort was to miniaturize these devices so they could be embedded into the structure with the actuators. This high-efficiency amplifier was specifically designed for capacitive loads that the small, "chiplet" actuators present to it (Fig. 6.39). These are high-efficiency devices: the measured efficiency is about 90% at 0.5 kHz, with a decrease down to about 20% at 40 kHz.

Another important feature could be the development of small, lightweight power supplies that can be placed with appropriate electronics for local control. For example, researchers at the USAF's Rome Laboratory and Johns Hopkins University have developed an all-plastic battery using polymers for electrode materials.[181] These batteries are useful in a wide variety of applications and can be recharged hundreds of times and operate under extreme conditions without serious performance degradation. The finished cells can be as thin as a business card and are malleable so they can be cut to fit. Specific energy densities of 30 to 75 W·h/kg are thought possible with these batteries.

Fig. 6.39 Miniature high-efficiency power supply and amplifiers (courtesy of Virginia Tech and Virginia Power Technologies).

Researchers at PSU have examined innovative antennas for use in smart skins. Specifically, barium strontium titanate (BST) materials are being examined for use as a tunable substrate in microwave phase shifters. The dielectric constant of these materials can be changed more than 50% by varying the composition and changing the applied bias voltage. Recent results[182] indicate that varying the applied direct current voltage can change the operating frequency of the antenna and incorporating BST phase shifters in the antenna array can provide electronic steerability. They were also combining these antenna concepts with advanced polymer materials and MEMS to develop a wireless telemetry system.[183,184]

D. Control Approaches and Algorithms

Control is one of the most challenging aspects in the smart materials and structures arena, in large part because of the complexities and uncertainties inherent to materials and structural dynamics of complex systems. It is an area that has received and continues to receive significant focus among academics and laboratory researchers.[185,186] One measure of controller performance is robustness. Robustness implies that the closed-loop stability and performance of the control system are insensitive to uncertainties stemming from modeling errors, nonlinearity, unmodeled dynamics, measurement errors, and other unknown disturbances. The more able a control system is to handle these uncertainties, the more robust it is considered to be. Inadequacy of modeling capabilities certainly increases the degree of difficulty in achieving structural control. The complex behavior of actuator materials, for example, presents a particularly difficult theoretical problem that may require development of new nonlinear control theories and algorithms.

The selection of the control approach is not straightforward. A significant problem lies in the control of large numbers of sensors and actuators, desirable for reasons of wide-band, spatially distributed structural control, redundancy, and robustness. Rapid, real-time computational capability at high bandwidth is required to satisfy structural control objectives—mapping sensor outputs into actuator inputs in real-time. The computational complexity is substantial and communication bottlenecks are expected. Current hardware limitations also prevent full interconnection of all of the devices. Control of these large device networks and the associated information management needs may, therefore, require development of new control approaches such as hierarchical control schemes, neural network schemes, or fuzzy logic controllers. While all are in development, none have been experimentally validated in complex systems to any substantial degree. SRI International developed a hierarchical control methodology to address the kilo-input/kilo-output problem.[187] Their technical approach was based on the wavelet transform, commonly used in signal processing and numerical analysis.[188] It is akin to a Fourier transform except that it is decomposed into time and scale rather than frequency. Figure 6.40 schematically illustrates a vibrating surface using this scaling approach. The pyramids above the surface are projections of a virtual surface at coarser and coarser scales (in this way, every device is connected to every other device, although not directly). In essence, data points are being thrown away as one moves away from the surface. Control can actually be implemented at any of these virtual surfaces—one way to get around the issue of local vs global control. This computationally efficient processing scheme also provides a method for device selection and placement. A particular advantage of this approach is that it does not require optimization of

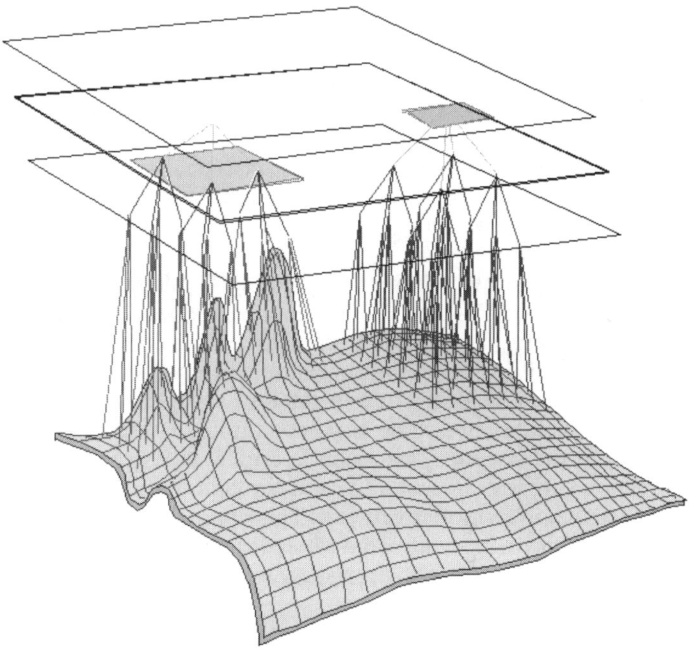

Fig. 6.40 Schematic of hierarchical control methodology to address the kilo-input/kilo-output control problem (courtesy of SRI).

placement of a small number of devices, a challenge when broadband control is desired.

A neural network (NN) is a data-processing scheme in which processing elements are fully interconnected.[189] Each processing element can receive data from many sources, but these data can only be sent out in one direction. The output can branch with other connections but each one would carry a signal based on the particular NN interpolating algorithm. Typically, each input is weighted, and all the weighted inputs are summed. Weighting factors are adjusted by the NN during "training" to minimize errors between the desired output and the NN output. These systems are not programmed but actually learn by training. The quality of the examples used in training determines how well or efficiently the network will perform. The primary advantage of a NN is the orders-of-magnitude faster computational speed. Of the several types of NN architectures, multilayer back-error propagation is the most common ("multilayer" refers to a feed forward arrangement while "backpropagation" refers to the method by which the NN is trained). These types of NNs are suitable for nonlinear control problems, such as system identification (the first step in design and implementation of active controllers),[190] damage detection,[191] and vibration control.[192]

A fuzzy logic controller is regarded as a set of heuristic (experience- or knowledge-based) decision rules that can be implemented on a computer.[193] Such control approaches are of interest to the smart structures community primarily because the control system can be designed without complete analytical knowledge of the structural system. It is believed that these control systems will be quite stable, reliable, and robust. They also allow for adaptive control to suit changing needs. The major issue in designing these systems is determining/selecting the appropriate rules. Geng et al.[194] have combined the fuzzy logic approach with a neural network scheme—Fuzzy Cerebellar Model Arithmetic Computer (FCMAC)—to allow for an adaptive, self-learning control system. Combining the two approaches offers the advantages of both, apparently without too much loss in performance (e.g., ability to learn and handle nonlinear behavior, rapid multiple input/output responses and no need for complete, detailed analytical models). The FCMAC controller hardware, MDSP-100, was successfully designed, built, and tested for controlling a magnetostrictive actuator to demonstrate active vibration isolation of a hexapod (Stewart platform).

One interesting technique, Vibration Control by Confinement™, or VCC™, has been developed by QRDC, Inc.[195] This approach takes advantage of the phenomenon of modal energy localization: geometric or material property variations in a structure can result in vibrations concentrated in particular locations of the structure. This is observed most often in rotating and periodic structures but it can occur in all structures under the right conditions. By redirecting energy to noncritical parts of the structure, vibration in critical areas can be reduced. This energy redirection can be achieved passively, through design, or actively via the use of a sensor/actuator/controller network. This approach yields simpler control systems—instead of having to control the whole structure one only needs to control selected regions. The approach has been successfully demonstrated on relatively simple structures.

Much of the university research in the area of smart structure control is focused on control of a few modes—using more conventional linear control approaches—in simple structures such as beams and plates with just a few sensors and actuators.

Some of the more complex demonstration structures are being used to evaluate the effectiveness of a variety of control laws. For example, 28 different control laws were evaluated in the series of tests on the PARTI wing.[84] The best was a linear quadratic Gaussian form using single-input/single-output with a strain gauge for feedback along with all 15 piezoceramic actuator groups operating. Even so, efforts to assess the state of control technology for more complex systems using smart materials and structures are inadequate.

E. Analytical Methods

Much progress in the area of smart structures has developed from interest in specific applications and devices with generally limited analysis. Models and analytical techniques that can reliably simulate and predict system responses are necessary to achieve effective and useful designs. One would like to begin with constitutive laws describing the active material/device behavior in terms of its inherent material properties. For example, Hom and Shankar[196] developed a constitutive model to describe electrostrictive ceramics. This fully coupled, three-dimensional model relates key variables (stress, strain, electric field, polarization, and temperature) based on particular material constants. As an additional example, five types of constitutive models (ferroelectric, internal variable, plasticity, hysteresis, and nonisothermal) have been developed to describe SMA behavior.[197] These nonlinear models rely on different characteristics and properties of SMAs to describe stress–strain relationships, the superelastic effect and its attendant energy dissipation, one-way and two-way shape memory effects, and so forth.

Ideally, material constitutive models like these would be combined with structural response models of the host structure and field equations to describe the system. Models must include all cross-coupling phenomena. Such an approach allows for a detailed understanding of the system.

Finite element modeling and validation for patch piezoelectric actuators has been examined at NASA. During the PARTI program, a finite element model of the PARTI wind-tunnel model was created and the natural frequency and mode shape results validated with ground vibration tests. Although global structural dynamic data (mode shape and natural frequency data) could generally be modeled with sufficient accuracy, local deformation and force data were difficult to model accurately with reasonable computational efficiency. It was found that considerably smaller finite element mesh sizes and a judicious use of finite elements were needed to capture the abruptly changing strain field in the area immediately next to the piezoelectric actuators.

Numerous researchers have developed analyses and numerical models to analyze piezoelectrically controlled structures.[198–203] These research efforts and many others clearly show that the key considerations for modeling in-plane piezoelectric actuation are

1) More refined mesh sizing (that may be accomplished through superelement modeling) and/or higher-order structural analysis theory to capture the widely varying strain field around the piezoelectric actuator.

2) Prudent modeling of damping to capture the dynamic piezoelectric effect more accurately.

3) Improved finite element modeling techniques to model larger, more complex structures with piezoelectric actuators.

In particular, Seeley's work[199] showed that the strain field in the host structure in the vicinity of the piezoelectric actuator is nonlinear. Most analyses used to model piezoelectric actuation (including the one used during the PARTI program) assume that the strain field through the thickness of the host structure is linear. Seeley used higher-order laminate theory, which was implemented using a finite element method, to model the piezoelectric actuation of composite plates. This research indicates that strain measurements (via traditional gauges) taken around a piezoelectric actuator may give unexpected results because most researchers assume that stress and strain fields are essentially linear.

Although previously mentioned analytical techniques can give very good theoretical results, they can also be complicated and difficult to implement, even for simple structures. The objectives of finite element modeling and validation research on patch piezoelectric actuators at NASA–Langley are to investigate and validate simple and accurate techniques for modeling structures containing piezoelectric actuators using MSC/NASTRAN.[204] MSC/NASTRAN was selected because it is a commonly used structural analysis code. Finite element models of structures of increasing complexity, such as composite plates and composite box beams with embedded actuators, will be developed and validated. Thus far, analytical models of a simple aluminum beam and a composite box beam with surface-bonded actuators have been validated with experimental results. The analytical technique[203] involves using a finite element approach to model the structure with actuators and a thermally induced strain to model the straining of the actuators with an applied voltage field.

During the PARTI program, Pototzky[205] used a finely meshed finite element model of a cantilevered beam to develop a method for modeling the aeroservoelastic response of a structure controlled with piezoelectric actuators. His method consisted of using a thermal mechanical analogy to create a static deflection shape of the beam actuated by the piezoelectric actuators. This shape was then appended to free vibration mode shapes, and the new augmented modal matrix was used in an aeroservoelastic analysis using the Interaction of Structures, Aerodynamics, and Controls (ISAC) Code.[206] Including this deflection shape as a mode in the aeroelastic analysis may allow for computation of the aerodynamic influence of piezoelectric actuation. In the aeroservoelastic analysis in ISAC, the appended static deflection shape was used as the control mode. When compared with experimental results, the transfer function data at zero airspeed were accurate. No air-on data were available for verification of aeroservoelastic results. Also, note that a very finely meshed finite element grid was used to ensure the accuracy of strain information. For larger, more complex models, this may be computationally difficult.

As in the area of finite element modeling, many researchers have developed methods to incorporate piezoelectric actuation to aeroservoelastic models.[207,208] Much of this work uses simplified structural models. The goals of the aeroservoelastic modeling and validation research in the NASA aircraft morphing program are to assess the state of the art in simulating piezoelectric actuation in aeroservoelastic modeling using simple and moderately complex models and, for the long term, to develop efficient methods for larger, more complex structures. For this research, doublet lattice aerodynamics and the ISAC code will be used to perform aeroservoelastic analyses. Existing data from the PARTI and ACROBAT programs will be used to assist in verifying analysis methods. Verification of analysis techniques will first be conducted at zero airspeed and then expanded to correspond to wind-on data. Data from ground modal tests conducted in the aircraft morphing program will be used to verify zero-airspeed characteristics further. Additional simplified models may be constructed and tested to provide more experimental data to compare with the aeroservoelastic modeling development.

To use devices such as THUNDER actuators in engineering applications, both modeling and characterization are essential. No simple analytical models have been available to understand static and dynamic behavior of THUNDER actuators. A NASTRAN nonlinear finite element model has been developed for predicting the dome heights of THUNDER piezoelectric actuators. A simple approach is used whereby temperature-induced expansion is used to simulate voltage actuation as described by Babuska and Freed.[209] To validate the finite element model analytically, a comparison was made with the nonlinear plate solution using von Kármán's approximation. The NASTRAN finite element model was also compared with experimental results.[210]

To be useable by a design engineer, however, faster, more efficient computational methods that retain important features of detailed physical models are necessary. The DARPA/industry-supported Synthesis and Processing of Intelligent, Cost-Effective Structures (SPICES) consortium, for example, addressed this efficiency issue via the use of superelements to describe piezoelectric device behavior within the context of a large ABAQUS finite element model.[149] Results showed good agreement between experimental and predicted broadband and modal responses for simple structures. Techniques to improve this finite element modeling capability have been identified for more complex structures. As another example involving complex constitutive behavior, Regelbrugge[211] modified Hom and Shankar's electrostrictor model to evaluate the performance of a vibration cancellation device more easily. He obtained excellent agreement between test data and the model for strain polarization as a function of induced strain and of electric field.

The USAF has supported work at CSA Engineering to develop finite element modeling techniques for smart structures, mostly considering vibration-suppression applications such as buffet load alleviation, flutter suppression, and component isolation. Presumably, having such models available will allow designers to characterize the collective system behavior and evaluate system performance enhancements (e.g., increased lift, decreased fuel consumption, and so forth) more easily.

System models are also being developed to describe improvements in helicopter performance. For example, to investigate the potential of active fiber composite rotors, two active twist rotor mathematical modeling methods have been developed at NASA–Langley. The first of these is a simple mathematical aeroelasticity model for composite helicopter rotor blades incorporating anisotropic embedded piezoceramic actuators. The computer implementation of this model, the Piezoelectric Twist Rotor Analysis (PETRA), has been created for use with the MATLAB[212] numerical analysis program and is ideally suited for conceptual active twist rotor design and optimization studies.[98,99,101] A procedure for using a commercially available comprehensive rotorcraft computer code (CAMRAD II[213]) for active twist rotor studies has also been developed. This allows active twist rotor numerical studies to be performed using a detailed state-of-the-art rotorcraft aerodynamics and structural dynamics model. Figure 6.41 compares the two codes in terms of relative twist as a function of frequency for a full-scale active twist rotor blade in hover. The twist amplitude and bandwidth behavior shown here would be sufficient

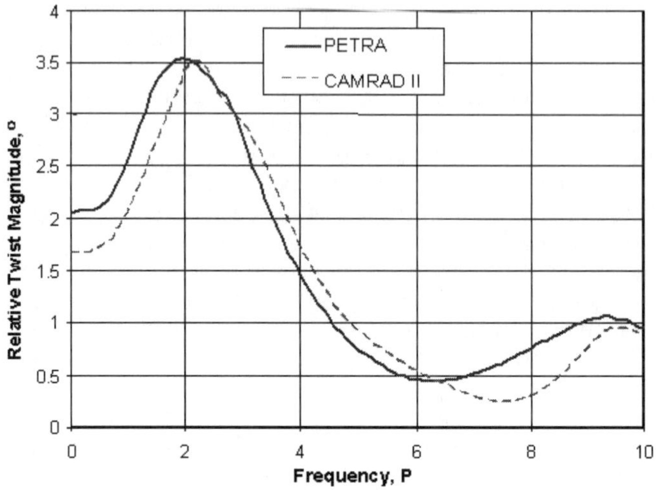

Fig. 6.41 Calculated hovering flight twist actuation frequency response for a full-scale active twist rotor blade concept.

for many IBC applications and vertical hub shear vibration reduction, in particular. Agreement between the two analytical methods is also extremely good and shows that fundamental active twist rotor dynamics are being modeled consistently.

Figure 6.42 shows an example of the predicted capability of active fiber composite rotor blades to alleviate stall on helicopter rotors.[101] Rapid buildup of torsional vibratory loads caused by stall severely restricts the maximum lift and forward flight speed capabilities of conventional helicopters. Vibration trends with

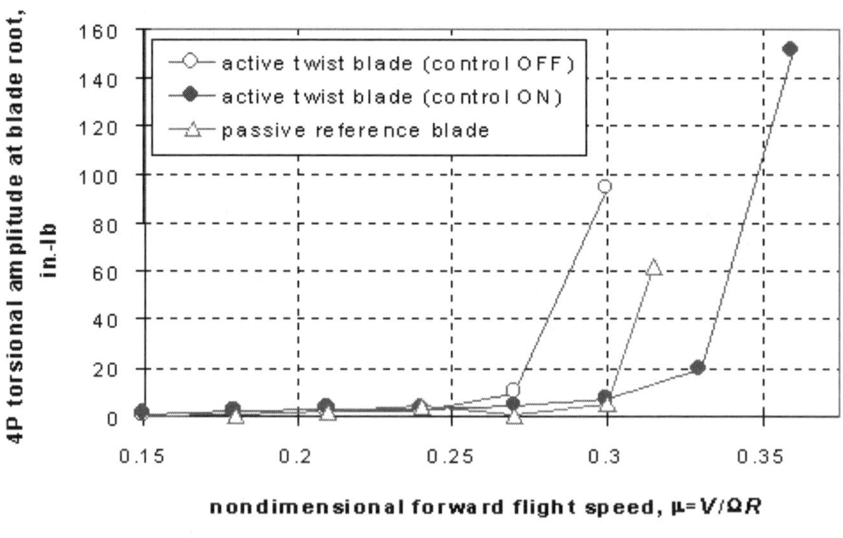

Fig. 6.42 Calculated suppression of dynamic stall induced torsional vibrations through active twist control.

increasing flight speed are shown for a conceptual full-scale active twist rotor blade with and without twist actuation control. Trends for a conventional, passive-structure blade are also shown as a reference. The rapid rise in dynamic stall-induced torsional vibratory loads for the active twist rotor blade using twist control has been effectively delayed by approximately 10% in nondimensional flight speed with respect to the baseline blade and by as much as 22% with respect to the active fiber composite blade with no twist control.

Another important design issue is that of controlling large structures with many sensors and actuators. Effective use of optimization models and schemes to determine the number and location of sensors and actuators needed to perform the desired task are being developed. For example, wind-tunnel results from the PARTI program showed that, at a given tunnel condition, one control law using all 15 actuators reduced gust loads by 75% while another control law used only six actuators to reduce gust loads by 72%. For many cases, such a minor loss of control effectiveness (3%) is inconsequential compared to the 60% reduction in the number of actuators required, which significantly reduces weight and complexity.[84] Two studies[214,215] investigated choosing optimal actuators from a set already in place on the PARTI model as opposed to choosing the optimal placement of actuators. Lin's study[215] focused on the selection of a set of effective actuators for improving the performance of active control at a single design point, namely, the onset of flutter. Simulation results using two control law designs showed that the optimal actuator set gave improved closed-loop performance, independent of the control law design selected.

Along with the general design approach and models noted above come a range of more specific concerns. These include

1) An inability to predict the life of the device in a free or integrated condition, especially its durability and reliability in terms of fatigue, fracture, and other mechanical performance criteria.

2) A lack of understanding of cross-coupling (electrical, magnetic, thermal, mechanical, fluidic, and so forth) among the various devices and the structure.

3) No fast, reliable method for determining the correct number and placement of individual sensors and actuators within or on the structure.

Efforts to address life prediction, in particular, may require further development of nonlinear mechanics models. Such efforts will also require extensive experimental tests to simulate the end-use operational conditions and verify models.

F. Integration

The application of actuators, sensors, and controllers to alter the performance of structures is becoming a viable tool for designers. Controlled structures technology has matured substantially over the last decade, and many new sensor and actuator types have been developed over the last 2 decades. Integration, therefore, can be approached from two perspectives. One is focused on practical manufacturing issues associated with actual integration of smart materials into real components. The second is focused on larger system considerations.

1. Practical Manufacturing Issues

Practical issues associated with building these structures and implementing them in real systems represent a significant barrier. To date, most examples of controlled structures involve surface mounting or bonding of sensors and actuators. This approach often carries the least risk because passive structural load paths are unaltered and failed components can be readily identified and repaired or replaced. Surface mounting is appropriate for some types of actuators, such as inertial proof-mass actuators and displacement/force actuators connecting the controlled structure to ground. However, strain actuators, which create a relative displacement within the structure, are usually most effective when they are embedded to the structural elements.[216] Except for simple structural geometries, such as beams and plates, surface mounting of strain actuators leads to less control authority than when they are embedded and their impedance is matched to the host structure. Structural integration of actuators, sensors, power and data electrical buses, and, perhaps, the controller electronics should also lead to more robust, durable, controlled structural systems. Numerous challenges arise when embedding smart materials to composite structures, including[216]

1) Electrical circuit failures due to dielectric breakdown and arcing,

2) Breaking of ceramic wafers and electrical leads (particularly in curved surfaces),

3) Low performance due to impedance mismatch or temperature changes, and

4) Compromised structural integrity due to microcracking and macrocracking in the host composite material.

These complications have been encountered in low-to-moderate strain-level applications but are even more problematic when high-stress, high-strain designs typical of aerospace applications are pursued.

Selection of the host material/structure strongly influences the type of manufacturing and fabrication processes actually used to create the smart structure. If devices are to be attached to a metal structure, typically achieved via adhesive bonding, the choice is relatively easy. When devices are to be embedded in the structure (this implies a polymer composite host material), selection of the process can be more difficult. Early efforts to embed piezoelectric sensors and actuators, fiber-optic sensors, and SMA actuator wires required hand layup and other special handling procedures,[7,8,217] a time-consuming and expensive process. Many current research programs are still using hand layup processes.[218] Thus, there has been and continues to be considerable user interest in developing cost-effective, automated methods to fabricate these structures.

One of the primary objectives of the SPICES program was the development and demonstration of cost-effective methods for fabrication of smart structures.[149,219] This consortium evaluated two potentially low-cost techniques for embedding ACX Quickpack™ piezoelectric plate actuators, frame-type piezoelectric stack actuators, fiber optics, and SMA wires: RTM for a flat plate component and advanced fiber placement for a trapezoidal rail component. Concerns about device compatibility with the host material, interfaces and interconnects between the devices and the structure, methods to fix devices in place during processing, and edge egress of wires/cables from the structure were addressed. For example, to prevent shorting out of the SMA in the rail component, a SMA tow-preg consisting of wires embedded in a thermoplastic sheath was developed. Special shielding methods were developed to protect piezoelectric actuators from the laser during fiber placement. For the RTM plate, a rigid substrate—SMA wires embedded in a thin, E-glass fiber-reinforced substrate—was developed. In this way, fibers could easily be prestrained. For the RTM component, registration holes were required to hold Quickpack^{TM} actuators in place. A unique method—a clean-trim utility conduit—eliminated the edge egress problem so no wires were hanging from the part. Lessons learned from this program are being applied in follow-on efforts.

While the SPICES program was very successful in identifying and solving problems unique to their selected processes, some others have not been addressed to any significant degree. Some of the modeling and design issues for materials have already been identified. Other concerns include a lack of methods to calibrate embedded sensors and actuators, very few reliable approaches to evaluate the manufacturability and cost of smart structures, and a limited understanding of the environmental and operational durability of structures with embedded devices.

2. *System Considerations*

For a number of years, researchers have touted the performance of new solid-state actuators, with each new material and actuator concept showing mostly incremental improvements in capabilities. This performance-driven competition spawned some novel concepts and provided the solid-state actuator community with a beneficial competitive environment. Unfortunately, until recently, other desirable characteristics of actuators, including weight, cost, reliability, and serviceability, have been downplayed. This, in turn, has led to a slow acceptance of solid-state actuators by designers, primarily because of real and perceived negative impacts on system weight, cost, and reliability. While these are considered secondary characteristics by some researchers, they are, in fact, first-order considerations for system designers. Hence, to increase the acceptability of solid-state actuator technology, the system performance benefits of including integrated actuators must be balanced against resource impacts and system reliability.

Power, mass, and cost are among the designer's primary considerations. Available electrical power is an especially important design consideration. Strain-actuation devices are usually controlled electrically, even though the actuator response may be due to an electric, magnetic, or thermal field. For command and control, actuators are integrated to an electronic circuit that also requires power, weight, and, perhaps, hardware and software interfaces with the system computer. While the electrical power needed by a few actuators may be small, highly distributed actuators can require significant system power caused by low electrical-to-mechanical energy conversion. Thus, electrical power consumption continues to be a major resource consideration for many applications.

Another feature of smart structures with serious system-level ramifications is the mass of actuators. Ferroelectric strain actuation materials often contain lead so that the actuators are usually more dense than the host structure. To offset the actuator mass, redesign of the host structure incorporating the performance benefits of active structural control can usually result in a neutral or beneficial effect on total system mass. However, simply bonding strain actuators onto the nominal host structure will always increase mass. It is more desirable to redesign the structural component with embedded actuators to take advantage of active strain actuation in reducing system mass.

Perhaps the most important resource that the designer must manage is cost. Because these actuators are usually produced in small quantities using labor-intensive processes, costs are typically high. Even though raw materials are inexpensive, current solid-state actuator fabrication processes result in nonnegligible costs. Electronics used for power and C2 also add expense as they are typically customized for a given application. Actuator and electronics parameters should be standardized where possible so that large-scale manufacturing techniques/methods can be employed to reduce the price of these advanced actuation systems.

Next, the designer considers reliability and, sometimes, serviceability. If these advanced actuators are an enabling aspect of the system, reliability must be very high because actuator failure could result in system failure. On the other hand, system performance may degrade yet be marginally acceptable if actuators fail. In this case, lower reliability can be tolerated, particularly if the system can be serviced to repair the actuator subsystem. To date, processes used to fabricate solid-state actuators are not well controlled and documented, a fact that has led to uneven quality of devices, variable device performance parameters, and large numbers of actuator failures, even in laboratory environments. Actuator fabrication processes must be documented and controlled to achieve high reliability. In addition, applications with graceful failure modes should be exploited first to build a database on operational reliability. Such a database could lead to acceptance by the designer to use strain actuation in critical applications.

Finally, the performance of the actuator must meet the demands of the application. Vibration control requirements for spacecraft are certainly much different than those for vibration control of an aircraft. Spacecraft are typically designed using stiffness criteria, whereas aircraft are designed using stress criteria (high strain). Although high-strain performance has been and will continue to be a highly desirable characteristic of strain actuators, the development community is encouraged not to ignore other characteristics in their quest for ever higher strain performance.

V. Concluding Remarks

Smart materials and structures are those that can sense external stimuli and respond with active control to that stimulus in real or near-real time. Numerous system demonstrations recently completed or currently underway indicate that smart technologies will likely provide new and innovative capabilities in future commercial and military aerospace systems: spacecraft and launch vehicles; aircraft, including fighter and transport aircraft and UAVs; and helicopters and tilt rotorcraft. Expected benefits include "maintenance on demand," increased passenger/crew comfort, increased system/component structural life, improved precision pointing and/or sensing capabilities, enhanced aircraft and rotorcraft handling, improved aerodynamics and, possibly, new flight profiles, improved low observable characteristics, and reduced manufacturing and assembly costs.

Good progress is being made in the development of appropriate electronics, control approaches, and analysis techniques although there are still issues to be resolved. More work is needed to develop fabrication techniques to make these smart structural systems affordable. Concurrent with that is a need for increased emphasis on integrated system design. Realizing many of the envisioned applications will depend critically on the development of higher authority, solid-state actuators. Recent discoveries in actuator materials indicate that such substantial improvements are possible, thus broadening the potential for successful implementation.

VI. Acknowledgments

We would like to acknowledge the assistance of the following individuals and organizations in the preparation of this paper (alphabetical by organization): Shoko Yoshikawa at Active Control experts, Inc.; Spencer Wu at the Air Force Office of Scientific Research; Keith Denoyer, Rory Ninneman, Brian Sanders, and James Tuss at the Air Force Research Laboratories; Gary Anderson at the Army Research Office; Jim Shoemaker at Ballistic Missile Defense Organization; Friedrich Straub at Boeing (Mesa); Mark Hopkins and Ed White at Boeing (St. Louis); Aaron Bent at Continuum Control Corp.; Ephrahim Garcia at DARPA; Porter Davis at Honeywell; Felicia Brady, Chuck Everett, and Dee Saunders at the Institute for Defense Analyses; Bernie Carpenter and Suraj Rawal at Lockheed Martin Astronautics; Kent Murphy and Mike Gunther at Luna Innovations; Steve Hall and Eric Prechtl at MIT; Don Krantz at MTS Corporation; Keith Belvin, Jennifer Florance, Joycelyn Harrison, Jennifer Heeg, Garnett Horner, Lucas Horta, Anna McGowan, Bob Moses, Sharon Padula, Darrel Tenney, and W. Keats Wilkie at NASA–Langley; Bob Badaliance, Lee Gause, and John Michopoulos at Naval Research Laboratory; Jay Kudva at Northrop Grumman; Jim Kelly, Wally Smith, and Kam Ng at the Office of Naval Research; Thomas Shrout at the Pennsylvania State University/Materials Research Laboratory and TRS Ceramics; Dave Flamm and Ken Chou at SRI International; Doug Lindner at Virginia Polytechnic Institute and State University; and Troy Schelling at Virginia Power Technologies.

The financial support of the Institute for Defense Analyses under project CRP-238 is gratefully appreciated.

References

[1]Clauser, H. R., "Modern Materials Concepts Make Structure Key to Progress," *Materials Engineering*, Vol. 68, No. 6, Nov. 1968, pp. 38–42.

[2]Clauser, H. R., "From Static to Dynamic Materials in Design," *Mechanical Engineering*, May 1975, pp. 20–26.

[3]Forward, R. L., "Electromechanical Transducer-Coupled Mechanical Structure With Negative Capacitance Compensation Circuit," U.S. Patent 4,158,787, 19 June 1979.

[4]Forward, R. L., and Peterson, R. W., "Cold Damping of Mechanical Structures," U.S. Patent 4,199,989, 29 April 1980.

[5]Forward, R. L., "Apparatus and Method for Electronic Damping of Resonances," U.S. Patent 4,352,481, 5 October 1982.

[6]Forward, R. L., Swigert, C. J., and Obal, M., "Electronic Damping of a Large Optical Bench," *The Shock and Vibration Bulletin, Part 4 Damping and Machinery Dynamics*, Bulletin 53, May 1983, pp. 51–61.

[7]Ikegami, R., Wilson, D. G., and Laakso, J. H., "Advanced Composites with Embedded Sensors and Actuators (ACESA)," Final Rept. AL-TR-90-022, Astronautics Laboratory/Air Force Space Technology Center, June 1990.

[8]Bronowicki, A. J., Mendenhall, T. L., and Manning, R. M., "Advanced Composites with Embedded Sensors and Actuators (ACESA)," Final Rept. AL-TR-89-086, Astronautics Laboratory/Air Force Space Technology Center, April 1990.

[9]Bronowicki, A. J., Betros, R. S., Dvorsky, G. R., Wyse, R. E., Innis, J. W., and Kuritz, S. P., "Advanced Composites with Embedded Sensors and Actuators (ACESA)," Phase III-V Final Rept. PL-TR-93-3017, Phillips Laboratory Space & Missiles Directorate, Kirtland AFB, Nov. 1993.

[10]Wada, B. K., Fanson, J. I., and Crawley, E. F., "Adaptive Structures," *Journal of Intelligent Material Systems and Structures*, Vol. 1, April 1990, pp. 157–174.

[11]Matsuzaki, Y., and Wada, B. K., (ed.), *Second Joint Japan/U.S. Conference on Adaptive Structures*, Technomic, Lancaster, PA, 1992.

[12]Loewy, R. G., "Recent Developments in Smart Structures with Aeronautical Applications," *Smart Materials and Structures*, Vol. 6, 1997, pp. R11–R42.

[13]DeCamp, R. W., Hardy, R., and Gould, D. K., "Mission-Adaptive Wing," *Proceedings of the SAE International Pacific Air and Space Technology Conference*, SAE-87-2419, Vol. A89-10627 01-01, Society of Automotive Engineers, Warrendale, PA, 1987, pp. 189–195.

[14]Powers, S. G., and Webb, L. D., " Flight Wing Surface Pressure and Boundary-Layer Data Report From the F-111," Rept. No. NASA-TM-4789, NASA Dryden Flight Research Center, Edwards, CA.

[15]Hall, J. M., "Executive Summary AFTI/F-111 Mission Adaptive Wing," WRDC-TR-89-3083, Wright Patterson AFB, OH, Sept. 1989.

[16]Wong, K. J., "AFTI/F-111 Mission Adaptive Wing Lift and Drag Flight Test Results," AFFTC-TR-86-42, Final Rept., March 1987.

[17]Kudva, J. N., Lockyer, A. J., and Van Way, C. B., "Structural Health Monitoring of Aircraft Components," *AGARD Lecture Series* 205, *Smart Structures and Materials: Implications for Military Aircraft of New Generation*, Paper 9, Oct. 1996, pp. 9-1–9-6.

[18]Scheuren, W. J., Caldwell, K. A., Goodman, G. A., and Wegman, A. K., "Joint Strike Fighter Prognostics & Health Management," *Proceedings of the International Powered Lift Conference*, Royal Aeronautical Society, London, UK, 1998, pp. 38.1–38.7.

[19]Fedele, P., Cafasso, G., and Leece, L., "A New Technique for Damage Identification and Health Monitoring of Structures Using Piezoelectric Sensors and Actuators," *7th International Conference on Adaptive Structures*, Technomic, Lancaster, PA, 1997, pp. 232–242.

[20]Concillio, A., and Leece, L., "A Survey on Italian Activities in Smart Structures," presented at *Engineered Adaptive Structures for Noise Control and Vibration Control II* (Castelvecchio, Italy, 16–21 May 1999), Engineering Foundation, New York, NY, 1999.

[21]Van Way, C. B., Kudva, J. N., Ziegler, M. L., Alper, J. M., West, M. N., and Edwards, T. G., "Development of an Automated Aircraft Structural Integrity Monitoring System—Overview of the Air Force/Navy Smart Metallic Structures Program," *Proceedings of the 37th Structures, Structural Dynamics, and Materials Conference*, AIAA-96-1615-CP, AIAA, Reston, VA, April 1996.

[22]Kudva, J. N., Grage, M. J., and Roberts, M. M., "Aircraft Structural Health Monitoring and Other Smart Structures Technologies—Perspectives on Development of Future Smart Aircraft," in publication.

[23]Kudva, J. N., Marantidis, C., Gentry, J., and Blazic, E., "Smart Structures Concepts for Aircraft Structural Health Monitoring," *Smart Structures and Intelligent Systems*, Paper 1917–92, Vol. 1917, Society of PhotoOptical Instrumentation Engineers, Bellingham, WA, 1993.

[24]Van Way, C. B., et al., "Integration of Smart Structures Concepts for Improved Structural Integrity Monitoring of the T-38 Aircraft," *Proceedings of the USAF ASIP Conference*, San Antonio, TX, Dec. 1993.

[25]Gentry, J. D., et al., "The Impact of Smart Structures on Aircraft Structural Integrity Programs," *Proceedings of the 1992 USAF Aircraft Structural Integrity Program Conference*, San Antonio, TX, Dec. 1992.

[26]Simpson, J. O., Wise, S. A., Bryant, R. G., Cano, R. J., Gates, T. S., Hinkley, J. A., Rogowski, R. S., and Whitley K.S., "Innovative Materials for Aircraft Morphing," *Industrial*

and Commercial Applications of Smart Structures Technologies, Paper 3326-20, Vol. 3326, Society of PhotoOptical Instrumentation Engineers, Bellingham, WA, 1998.

[27]Lichtenwalner, P., Dunne, J. P., Becker, R. S., and Baumann, E. R., "Active Damage Interrogation System for Structural Health Monitoring," *Industrial and Commercial Applications of Smart Structures Technologies*, Vol. 3044, Society of PhotoOptical Instrumentation Engineers, Bellingham, WA, 1997, pp. 186–194.

[28]Obermeier, E., and Haslam, B., "Eurofighter Technology for the 21st Century," *Proceedings of the 21st ICAS Congress*, ICAS Paper 98-0,4, AIAA, Reston, VA, Sept. 1998.

[29]Hunt, S. R., and Hebden, I. G., "Eurofighter 2000: An Integrated Approach to Structural Health Monitoring," *Proceedings 19th ICAF Symposium: Fatigue In New and Aging Aircraft*, Cradley Health Engineering Materials Advisory Services, LTD, UK, 1997, pp. 481–498.

[30]Kiefer, K., "Applications of Wireless Data Acquisition and Control Techniques," presented at *Engineered Adaptive Structures for Noise Control and Vibration Control II* (Castelvecchio, Italy, 16–21 May 1999), Engineering Foundation, New York, NY, 1999.

[31]Krantz, D., Belk, J., Biermann, P. J., Dubow, J., Gause, L. W., Harjani, R., Mantell, S., Polla, D., and Troyk, P., "Project Update: Applied Research on Remotely-Queried Embedded Microsensors," *Smart Electronics and MEMS*, Vol. 3328, Society of PhotoOptical Instrumentation Engineers, Bellingham, WA, 1999, pp. 124–132.

[32]Baumann, E. R., Becker, R. S., Ellerbrock, P. J., and Jacobs, S. W., "DC-XA Structural Health Monitoring," *Industrial and Commercial Applications of Smart Structures Technologies*, Vol. 3044, Society of PhotoOptical Instrumentation Engineers, Bellingham, WA, 1997, pp. 195–206.

[33]Ellerbrock, P. J., "DC-XA Structural Health Monitoring Fiber Optic-Based Strain Measurement System," *Industrial and Commercial Applications of Smart Structures Technologies*, Vol. 3044, Society of PhotoOptical Instrumentation Engineers, Bellingham, WA, 1997, pp. 206–218.

[34]Miller, D., et al., "The Middeck Active Control Experiment (MACE): Summary Report," MIT Space Engineering Research Center, SERC Rept. #7-96, Massachusetts Institute of Technology, Cambridge, MA, June 1996.

[35]Davis, L. D., and Hyland, D. C., "Adaptive Neural Control for Space Structure Vibration Suppression," PL-TR-96-1133, Phillips Laboratory Space & Missiles Directorate, Kirtland AFB, Aug. 1996.

[36]Denoyer, K. K., and Ninneman, R. R., "Increasing Autonomy of Spacecraft Using Neural Network Adaptive Control," *Proceedings of the Space Technology and Applications International Forum 1999*, Albuquerque, NM, Feb. 1999, pp. 636–641.

[37]Manning, R. A., and Foley, C. N., "Identification of Experiment Dynamics and Environmental Effects on the ACTEX I Flight Experiment," *Proceedings of the AIAA Guidance, Dynamics, and Controls Conference*, AIAA-99-3591, AIAA, Reston, VA, 1999.

[38]Obal, M., and Sater, J. M., "Multi-Functional Structures: The Future of Spacecraft Design?," *5th International Conference on Adaptive Structures*, Technomic, Lancaster, PA, 1995, pp. 720–734.

[39]Cobb, R. G., Sullivan, J. M., Das, A., Davis, L. P., Hyde, T. T., Davis, T., Rahman, Z. H., and Spanos, J. T., "Vibration Isolation and Suppression System for Precision Payloads in Space," *Smart Materials and Structures*, Vol. 8, 1999, pp. 1–15.

[40]Sciulli, D., and Griffin, S. F., "Hybrid Launch Isolation System," *Industrial and Commercial Applications of Smart Structures Technologies*, SPIE 3674-40, Vol. 3674, Society of PhotoOptical Instrumentation Engineers, Bellingham, WA, 1999, pp. 352–359.

[41]Wilke, P. S., Johnson, C. D., and Grosserode, P. J., "GFO/Taurus Whole Spacecraft Isolation System," *Proceedings of the 12th AIAA/USU Conference on Small Satellites*, SSC-98-III-1, AIAA, Reston, VA, Sept. 1998.

[42]Carpenter, B. F., Clark, C., and Weems, W., "Shape Memory Actuated Release Mechanisms," *Industrial and Commercial Applications of Smart Structures Technologies*, Paper 2721-31, Vol. 2721, Society of PhotoOptical Instrumentation Engineers, Bellingham, WA, 1996, pp. 420–426.

[43]Private communication with Dr. Suraj Rawal, Lockheed Martin Astronautics, 31 July 1999.

[44]Fuller, C. R., and Von Flotow, A. H., "Active Control of Sound and Vibration," *IEEE Control Systems*, Dec. 1995, pp. 9–19.

[45]Lyle, K. H., and Silcox, R. J., "A Study of Active Trim Panels for Noise Reduction in an Aircraft Fuselage," *General, Corporate and Regional Aviation Meeting and Exposition*, Paper SAE-95-1179, Society of Automotive Engineers, Warrendale, PA, 1995.

[46]Silcox, R. J., Lester, H. C., and Coats, T. J., "An Analytical Study of Intensity Flow for Active Structural Acoustic Control," *Proceedings of the SAE Noise and Vibration Conference*, SAE-93-1284, Society of Automotive Engineers, Warrendale, PA, 1993.

[47]Ruckman, C. E., and Fuller, C. R., "Optimizing Actuator Locations in Active Noise Control Systems Using Subset Selection," *Journal of Sound and Vibration*, Vol. 186, No. 3, Sept. 1995, pp. 395–406.

[48]Padula, S. L., and Kincaid, R. K., "Aerospace Applications of Integer and Combinatorial Optimization," NASA TM-110210, NASA Langley Research Center, Hampton, VA, Oct. 1995.

[49]Hirsch, S. M., Jayachandran, V., and Sunn, J. Q., "Structural-Acoustic Control for Quieter Aircraft Interior—Smart Trim Technology," *Composite Structures*, Vol. 42, 1998, pp. 189–202.

[50]Fuller, C. R., Guigou, C., and Gentry, C. A., "Foam-PVDF Smart Skin for Active Control of Sound," *Industrial and Commercial Applications of Smart Structures Technologies*, Vol. 2721, Society of PhotoOptical Instrumentation Engineers, Bellingham, WA, 1996, pp. 26–37.

[51]Sollo, A., Leece, L., Guaranta, V., Doelman, N., and Doppenberg, E., "Active Noise Control on ATR Fuselage Mock-Up by Piezoceramic Actuators," *Proceedings of the AIAA/CEAS Aeroacoustics Conference, 4th*, AIAA-98-2230, AIAA, Reston, VA, 1998, pp. 174–183.

[52]Leece, L., and Viscardi, M., "Active Control of Cabin Noise," presented at *Engineered Adaptive Structures for Noise Control and Vibration Control II* (Castelvecchio, Italy, 16–21 May 1999), Engineering Foundation, New York, NY, 1999.

[53]Kincaid, R. K., and Padula, S. L., "Quelling Cabin Noise in Turboprop Aircraft via Active Control," *Journal of Combinatorial Optimization*, Vol. 1, 1997, pp. 229–250.

[54]Mathur, G., Tran, B., and Simpson, M., "Broadband Active Structural Acoustic Control of Aircraft Cabin Noise—Lab Tests," *AIAA/CEAS 3rd Aeroacoustics Conference*, AIAA-97-1636, AIAA, Reston, VA, May 1997.

[55]Cruz, J. D., "Active Suppression of Aircraft Panel Vibration with Piezoceramic Strain Actuators," *Journal of Aircraft*, Vol. 35, No. 1, 1998, pp. 139–144.

[56]Larson, C. R., Rosenthal, J. S., Neurgoankar, R. R., Falangas, E., Nelson, J. G., Dobbs, S. K., Hustedde, C. L., and McGrath, S. F., "Piezoceramic Active Vibration Suppression Flight Demonstration Program on the B-1B Aircraft," *Proceedings of the 39th Structures, Structural Dynamics and Materials Conference*, AIAA-98-1926, AIAA, Reston, VA, 1998, pp. 1847–1856.

[57]Larson, C. R., Falangas, E., Dobbs, S. K., Neurgoankar, R. R., Nelson, J. G., Rosenthal, J. S., Hustedde, C. L., and McGrath, S. F., "Piezoceramic Active Vibration Suppression Control System Development for the B–1B Aircraft," *Industrial and Commercial Applications of Smart Structures Technologies*, Vol. 3326, Society of PhotoOptical Instrumentation Engineers, Bellingham, WA, 1998, pp. 294–305.

[58]Finn, E. D., "Lowering the Volume on Helicopters," *Aerospace America*, Jan. 1997, pp. 24–25.

[59]Moses, R. W., "Contributions to Active Buffeting Alleviation Programs by the NASA Langley Research Center," *Proceedings of the 40th Structures, Structural Dynamics, and Materials Conference*, AIAA-99-1318, AIAA, Reston, VA, 1999.

[60]Ryall, T. G., Moses, R. W., Hopkins, M. A., Henderson, D., Zimcik, D. G., and Nitzsche, F., "Buffet Load Alleviation," *Proceedings of the Second Australasian Congress on Applied Mechanics*, Paper R-017, Canberra, Australia, 1999, pp. 126–131.

[61]Moses, R. W., "Vertical Tail Buffet Alleviation Using Piezelectric Actuators—Some Results of the Actively Controlled Response of Buffet—Affected Tails (ACROBAT) Program," *Industrial and Commercial Applications of Smart Structures Technologies*, Vol. 3044, Society of PhotoOptical Instrumentation Engineers, Bellingham, WA, 1997, pp. 87–98.

[62]Zimmerman, N. H., and Ferman, M. A., "Prediction of Tail Buffet Loads for Design Application," Vols. I and II, Rept. No. NADC-88043-60, July 1987.

[63]Lee, B. H. K., Brown, D., Zgela, M., and Poirel, D., "Wind Tunnel Investigation and Flight Tests of Tail Buffet on the CF-18 Aircraft", *Aircraft Dynamic Loads Due to Flow Separation*, N91-14324 06-05, AGARD-CP-483, NATO Advisory Group for Aerospace Research and Development, April 1990.

[64]Pettit, C. L., Banford, M., Brown, D., and Pendleton, E., "Pressure Measurements on an F/A-18 Twin Vertical Tail in Buffeting Flow," Volumes 1-4, United States Air Force, Wright Laboratory, TM-94-3039, Wright Patterson AFB, OH, Aug. 1994.

[65]Meyn, L. A., and James, K. D., "Full-Scale Wind-Tunnel Studies of F/A-18 Tail Buffet," *Journal of Aircraft*, Vol. 33, No. 3, July 1996, pp. 589–595.

[66]Moses, R. W., and Pendleton, E., "A Comparison of Pressure Measurements Between a Full-Scale and a 1/6-Scale F/A-18 Twin Tail During Buffet," *83rd Structures And Materials Panel (SMP) Meeting, Loads and Requirements for Military Aircraft*, AGARD-R-815, NATO Advisory Group for Aerospace Research and Development, Sept. 1996.

[67]Ashley, H., Rock, S. M., Digumarthi, R., Chaney, K., and Eggers, A. J., Jr., "Active Control For Fin Buffet Alleviation," WL-TR-93-3099, Air Force Wright Laboratory, Wright Patterson AFB, Dayton, OH, Jan. 1994.

[68]Hauch, R. M., Jacobs, J. H., Ravindra, K., and Dimas, C., "Reduction of Vertical Tail Buffet Response Using Active Control," *Proceedings of the 36th Structures, Structural Dynamics, and Materials Conference*, AIAA-95-1080-CP, Part 4, AIAA, Reston, VA, April 1995, pp. 2281–2288.

[69]Hauch, R. M., Jacobs, J. H., Dima, C., and Ravindra, K., "Reduction of Vertical Tail Buffet Response Using Active Control," *Journal of Aircraft*, Vol. 33, No. 3, May–June 1996, pp. 617–622.

[70]Lazarus, K. B., Saarmaa, E., and Agnes, G. S., "An Active Smart Material System for Buffet Load Alleviation," *Industrial and Commercial Applications of Smart Structures Technologies*, Vol. 2447, Society of PhotoOptical Instrumentation Engineers, Bellingham, WA, 1995, pp. 179–192.

[71]Hopkins, M., Moses, R., Zimcik, D., Henderson, D., Ryall, T., and Spangler, R., "Active Vibration Suppression Systems Applied to Twin Tail Buffeting," *Industrial and Commercial*

Applications of Smart Structures Technologies, Vol. 3326, Society of PhotoOptical Instrumentation Engineers, Bellingham, WA, 1998, pp. 27–33.

[72]Spangler, R. L., and Jacques, R. N., "Testing of an Active Smart Material System for Buffet Load Alleviation," *Proceedings of the 40th Structures, Structural Dynamics, and Materials Conference*, AIAA-99-1316, AIAA, Reston, VA, 1999.

[73]Hanagud, S., Bayon de Noyer, M., Lou, H., Henderson, D., and Nagaraja, K. S., "Tail Buffet Alleviation of High Performance Twin Tail Aircraft Using Piezo-Stack Actuators," *Proceedings of the 40th Structures, Structural Dynamics, and Materials Conference*, AIAA-99-1320, AIAA, Washington, DC, 1999.

[74]Thorby, D. C., "Can Piezoelectric Methods Be Used to Control Aircraft Buffet Response?," *CEAS International Forum on Aeroelasticity and Structural Dynamics*, Paper A97-3665A 09-31, Vol. 2, Associazione Italiana di Aeronautica ed Astronautica, Rome, Italy, 1997, pp. 351–358.

[75]Simpson, J., and Schweiger, J., "An Industrial Approach to Piezo Electric Damping of Large Fighter Aircraft Components," *Industrial and Commercial Applications of Smart Structures Technologies*, Vol. 3326, Society of PhotoOptical Instrumentation Engineers, Bellingham, WA, 1998, pp. 34–46.

[76]Becker, J., Schröder, W., Dittrich, K., Bauer, E., and Zippold, H., "The Advanced Aircraft Structures Program—An Overview," *Industrial and Commercial Applications of Smart Structures Technologies*, Vol. 3674, Society of PhotoOptical Instrumentation Engineers, Bellingham, WA, 1999, pp. 2–12.

[77]Zaglauer, H. W., Dürr, J. K., Flöth, E., Ihler, E., Herold–Schmidt, U., Dittrich, K., Simpson, J., and Becker, J., "Fin-Buffet Alleviation via Distributed Piezoelectric Actuators—Materials Qualification Program," *Industrial and Commercial Applications of Smart Structures Technologies*, Vol. 3674, Society of PhotoOptical Instrumentation Engineers, Bellingham, WA, 1999, pp. 22–30.

[78]Manser, R., Simpson, J., Becker, J., Dürr, J. K., Flöth, E., Herold–Schmidt, U., Stark, H., and Zaglauer, H.W., "Fin-Buffet Alleviation via Distributed Piezoelectric Actuators—Full Scale Demonstrator Tests," *Industrial and Commercial Applications of Smart Structures Technologies*, Vol. 3674, Society of PhotoOptical Instrumentation Engineers, Bellingham, WA, 1999, pp. 13–21.

[79]Becker, J., and Wolfgang, W., "Comparison of Piezoelectric Systems and Aerodynamic Systems for Aircraft Vibration Alleviation," *Industrial and Commercial Applications of Smart Structures Technologies*, Vol. 3326, Society of PhotoOptical Instrumentation Engineers, Bellingham, WA, 1998, pp. 13–26.

[80]Stüwing, M., Sachau, D., and Breitbach, E., "Adaptive Vibration Damping of Fin Structures," *Industrial and Commercial Applications of Smart Structures Technologies*, Vol. 3674, Society of PhotoOptical Instrumentation Engineers, Bellingham, WA, 1999, pp. 31–38.

[81]Gade, P. V. N., and Inman, D. J., "Two-Dimensional Active Wing/Store Flutter Suppression Using H_∞ Theory," *Journal of Guidance, Control, and Dynamics*, Vol. 20, 1997, pp. 949–955.

[82]McGowan, A. R., Wilkie, W. K., Moses, R. W., Lake, R. C., Pinkerton–Florance, J., Wieseman, C.D., Reaves, M.C., Taleghani, B. K., Mirick, P. H., and Wilbur, M. L., "Aeroservoelastic and Structural Dynamics Research on Smart Structures Conducted at NASA Langley Research Center," *Industrial and Commercial Applications of Smart Structures Technologies*, Paper 3326-20, Vol. 3326, Society of PhotoOptical Instrumentation Engineers, Bellingham, WA, 1998.

[83]Heeg, J., and McGowan, A., "The Piezoelectric Aeroelastic Response Tailoring Investigation: A Status Report," *Industrial and Commercial Applications of Smart Structures Technologies*, Vol. 2447, Society of PhotoOptical Instrumentation Engineers, Bellingham, WA, 1995, pp. 2–13.

[84]McGowan, A. R., Heeg, J., and Lake, R. C., "Results of Wind-Tunnel Testing from the Piezoelectric Aeroelastic Response Tailoring Investigation," *Proceedings of the 37th Structures, Structural Dynamics, and Materials Conference*, AIAA-96-1511, AIAA, Reston, VA, 1996.

[85]"The Langley Transonic Dynamics Tunnel," LWP-799, Sept. 1969.

[86]Derham, R. C., and Hagood, N. W., "Rotor Design Using Smart Materials to Actively Twist Blades," *Proceedings of the American Helicopter Society 52nd Annual Forum*, Vol. 2, American Helicopter Society, Alexandria, VA, 1996, pp. 1242–1252.

[87]Straub, F. K., and King, R. J., "Application of Smart Materials to Control of a Helicopter Rotor," *Industrial and Commercial Applications of Smart Structures Technologies*, Vol. 2721, Society of PhotoOptical Instrumentation Engineers, Bellingham, WA, 1996, pp. 66–76.

[88]Prechtl, E. F., and Hall, S. R., "An X-Frame Actuator Servo-Flap Actuation System for Rotor Control," *Smart Structures and Integrated Systems*, SPIE 3329-33, Vol. 3329, Society of PhotoOptical Instrumentation Engineers, Bellingham, WA, 1998.

[89]Hall, S. R., and Prechtl, E. F., "Preliminary Testing of a Mach-Scaled Active Rotor Blade with a Trailing Edge Servo-Flap," *Smart Structures and Integrated Systems*, Vol. 3668, Society of PhotoOptical Instrumentation Engineers, Bellingham, WA, 1999, pp. 14–21.

[90]Prechtl, E. F., and Hall, S. R., "Design of a High Efficiency, Large Stroke, Electromechanical Actuator," *Smart Materials and Structures*, Vol. 8, 1999, pp. 13–30.

[91]Bent, A., Hagood, N., and Rodgers, J., "Anisotropic Actuation with Piezoelectric Fiber Composites," *Journal of Intelligent Material Systems and Structures*, Vol. 6, May 1995, pp. 338–349.

[92]Bent, A., and Hagood, N., "Improved Performance in Piezoelectric Fiber Composites Using Interdigitated Electrodes," *Smart Materials*, Vol. 2441, Society of PhotoOptical Instrumentation Engineers, Bellingham, WA, Feb. 1995, pp. 196–212.

[93]Hagood, N., Kindel, R., Ghandi, K., and Gaudenzi, P., "Improving Transverse Actuation of Piezoceramics Using Interdigitated Surface Electrodes," *Smart Structures and Intelligent Systems*, Vol. 1917, Part 1, Society of PhotoOptical Instrumentation Engineers, Bellingham, WA, Feb. 1993, pp. 341–352.

[94]Chen, P. C., and Chopra, I., "Hover Testing of Smart Rotor with Induced-Strain Actuation of Blade Twist," *AIAA Journal*, Vol. 35, No. 1, Jan. 1997, pp. 6–16.

[95]Bernhard, A. P., and Chopra, I., "Trailing Edge Flap Activated by a Piezo-Induced Bending-Torsion Coupled Beam," *Journal of the American Helicopter Society*, Vol. 44, No. 1, Jan. 1999, pp. 3–15.

[96]Koratkar, N., and Chopra, I., "Development of a Smart Rotor with Piezoelectric Bender Actuated Trailing Edge Flaps," *Proceedings of the 4th ARO Workshop on Smart Structures*, Pennsylvania State University, State College, PA, 16–18 Aug. 1999.

[97]du Plessis, A., and Hagood, N., "Performance Investigation of Twist Actuated Single Cell Composite Beams for Helicopter Blade Control," *6th International Conference on Adaptive Structures Technology*, Technomic, Lancaster, PA, 1996, pp. 191–216.

[98]Wilkie, W. K., Belvin, W. K., and Park, K. C., "Aeroelastic Analysis of Helicopter Rotor Blades Incorporating Anisotropic Piezoelectric Twist Actuation," *Proceedings of the*

ASME World Congress and Exposition, Adaptive Structures Symposium, American Society of Mechanical Engineers, New York, NY, 1996, pp. 423–433.

[99]Wilkie, W. K., "Anisotropic Piezoelectric Twist Actuation of Helicopter Rotor Blades: Aeroelastic Analysis and Design Optimization," Ph.D. Dissertation, University of Colorado, Boulder, 1997.

[100]Rodgers, J. P., Hagood, N. W., and Weems, D. A., "Design and Manufacture of an Integral Twist-Actuated Rotor Blade," *Proceedings of the 38th Structures, Structural Dynamics and Materials Conference and Adaptive Structures Forum*, Paper AIAA-97-1264, AIAA, Reston, VA, April 1997.

[101]Wilkie, W. K., Wilburn, M. L., Mirick, P. H., Cesnik, C. E. S., and Shin, S. J., "Aeroelastic Analysis of the NASA/Army/MIT Active Twist Rotor," American Helicopter Society 55th Annual Forum, Montreal, Canada, 25–27 May 1999.

[102]Ham, N., "Helicopter Individual-Blade-Control Research at MIT 1977-1985," *Vertica*, Vol. 11, 1987, pp. 109–122.

[103]Ormiston, R. A., "Smart Materials and Structures for Rotor Dynamics Applications," *Proceedings of the 4th ARO Workshop on Smart Structures*, Pennsylvania State University, State College, 16–18 Aug. 1999.

[104]Yeager, W., Mirick, P., Hamouda, M. N., Wilbur, M., Singleton, J., and Wilkie, W. Keats, "Rotorcraft Aeroelastic Testing in the Langley Transonic Dynamics Tunnel," *Journal of the American Helicopter Society*, Vol. 38, No. 3, July 1993, pp. 73–82.

[105]Barrett, R. M., and Gross, R. S., "Adaptive Aerostructures Demonstrations of Flightworthy Hardware," *Proceedings of the 4th European Smart Structures and Materials and 2nd Micromechanics, Intelligent Materials and Robotics Conference*, Inst. of Physics Publishing, London, 1998, pp. 749–753.

[106]Ehlers, S. M., and Weisshaar, T. A., "Static Aeroelastic Behavior of Adaptive Laminated Piezoelectric Composite Wing," *AIAA Journal*, Vol. 28, No. 4, 1990.

[107]Barrett, R., "Active Plate and Wing Research Using EDAP Elements," *Journal of Smart Materials and Structures*, Vol. 1, No. 3, pp. 214–226.

[108]Barrett, R., Gross, R. S., and Brozoski, F. T., "Design and Testing of Subsonic All-Moving Smart Flight Control Surfaces," *Proceedings of the 36th Structures, Structural Dynamics, and Materials Conference*, AIAA-95-1081, AIAA, Reston, VA, 1995, pp. 2289–2296.

[109]Barrett, R., and Stutts, J., "Design and Testing of a 1/12th Scale Solid State Adaptive Rotor," *Journal of Smart Materials and Structures*, Vol. 6, No. 4, 1997, pp. 491–497.

[110]Barrett, R., and Stutts, J., "Modeling, Design and Construction of a Barrel-Launched Adaptive Munition," *Smart Structures and Integrated Systems*, Vol. 1997, Society of PhotoOptical Instrumentation Engineers, Bellingham, WA, 1997, pp. 578–589.

[111]Monner, H. P., Hanselka, H., and Breitbach, E., "Development and Design of Flexible Fowler Flaps for an Adaptive Wing," *Industrial and Commercial Applications of Smart Structures Technologies*, Vol. 3326, Society of PhotoOptical Instrumentation Engineers, Bellingham, WA, 1998, pp. 60–70.

[112]Kudva, J. N., Lockyer, A.J., and Appa, K., "Adaptive Aircraft Wing," *AGARD Lecture Series 205, Smart Structures and Materials: Implications for Military Aircraft of New Generation*, Paper 10, Oct. 1996, pp. 10-1–10-5.

[113]Kudva, J. N., Martin, C. A., Scherer, L. B., Jardine, A. P., McGowan, A. R., Lake, R. C., Sendeckyj, G., and Sanders, B., "Overview of the DARPA/AAFRL/NASA Smart Wing Program," *Industrial and Commercial Applications of Smart Structures Technologies*, SPIE 3674-28, Vol. 3674, Society of PhotoOptical Instrumentation Engineers, Bellingham, WA, pp. 230–236.

[114]Kudva, J., Appa, K., Martin, C., Jardine, P., Sendeckyj, G., Harris, T., McGowan, A., and Lake, R., "Design, Fabrication and Testing of the DARPA/WL Smart Wing Wind-Tunnel Model," *Proceedings of the 38th Structures, Structural Dynamics, and Materials Conference and Adaptive Structures Forum*, AIAA-97-1198, AIAA, Reston, VA, 1997.

[115]Martin, C., Bartley–Cho, J., Flanagan, J., and Carpenter, B. F., "Design and Fabrication of Smart Wing Wind Tunnel Model and SMA Control Surfaces," *Industrial and Commercial Applications of Smart Structures Technologies*, SPIE 3674-28, Vol. 3674, Society of PhotoOptical Instrumentation Engineers, Bellingham, WA, 1999, pp. 237–248.

[116]Jardine, A. P., Bartley–Cho, J., and Flanagan, J., "Improved Design and Performance of the SMA Torque Tube for the DARPA Smart Wing Program," *Industrial and Commercial Applications of Smart Structures Technologies*, SPIE 3674-29, Vol. 3674, Society of PhotoOptical Instrumentation Engineers, Bellingham, WA, 1999, pp. 260–269.

[117]Scherer, L. B., Martin, C. A., Appa, K., Kudva, J., and West, M. N., "Smart Wing Wind-Tunnel Test Results," *Industrial and Commercial Applications of Smart Structures Technologies*, Paper 3044-05, Vol. 3044, Society of PhotoOptical Instrumentation Engineers, Bellingham, WA, 1997.

[118]Scherer, L. B., Martin, C. A., West, M., Florence, J. P., Wiseman, C. D., Burner, A. W., and Fleming, G. A., "DARPA/AFRL/NASA Smart Wing Wind Tunnel Test Results," *Industrial and Commercial Applications of Smart Structures Technologies*, SPIE 3674-28, Vol. 3674, Society of PhotoOptical Instrumentation Engineers, Bellingham, WA, 1999, pp. 249–259.

[119]Austin, F., Van Nostrand, W. C., Siclari, M., Aidala, P., and Clifford, R., "Design and Performance Predictions of Smart Wing for Transonic Cruise," *Industrial and Commercial Applications of Smart Structures Technologies*, Vol. 2721, Society of PhotoOptical Instrumentation Engineers, Bellingham, WA, 1996, pp. 17–25.

[120]Yurkovich, R., "Optimum Wing Shape for an Active Flexible Wing," *Proceedings of the 36th Structures, Structural Dynamics, and Materials Conference*, AIAA-95-1220-CP, Part 1, AIAA, Reston, VA, April 1995, pp. 520–530.

[121]Pendleton, E., Griffin, K., Kehoe, M., and Perry, B., "A Flight Research Program for Active Aeroelastic Wing Technology," *Proceedings of the 37th Structures, Structural Dynamics, and Materials Conference*, AIAA-96-1574, AIAA, Reston, VA, 1996.

[122]Pendleton, E., Bessette, D., Field, P., Miller, G., and Griffin, K., "The Active Aeroelastic Wing Flight Research Program," *Proceedings of the 39th Structures, Structural Dynamics, and Materials Conference*, AIAA-98-1972, AIAA, Reston, VA, 1998.

[123]Hopkins, M. A., Dunne, J. P., Baumann, E. W., and White, E. V., "Adaptive Fighter Engine Inlet," *Proceedings of the 40th Structures, Structural Dynamics, and Materials Conference*, AIAA-99-1512, AIAA, Reston, VA, 1999.

[124]Park, M., Green, L., Montgomery, R., and Raney, D., "Determination of Stability and Control Derivatives Using Computational Fluid Dynamics and Automatic Differentiation," *Proceedings of the 17th Applied Aerodynamics Conference*, AIAA-99-3136, AIAA, Reston, VA, 1999.

[125]Heigl, H., Hetzler, W., and Lenz, E., "Guiding Guided Missiles," *Aerospace*, Jan. 1998, pp. 41–46.

[126]Barrett, R., "Active Plate and Missile Wing Development Using DAP Elements," *AIAA Journal*, Vol. 32, No. 3, 1994, pp. 601–609.

[127]Barrett, R., "Advanced Low-Cost Smart Fin Technology Evaluation," Final Rept. to Wright Laboratory, USAF Armament Directorate, Contract No. F08630-93-C-0039, Eglin AFB, Dec. 1993.

[128]Priou, A., "Electromagnetic Antenna and Smart Structures," *AGARD Lecture Series 205, Smart Structures and Materials: Implications for Military Aircraft of New Generation*, Paper 11, Oct. 1996, pp. 11-1–11-5.

[129]Bartley–Cho, J., Alt, K., Coughlin, D., Lockyer, A., Kudva, J. N., Tuss, J., and Goetz, A., "Development and Testing of a Conformal Load-Bearing, Smart Skin Antenna Structure," *Proceedings of the 40th Structures, Structural Dynamics, and Materials Conference*, AIAA-99-1515, AIAA, Reston, VA, 1999.

[130]Hopkins, M. A., Tuss, J. M., Lockyer, A. J., and Kudva, J. N., "Smart Skin Conformal Load-Bearing Antenna and Other Smart Structure Developments," NATO *Workshop on Smart Electromagnetic Antenna Structures*, 25–26 Nov. 1996.

[131]Alt, K. H., Lockyer, A. J., Martin, C. A., and Kudva, J. N., "Application for Smart Skin Technologies to the Development of a Conformal Antenna Installation in the Vertical Tail of a Military Aircraft," *Smart Electronics*, Vol. 2448, Society of PhotoOptical Instrumentation Engineers, Bellingham, WA, 1995, pp. 42–52.

[132]Polites, M., "Digital Avionics," *Aerospace America*, Dec. 1996, pp. 38–39.

[133]Barnett, D. M., and Rawal, S. P., "Multifunctional Structures Technology Experiment on Deep Space 1 Mission," *Proceedings of the IEEE Aerospace and Electronic Systems Conference*, Institute of Electrical and Electronics Engineers, Piscataway, NJ, 1999, pp. 13–19.

[134]Barnett, D. M., Rawal, S., and Rummel, K., "Multifunctional Structures for Advanced Spacecraft," in preparation.

[135]Private communication with Mr. R. Hanson, ITN, Jan. 1999.

[136]Huber, J. E., Fleck, N. A., and Ashby, M. F., "The Selection of Mechanical Actuators Based on Performance Indices," *Proeedings of the Royal Society of London*, Volume A 453, Royal Society of London, London, 1997, pp. 2185–2205.

[137]Simpson, J. O., Ounaies, Z., and Fay, C., "Polarization and Piezoelectric Properties of a Nitrile Substituted Polyimide," *Materials Research Society Proceedings: Materials for Smart Systems II*, Materials Research Society, Pittsburgh, PA, MRS A98-19626 04-23, Vol. 459, 1997, pp. 59–64.

[138]Ounaies, Z., Young, J., Simpson, J. O., and Farmer, B., "Dielectric Properties of Piezoelectric Polyimides," *Materials Research Society Proceedings: Materials for Smart Systems II*, Materials Research Society, Pittsburgh, PA, MRS A98-19626 04-23, Vol. 459, 1997, pp. 53–58.

[139]TRS Ceramics, Inc., company literature (2820 East College Avenue, Suite J, State College, PA 16801), 1999.

[140]Pan, M.–J., Pertsch, P., Yoshikawa, S., Shrout, T. R., and Vedula, V., "Electroactive Actuator Materials: Investigations on Stress and Temperature Characteristics," *Smart Materials*, Vol. 3324, Society of PhotoOptical Instrumentation Engineers, Bellingham, WA, 1998, pp. 145–153.

[141]Tickle, R., and James, R. D., "Magnetic and Magnetomechanical Properties of Ni_2MnGa," *Journal of Magnetism and Magnetic Materials*, in press.

[142]James, R. D., and Wuttig, M., "Magnetostriction of Martensite," *Philosophical Magazine A: Physics of Condensed Matter Structure, Defects and Mechanical Properties*, Vol. A77, 1998, pp. 1273–1299.

[143]Furuya, Y., Hagood, N., Kimura, H., and Watanabe, T., "Shape Memory Effect and Magnetostriction in Rapidly Solidified Fe-29.6 at % Pd Alloy," *Materials Transactions, Japan Institute of Metals*, Vol. 39, 1998, pp. 1248–1254.

[144]West, J., "Basics of Actuator Technology," *Lasers & Optronics*, Sept. 1993, pp. 21–22.

[145]Claus, R. O., "Sensor Instrumentation for Smart Materials and Structures," *Proceedings of the American Society of Mechanical Engineers (ASME) Winter Annual Conference*, AD-Vol. 52, American Society of Mechanical Engineers, New York, 1996, pp. 463–469.

[146]Murphy, K. A., Gunther, M. F., May, R. G., Claus, R. O., Tran, T. A., Greene, J. A., and Duncan, P. G., "EFPI Sensor Manufacturing and Applications," *Industrial and Commercial Applications of Smart Structures Technologies*, Vol. 2721, Society of PhotoOptical Instrumentation Engineers, Bellingham, WA, 1996, pp. 476–482.

[147]Dunphy, J. A., "Feasibility Study Concerning Optical Fiber Sensor Vibration Monitoring Subsystem," *Industrial and Commercial Applications of Smart Structures Technologies*, Vol. 2721, Society of PhotoOptical Instrumentation Engineers, Bellingham, WA, 1996, pp. 483–492.

[148]Brown, T., Wood, K., Childers, B., Cano, B., Jensen, B., and Rogowski, R., "Fiber Optic Sensors for Health Monitoring of Morphing Aircraft," *Industrial and Commercial Applications of Smart Structures Technologies*, Vol. 3674, Society of PhotoOptical Instrumentation Engineers, Bellingham, WA, 1999, pp. 60–71.

[149]Jacobs, J. H., "Synthesis and Processing of Intelligent Cost-Effective Structures: A Final Review of the ARPA SPICES Program," *Industrial and Commercial Applications of Smart Structures Technologies*, Vol. 2721, Society of PhotoOptical Instrumentation Engineers, Bellingham, WA, 1996, pp. 167–188.

[150]Bridger, K., Jones, L., Poppe, F., Brown, S., and Winzer, S. R., "High-Force, Cofired Multilayer Actuators," *Industrial and Commercial Applications of Smart Structures Technologies*, Vol. 2721, Society of PhotoOptical Instrumentation Engineers, Bellingham, WA, 1996, pp. 341–352.

[151]Gentilman, R. L., Fiore, D., Pham-Nguyen, H., Serwatka, W., Pazol, B. G., Near, C. D., McGuire, P., and Bowen, L., "1-3 Piezocomposite Smart Panels for Active Surface Control," *Industrial and Commercial Applications of Smart Structures Technologies*, Vol. 2721, Society of PhotoOptical Instrumentation Engineers, Bellingham, WA, 1996, pp. 234–239.

[152]French, J. D., Weitz, G. E., Luke, J. E., and Cass, R. B., "Production of Piezoelectric Fibers for Smart Materials and Active Control Devices," *Industrial and Commercial Applications of Smart Structures Technologies*, Paper 3044-39, Vol. 3044, Society of PhotoOptical Instrumentation Engineers, Bellingham, WA, 1997.

[153]Haertling, G. H., "RAINBOW Ceramics—A New Type of Ultra-High-Displacement Actuator," *American Ceramic Society Bulletin*, Vol. 73, 1994, pp. 93–96.

[154]Hooker, M. W., "Properties and Performance of RAINBOW Piezoelectric Actuator Stacks," *Industrial and Commercial Applications of Smart Structures Technologies*, Vol. 3044, Society of PhotoOptical Instrumentation Engineers, Bellingham, WA, 1997, pp. 413–420.

[155]Dausch, D. E., and Wise, S. A., "Compositional Effects on Electromechanical Degradation of RAINBOW Actuators," NASA-TM 206282, NASA Langley Research Center, Hampton, VA, 1998.

[156]Wise, S. A., Hardy, R. C., and Dausch, D. E., "Design and Development of an Optical Path Difference Scan Mechanism for Fourier Transform Spectrometers Using High Displacement RAINBOW Actuators," *Industrial and Commercial Applications of Smart Structures Technologies*, Vol. 3044, Society of PhotoOptical Instrumentation Engineers, Bellingham, WA, 1997, pp. 342–349.

[157]Hellbaum, R. F., Bryant, R. G., Fox, R. L., Jalink, A., Rohrbach, W. W., and Simpson, J. O., "Thin Layer Composite Unimorph and Ferroelectric Driver and Sensor," U.S. Patent 5,632,840, 27 May 1997.

[158]Bryant, R. G., "Thunder Actuators," *5th Annual Workshop: Enabling Technologies for Smart Aircraft Systems*, NASA Langley Research Center, Hampton, VA, 14–16 May 1996.

[159]Li, G., and Haertling, G. H., "The Piezoelectric, Pyroelectric and Photoelectronic: Properties of PLZT Rainbow Ceramics," *Proceedings of the 10th IEEE International Symposium on Application of Ferroelectrics*, Vol. 2, Institute of Electrical and Electronic Engineers, Piscataway, NJ, 1996, pp. 907–910.

[160]Dausch, D. E., "Ferroelectric Polarization Fatigue in PZT-Based RAINBOWs and Bulk Ceramics," *Journal of the American Ceramics Society*, Vol. 80, No. 9, 1997, pp. 2355–2360.

[161]Pinkerton, J. L., and Moses, R. W., "A Feasibility Study To Control Airfoil Shape Using THUNDER," NASA TM-4767, NASA–Langley Research Center, Hampton, VA, Nov. 1997.

[162]Pinkerton, J. L., McGowan, A. R., Moses, R. W., Scott, R. C., and Heeg, J., "Controlled Aeroelastic Response and Airfoil Shaping Using Adaptive Materials and Integrated Systems," *Industrial and Commercial Applications of Smart Structures Technologies*, Paper 3326-20, Vol. 3326, Society of PhotoOptical Instrumentation Engineers, Bellingham, WA, 1998.

[163]Kim, C., Glazounov, A., Flippen, D., Pattnaik, A., Zhang, Q., and Lewis, D., "Piezoelectric Ceramic Assembly Tubes for Torsional Actuators," *Smart Materials*, Vol. 3675, Society of PhotoOptical Instrumentation Engineers, Bellingham, WA, 1999, pp. 53–62.

[164]Van Humbeeck, J., Reynaerts, D., and Peirs, J., "New Opportunities for Shape Memory Alloys for Actuators, Biomedical Engineering, and Smart Materials," *Materials Technology*, Vol. 11, No. 2, 1996, pp. 55–61.

[165]Girshovich, S., Shikhamanter, L., and Weissberg, V., "The Application of Shape Memory Actuators in Adaptive Wings," *Proceedings of the 37th Israel Annual Conference on Aerospace Sciences*, A97-25510 06-01, Technion-Israel Institute of Technology, Haifa, Israel, pp. 152–153.

[166]Schetky, L. M., and Steinetz, B. M., "Shape Memory Alloy Adaptive Control of Gas Turbine Engine Compressor Blade Tip Clearance," *Industrial and Commercial Applications of Smart Structures Technologies*, Paper 3326-36, Vol. 3326, Society of PhotoOptical Instrumentation Engineers, Bellingham, WA, 1998, pp. 346–354.

[167]Turner, T. L., Zhong, Z. W., and Mei, C., "Finite Element Analysis of the Random Response Suppression of Composite Panels at Elevated Temperatures Using Shape Memory Alloy Fibers," *Proceedings of the 35th Structures, Structural Dynamics, and Materials Conference*, AIAA-94-1324 CP, AIAA, Reston, VA, 1994, pp. 136–146.

[168]Mei, C., Zhong, Z. W., and Turner, T. L., "Control of Sonic Fatigue for High Speed Flight Vehicles Using Shape Memory Alloy," *Industrial and Commercial Applications of Smart Structures Technologies*, Paper 3326-53, Vol. 3326, Society of PhotoOptical Instrumentation Engineers, Bellingham, WA, 1998.

[169]Weisensel, G. N., and Pierce, T. D., "High-Authority Smart Material Integrated Electric Actuator," *Industrial and Commercial Applications of Smart Structures Technologies*, Paper 3044-37, Vol. 3044, Society of PhotoOptical Instrumentation Engineers, Bellingham, WA, 1997.

[170]Weisensel, G. N., Hanses, T. T., and Hrbek, W. D., "High-Power Ultrasonic TERFENOL-D Transducers Enable Commercial Applications," *Industrial and Commercial Applications of Smart Structures Technologies*, Paper 3326-46, Vol. 3326, Society of PhotoOptical Instrumentation Engineers, Bellingham, WA, 1998, pp. 450–458.

[171]Weisensel, G. N., McMasters, O. D., and Chave, R. G., "Cryogenic Magnetostrictive Transducers and Devices for Commercial, Military and Space Applications," *Industrial*

and Commercial Applications of Smart Structures Technologies, Paper 3326-47, Vol. 3326, Society of PhotoOptical Instrumentation Engineers, Bellingham, WA, 1998, pp. 459–471.

[172]Rusovici, R., Giurgiutiu, V., and Rogers, C. A., "Experimental Study of Hydraulically-Amplified, High Displacement, Induced-Strain Actuators, Proof of Concept Demonstrator," *Proceedings of the American Society of Mechanical Engineers (ASME) Winter Annual Conference*, AD-Vol. 52, American Society of Mechanical Engineers, New York, NY, 1996, pp. 567–574.

[173]Lee, G. B., Ho, C. M., Jiang, F., Liu, C., Tsao, T., Tai, Y. C., and Scheuer, F., "Control of Roll Moment by MEMS," *Proceedings of the American Society Of Mechanical Engineers (ASME) Winter Annual Conference*, AD-Vol. 52, American Society of Mechanical Engineers, New York, NY, 1996, pp. 797–803.

[174]Glezer, A., "Synthetic Jet Actuators for Shear Flow Control, Part I: Synthetic Jet Technology," *5th Annual Workshop: Enabling Technologies for Smart Aircraft Systems*, NASA–Langley Research Center, Hampton, VA, 14–16 May 1996.

[175]McGowan, A. R., "A Feasibility Study on Using Shunted Piezoelectrics to Reduce Aeroelastic Response," *Industrial and Commercial Applications of Smart Structures Technologies*, SPIE 3674-20, Vol. 3674, Society of PhotoOptical Instrumentation Engineers (SPIE), Bellingham, WA, 1999.

[176]Wu, S. Y., and Bicos, A. S., "Structural Vibration Damping Experiments Using Improved Piezoelectric Shunts," *Passive Damping and Isolation*, Vol. 3045, Society of PhotoOptical Instrumentation Engineers, Bellingham, WA, 1997, pp. 40–50.

[177]Brennan, M. C., and McGowan, A. R., "Piezoelectric Power Requirements for Active Control," *Mathematics and Control in Smart Structures*, Paper 3039-69, Vol. 3039, Society of PhotoOptical Instrumentation Engineers, Bellingham, WA, 1997.

[178]Brennan, M. C., "Piezoelectric Power Consumption for Active Vibration Control," Master's Thesis, The George Washington University, July 1997.

[179]Nye, T., Casteel, S., Navarro, S., and Kraml, B., "Experiences with Integral Microelectronics on Smart Structures for Space," *Smart Electronics*, Vol. 2448, Society of PhotoOptical Instrumentation Engineers, Bellingham, WA, 1995, pp. 193–203.

[180]Zvonar, G. A., Luan, J., Lee, F. C., Lindner, D. K., Kelly, S., Sable, D., and Schelling, T., "High-Frequency Switching Amplifiers for Electrostrictive Actuators," *Industrial and Commercial Applications of Smart Structures Technologies*, Vol. 2721, Society of PhotoOptical Instrumentation Engineers, Bellingham, WA, 1996, pp. 465–475.

[181]"All-Plastic Battery," *NASA Tech Briefs*, Oct. 1996, p. 26.

[182]Jose, K. A., Varadan, V. K., and Varadan, V. V., "Tunable and Electronically Steerable Microstrip Antenna," *Smart Electronics*, Vol. 3673, Society of PhotoOptical Instrumentation Engineers, 1999, pp. 195–200.

[183]Piscotty, D., Jose, K. A., Varadan, V. V., and Varadan, V. K., "Design and Development of 150 MHz Wireless Telemetry System for MEMS-IDT Based Sensor," *Smart Electronics*, Vol. 3673, Society of PhotoOptical Instrumentation Engineers, Bellingham, WA, 1999, pp. 165–172.

[184]Varadan, V. K., and Varadan, V. V., "Smart Structures, MEMS and Smart Electronics for Aircraft," *AGARD Lecture Series 205, Smart Structures and Materials: Implications for Military Aircraft of New Generation*, Oct. 1996, Paper 8, pp. 8-1–8-19.

[185]Varadan, V. V. (ed.), *Mathematics and Control in Smart Structures*, Vol. 3323, Society of PhotoOptical Instrumentation Engineers, Bellingham, WA, 1998.

[186]Varadan, V. V. (ed.), *Mathematics and Control in Smart Structures*, Vol. 3667, Society of PhotoOptical Instrumentation Engineers, Bellingham, WA, 1999.

[187]Chou, K. C., Flamm, D. S., Guthart, G. S., and Ueberschaer, R. M, "Multiscale Approach to the Control of Smart Structures," *Industrial and Commercial Applications of Smart Structures Technologies*, Vol. 2721, Society of PhotoOptical Instrumentation Engineers, Bellingham, WA, 1996, pp. 94–105.

[188]Studt, T., "Wavelet Technology Offers Designers Alternative to Fourier Analysis," *R&D Magazine*, May 1996, p. 51.

[189]Studt, T., "Neural Networks: Computer Toolbox for the 90's," *R&D Magazine*, Sept. 1991, pp. 36–42.

[190]Damle, R., Rao, V., and Kern, F., "Multivariable Neural Network Based Controllers for Smart Structures," *Journal of Intelligent Material Systems and Structures*, Vol. 6, July 1995, pp. 516–528.

[191]Manning, R. A., "Damage Detection in Adaptive Structures Using Neural Networks," *Proceedings of the Adaptive Structures Forum*, AIAA-94-1752-CP, AIAA, Reston, VA, 1994, pp. 160–172.

[192]Carneal, J. P., and Fuller, C. R., "A Biologically Inspired Controller for Sound and Vibration Control," *Proceedings of the Adaptive Structures Forum*, AIAA-94-1752-CP, AIAA, Reston, VA, 1994, pp. 474–484.

[193]Kwak, M. K., and Sciulli, D., "Fuzzy-Logic Based Vibration Suppression Control Experiments on Active Structures," *Proceedings of the 36th Structures, Structural Dynamics, and Materials Conference*, AIAA-95-1085-CP, Part 4, AIAA, Reston, VA, April 1995, pp. 2319–2327.

[194]Geng, Z. J., Haynes, L., Wada, B. K., and Garba, J., "Active Vibration Isolation Using Fuzzy CMAC Neural Networks," *Proceedings of the 36th Structures, Structural Dynamics, and Materials Conference*, AIAA-95-1084-CP, Part 4, AIAA, Reston, VA, April 1995, pp. 2308–2318.

[195]QRDC, Inc., company literature (P.O. Box 562, Excelsior, MN 55331), 1996.

[196]Hom, C. L., and Shankar, N., "A Fully Coupled Constitutive Model for Electrostrictive Ceramic Materials," *Journal of Intelligent Material Systems and Structures*, Vol. 5, Nov. 1994, pp. 795–801.

[197]Jia, H., Lalande, F., and Rogers, C. A., "Review of Constitutive Modeling of Shape Memory Alloys," *Proceedings of the American Society Of Mechanical Engineers (ASME) Winter Annual Conference*, AD-Vol. 52, American Society of Mechanical Engineers, New York, NY, Nov. 1996, pp. 585–591.

[198]Hauch, R., "An Industrial Approach to Static and Dynamic Finite Element Modeling of Composite Structures with Embedded Actuators," *Smart Structures and Integrated Systems*, Vol. 2443, Society of PhotoOptical Instrumentation Engineers, Bellingham, WA, 1995, pp. 458–469.

[199]Seeley, C. E., "Analysis and Optimization of Smart Composite Structures Including Debonding," Ph.D. Dissertation, Arizona State University, May 1997.

[200]Freed, B. D., and Babuska, V., "Finite Element Modeling of Composite Piezoelectric Structures with MSC/NASTRAN," *Smart Structures and Integrated Systems*, Paper 3041-60, Vol. 3041, Society of PhotoOptical Instrumentation Engineers, Bellingham, WA, 1997.

[201]Crawley, E. F., and de Luis, J., "Use of Piezoelectric Actuators as Elements of Intelligent Structures," *AIAA Journal*, Vol. 25, No. 10, 1987, pp. 1373–1385.

[202]Ha, S. K., Keilers, C. H., and Chang, F.K., "Finite Element Analysis of Composite Structures Containing Distributed Piezoceramic Sensors and Actuators," *AIAA Journal*, Vol. 30, No. 3, March 1992, pp. 2323–2330.

[203]Belvin, W. K., "Spacecraft Jitter Attenuation Using Embedded Piezoelectric Actuators," *Proceedings of the International Forum on Aeroelasticity and Structural Dynamics*, Confederation of European Aerospace Societies, 1995.

[204]Sang, L. H., *MSC/NASTRAN Handbook for Non-Linear Analysis: Based on Version 67*, The MacNeil Schwendler Corporation, Los Angeles, CA, 1992.

[205]Pototzky, T. S., "Analytical Model Development for Piezoelectric-Structural Interactions," NASA Contractor Rept.

[206]Adams, W. M., Jr., and Hoadley, S. T., "ISAC: A Tool for Aeroservoelastic Modeling and Analysis," NASA-TM-109031, NASA Langley Research Center, Hampton, VA, Dec. 1993.

[207]Nam, C., and Youdan, K., "Optimal Design of Composite Lifting Surface for Flutter Suppression with Piezoelectric Actuators," *AIAA Journal*, Vol. 33, No. 10, Oct. 1995, pp. 1897–1904.

[208]Heeg, J., "Analytical and Experimental Investigation of Flutter Suppression via Piezoelectric Actuation," NASA TP 3241, NASA Langley Research Center, Hampton, VA, March 1993.

[209]Babuska, V., and Freed, B. D., "Finite Element Modeling of Composite Piezoelectric Structures with MSC/NASTRAN," *Smart Structures and Integrated Systems*, Paper 3041-60, Vol. 3041, Society of PhotoOptical Instrumentation Engineers, Bellingham, WA, 1997.

[210]Taleghani, B. K., and Campbell, J. F., "Non-Linear Finite Element Modeling of THUNDER Piezoelectric Actuator," *Smart Structures and Integrated Systems*, Paper 3668-52, Vol. 3668, Society of PhotoOptical Instrumentation Engineers, Bellingham, WA, 1999, pp. 555–566.

[211]Regelbrugge, M. E., "Design Issues for Electrostrictive Actuators," *5th International Conference on Adaptive Structures*, Technomic, Lancaster, PA, 1995, pp. 555–568.

[212]MATLAB, High Performance Numeric Computation and Visualization Software: Reference Guide, The MathWorks, Inc., Natick, MA, 1992.

[213]Johnson, W., "CAMRAD II, Comprehensive Analytical Model of Rotorcraft Aerodynamics and Dynamics," Johnson Aeronautics, Palo Alto, CA, 1994.

[214]Lim, K. B., Lake, R. C., and Heeg, J., "Selection of Piezoceramic Actuators on an Experimental Flexible Wing," *Proceedings of the AIAA Guidance, Navigation and Control Conference*, AIAA-96-3758, AIAA, Reston, VA, 1996.

[215]Lin, C. Y., "Strain Actuated Aeroelastic Wing," Ph.D. Dissertation, Massachusetts Institute of Technology, May 1997.

[216]Belvin, K., Horner, G., Hardy, R., Armstrong, D., and Rosenbaum, D., "Integration Issues for High-strain Actuation Applications," *Industrial and Commercial Applications of Smart Structures Technologies*, Paper 3326-20, Vol. 3326, Society of PhotoOptical Instrumentation Engineers, Bellingham, WA, 1998.

[217]Joshi, S. P., and Chan, W. S., "Fabrication and Curing of Laminates with Multiple Embedded Piezoceramic Sensors and Actuators," *Proceedings of the ADPA/AIAA/ASME/SPIE Conference on Active Materials and Adaptive Structures*, Institute of Physics Publishing, Bristol, UK, 1991, pp. 701–705.

[218]Rodgers, J. P., and Hagood, N. W., "Manufacture of Adaptive Composite Plates Incorporating Piezoelectric Fiber Composite Plies," *Proceedings of the 36th Structures, Structural Dynamics, and Materials Conference and Adaptive Structures Forum*, AIAA-95-1096-CP, AIAA, Reston, VA, 1995, pp. 2824–2835.

[219]Trottier, M., Millet Day, F., Duryea, H. E., Dunphy, J. R., Koopman, G., and Carpenter, B. F., "Performance of Integrated Fiber Optic, Piezoelectric, and Shape Memory Alloy Actuator/Sensors in Thermoset Composites," *Industrial and Commercial Applications of Smart Structures Technologies*, Paper 2447-27, Vol. 2447, Society of PhotoOptical Instrumentation Engineers, Bellingham, WA, 1995.

Chapter 7

Computational Structures Technology

Ahmed K. Noor
Center for Advanced Computational Technology
University of Virginia
NASA Langley Research Center
Hampton, VA 23681 U.S.A.

I. Introduction

COMPUTATIONAL structures technology (CST) blends the insightful modeling of structural response with the development of computational methods. CST is an outgrowth of matrix and finite element methods of structural analysis that were developed over the past five decades. Brief historical reviews of the finite element method are given in two articles.[1,2] Computing technology is the technological foundation of CST. The advances made in computer hardware, firmware, and software have significantly impacted all aspects of CST development. Current CST activities include computational material modeling; computational methods for predicting the response, performance, failure and life of structures and their components; and automated methods of structural synthesis and optimization.

Application of CST to contemporary structures problems typically involves a sequence of five steps: observation of the response phenomena of interest; development of computational models for the numerical simulation of these phenomena; development and assembly of software and/or hardware to implement the computational models; post-processing and interpretation of the predictions of the computational models; and utilization of the computational models in the analysis and design of structures (Fig. 7.1).

Development of a computational model includes selection of the mathematical model that describes the phenomena; mathematical analysis of the model to ensure that the problem is properly formulated; testing the range of validity of the model; and development of a discrete model, computational strategy, and numerical algorithms to approximate the mathematical model. Successful computational models for structures are those based on thorough familiarity with the response phenomena being simulated and a good understanding of the mathematical models available to describe them.

Within the aforementioned general framework, CST is currently being used in a broad range of applications including aerospace, automotive, naval, and nuclear structures. Reviews of some of the activities in the U.S. and European aerospace

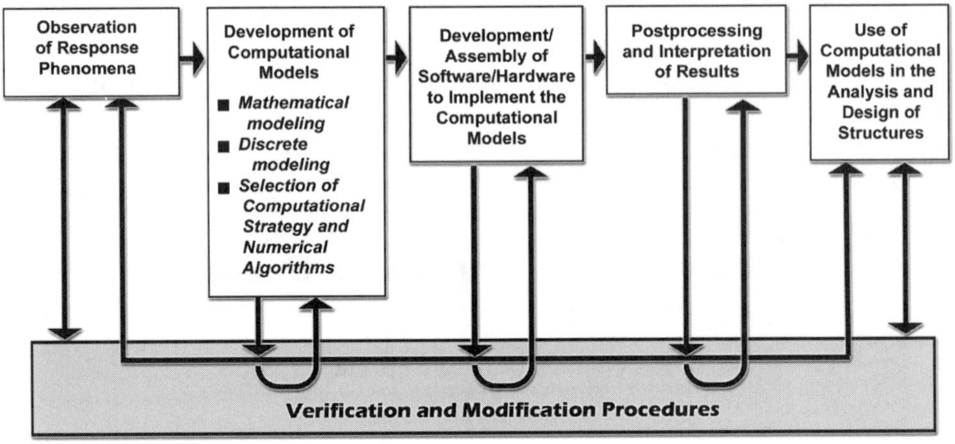

Fig. 7.1 Application of CST to practical structural problems. (See color section.)

industries are given in Refs. 3 and 4. Other industrial applications of CST are described in Refs. 5 and 6. Large structural calculations performed to date account for complicated geometry, complex loading history, and material behavior.

CST represents one of the most significant developments in the history of the structures field. Its use has transformed much of theoretical structural mechanics and materials science into practical tools that affect all phases of the design, fabrication, testing, and operation of engineering systems. Although CST led the way among computational aerosciences until the 1970s, the emphasis has shifted in the early 1980s to other disciplines, particularly computational fluid dynamics. However, since the late 1980s, there has been renewed interest in CST. There are three compelling motivations for vigorously pursuing CST development: the practical problems awaiting solutions; the need for reducing dependence on extensive and costly testing; and the desire to use the power of a new and emerging computing paradigm in the design and operation of future engineering structures.

A number of survey papers and monographs have been written on various aspects of computational structures technology (see, for example, Refs. 7–10). Also, a number of workshops and symposia have been devoted to CST and proceedings have been published (for example, Refs. 11–16). The objectives of this chapter are to present a brief review of the history of the development of CST software, summarize some of the recent developments and trends in CST, and identify research areas in CST that have high potential for meeting future technological needs. The number of publications on CST has been steadily increasing, and a vast amount of literature currently exists on various aspects and applications of CST. The cited references are selected for illustrating the ideas presented and are not necessarily the only significant contributions to the subject. The discussion in this chapter is kept on a descriptive level; for details, the reader is referred to the cited literature.

II. Brief History of the Development of CST Software

Development of CST software spans a period of approximately 50 years and may be divided into six stages, each of the first four stages lasting approximately

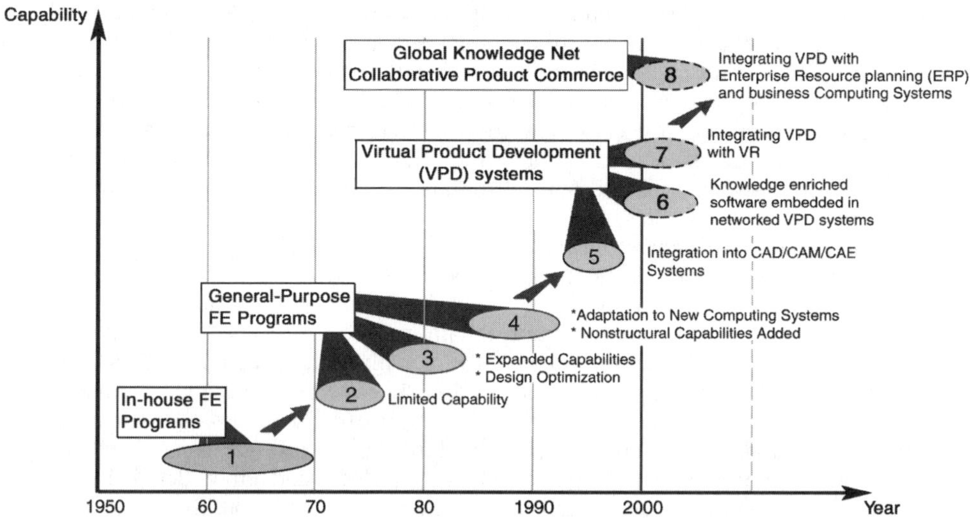

Fig. 7.2 Evolution of CST software.

8 to 12 years, the fifth stage lasting 3 to 4 years, and the sixth stage lasting 2 to 3 years (Fig. 7.2). In the first stage (during the 1950s and 1960s) the aircraft industry pioneered development of in-house finite element programs. These programs were generally based on the force method of analysis and were used to automate analysis of highly redundant structural components. Subsequent efforts in industry and academia led to development of simple two- and three-dimensional finite elements based on the displacement formulation. The variational process for formulating the elemental matrices was also introduced in this period.

In the second stage, general-purpose finite element programs such as NASTRAN, ASKA, ANSYS, STARDYNE, MARC, SAP, SESAM, and SAMCEF were released for public use in the U.S. and Europe. These programs brought a significant technology base that led to the development of numerous commercial finite element software systems. This development included mixed and hybrid finite element models, with the fundamental unknowns consisting of stress and displacement parameters; efficient numerical algorithms for solution of algebraic equations and extraction of eigenvalues; and substructuring and modal synthesis techniques for handling large problems. The finite element method's success in linear static problems has encouraged bolder applications to nonlinear and transient response problems.

The third stage involved refining the commercial software codes and expanding their technology base. Design optimization techniques were also developed in this stage, as were pre- and post-processing software and computer-aided design systems. The technology development included singular elements for fracture mechanics applications; boundary element techniques; coupling of finite elements with other discretization techniques such as boundary elements; and quality assurance methods for both software and finite element models.

The fourth stage included the adaptation of CST software to new computing systems (vector, multiprocessor, and parallel machines); development of efficient computational strategies and numerical algorithms for these machines; widespread availability of CST software on workstations, personal computers, and laptop computers; and the addition of substantial nonstructural modeling and analysis capabilities to CST software systems such as MSC/NASTRAN and ANSYS. The latter capabilities were added because current high-performance engineering systems (such as high-performance aircraft and microsized spacecraft) require significant interactions between CST and other disciplines such as aerodynamics, controls, acoustics, electromagnetics, and optics.

Technology development in the fourth stage included the introduction of advanced material models; development of stochastic models and probabilistic methods for structural analysis and design; and development of facilities for quality assessment and control of numerical simulations.

The fifth stage started in the mid-1990s and involved the integration of CST and other simulation software into CAD/CAM/CAE systems to provide a complete virtual product development (VPD) facility. Parametric, variational, and feature-based solid modeling methods are combined to generate a single "smart" product model. This eliminates data transfers and data interfaces and allows detailed analyses from within the CAD system on the latest model throughout the design process. Technology development included mesh automation, object-oriented tools and facilities, particularly in the user interfaces and databases. The object-oriented interface lets the user rapidly generate models by assembling and resizing components and then get immediate predictions of response, risk, cost, and performance information.

The sixth stage began in the late 1990s and included the use of the Internet to provide a collaborative networked (web-based) VPD facility, adding intelligent software agents and expert systems into VPD, and integrating the VPD with virtual reality facilities.

The six stages of CST software development parallel the six stages of the computing environment's evolution: noncentralized mainframes; centralized mainframes; mainframe computing with timesharing; parallel computing; heterogeneous meta computing and meta-centers (distributed computing and networking); and web-based collaborative computing. A summary of the major characteristics of finite element systems used in the late 1980s is given in Ref. 17, and a guide to information sources on finite element methods is given in Ref. 18. A finite element bibliography, including books and conference proceedings published since 1967, is available on the web. The URL address is http://www.solid.ikp.liu.se/fe/index.html.

III. Goals of CST Activities

In general, five major goals can be identified for CST activities. The five goals are described subsequently for aerospace-related CST activities. However, most of the goals described apply also to structures in fields other than aeronautics and space. The first goal is to predict the response, performance, failure, and life of structures and their components at actual operating conditions. Major advances in support of this goal continue to occur on a broad front. They include development of sophisticated computational models to simulate the mechanical, thermal, and electromagnetic responses of structures; efficient discretization techniques (namely, improved finite elements, boundary elements, and hybrid analytical/numerical methods); computational strategies for nonlinear static, dynamic, and fracture problems; quality assessment and adaptive refinement of numerical simulations; and versatile, powerful software systems for structural analysis.

Life prediction methods are required for performing structural integrity assessments of engineering systems. For metallic structures, empirical methods are being replaced by fracture-mechanics-based computational methods. For composite structures, mechanistic approaches are being developed for predicting strength and life under sustained and cyclic mechanical, thermal, and chemical environments. Because of the difficulty of understanding and modeling damage and failure phenomena, computational methods for strength and failure predictions lag behind those for response predictions.

CST's second goal is to complement experiments and other tests involving structures and their components and to aid in the design of such experiments. The need for reducing dependence on extensive and costly testing gives this goal added importance. In some cases, computational techniques are the only tools available to the designer. Examples include hypersonic aerospacecraft, outer planetary entry probes, large space structures, and extraterrestrial bases.

Another important area pertaining to the second goal is the development of computational models and methods used in support of nondestructive evaluation. Still another is the system identification (ID) method that allows the use of experiments to validate and improve the quality of numerical models through the use of inverse methods. The ID methods have become particularly important in the control–structure area because they promise a control system that can automatically identify the structure it is trying to control.

Inverse methods for structural dynamics have evolved over time from the ad hoc adjustment of parameters to sophisticated, statistically based parameter estimation methods using large quantities of experimentally estimated modal parameters, sensitivity analysis, and optimization techniques. Computational-analysis-based reliability assessment is now possible with the use of these techniques.

The third goal is to study phenomena that are difficult to understand by means other than numerical simulations. Such studies involve the implications of various microstructural damage mechanisms on the macroscopic response and performance of structural components. Also included in this goal are nonlinear chaotic dynamics of structural systems whose history indicates sensitivity to initial conditions.

The fourth goal is to give material developers guidance, enabling them to improve material systems and thereby meet performance and design requirements. Examples of computations supporting this goal include simulating the effect of a fiber–matrix interface on residual stresses, and mechanical response of metal–matrix composites.

The fifth goal of CST is to aid in the design process of engineering systems. Activities here include the development of efficient techniques for large-scale design optimization of structures; techniques for calculating the sensitivity of the structures' responses to design modifications; and inverse design problems. Sensitivity calculations are now part of many commercial software systems. Optimization techniques are gaining more acceptance in aerospace and automotive engineering[19,20] and are slowly but steadily changing all phases of the structural design process, from concept identification to the detailed design of components and flight vehicles. Inverse design problems attempt to identify design characteristics that produce the desired performance.

Current CST activities include the study of phenomena occurring at disparate spatial and time scales. Length scales range from the atomic and molecular levels to the structural level. Today no important design can be completed without CST, nor can any new theory be validated without it. Commercial programs for structural analysis have a rich variety of elements, and are widely used for performing structural calculations on large components and/or entire structures.

IV. Recent Advances in CST

The development of CST cuts across a number of disciplines including structural mechanics, materials science, discretization techniques, numerical analysis, and computer science. Advances in each of these disciplines can have a strong impact on CST. Some of the recent advances which have had, or promise to have, a strong impact on CST are described subsequently. The advances are grouped into the following categories: element technology and discretization techniques; computational material modeling; computational modeling of composite, sandwich, and smart structures; life management computational tools and methodologies; transient response analysis; nonlinear analysis; numerical simulation of frictional contact/impact response; computational methods for articulated structural dynamics; probabilistic analysis and stochastic modeling; qualitative analysis and simulation; neuro-computing; hybrid techniques; error estimation and adaptive improvement strategies; strategies for solution of coupled problems; sensitivity analysis; integrated analysis and design; strategies and numerical algorithms for new computing systems; model generation facilities; and application of object-oriented technology. This list is by no means complete or exhaustive. The intention is to stimulate the interest of researchers for realizing the full potential of CST in the analysis and design of engineering structures. Among the advances not covered herein are: computational methods in fracture and damage mechanics (see, for example, Refs. 21–26); computational methods for inverse problems (e.g., parameter/system identification);[27,28] computational methods for material process modeling; and multidisciplinary optimization problems.

A. Element Technology and Discretization Techniques

The establishment of reliable and efficient finite elements, boundary elements, and other discretization techniques for modeling structures with complicated geometry has been, and continues to be, the focus of intense research effort. In this section, some of the recent advances are reviewed. These advances are grouped into six categories: formulative aspects of finite elements; spectral and wavelet and other nonclassical elements; meshless methods (also referred to as element-free Galerkin methods or nodes without elements); finite volume method; generalized finite element methods; and finite elements for shell analysis. Recent advances in boundary element methods are reviewed in recent survey articles,[29–31] in a monograph[32] and in a special issue of a journal.[33]

1. Formulative Aspects of Finite Elements

The desire to establish a variational basis for high-performance finite elements (viz., simple elements with few degrees of freedom—all physical, which delivers results of engineering accuracy on coarse meshes) has led to the development of generalized mixed variational principles (GMVPs) in the early 1980s,[34,35] which were referred to as parametrized variational principles (PVP) by Felippa[36,37] in 1989. For a discussion of the attributes of high-performance elements see Ref. 36. The functional of a generalized multifield variational principle (GMVP) is a parametrized combination of canonical functionals, with each of the canonical functionals represented by a point in the parametric continuum.

Generalized multifield variational principles provide a framework by which classical variational principles may be embedded into a parametrizable continuum of functions. To date, GMVPs have been constructed for finite deformation elasticity, incompressible solids, micropolar elasticity, and linear electrodynamics. Two of the important applications of GMVPs are selection of variational functionals for ill-conditioned problems,[35] and development of finite element templates. A template is a parametrized algebraic form that yields a continuum of finite elements for a specific application, with a given element configuration, and a given number of degrees of freedom.[37] GMVPs provide a variational framework for some of the elements that lack such framework. Two notable examples are the scaled free formulation (SFF) elements, and the assumed natural deviatoric strains (ANDES) elements. The SFF element was shown to be associated with a multifield variational principle with one free parameter that interpolated between hybrid versions of the potential energy principles and Hellinger–Reissner principle. The ANDES element is associated with a one-parameter hybrid multifield functional that interpolated between the potential energy and the Hu–Washizu functionals. The development of GMVPs and the widespread availability of powerful computerized symbolic manipulation systems can help in the systematic construction of optimal finite elements by the template approach.

2. *Spectral, Wavelet, and Other Nonclassical Elements*

During the past decade, spectral element (p-type finite element) methods have proven to be an efficient and accurate way to solve partial differential equations, especially in the fluid mechanics area. Spectral element methods use the fast convergence and good phase properties inherent in singular Sturm–Lionville approximations, while allowing the solution domain to include complex geometries. Developments in the last 7 years included extension of spectral methods to triangular and tetrahedral elements in two- and three-dimensions, respectively (see, for example, Refs. 38–40).

Wavelets are functions with local support (i.e., functions that are local in time and frequency, or in space and wave number–see, for example, Refs. 41 and 42). Several engineering applications of wavelets have been reported in the literature, especially in signal and image processing. However, only in the last 5 years have finite element application of wavelets to model localization phenomena become an active research area (see, for example, Ref. 43).

Recent work included evaluating the performance of hierarchical/wavelet formulations in terms of convergence rate, dispersive behavior, and computational complexity; and using multiple scaling functions for multiresolution analysis. The resulting techniques are referred to as multiwavelet methods.[44] With multiwavelet methods no special quadratures are needed in the generation of wavelet elements, and the difficulties associated with the application of wavelets to closed bounded domains are greatly alleviated.[45]

Wavelet elements have the potential of computing multiscale solutions to partial differential equations with higher convergence rates than conventional finite element methods. Their built-in adaptive nature offers the possibility of automatic adaptivity.[46] They are well suited to preconditioning finite element approximations of boundary value problems. They also yield methodologies for obtaining sparse representations of boundary element formulations.

In addition to the spectral and wavelet elements, a number of other nonclassical elements have been developed using a variety of approximation functions, including digital Walsh functions, orthogonal and nonorthogonal polynomials, spline functions, and fundamental solutions of elasto-static problems.[47]

3. *Meshless Methods*

The term "meshless methods" encompasses a class of techniques striving to simplify the preparation of computational models by eliminating or reducing the effort required to generate finite element meshes. This includes a variety of methods, such as generalized finite differences, element-free Galerkin, hp clouds, wavelets, external approximation, R-functions, and others. While virtually all of these methods require some kind of discretization to solve the problem, generation of such a discretization typically lends itself to a high level of automation, allowing the user to perform meshless analysis. Besides simplifying model preparation, these methods are also advantageous in problems with changing domains, such as crack propagation, where classical finite element techniques experience difficulties in following the changing domain. Meshless methods are also suited for capturing thin zones of steep strain gradients, particularly when high-order spatial gradients are included in the constitutive relations, as in the case of gradient-dependent plasticity.

One of the advantages of the methodology of meshless methods is that the model description is the same as that required by CAD systems. Recent work included the development of an element-free Galerkin method that uses a moving least square interpolant.[48] The method has been applied to crack propagation problems and was found to possess the following advantages: avoiding the locking associated with compressible materials; providing high rates of convergence with low-order polynomials; and facilitating adaptivity, because it is only necessary to add extra nodes in the area where extra resolution is needed. Recent work also includes reproducing kernel particle methods for structural dynamics problems,[49] which are extensions of the smooth particle hydrodynamic method developed by Gingold and Monaghan to provide a mesh-free environment.[50] It can be viewed as a continuous least-square polynomial. Recently two meshless approaches have been developed that do not involve a mesh either for interpolation purposes or for the purpose of performing numerical integrations, to form the system of algebraic equations. The first is based on a local boundary integral equation (LBIE) approach.[51] The second is based on a meshless local Petrov–Galerkin (MLPG) approach and uses the moving least-squares approximation. The integrals in the MLPG approach are evaluated only over regularly shaped subdomains and their boundaries. The local boundary is selected to be the surface of a sphere centered at the node in question.[51]

4. *Finite Volume Method*

Although attempts were made in the 1960s to use an integration technique (cell method) for generating finite difference approximations for plane elasticity, plates and elasto-plastic flow, it was not until the early 1970s that the finite volume method evolved as a finite difference approximation on nonorthogonal grids. The method has been widely used in fluid dynamics and heat transfer applications, but it has not gained acceptance in structures applications. The method was shown to be a special case of the finite element method with non-Galerkin weighting.[52–54] The method has been applied to structural analysis of complex domains,[55] modeling the behavior of nonlinear materials,[56,57] and dynamic fracture mechanics problems.[58,59] One motivation for developing this method for structures was to solve fluid-structure interaction problems (such as simulation of flow-induced vibration phenomena) using commercially popular finite volume-based CFD software.[60] The method has the advantage of computing element matrices and vectors using boundary integrals along the control volume sides.[61,62] Also, the potential of mixing finite element and finite volume methods for the discretization of complex domains needs to be investigated.

5. *Generalized Finite Element Methods*

The high-fidelity modeling of complex three-dimensional structural components in high-tech industries by conventional finite element methods often requires an excessive number of elements to capture high gradients in the response quantities. To alleviate this problem several methods were proposed under the names finite element partition of unity, cloud-based hp finite element methods, and special finite element method.[63–65] Recently, it was shown that these methods, and several of the meshless methods, can be considered as special cases of a generalized finite element method based on the use of a set of functions whose values sum to the unity at each point in the domain. The functions are defined over element patches.

6. *Finite Elements for Shell Analysis*

Attempts to develop finite elements for shell analysis date back to the late 1960s. Since that time a significant capability has been developed for modeling shell behavior. Fundamental considerations for, and reviews of, the development of some of the finite elements for shell analysis are included in two recent papers[66,67] and a monograph.[68] Mixed shell elements, which satisfy the fundamental condition for stability and optimality, are reported in Refs. 69–71. These elements are particularly useful for membrane-dominated and bending-dominated behavior. Despite the progress made, to date there is no mathematically validated approach for overcoming the difficulties associated with finite element shell analysis (such as complete removal of shear and membrane locking).

B. Computational Material Modeling

Considerable attention has recently been devoted to computational material modeling. In the context of CST, the objectives of this activity are to model realistic constitutive behavior by means of equations that represent the main physics and are capable of predicting results consistent with macroscopic observations (see, for example, Refs. 72–77); and to develop a hierarchy of material models (multilevel/multiscale) to describe the different phenomena associated with response, life, and failure of structures.

The first activity includes mesoscale continuum mechanics formulations, which are intermediate between a direct atomistic and an unstructured continuum description of deformation processes. They provide frameworks for describing fracture processes and material instabilities such as shear bands.[77]

A major consideration in the application of constitutive equations is the determination of material parameters from test data. This can be a difficult undertaking for the more complex equations with a large number of parameters. In most formulations for metals, the parameters divide into a few that govern the overall strain rate and temperature dependence of inelastic flow and would generally appear in the kinetic equation. A larger number of material parameters usually arise in the evolution equations for the hardening or possible softening properties. Attempts to rationalize and formalize the procedure have been made.

The multiscale material modeling activity has resulted in a gradual evolution from empirical models to physics-based models that predict the processing response and properties. The hierarchy of material models that has been developed is shown in Fig. 7.3. The models are arranged according to the phenomena they describe and the length scale at which the phenomena are studied (from 10^{-10} m to 1 m). The disciplines involved include computational chemistry that covers molecular dynamics and quantum mechanics, computational material science, and computational structural mechanics. The models used range from atomistic to single crystals to polycrystals to micromechanical and macromechanical models.[78] However, many gaps still exist in the hierarchy of models, and to date no rational way exists to integrate these models and to couple them with experiments to relate the phenomena at the very small length scales with the macroscopic behavior.

The hierarchy of material models is an important element in the computationally driven material development activity that has been pursued at a number of research labs for more than 15 years (see, for example, Refs. 79 and 80). The central paradigm of the computationally driven material development activity is the sequential interrelation of processing, structure, properties, and performance

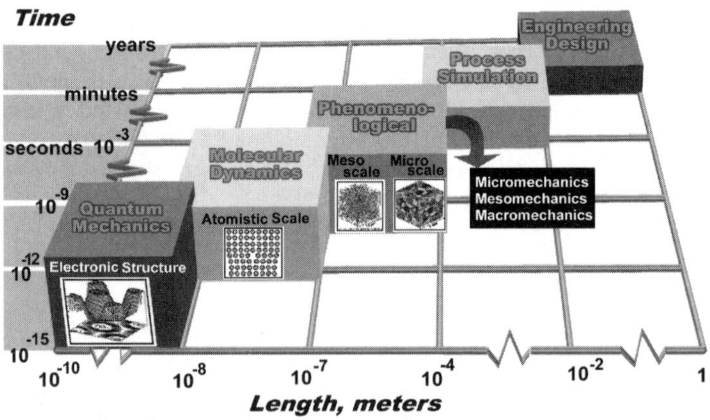

Fig. 7.3 Multi-scale phenomena associated with computationally driven material development. (See color section.)

of materials. Ideally, the inner structure of the material is used to predict properties, then the properties are used to predict performance, and finally, a sequence of processing procedures is selected to yield the desired material and inner structure at a reasonable cost.

C. Computational Modeling of Composite, Sandwich, and Smart Structures

Recently, increasing work has focused on the development of computational models for studying various phenomena associated with the response, life, failure, and performance of composite and sandwich structures. The phenomena involved cover a wide range of length scales, from microstructural to structural response. Recent noteworthy contributions can be grouped into the following five categories: accurate determination of inter-laminar stresses; multiscale modeling approaches; thermal and thermomechanical buckling and postbuckling; heat transfer analysis; specialized software for predicting the response and failure of high-temperature sandwich and composite structures; and modeling of smart structures. These categories are described subsequently.

1. Accurate Determination of Inter-laminar Stresses

The importance of inter-laminar stresses in predicting the onset of some of the damage mechanisms in multilayered composite structures has long been recognized. Therefore, since the 1970s various techniques have been proposed for the accurate determination of transverse stresses in laminated composites. These include using three-dimensional or quasi-three-dimensional finite elements; two-dimensional finite elements based on higher-order shear deformation theories with either nonlinear or piecewise linear approximations for the displacements in the thickness direction; and post-processing techniques, used in conjunction with two-dimensional finite elements based on the classical or first-order shear deformation theory (with linear displacement approximation through the thickness of the entire laminate).[81,82]

Experience with most of the three-dimensional finite elements and two-dimensional finite elements based on higher-order shear deformation theories has shown that unless the three-dimensional equilibrium equations are used in evaluating the thickness distribution of the transverse stresses, the resulting stresses are inaccurate. Because the finite element models based on the first-order shear deformation theory are considerably less expensive than those based on three-dimensional and higher-order, two-dimensional theories, their use in conjunction with post-processing techniques has received increasing attention in recent years. The post-processing techniques proposed for the evaluations of transverse stresses are based on the use of three-dimensional equilibrium equations; predictor–corrector approaches; and simplifying assumptions.[83–86]

2. Multiscale Modeling Approaches

The global response and performance of composite and sandwich structures is governed by a number of local (constituent scale) phenomena such as fiber–matrix interface bonding, core-free sheet interface bonding, progressive damage and failure processes, and constitutive material behavior that exhibits cyclic nonlinear history dependence. An attempt has been made to develop an integrated multiscale modeling approach that incorporates nonlinear constituent material models and failure models, composite micromechanical models, and finite element global structural analysis models. The multiscale approach has been used to assess the performance and structural integrity of aircraft engine components.

3. *Buckling and Postbuckling*

In recent years, considerable work has been devoted to the study of buckling and postbuckling responses of composite and sandwich plates and shells. Both thermal as well as combined thermal and mechanical loading cases have been considered. Attempts have been made to identify the differences between the isothermal and thermal responses. Reviews of some of the recent contributions are contained in three survey papers[81,82,87] and a monograph.[88] Of special interest is the study of the sensitivity of the buckling and postbuckling responses to variations in the material and lamination parameters of the composite.[89–94]

4. *Heat Transfer*

A number of approaches have been developed for the determination of the heat transfer characteristics and thermal response of composite structures. Some of these approaches are similar to the ones used in predicting the deformational response of composite structures. Recent contributions include the development of two-dimensional finite element models for heat transfer in multilayered composite panels.[95,96]

Moderate intensity, short-duration radiant exposures of composites with organic matrices may result in processes such as pyrolysis, ablation, and diffusion of moisture or vapor. Only a few publications have addressed this subject.

5. *Special-Purpose Analysis and Design Codes*

A variety of special-purpose codes have been developed for assessing the structural integrity/durability and reliability of the high-temperature structures used in advanced propulsion systems. The codes developed fall into a number of categories, including: probabilistic structural analysis; progressive fracture of composite structures; coupled structural/thermal/electromagnetic tailoring; and micromechanics analysis tools for multiphase composite materials subjected to complex histories.[97–100]

6. *Smart Structures*

Smart structures sense external stimuli, process the sensed information, and respond with active control. The response can consist of deforming or deflecting the structure, or communicating the information to another control center.[101] Smart materials deform or deflect the structure by changing their physical properties when subjected to electric, magnetic, or thermal loads. The active elements in smart structures can be embedded in or attached to the structure. The computational models developed for predicting the thermomechanical response of laminated composites were extended to predict the electro–thermo mechanical response of anisotropic composite structures made of smart materials as well as hybrid fibrous composite/smart materials.[102–106] Review of the finite element applications to smart materials and structures is given in Ref. 107.

D. Computational Tools and Methodologies for Life Management

Life management tools range from those used in the initial design phase of new systems to those required to make technology insertion and retain-or-retire decisions of aging systems. Among the computational needs for both new and aging aerospace systems are life prediction methods for performing structural integrity assessments. At high temperatures, life prediction involves a complex combination of processes that change the stress and material states of a structural component.

Structural life prediction procedures use material life models in conjunction with thermal, acoustic, and structural analysis to quantify material response in identified local damage initiation-prone regions of a structure. A large number of high-temperature material life models have been proposed (see, for example, Refs. 108–110). Acoustic fatigue life estimates have been made for simple structures only (see, for example, Refs. 111 and 112).

The complexity of life prediction for high-temperature structures may be attributed to a number of factors. The processes experienced by the structural component are multidisciplinary in character. The mechanics, thermodynamics, and chemistry aspects are coupled in such a way that classical representations such as fatigue (which attempts to isolate cycle-dependent mechanical behavior) and creep rupture (which attempts to isolate time-dependent behavior alone) are inadequate representations of the total performance of the structure.

Some of the inputs to the life-prediction methods are random in time or are stochastically distributed. Cyclic or fatigue loading of high-temperature structural components is applied in a fairly random or stochastic sequence. Moreover, other variables such as corrosion, oxidation, and chemical activities may occur in a random fashion.

The rate equations that are used in representing the evolution of material properties are difficult to establish experimentally. Because of the complexity of the life prediction problem, it is common to represent changes in life by empirical or phenomenological descriptions that are limited to correlations of experimental observations. Recent work has focused on establishing mechanistic models for predicting remaining strength and life of structures. However, before these models can be used they have to be validated and calibrated by macroexperiments.

Performance simulation has been proposed as an approach for predicting the remaining strength and life of material systems.[113–114] The performance simulation approach is based on experimental determination of failure and damage modes and the use of that information to establish critical elements of the materials that control the failure event. These representations are combined into a single integral equation that represents the remaining strength as a function of time and loading history. The integral equation is discretized in time, and the physical behavior of the system at different times is estimated by accounting for the changes in the material state and the stress state during the loading history. The approach has the potential of representing highly nonlinear and coupled phenomena such as creep, damage, oxidation, corrosion, and thermodynamic variations.

Extensive analytical and experimental work is being carried out at NASA Langley and NASA Glenn to provide a basic understanding of the damage mechanisms and fatigue of composites at high temperatures. The work at Langley is primarily focused on continuous-fiber-metal matrix composites (see a review of this work in Ref. 115). The controlling parameters identified for fatigue damage include relative strains to fatigue failure of the fiber and matrix, fiber/matrix interface strength, and thermal residual stresses.

The NASA Glenn activities are part of the HiTEMP and EPM Projects and include many tasks. First, the projects assess life and fracture prediction methods for metal-matrix composites (MMCs consisting of a titanium alloy matrix and ceramic reinforcement) and evaluate different means of nondestructive evaluation (NDE). This work was done jointly with Pratt and Whitney, a division of United Technologies, Inc., and included determination of actual fracture strength and lives of MMC rings.[116]

NASA Glenn projects also predict the probability of failure of monolithic ceramic components as a function of time in service. A computer program, CARES/Life, has been developed that couples commercial finite element programs (used for determining the temperature and stress distributions) to reliability evaluation and fracture mechanics routines for modeling strength-limiting defects.[117] The code has been applied to aeronautics and space propulsion systems, including ultra-high temperature ceramics for the leading edges of reusable space vehicles.

Finally, the NASA Glenn projects assess the integrity of ceramic matrix composites (CMC). A computer program (PCEMCAN) has been developed for predicting the probabilistic CMC properties and strength from the constituent properties and their probabilistic distributions. The program has the capability to perform two-dimensional woven CMC analysis.[118]

For aging air vehicles, fatigue damage, loss of corrosion protection, and the subsequent occurrence of damaging corrosion, can become serious threats to their operational integrity. Better life prediction schemes (for example, those based on probabilistic analysis and stochastic modeling) are required in support of the alternate usage schedules, management techniques, corrosion protection refurbishment, and affordable inspection and maintenance approaches. There is also a need for computational simulation of technology insertion processes and repair technologies.

E. Transient Response Analysis

The development of efficient computational strategies and numerical algorithms for simulating the transient response of structures has been, and continues to be, an active research area. Some of the advances made in this area are contained in Ref. 119. Among the recent developments are variable multi-step integration algorithms for first-order equations, which are extensions of multi-time step (subcycling) algorithms based on nodal or element partitions allowing each node to be integrated with a different time step.[120] A stability analysis for the algorithms is presented in Ref. 121.

Also, space-time finite elements, which include discontinuity-capturing operators to accurately resolve sharp gradients or discontinuities in wave propagation and impact problems, have been developed.[122–125] Also, the time-discontinuous Galerkin method has been proposed to provide controllable numerical dissipation in the high-frequency response domain without introducing excessive algorithmic damping in the important low-frequency modes.

Finally, multiple-scale finite element methods for linear structural dynamics problems, based on decomposing the response into fast and slow time scales using temporal shifting, have recently been developed.[126] The slow time scale is solved by the direct time integration method with a reasonably large time step. The total response is then recovered via a sampling theorem and superposition.

F. Nonlinear Analysis

Significant and far-reaching advances have been made in the development and application of computational methods to analyze nonlinear behavior of structures and solids. Special strategies and techniques have been proposed for handling the difficulties associated with the roughness (lack of smoothness) and instabilities that can occur both in time and space. Smoothness in time depends on factors such as contact–impact, the character of loading, and the smoothness of the constitutive response. Smoothness in space depends also on the character of the loading, but is impaired dramatically by failure phenomena such as cracking. Instabilities can be classified as geometric instabilities and material instabilities. Material instabilities may result in localization of deformation, which dramatically increases the need for high-resolution numerical techniques, or a discontinuous treatment of the motion. Some of the advances in nonlinear analysis are described in other subsections.

G. Numerical Simulation of Frictional Contact/Impact Response

The numerical simulation of the nonlinear response of structures subject to impact loads has been the focus of intense efforts because of the pressing need for collision protection (or crashworthiness) of transportation vehicles, nuclear reactors, and containment vessels. Several software systems are currently available for frictional contact/impact analysis of structures, including ABAQUS, ADINA, DYCAST, DYNA, KRASH, LS/DYNA, McALOG/RADIOSS, MSC/DYTRAN, NIKE, PAMCRASH, and PRONTO. Recent activities included study of the mechanics of large dynamic deformation, and of material strain-rate sensitivity, and development of friction models, contact/impact algorithms, and semi-empirical damage and failure mechanisms (see, for example, Refs. 127–136).

H. Computational Methods for Articulated Structural Dynamics

Flexible multi-body systems (FMS) have important applications in aircraft and space structures, including deployable space structures, space robots and manipulators, and aircraft landing gears. The response of FMS involves coupled effects of deformation, angular velocity and acceleration relative to an inertial reference frame. Considerable progress has been made in predicting the response of FMS (see, for example, Refs. 137–140). The progress is manifested by the development of efficient methods for solving differential–algebraic equations of motion of rigid and flexible multi-body systems; and software systems and object-oriented tools for the numerical simulation of the transient dynamic response of FMS.

I. Nondeterministic Modeling and Analysis Methods

In the past two decades there has been a growing realization among engineers that unavoidable uncertainties in geometry, material properties, boundary conditions, loading, operational environment, modeling, and analysis assumptions must be taken into account to produce meaningful designs. Three possible approaches for handling uncertainty can be identified, depending on the type of uncertainty and the amount of information available about the structural characteristics and the operational environment. The three approaches are probabilistic analysis, fuzzy-logic approach, and set theoretical, convex (or anti-optimization) approach. A discussion of the three approaches and their combinations is given in Refs. 141–143. In probabilistic analysis, the structural characteristics and/or the source variables are assumed to be random variables (or functions), and the joint probability density functions of these variables are selected. The main objective of the analysis is the determination of the reliability of the system. Reliability is defined as the probability that the structure will adequately perform its intended mission for a specified interval of time when operating under specified environmental conditions. Possible limitations of probabilistic methods are discussed in Ref. 144.

Southwest Research Institute, under NASA funding, developed a code that combines a fast probability integration algorithm with general-purpose stress, free vibration, buckling and dynamic structural analysis programs, based on finite element and boundary element techniques. Output of the analysis consists of a user-defined region for the probability distribution function (or cumulative distribution function), together with confidence bands.[145–146] A comparative study of probabilistic finite element methods of structures was reported in a journal article[147] and in a monograph.[148] A state-of-the-art survey on stochastic modeling is given in Ref. 149, and some of the recent applications are given in Ref. 150.

If the uncertainty is because of vaguely defined structural and/or operational characteristics, imprecision of data, and subjectivity of opinion or judgment, fuzzy logic-based treatment is appropriate (see, for example, Refs. 151–153). Randomness describes the uncertainty in the occurrence of an event (such as damage or failure of the structure). Fuzziness describes the ambiguity of the event (such as imprecisely defined criteria for failure or damage). When the information about the structural and/or operational characteristics is fragmentary (that is, only a bound on a maximum possible response function is known), then convex modeling can be used. Convex modeling produces the maximum or least-favorable response and the minimum or most-favorable response of the structure under the constraints within the set-theoretical description.[154]

J. Qualitative Analysis and Simulation

Many engineering structures are made up of several interconnected components. The complexity of the response of the components and/or the incomplete knowledge about the structure characteristics can make the prediction of the response of the system by traditional computational (quantitative) methods intractable. Qualitative analysis and simulation, a branch of AI, can help in this situation by building models that reflect the phenomena associated with the response of the actual system. Application of qualitative analysis to snap-through mechanics problems is described in Ref. 155. Recent activities have focused on the application of interval methods to estimate the range of possible solutions, and on developing hybrid approaches integrating qualitative, quantitative and rule-based analysis.[156,157]

K. Neuro-computing

Neuro-computing is rooted in various disciplines, but the concept of artificial neural networks was inspired by biological network. Neural networks are pattern computers—information processing devices (either algorithms or actual hardware). They use simplified mathematical functions to approximate the behavior of neurons in the brain. Biological neurons, as the structural constituents of the brain, are much slower than silicon logic gates. However, inferencing in biological neural networks is faster than the fastest computer. The brain compensates for the relatively slower electro-chemical operation by having an enormous number of massively interconnected neurons.

The application of artificial neural networks (ANNs) in structural mechanics has been gaining support in recent years. Operations performed by neural networks include classification, pattern matching, optimization, control, and noise removal. Neural networks are particularly useful in situations where good analytic models are either unknown or extremely complex. When analytic models are either missing or incomplete, neural nets can be used to estimate a model from empirical data. They essentially work as interpolators—using partial input/output data about a system, a neural net can be used to estimate other output values. When good analytic models are available, neural networks are likely to be less accurate than numerical solution techniques. ANNs have been used as computational tools in various areas of structural mechanics, some of which are analysis, identification, simulation, design, and optimization (see, for example, Ref. 158).

L. Hybrid Techniques

The application of hybrid techniques to complicated structural problems can result in a dramatic savings in computational effort. Four groups of hybrid techniques can be identified in relation to the four groups of numerical, experimental, analytical, and neural techniques: hybrid analytical techniques;[159] hybrid analytical/numerical techniques (see, for example, Ref. 160); hybrid numerical/experimental techniques;[161] and hybrid neuro-computing/numerical techniques. Examples of the four groups are shown in Fig. 7.4a and 7.4b. Recent work has focused on the coupling of different methods to generate effective hybrid techniques; specifically, applications of neuro-computing/numerical techniques including nonlinear structural and sensitivity analyses;[162,163] and micromechanics based constitutive damage models.[164] In the second application, the finite element results, in the form of equivalent inelastic stress–strain response, are used to train ANN. The finite element analyses are performed for a relatively small number of selected strain-paths, geometric and damage parameters. The ANN is used to construct a micromechanical damage model that can be applied in nonlinear analysis of composite structures.

M. Error Estimation and Adaptive Improvement Strategies

Assessment of the reliability of computational models has been the focus of intense efforts in recent years. These efforts can be grouped into three general categories (see, for example, Refs. 165–170): *a posteriori* error estimation, super-convergent recovery techniques, and adaptive improvement strategies. *A posteriori* error estimates use information generated during the solution process to assess the discretization errors. In super-convergent recovery techniques more accurate values of certain response quantities (e.g., derivatives of fundamental unknowns) are calculated than those obtained by direct finite element calculations. Adaptive

a)

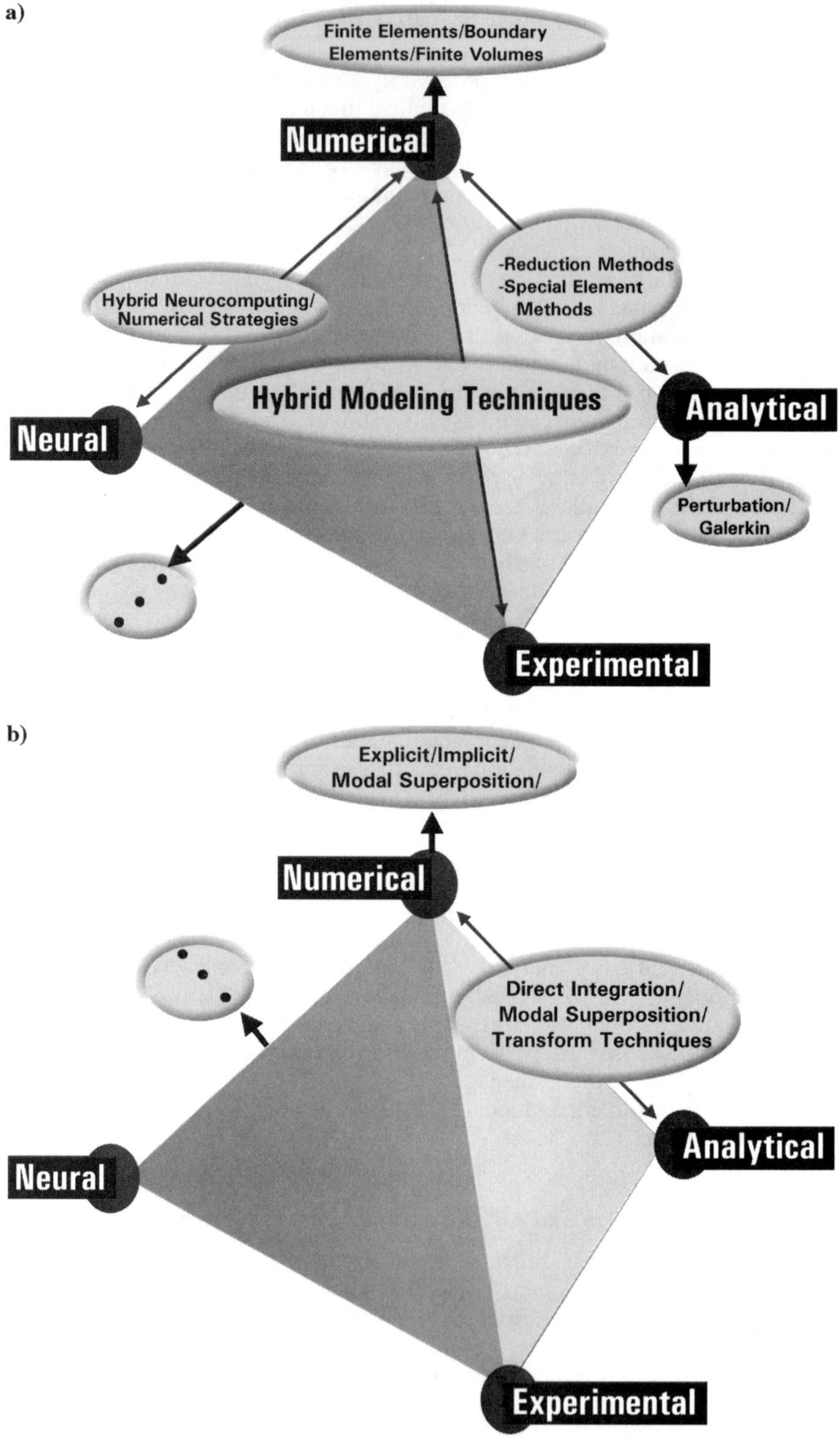

Fig. 7.4 Examples of hybrid analysis techniques. (See color section.)

strategies attempt to improve the reliability of the computational model by controlling the discretization error.

1. *Error Estimation*

Two broad classes of error estimation schemes are currently used—residual methods and interpolation methods. Residual methods involve the use of local residuals, usually as data in a local auxiliary problem designed to generate the local error to an acceptable accuracy. A significant amount of computation may be required in implementing these methods. In interpolation methods the available approximate solution for a given mesh (or time step) is used to estimate higher derivatives locally (that is, local gradients or second derivatives). The higher derivatives are used, in turn, to determine the local error. Although these error estimates can be very crude, they are portable: a subroutine for computing local estimates can be added to virtually any existing code that operates on unstructured meshes with some effort.

Recent work included the development of alternative methods to estimate and control the approximation error with respect to measures other than the usual energy norm. The measures are expressed in terms of a linear functional of the solution and represent a physical quantity of interest such as point-wise or averaged values of the stress, displacement, strain, velocity, or pressure. The starting point of the alternative methods is the introduction of influence functions that indicate how the sources of error influence the numerical error in the quantity of interest. It is shown to be the solution of an adjoint problem to the solution problem.[171]

Although significant progress has been made in developing *a posteriori* error estimates for linear elliptic problems (see, for example, Refs. 172–175), the error estimates for nonlinear and time-dependent problems (parabolic and hyperbolic) are considerably less developed. This is particularly true for bifurcation problems, problems with multiple scales, and problems with resonance. Work on error estimation for highly nonlinear problems has mainly been a subject of ad hoc experimentation.

2. *Super-Convergent Recovery Techniques*

Super-convergent recovery techniques refer to simple post-processing techniques that provide increased accuracy of the sought quantities at some isolated points (for example, Gauss–Legendre, Jacobi, or Lobatto) in a sub-domain or even in the whole domain.[176,177] In the latter two cases, the techniques are referred to as local and global super-convergent recovery techniques, respectively. Recent work included the development of local super-convergent patch derivative techniques for both interior and boundary (or material interface) points.[178–185] It was shown in Refs. 178 and 179 that the super-convergent recovery technique can be used to obtain *a posteriori* error estimates for the finite element solution.

3. *Adaptive Strategies*

Different strategies have been used for adaptive improvement of the numerical solutions (see, for example, Refs. 170 and 186–190), including mesh refinement (or de-refinement) schemes (h methods); moving mesh (no redistribution) schemes (r methods); subspace enrichment schemes (selection of the local order of approximation, p methods); mesh superposition schemes (overlapping local finite element meshes on the global one, s methods); and hybrid (or combined) schemes.

Examples of these schemes are simultaneous selection of the meshes and local order of approximation (h-p methods). Recent theoretical results have shown that the fastest possible convergence rates can be attained by optimally decreasing the mesh size h and increasing the degree of the polynomial degree p in a special way, or by simultaneous selection of the meshes and node redistribution (h-r method). These methods can be effective in shock problems because an r-method might align the mesh along discontinuities prior to a mesh refinement.

Although all the aforementioned strategies can be applied to structural and nonstructural problems, only the first strategy (h-method) has been applied, in conjunction with unstructured grids, to thermal stress analysis.[191]

N. Strategies for Solution of Coupled Problems

Two general strategies are used for solution of multifield problems, such as fluid flow/acoustic/thermal/structural problems: namely, the staged solution strategy, and the coupled solution strategy, which are briefly outlined subsequently.

1. Staged Solution Strategy

The multiple fields are treated separately. The discrete models for each of the fields may be developed separately. Coupling effects are viewed as information that must be transferred between the discrete models of the different fields. For thermal/fluid/structural problems, the heating rates are first predicted on aerodynamic surfaces using coupled fluid flow/thermal analysis. Next, the structural temperatures are determined via thermal analysis. The structural response is then computed using the structural equations of motion and constitutive relations. The structural deformations can be used to correct the shape of the body, and the whole process is repeated. This is depicted in an integrated fluid flow/thermal/structural analysis capability as presented in Refs. 192 and 193. A more recent application of the strategy is presented in Ref. 194.

A modification of this strategy is the multistagger solution strategy in which a partial decoupling is made of the full system of equations. The full system of coupled equations is partitioned into smaller subsystems of equations. Each subsystem is solved separately under the assumption that the variables of the other subsystems are frozen (temporarily).[194] A staggered solution process is presented in Refs. 197 and 198 to analyze the coupling of incompressible and compressible fluid flows with structural motions.

2. Coupled Solution Strategy

The multiple-field problem is treated as an indivisible whole. The discrete models of the different fields are tightly coupled.

O. Sensitivity Analysis

Sensitivity analysis refers to methods of calculating the rates of change of response quantities (such as displacements, stresses, vibration frequencies, and buckling loads) with respect to changes in the structure characteristics (such as geometric and material parameters of the structure); and the optimum design variable values with respect to changes in the structure parameters (such as applied loads and allowable stresses). The two types of calculations are usually designated by response and optimum design sensitivity analysis, and the rates of changes are referred to as sensitivity coefficients.

A number of techniques have been developed for evaluating the sensitivity coefficients of response quantities using either the governing discrete (such as finite element) equations or continuum equations of the structure. The techniques used with the discrete equations can be grouped into three categories: analytical direct differentiation methods, semi-analytical or quasi-analytical methods, and finite difference methods.[199,200] To avoid the subtractive cancellation errors inherent in the classical finite difference method, complex variables have been used to estimate derivatives of real functions.[201]

Methods for computing sensitivity coefficients for linear structural response have been developed for more than 20 years (see, for example, Refs. 202 and 203). However, only in the last few years have attempts been made to extend the domain of sensitivity analysis to nonlinear structural response and to path-dependent problems, for which the sensitivity coefficients depend also on the deformation history (for example, visco-plastic response and frictional contact, see Refs. 204–207); and to structural systems exhibiting probabilistic uncertainties.

Because of the importance of its role in structural optimization and in assessing the effect of uncertainties in the input parameters on the structural response, some commercial software systems have incorporated response sensitivity analysis into the systems. Also, algorithmic (automatic) differentiation facilities have been developed for evaluating the derivatives of functions defined by computer programs, exactly to within machine precision.[208] One of the facilities developed for algorithmic differentiation, ADIFOR (Automatic DIfferentiation of FORtran), is described in Refs. 209 and 210. The use of ADIFOR to evaluate the sensitivity coefficients from incremental/iterative forms of three-dimensional fluid flow problems is discussed in Ref. 211, and the additional facilities needed for ADIFOR to become competitive with hand-differentiated code are listed in Ref. 212.

P. Integrated Analysis and Design

The design of vehicular and propulsion systems for future aerospace missions requires integration between the structures discipline and other traditionally separate disciplines such as aerodynamics, heat transfer, chemistry, materials, and control; integrated system analysis to couple subsystems, components, and subcomponents at an appropriate level of detail; and a simulation environment that provides a user-friendly interface between the analyst and the multitude of complex codes and computing systems that are required to perform the simulations. NASA Glenn has built a numerical propulsion system simulator (NPSS) to determine the system attributes such as performance, reliability, stability, and life.[212]

Q. Strategies and Numerical Algorithms for New Computing Systems

In recent years, intense efforts have been devoted to the development of efficient computational strategies and numerical algorithms that exploit the capabilities of new computing systems; in particular, the vector and parallel processing of the powerful high-performance computers (see, for example, Refs. 14 and 213). Efficient direct and iterative numerical algorithms have been developed for solving large sparse linear systems of equations. Parallel strategies and software tools have been developed for static partitioning of large problems on parallel computers. The software tools are often run as preprocessors to large-scale applications.

Most parallel strategies are related to the "divide and conquer" paradigm based on breaking a large problem into a number of smaller sub-problems that may be solved separately on individual processors. The degree of independence of the sub-problems is a measure of the effectiveness of the algorithm because it determines the amount and frequency of communication and synchronization. The numerical algorithms developed for structural analysis can be classified into three major categories, namely element-wise algorithms, node-wise algorithms, and domain-wise algorithms.

The element-wise parallel algorithms include element-by-element equation solvers and parallel frontal equation solvers. The node-wise parallel equation solvers include node-by-node iterative solvers and column-oriented direct solvers. The domain-wise algorithms include nested dissection-based (substructuring) techniques and domain decomposition methods. The first two categories of numerical algorithms allow only small granularity of the parallel tasks, and require frequent communication among the processors. By contrast, the third category allows a larger granularity that can result in improved performance for the algorithm.

Nested dissection ordering schemes have been found to be effective in reducing both the storage requirements and the total computational effort required for direct factorization. The performance of nested dissection-based linear solvers depends on balancing the computational load across processors in a way that minimizes inter-processor communication. Several nested dissection-ordering schemes have been developed that differ in the strategies used in partitioning the structure and selecting the separators. Among the proposed partitioning strategies are recursive bisection strategies (for example, spectral graph bisection, recursive coordinate bisection, and recursive graph bisection);[214,215] combinatorial and design-optimization based strategies (for example, simulated annealing algorithm, genetic algorithm, and neural-network-based techniques);[216] heuristic strategies (such as methods based on geometric projections and mappings); and algorithms (based on embedding the problem in Euclidean space). For highly irregular and/or three-dimensional structures, the effectiveness of nested dissection-based schemes may be reduced. However, this is also true for most other parallel numerical algorithms. Scalable parallel computational strategies for nonlinear, postbuckling and dynamic contact/impact problems are presented in Refs. 217 and 218.

Recent work has been devoted to dynamic partitioning and load balancing techniques for distributed adaptive grid hierarchies. The techniques seek to improve the accuracy of the solution by dynamically refining the computational grid in regions of high local solution error. Although distributed implementations have the potential of solving very large complex problems, they lead to challenges in dynamic resource allocation, data distribution and load balancing, communications and coordination and resource management.[219] To date, no general purpose tools for dynamic load balancing are available.

R. Model Generation Facilities

It has long been recognized that for complex structures the task of generating the finite element model is both tedious and error prone. Moreover, the cost of

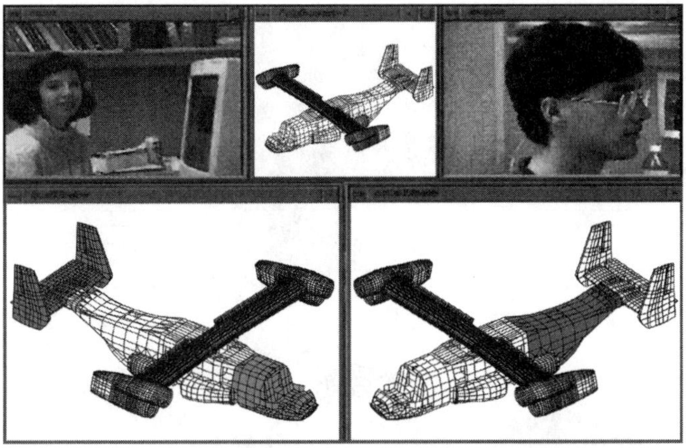

Fig. 7.5 Collaborative model development. (See color section.)

the analysis directly depends on the size, shape, and number of elements. Extensive research has been devoted to automatic and semi-automatic generation of finite element grids (see, for example, Refs. 220 and 221). The techniques proposed for model generation include two-dimensional quadtree, three-dimensional octree, triangulation parameter space mapping, and substructuring. Among the recent activities on model generation are application of knowledge-based analysis assistance tools that allow a simple description of the analysis objectives and generate the corresponding discrete models appropriate for these objectives, and the development of a solid modeler-based preprocessor CUBIT for the automatic generation of well-formed quadrilateral and hexadedral finite element grids (with reasonably small distortion metric). CUBIT provides a combination of techniques, including paving, mapping, sweeping, and various other algorithms for discretizing the geometry into a finite element grid. Recent activities on model generation also include the use of visual object-oriented technology facilities, development of collaborative model generation (Fig. 7.5) (this also includes interactive assembly of finite element component models and stock parts), and real-time model generation using immersive technologies, such as special gloves and VR facilities (Fig. 7.6). Such technologies enable interaction with complete or incomplete models, and allow interactive feedback to the model generation process. The capability is particularly useful for examining the finite element topology in certain spatial localities where the automated model generation facility may have produced certain complexities or anomalies.

S. Application of Object-Oriented Technology

Because of the increasing size and complexity of analysis and design software systems, change in programming paradigm and a shift in software design strategy have occurred for developing easily readable, expandable, and maintainable software systems. The change is from procedure-oriented to object-oriented programming, and the shift is toward object-oriented system development methodologies. The use of object-oriented technology can lead to smaller programs; provide better

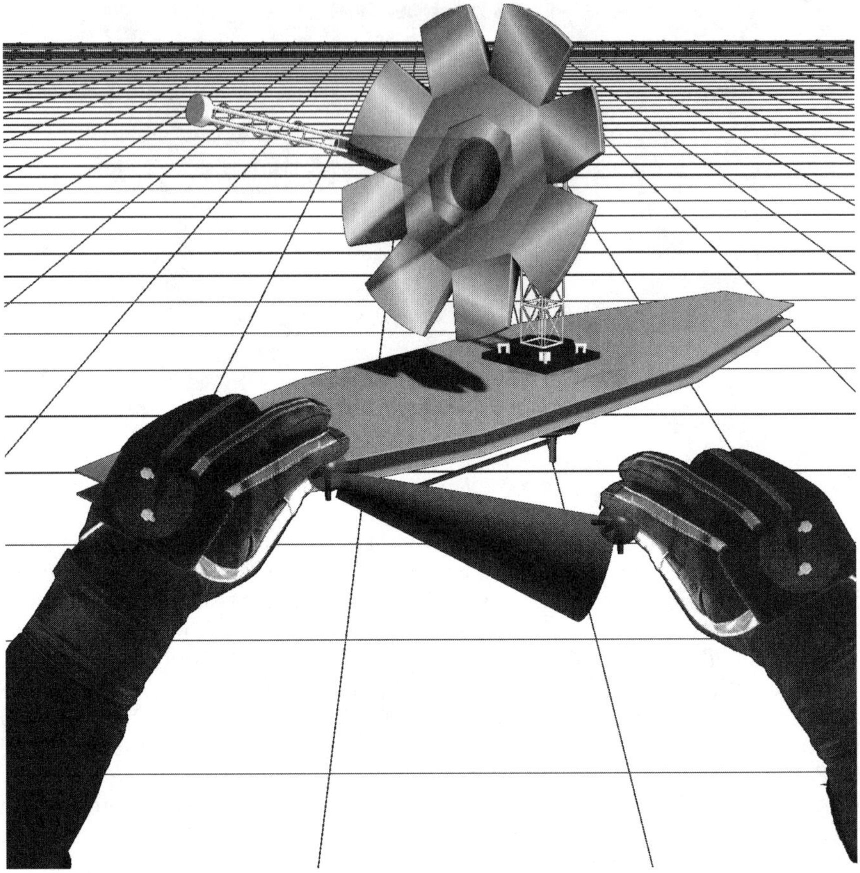

Fig. 7.6 Real-time model generation. (See color section.)

management of data and procedures; and simplify extensions of the system to cope with the advances in analysis methods, new materials and structural components, and new computer architectures. Some of the recent structural and dynamic applications of object-oriented technology are reported in Refs. 222–229. The efficiency issues in an object-oriented programming environment are discussed in Ref. 230.

The current trend is to move from object-oriented to component-based software development (CBD). CBD allows developers to use and reuse code in applications that are written in any language and that run on any platform, almost anywhere on most networks. A prerequisite for the success of CBD is a standard infrastructure, with three main elements for the components. The three elements are uniform design notation (which can be achieved by using a uniform modeling language or UML); a standardized interface that enables any application (in any language) to access component features by binding to the component model or interface definition language; and repositories, which are catalogues of available components with a description of the features that enable the developer to find the components appropriate for an application.

V. CST at Universities and Industry

Through their dual mission of teaching and research, universities have aided CST development. Their contributions started in the early days of finite element development and included new finite elements; numerical algorithms for solving equations and extracting eigenvalues; computational strategies for contact/impact, large deformation, buckling, postbuckling, and dynamic problems; and optimization techniques. Until recently, however, when high-performance computers and advanced workstations became widely available to university researchers, most university CST research focused on fundamental problems and did not involve very large-scale numerical simulations.

Universities began teaching finite element courses in the mid-1960s; since then a variety of CST-related and computer-aided engineering courses have been added to many undergraduate and graduate curricula. CST can significantly enhance structures education by allowing students to apply the basic principles to real (though scaled-down) structures. It can also provide a virtual experimental facility for simulating structural, dynamic, and wind tunnel experiments. Although observations of real tests and experiments are invaluable, physical insight is rarely derived from a single experiment; CST provides a facility in which the student may change the design variables and study such changes' effects on the response characteristics of the structure. University of Virginia researchers are developing a virtual structural laboratory as part of a distributed learning environment (Fig. 7.7), that

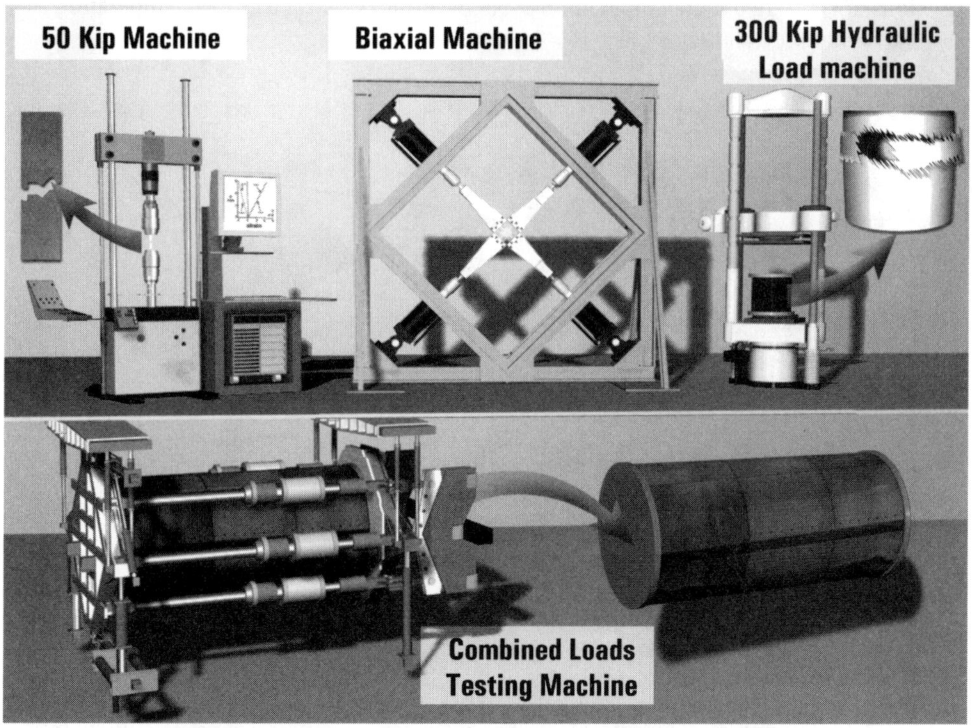

Fig. 7.7 Virtual structural laboratory. (See color section.)

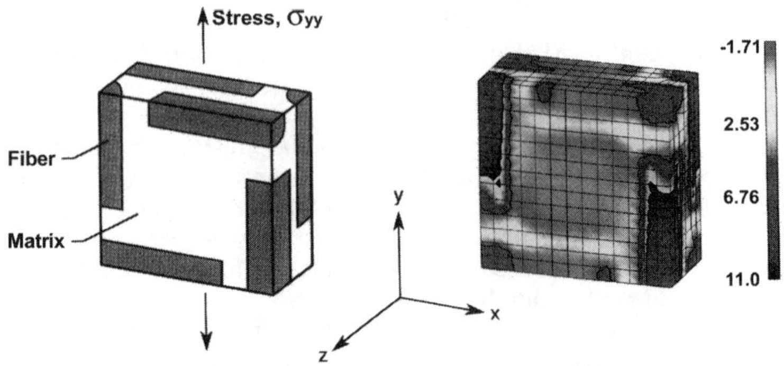

Fig. 7.8 Microstress σ_{yy} distribution in a typical cell of a three dimensional short fiber-reinforced metal matrix composite loaded in the y-direction (courtesy of the Mechanical Engineering Department, University of California at Santa Barbara). (See color section.)

will be Internet-centric and web-based with a virtual reality modeling language, multimedia, immersive facilities, and multisensory interfaces.

Examples of university studies include micromechanical analysis of short-fiber composites (Fig. 7.8), and thermomechanical buckling of composite panels with cutouts (Fig. 7.9). The second example showed that, contrary to intuitive expectations, for certain boundary conditions both the critical load and critical temperature can increase as the size of the cutout increases.

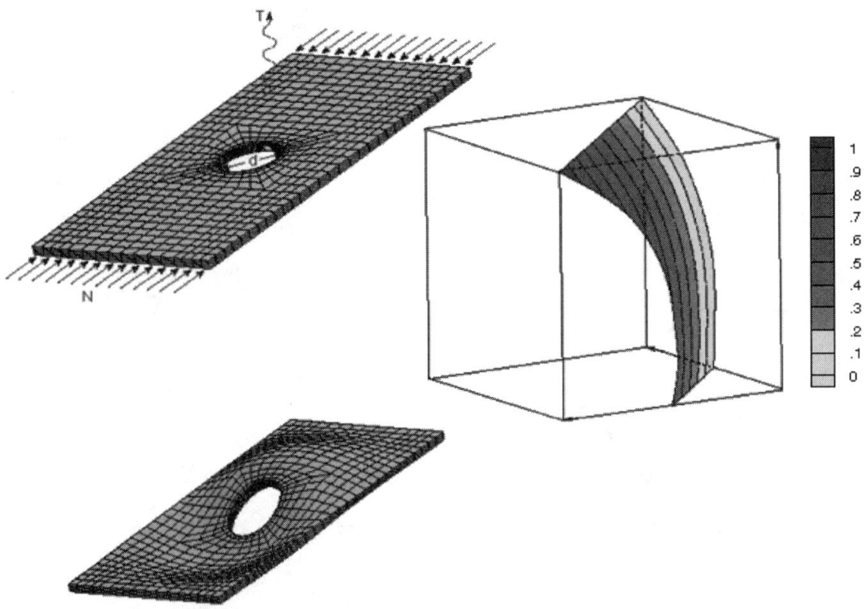

Fig. 7.9 Thermo-mechanical stability boundary for a composite panel with circular cutout. (See color section.)

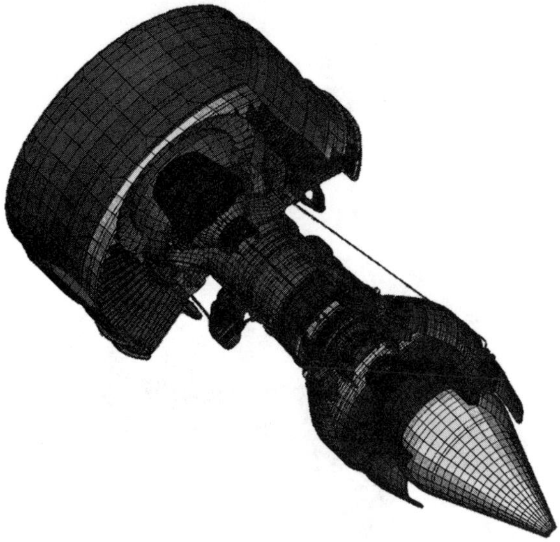

Fig. 7.10 MSC/NASTRAN model of GE engine—180,000 degrees of freedom (courtesy of GE Aircraft Engines). (See color section.)

In industry, recent CST activities include development of pre- and postprocessors. The goal of this work is to reduce the efforts needed for creating the finite element model and to cut the time required for interpretation and visualization of results. Also, in-house special-purpose programs have been developed for detailed analysis of structural connections and joints.

Although the sophistication of computational models in industry has increased in the last decade, the traditional role of computational structural analysis has not changed significantly in industry. In the aircraft industry, computational structural analysis is used to establish load paths and local stress levels. These are fed into separate analyses, or component tests, to predict strength, life, or other failure modes, and the whole process is then validated by full-scale tests. The level of detail in the finite element modeling currently used in the aircraft, nuclear, and automotive industries is provided by the five finite element models shown in Figs. 7.10–7.18 (GE-90 engine model, subassembly of an engine, B-2 Bomber model, V-22 Osprey tilt rotor model, landing gear and structural joint models, internal loads model of Boeing 777, car model used in crush simulation, and a fusing assembly). Today many of these calculations can be performed on the powerful workstations in a matter of hours.

VI. A Look at the Future

Economic stresses are forcing many industries to reduce cost and time to market, and to insert emerging technologies into their products. This trend is likely to continue and will require engineers to design faster, ever more complex systems. They must find globally optimal designs that take uncertainties and risk into consideration.

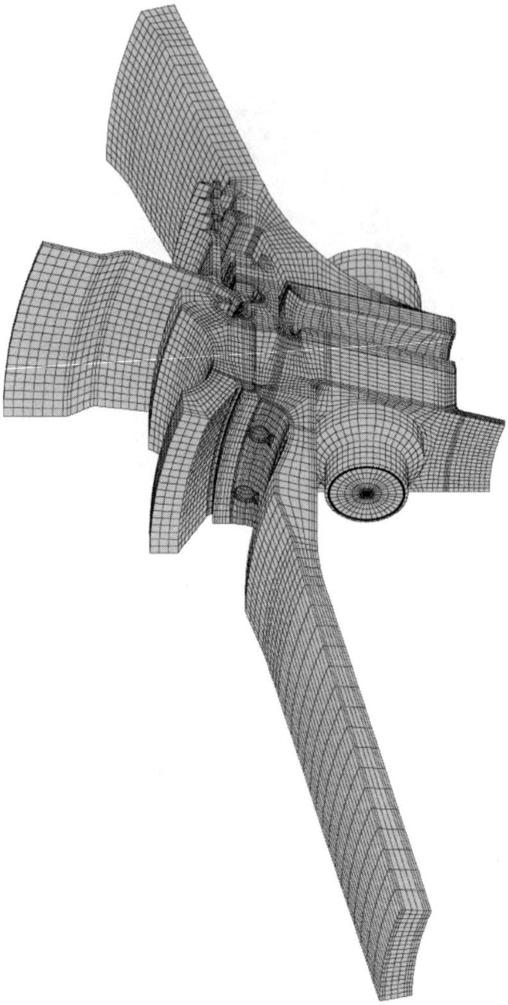

Fig. 7.11 ANSYS model of a subassembly of a high by-pass aircraft gas turbine engine containing the fan-booster bearing race. The model has 97,250 eight-noded elements and 115,369 nodes. Gap elements are used at the contact surfaces (courtesy of GE Aircraft Engines). (See color section.)

The realization of future cost-effective, high-performance engineering systems requires technology advances in the materials and structures areas; computational technologies and tools for solving complex problems with system uncertainties (for example, computational intelligence and its associated tools, also called soft computing); and distributed integrated virtual environment for simulating the entire lifecycle of the engineering system, from concept development, detailed design, and prototyping to qualification testing and operations (see Fig. 7.19). The environment links scientists, design teams, manufacturers, suppliers, and consultants in the creation and operation of engineering systems. It enables a seamless integration of teams, processes, and disciplines.

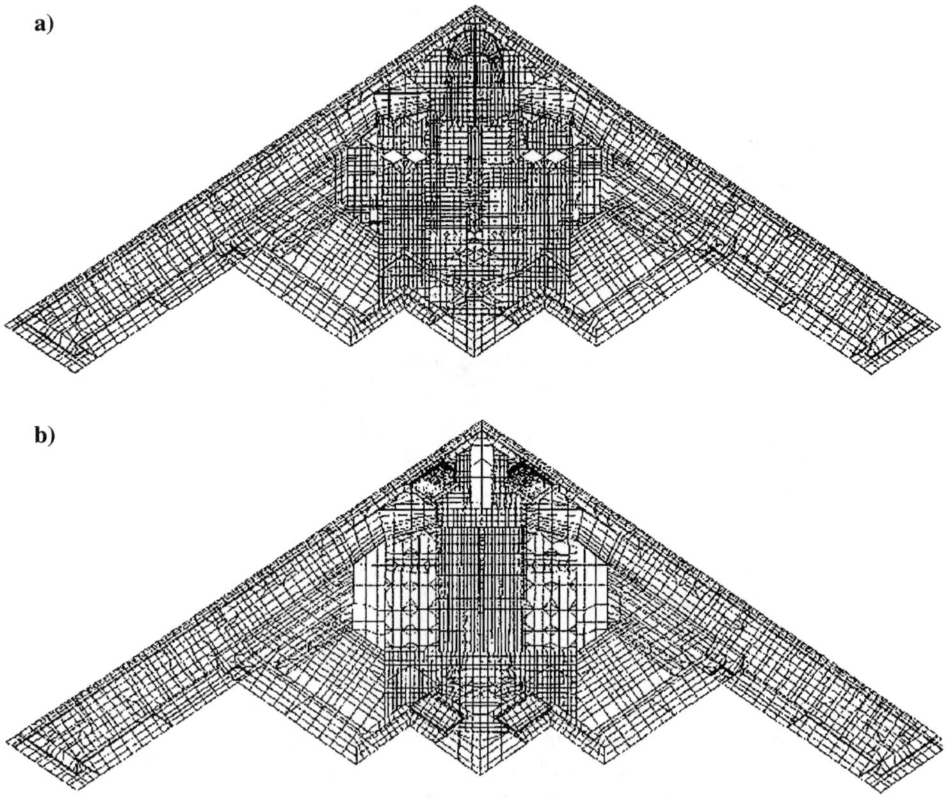

Fig. 7.12 A finite element model of the B-2 bomber with over 100,000 degrees of freedom: a) upper surface description; b) lower surface description (courtesy of J. Kudva and M. Lonner, Northrop Grumman Corp.).

In order for CST to play a significant role in the future development of structures technology, as well as in the design and certification of future engineering systems, major advances and new computational tools are needed in several key areas. To this end the research community must address several primary pacing items and related tasks. In identifying these pacing items, researchers must consider the characteristics of future engineering systems and their implications for CST; future computing paradigm and environment; and developments in other technologies that can impact CST (virtual product development, information technology and knowledge-based engineering) and other fields of computational science and engineering that can be adapted to CST.

A. Characteristics of Future Engineering Systems

The demands that future high-performance engineering systems place on CST differ somewhat from those of current systems. The trend toward miniaturization, and the radically different and more unpredictable operational environments for

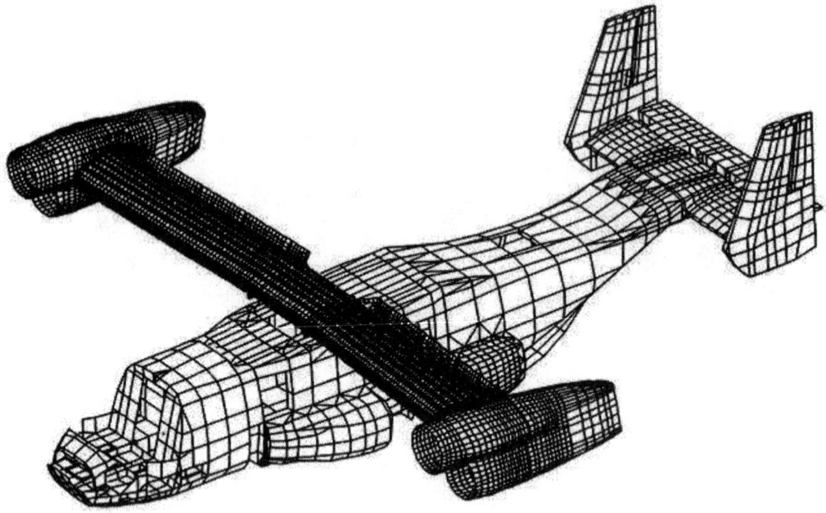

Fig. 7.13 MSC/NASTRAN finite element dynamics model of the V-22 Osprey Tiltrotor—134,982 degrees of freedom (22,497 grid points), 44,006 elements (courtesy of Bell-Boeing).

many of the future systems, are two reasons for this difference. Another is the stringency of design requirements for economy, high performance, light weight, and rapid prototyping. The twenty-first century will see engineering systems made from computationally designed materials at the atomic/molecular level to accomplish a variety of missions. Programmable, multifunctional materials will be able to adjust their shape and their mechanical, electromagnetic, optical and acoustic properties on demand. The technical needs for future systems in the materials and structures areas include:

a) Development of new high-performance material systems such as smart/intelligent, multifunctional and bio-mimetic material systems;

b) Development of innovative materials processing technologies that enable low-cost fabrication;

c) Development of novel structural concepts, such as structural tailoring and intelligent/smart structures, with active and/or passive adaptive control of dynamic deformations; and

d) Investigation of more complex phenomena and interdisciplinary couplings such as fluid flow/acoustics/thermal/control/electromagnetic/optics and structural couplings (see Fig. 7.20).

B. New and Emerging Computing Paradigm and Environment

Significant advances have been made and continue to be made in the entire spectrum of computing and communication technologies. The advances included fast microprocessors; massively parallel computer architectures; high-speed, high-capacity optical and wireless networks; communication protocols; distributed software structures; and security mechanisms. A partial list of some the recent and projected advances in computing technology are given in Refs. 231 and 232.

The impact of these advances on the structures technology is given in Ref. 233, and are summarized subsequently. At one end of the spectrum are heterogeneous meta-computers and meta-centers. Meta-computers refer to an integrated (virtual) computing environment with geographically separated resources connected by one or more high-speed inter-connectors. The middle-wire (a software layer) transforms a collection of independent hosts into a single, coherent virtual machine. Use of meta-computers has significantly affected CST by greatly alleviating the size limitations that current memory capacities impose on numerical simulations. Moreover, highly parallel systems and meta-computers have achieved sustained speeds in the tera-flop (trillion floating-point operations per second) range. The meta-computer hardware components are described in Refs. 232–234.

At the other end of the computing spectrum, small systems such as handheld computers (referred to as personal digital assistants), and wearable computers have emerged with cellular technology to support portable computing. Also, new powerful microprocessors such as Intel Titanium IA-64 and other GHz-scale

a)

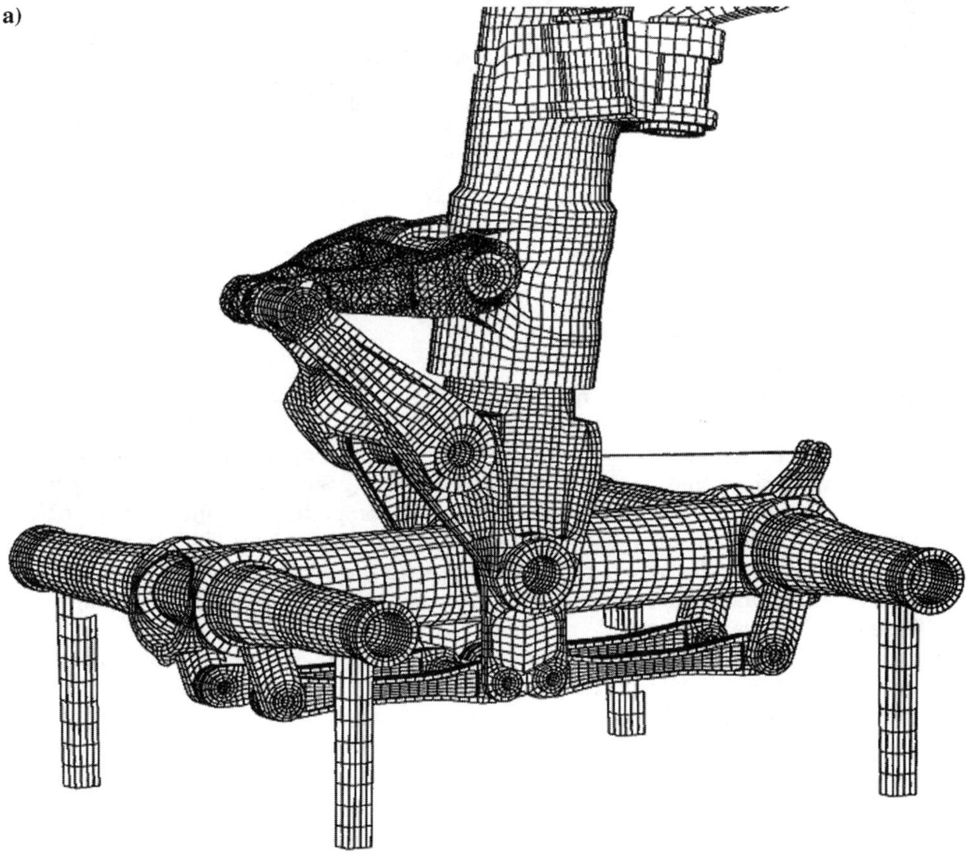

Fig. 7.14 CATIA ELFINI model of the Boeing 767-400 main landing gear. a) 320,000 degrees of freedom finite element model, b) stress results (courtesy of Rodney L. Dreisbach, the Boeing Company, Seattle, WA). (See color section.)

b)

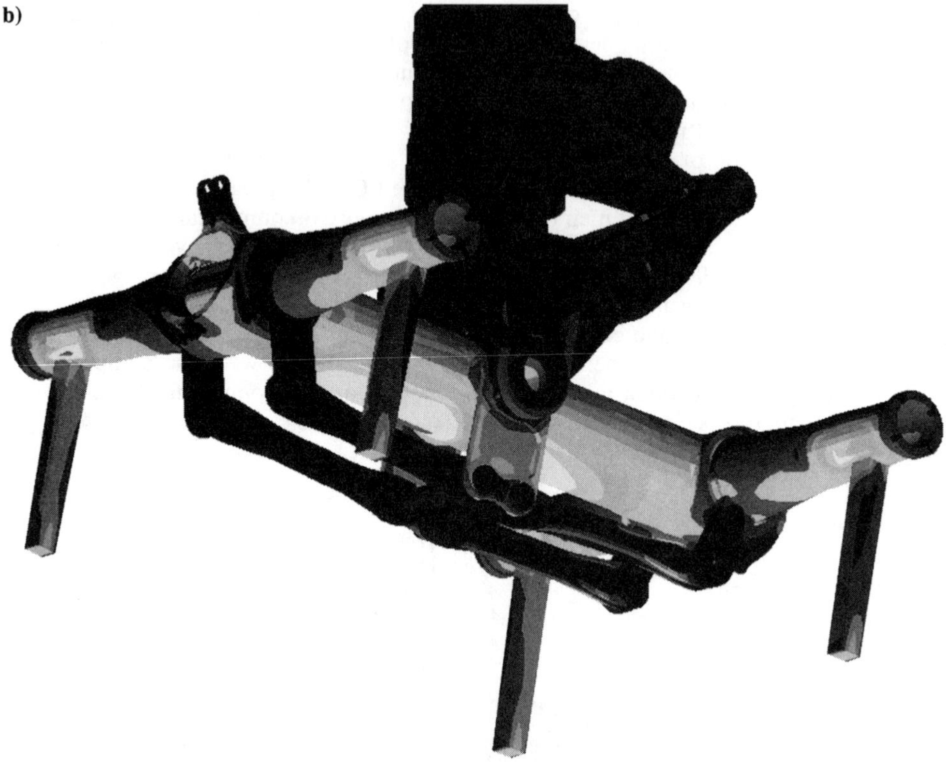

Fig. 7.14 (Continued).

processors will be used as structural health monitoring systems for detecting and recording damage. The aforementioned advances have resulted in an explosive growth in computer power and connectivity that is reshaping relationships among researchers and organizations, and the emergence of the new paradigm of parallel, distributed, collaborative, immersive computing. The implications of the new paradigm for engineering practice in general, and for CST in particular, are immense. The new paradigm, along with the advances in modeling and simulation, virtual product development, information technology, and knowledge-based engineering, will enable dramatic changes in which engineering systems are designed, produced, operated, maintained, and disposed. Future environments will allow diverse, geographically dispersed teams to share and transform information into knowledge by combining and analyzing it in new ways. The interactive collaboration among experts in several disciplines will enable them to attack heretofore-intractable multiphysics problems involving phenomena at disparate length and time scales.

Realizing the full potential of new computing technology, however, requires new approaches for human–machine interfaces. The full power of creative artificial intelligence research should be brought to bear on the development of these interfaces, which should effectively link to as many of the human sensing and communication mechanisms as possible.

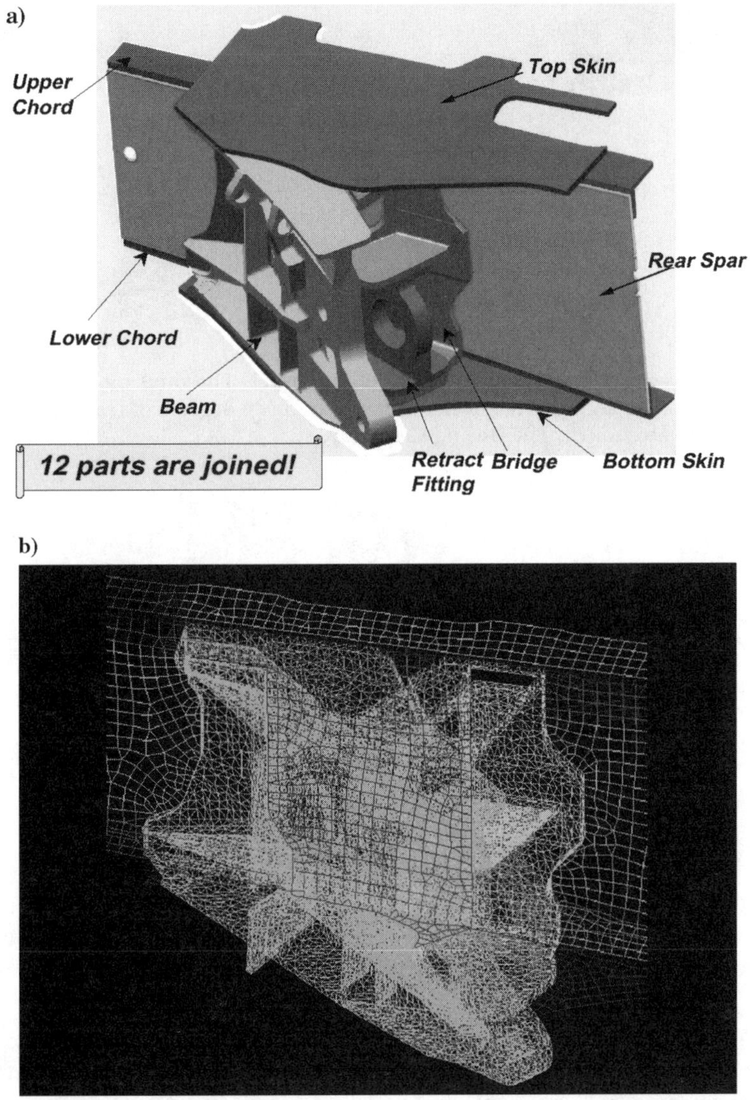

Fig. 7.15 CATIA ELFINI model of a structural joint joining 12 parts a) CAD model, b) Finite element model. (Courtesy of Rodney L. Dreisbach, The Boeing Company, Seattle, WA). (See color section.)

C. Virtual Product Development Systems and Information Technology

Current virtual product development (VPD) systems have embedded simulation capabilities for the entire lifecycle of the product. A description of the capabilities of over 120 CAD systems (including some VPD systems) is given in Ref. 235. The addition of intelligent agents and networking facilities will transform the VPD systems into knowledge-enriched networked VPD systems. The benefits obtained include higher productivity, better product quality, and a broader design to provide

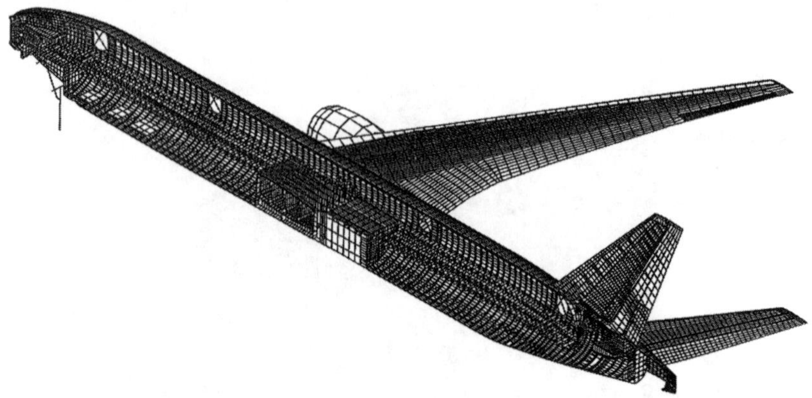

Fig. 7.16 Boeing 777 Airplane internal loads model. The total model has 40 substructures, 70,000 nodes, 175,000 elements and 275,000 degrees of freedom (courtesy of Rodney L. Dreisbach, The Boeing Company, Seattle WA).

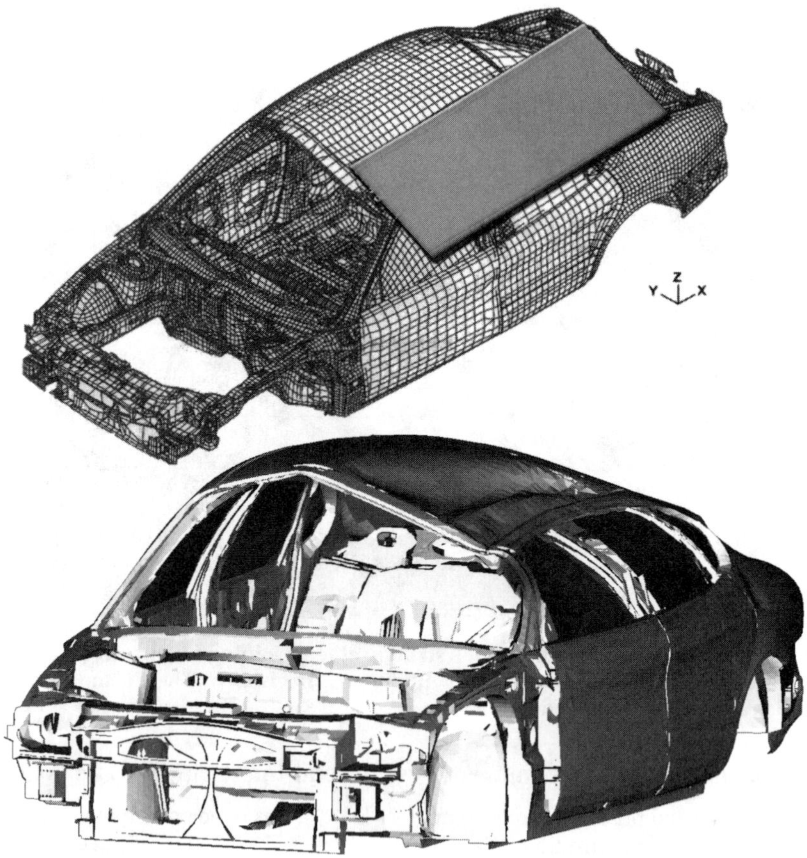

Fig. 7.17 Crush analysis of a Taurus car model performed by implicit dynamic finite element analysis (corresponding to an incremental static solution). A rigid plate moving at 10 mm/sec is crushing the car (courtesy of ADINA R&D, Inc.). (See color section.)

an integrated system solution. Current work focuses on integrating VPD with other enterprise software facilities, such as enterprise resource planning (ERP), supply chain management (SCM), and customer relationship management (CRM), to create a global knowledge net for collaborative product commerce (see Fig. 7.2)

Information technology (IT) is concerned with the dissemination, processing, storage, retrieval, and use of information.[236] IT will change the product development from a sequence of distinct phases into a continuous process covering the

a)

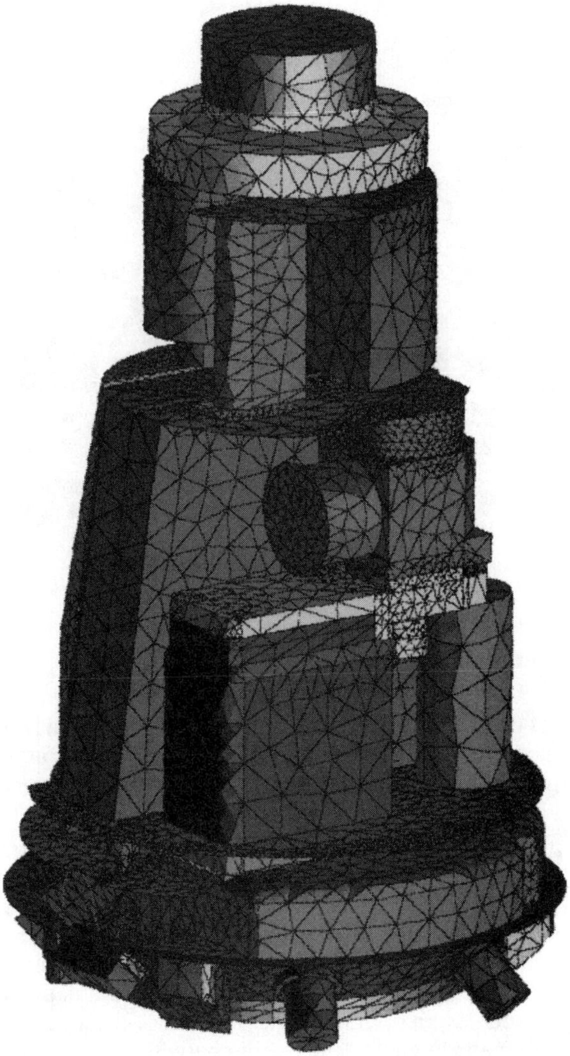

Fig. 7.18 Fusing assembly modeled using SALINAS code and 250,000 tetrahedral elements: a) finite element model; b) first vibration mode. Colors represent subcomponents within the system (courtesy of J. S. Peery, Sandia National Laboratory). (See color section.)

b)

Fig. 7.18b (See color section.)

entire life cycle of the product with full interplay of information from beginning to end and everywhere throughout. IT will also be an important component of future distributed learning environments used for engineering training and education.

D. Primary Pacing Items

The primary pacing items for CST are high-fidelity modeling of the structure and its components; failure and life prediction methodologies; hierarchical, integrated methods and adaptive modeling techniques; nondeterministic analysis, soft computing and risk assessment; validation of numerical simulations; and multidisciplinary analysis and design optimization. For each of the aforementioned items, attempts should be made to exploit the major characteristics of high-performance computing technologies, as well as future computing environments. The six primary pacing items are described subsequently. Note that some of the tasks within the pacing items are of a generic nature. Others are specific to either certain components of future engineering systems (e.g., propulsion systems or airframes of flight vehicles).

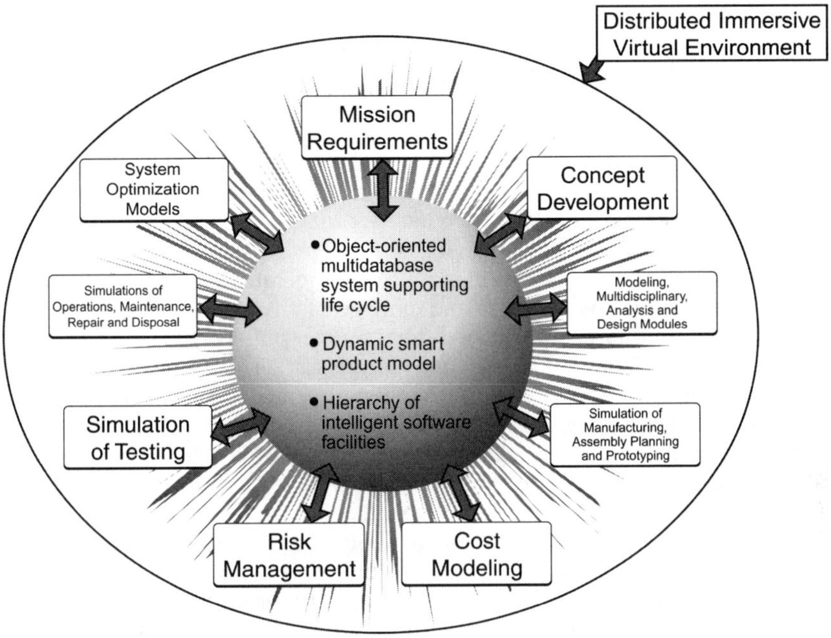

Fig. 7.19 Distributed virtual environment for simulating the entire life cycle of the engineering system.

1. *High-Fidelity Modeling of the Structure*

The reliability of the predictions of response, failure and life of structures is critically dependent on the accurate characterization and modeling of material behavior and high-fidelity modeling of the critical details of the structure and its components (such as joints, damping, and, for large deformations, frictional contact between the different parts of the structure).

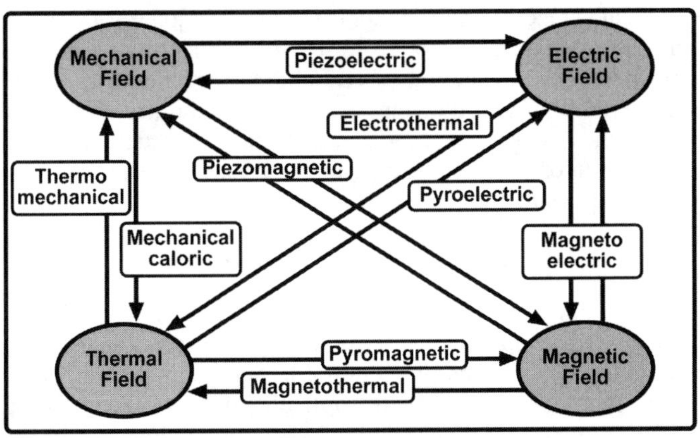

Fig. 7.20 Couplings between mechanical, electric, magnetic and thermal fields in smart materials.

As modeling strives toward a higher, physics-based predictive capability based on first principles, validation methodology and inverse methods must increasingly focus on the phenomenological identification and validation of sub-grid physics such as material constitutive and interfacial microslip models. While experiment design can help to isolate sub-grid models for identification and validation purposes, new inverse methods need to be developed to help isolate, identify, and validate the contributing physics in complex structural applications.

The simple material models used to date are inadequate for many of the future applications, especially those involving severe environment (for example, high temperatures). Needed work on material modeling can be grouped in two general areas—modeling the response and damage of advanced material systems in the actual operating environment of future engineering systems, and numerical simulation of manufacturing (fabrication) processes.

Advanced material systems include functionally graded materials (FGMs), multifunctional materials, smart materials, bio-mimetic materials, textile and high-temperature composites, and advanced metallics. For FGMs, novel processing techniques are used to produce engineered, gradual transitions in microstructure, composition, and properties to satisfy spatially varying functional performance requirements within a single component. Use of FGMs can alleviate the high gradients of internal stresses and strains resulting from the different local deformation fields induced when dissimilar materials are joined to form a component. This technology could allow researchers to develop complicated structures without using conventional methods of joining. Besides being less costly, jointless gradient materials are much more robust because stresses are distributed throughout the volume of the material rather than being concentrated at the joint. Multifunctional materials, in addition to supporting mechanical loads, may incorporate sensors to detect and evaluate loads or failure, and to interact with the surrounding electromagnetic environment. By contrast, materials as we know them and use them at the present are largely "mono-functional."

Bio-mimetic technology aims at producing new materials by mimicking the synthesis, processing, and properties of materials found in biological systems. Among the unique and useful characteristics of such systems are multifunctionality, hierarchical organization, self-repair, adaptability, and durability. Moreover, biological structural systems do not distinguish between materials and structures.

The length scale selected in the model must be adequate for capturing the response phenomena of interest (for example, micromechanics, mesomechanics, and macromechanics). For materials used in high-temperature applications, work is needed on the modeling of damage accumulation and propagation to fracture; modeling of thermoviscoplastic response, thermal-mechanical cycling and ratcheting; and prediction of long-term material behavior from short-term data, which are particularly important.

2. *Failure and Life Management Computational Tools and Methodologies*

Practical numerical techniques are needed for predicting the life as well as the failure initiation and propagation in structural components made of new, high-performance materials in terms of measurable and controllable parameters. Examples of these materials are electronic, optical and smart materials for space applications; high-temperature materials for hypersonic vehicles; and piezoelectric composites. For some of the materials, accurate constitutive descriptions, failure criteria, damage theories, and fatigue data are needed, along with more realistic characterization of interface phenomena (such as contact and friction). The constitutive descriptions may require investigations at the microstructure level or even the atomic level, as well as carefully designed and conducted experiments. Failure and life prediction of structures made of these materials is difficult and numerical models often constructed under restricting assumptions may not capture the dominant and underlying physical failure mechanisms. Moreover, material failure and structural response (such as instability) often couple in the failure mechanism. The coupling of nondestructive evaluation (NDE) techniques with numerical simulations can significantly enhance the capability of NDE for detecting flaws during manufacturing, design certification, maintenance, inspection and repair.[237]

3. Hierarchical, Integrated Multiple Methods and Adaptive Modeling Techniques

The effective use of numerical simulations for predicting the response, life, performance, and failure of future engineering structures requires strategies for treating phenomena occurring at disparate spatial and time scales, using reasonable computer resources. The strategies are based on using multiple mathematical models in different regions of the structure to take advantage of efficiencies gained by matching the model to the expected response in each region. To achieve the full potential of hierarchical modeling, there should be minimal reliance on *a priori* assumptions about the response. This is accomplished by adding adaptivity to the strategy. The key tasks of the research in this area are simple design-oriented models for use in the early stages of the design process; rational selection of a set of nested mathematical models for different regions, and discretization techniques for use in conjunction with the mathematical models (this, in turn, requires the availability of a capability for holistic modeling from micro to structural response with varying degrees of accuracy); simulation of local phenomena through global/local methodologies; automated (or semi-automated) coupling of different mathematical/discrete models; error estimation and adaptive modeling strategies; efficient methods for coupling different components of structures (such as, for flight vehicles, coupling the engine, airframe and rotor/engine-frame); and sensitivity analysis to assess the sensitivity of the response to each of the parameters neglected in the current mathematical model.

A variational multiscale method was presented for solution of multiscale problems in fluid mechanics, acoustics and electromagnetics.[238,239] The method is based on decomposing the solution into two parts: a course scale part represented by finite elements, and a fine scale part represented by Green's function. The two parts may overlap or be disjointed, and the fine scale part may be global or local. Application of this method to structural mechanics problems deserves attention.

4. Nondeterministic Analysis, Soft Computing and Risk Assessment

The novel modes of computation that exploit tolerance for imprecision and uncertainty in geometry, material properties, boundary conditions, loading, and operational environment to achieve tractability, robustness, and low cost need to be extended to structural design. These facilities include neuro-computing, fuzzy logic and genetic algorithms, and have been collectively referred to as "soft computing."[240] The distinction between expert systems (the most mature and resilient product of AI), soft computing paradigms, and conventional computational

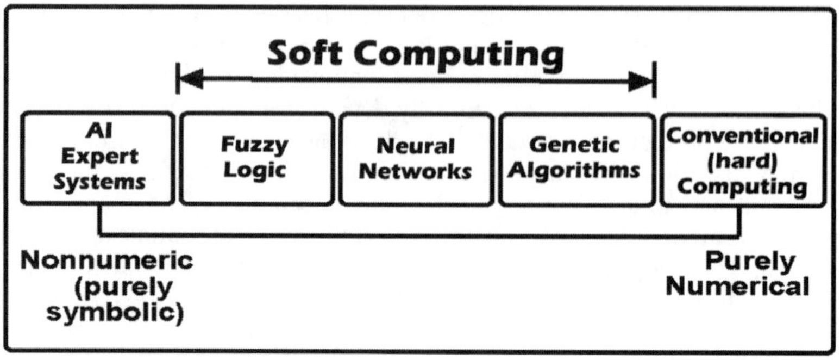

Fig. 7.21 Relationship between AI expert systems, soft computing paradigms and conventional (hard) computing.

methods can be illustrated by the extent to which they use nonnumeric (symbolic) and numerical computations (see Fig. 7.21).

The ability to quantify model biases and the inherent uncertainties in the response of the structure is obviously of great value. However, the principal benefit of using any of the soft computing tools consists of the insights into engineering, safety, and economics that are gained in the process of arriving at those quantitative results and carrying out reliability analyses. A review of recent advances in structural reliability is presented in Ref. 241. As future engineering structures become more complicated, modeling of failure mechanisms will account for uncertainties from the beginning of the design process, and potential design improvements will be evaluated to assess their effects on reducing overall risk. The results, combined with economic considerations, will be used in systematic cost-benefit analyses (perhaps also done on a nondeterministic basis) to determine the structural design with the most acceptable balance of cost and risk.

5. *Validation of Numerical Simulations*

In addition to selecting a benchmark set of structures for assessing new computational strategies and numerical algorithms, a high degree of interaction and communication is needed between computational modelers and experimentalists. This is done on four different levels, namely: laboratory tests on small specimens to obtain material data; component tests to validate computational models; full-scale tests to validate the modeling of details; and, for flight vehicles, flight tests to validate the entire modeling process. At each level, analysis of the computational model can be used to help set up the test, and the test data to help construct a more accurate computational model. For the first two levels, the use of flexible multibody system codes allows the testing machine to be included in the analysis.

6. *Multi-disciplinary Analysis and Design Optimization*

The realization of new complex engineering systems requires integration between the structures discipline and other traditionally separate disciplines such as aerodynamics, propulsion, and control. This is mandated by significant interdisciplinary interactions and couplings that need to be accounted for in predicting response, as well as in optimal design of these vehicles. Examples are the couplings between the aerodynamic flow field, structural heat transfer, and structural response of high-speed aircraft and propulsion systems; and the couplings between the control system and structural response in control configured aircraft and spacecraft. This activity also includes design optimization with multiobjective functions (such as performance, durability, integrity, reliability, and cost), and multiscale structural tailoring (micro, local and global levels). For propulsion systems of flight vehicles it also includes design with damping for high-cycle fatigue, low-cycle fatigue, and acoustic fatigue.

Typically, in the design process questions arise regarding the influence of design variable changes on system behavior. Answers to these questions, quantified by the derivatives of behavior with respect to the design variables or by parametric studies, guide design improvements toward a better overall system. In large applications this improvement process is executed by numerical optimization, combined with symbolic/AI techniques and human judgment aided by data visualization. Efficiency of the computations that provide data for such a process is decisive for the depth, breadth, and rate of progress achievable, and hence, ultimately, is critical for the final product quality.

A recent trend in multidisciplinary optimization is the use of response surface approximations in which the objective and/or constraint functions are replaced with simple functions, often polynomials, which are fitted to data at a set of carefully selected design points.[242] Response surface techniques can facilitate communication between specialists from different disciplines on the design team. The coupling of genetic algorithms with response surface techniques can produce a minimum-bias based design.

E. Related Tasks

For CST to impact the product development process, the research community needs to adapt of CST software to the emerging parallel, distributed, collaborative, immersive computing paradigm. This includes exploiting the existing infrastructure for distributed visual problem solving environments (consisting of an integrated set of high level facilities and tools providing intelligence and expert assistance for solving problems from a prescribed domain); and development of real-time collaborative model, grid generation and computational steering facilities; interfaces between CST software and advanced visualization facilities (such as multimedia and virtual reality); and infrastructure for component-based software development for facilitating the integration and reuse of diverse codes. Also, the research community must develop intelligent software agents (simulation advisors) to enable the designers to use the CST software (embedded in the virtual product development systems) early in the design process.

CST's wide acceptance can affect the design and operation of future engineering systems and structures in three ways. It can provide a better understanding of the phenomena associated with response, failure and life, thereby identifying desirable structural design attributes. CST can verify and certify designs and also allow low-cost modifications to be made during the design process. Finally, it can improve the design team's productivity and allow major improvements and innovations during the design phase, enabling fully integrated simulation-based design environment. Such an environment allows computer simulation of the entire life cycle of the structure, including material selection and processing, multidisciplinary design, automated manufacturing and fabrication, quality assurance, certification, operation, health monitoring and control, retirement, and disposal.

VII. Concluding Remarks

The history, recent developments, and trends in computational structures technology have been summarized. Discussion has focused on the development of CST software and goals of CST activities. Some recent advances in a number of CST areas have been described, including discretization techniques and element technology; computational material modeling; modeling of composite, sandwich and smart structures; computational tools and methodologies for life management; transient response analysis; nonlinear analysis; numerical simulation of frictional contact/impact response; articulated structural dynamics; nondeterministic modeling and analysis methods; qualitative analysis and simulation; neuro-computing; hybrid techniques; error estimation and adaptive improvement strategies; strategies for solution of coupled problems; sensitivity analysis; integrated analysis and design; strategies and numerical algorithms for new computing systems; model generation facilities; and application of object-oriented technology. Research areas in CST that have high potential for meeting the future technological needs have been identified.

The outlook for CST is bright. Innovative structural concepts, advanced computational material models, intelligent computational tools for analysis and synthesis, along with the rapid convergence of computing, communications, and information technologies should significantly enhance CST development and improve the performance of engineering systems.

Acknowledgment

The present work is partially supported by NASA Cooperative Agreement NCC-1-263. The author acknowledges useful discussions and suggestions by many of his colleagues including S. Arnold, I. Babuska, J. Bathe, T. Belytschko, L. Berke, C. Chamis, R. Dreisbach, C. Felippa, R. Haftka, G. Halford, C. Harris, D. Herting, D. Hopkins, J. Housner, R. Johns, W. Johnson, J. Kudva, A. Kurdila, J. Mackerle, R. McKnight, M. Mohagheh, J. Peery, L. Pinson, J. Sobieski, A. Weiting, and E. Wilson.

References

[1]Clough, R. W., "Original Formulation of the Finite Element Method," *Finite Elements in Analysis and Design*, Vol. 7, 1991, pp. 89–101.

[2]Wilson, E. L., "Automation of the Finite Element Method—A Personal Historical Review," *Finite Elements in Analysis and Design*, Vol. 13, Nos. 2 and 3, June 1993, pp. 91–104.

[3]*Focus '93—Computational Structures Technology, Aerospace America*, Feb. 1993.

[4]Noor, A. K., Housner, J. M., Starnes, Jr., J. H., Hopkins, D. A., and Chamis, C. C. (compilers), "Computational Structures Technology for Airframes and Propulsion Systems," NASA CP-3142, May 1992.

[5]Goudreau, G. L., "Computational Structural Mechanics: From National Defense to National Resource," *IEEE Computational Science and Engineering*, Vol. 1, No. 1, Spring 1994, pp. 33–42.

[6]Argyris, J., St. Doltsinis, I., Frik, G., and Tenek, L., "Computational Structures Technology in Europe—Past, Present and Future," *IEEE Computational Science and Engineering*, Vol. 1, No. 1, Spring 1994, pp. 43–54.

[7]Noor, A. K., and Atluri, S. N., "Advances and Trends in Computational Structural Mechanics," *AIAA Journal*, Vol. 25, No. 7, July 1987, pp. 977–995.

[8]Noor, A. K., and Venneri, S. L., "A Perspective on Computational Structures Technology," *IEEE Computer Journal*, Vol. 26, No. 10, Oct. 1993, pp. 38–46.

[9]Noor, A. K., and Venneri, S. L. (editors), "Computational Structures Technology," *Flight Vehicle Materials, Structures and Dynamics*, American Society of Mechanical Engineers, New York, Vol. 6, 1995.

[10]Ju, J. W. (editor), *Numerical Methods in Structural Mechanics*, ASME Press, New York, 1995.

[11]Ladvese, P., and Zienkiewicz, O. C. (editors), *New Advances in Computational Structural Mechanics*, Proceedings of the European Conference on New Advances in Computational Structural Mechanics, Giens, France, April 2–5, 1991, Elsevier, Amsterdam, 1992.

[12]Stein, E. (editor), *Progress in Computational Analysis in Inelastic Structures*, Springer Verlag, Vienna, 1993.

[13]Topping, B. H. V. (editor), *Advances in Computational Structural Mechanics*, Civil-Comp Press, Edinburgh, U.K., 1998.

[14]Storaasli, O. O. (editor), Large Scale Analysis and Design on High Performance Computers and Workstations, (special issue) *Advances in Engineering Software*, Vol. 29, No. 3–6, 2000.

[15]Mang, H. A., and Rammerstorfer, F. G. (editors), *IUTAM Symposium on Discretization Methods in Structural Mechanics*, Proceedings of the Symposium, Vienna, Austria, June 1997, Kluwer, The Netherlands, 1999.

[16]Topping, B. H. V. (editor), *Developments in Computational Mechanics with High Performance Computing*, Proceedings of the Seventh International Conference, Oxford, UK, Sept. 1999, Civil-Comp, Edinburgh, U.K., 1999.

[17]Mackerle, J., and Fredriksson, B., *A Handbook of Finite Element Software. Supercomputers, Main-Frames, Minicomputers and Microcomputers*, Studentlitteratur, Lund, Sweden, 1988.

[18]Mackerle, J., *Finite Element Methods, A Guide to Information Sources*, Elsevier, Amsterdam, 1991.

[19]*Proceedings of the Seventh AIAA/USAF/NASA/ISSMO Symposium on Multidisciplinary Analysis and Optimization*, Sept. 2–4, 1998, St. Louis, MO, A Collection of Technical Papers, 3 volumes, AIAA, Reston, VA, 1998.

[20]Papadrakakis, M., Tsompanakis, Y., Lagaros, N. D., Hinton, E., Sienz, J., Thierauf, G., and Cai, J., "Innovative Computational Methods for Structural Optimization," *Innovative Computational Methods for Structural Mechanics*, Papadrakakis, M. and Topping, B. H. V. (editors), Saxe-Coburg Publications, U.K., 1999.

[21]Nishioka, T., "Computational Dynamic Fracture Mechanics," *International Journal of Fracture*, Vol. 86, No. 1/2, 1997, pp. 127–159.

[22]Aliabadi, M. H. (editor), "Nonlinear Fracture and Damage Mechanics," *Advances in Fracture Mechanics*, WIT Press, Southampton, U.K., Vol. 4, June 1999.

[23]Skrzypek, J. J., and Ganczarski, A., "Modeling of Material Damage and Failure of Structures: Theory and Applications," *Foundations of Engineering Mechanics Series*, Springer Verlag, Berlin, 1999.

[24]Krajcinovic, D., "Damage Mechanics: Accomplishments, Trends and Needs," *International Journal of Solids and Structures*, Vol. 37, 2000, pp. 267–277.

[25]Voyiadjis, G. Z., and Ju, J.-W. W. (editors), Advances in Computational Methods for Damage Mechanics (special issue), *Computer Methods in Applied Mechanics and Engineering*, Vol. 183, Nos. 1–2, 2000.

[26]Voyiadjis, G. Z., and Ju, J.-W. W. (editors), Advances in Computational Methods for Damage Mechanics (special issue), *Computer Methods in Applied Mechanics and Engineering*, Vol. 183, Nos. 3–4, 2000.

[27]Trujillo, D. M., and Busby, H. R., *Practical Inverse Analysis in Engineering*, CRC Press, Boca Raton, Florida, 1997.

[28]Tanaka, M., and Dulikravich, G. S. (editors), *Inverse Problems in Engineering Mechanics*, Elsevier Science, New York, 1998.

[29]Tanaka, M., Sladek, V., and Sladek, J., "Regularization Techniques Applied to Boundary Element Methods," *Applied Mechanics Reviews*, Vol. 47, No. 10, Oct. 1994, pp. 457–499.

[30]Beskos, D. E., "Boundary Element Methods in Dynamic Analysis: Part II (1986–1996)," *Applied Mechanics Reviews*, Vol. 50, 1997, pp. 149–197.

[31]Aliabadi, M. H., "Boundary Element Formulations in Fracture Mechanics," *Applied Mechanics Reviews*, Vol. 50, No. 2, Feb. 1997, pp. 83–96.

[32]Banerjee, P. K., *Boundary Element Methods in Engineering*, McGraw-Hill, New York, 1994.

[33]Cruse, T. A. (editor), Advances in Boundary Element Methods (special issue), *Computer Methods in Applied Mechanics and Engineering*, Vol. 173, Nos. 3–4, 1999.

[34]Rong, T.-Y., "Generalized Mixed Variational Principles and New FEM Models in Solid Mechanics," *International Journal of Solids and Structures*, Vol. 24, No. 11, 1988, pp. 1131–1140.

[35]Rong, T.-Y., and Lu, A.-Q., "Parametrized Lagrange Multiplier Method and Construction of Generalized Mixed Variational Principles for Computational Mechanics," *Computer Methods in Applied Mechanics and Engineering*, Vol. 164, Nos. 3–4, 1998, pp. 287–296.

[36]Felippa, C. A., "A Survey of Parametrized Variational Principles and Applications to Computational Mechanics," *Computer Methods in Applied Mechanics and Engineering*, Vol. 113, 1994, pp. 109–139.

[37]Felippa, C. A., Haugen, B., and Militello, C., "From the Individual Element Test to Finite Element Templates: Evolution of the Patch Test," *International Journal for Numerical Methods in Engineering*, Vol. 38, 1995, pp. 199–239.

[38]Selected Papers From the Second International Conference on Spectral and High-Order Methods (special issue), *Computer Methods in Applied Mechanics and Engineering*, Vol. 116, Nos. 1–4, June 1994.

[39]Sherwin, S. J., and Karniadakis, G. E., "A Triangular Spectral Element Method; Applications to the Incompressible Navier-Stokes Equations," *Computer Methods in Applied Mechanics and Engineering*, Vol. 123, 1995, pp. 189–229.

[40]Sherwin, S. J., and Karniadakis, G. E., "Tetrahedral hp Finite Elements: Algorithms and Flow Simulations," *Journal of Computational Physics*, Vol. 124, 1996, pp. 14–45.

[41]Strang, G., "Wavelets and Dilation Equations: A Brief Introduction," *SIAM Review*, Vol. 31, No. 4, Dec. 1989, pp. 614–627.

[42]Williams, J. R., and Amaratunga, K., "Introduction to Wavelets in Engineering," *International Journal for Numerical Methods in Engineering*, Vol. 37, No. 14, July 1994, pp. 2365–2388.

[43]Ko, J., Kurdila, A. J., and Pilant, M., "A Class of Wavelet-Based Finite Element Methods for Computational Mechanics," *Proceedings of the 35th AIAA/ASME/ASCE/AHS/ASC Structures, Structural Dynamics and Materials Conference*, Hilton Head, South Carolina, April 18–20, 1994, Part 2, AIAA Paper 94-1388-CP.

[44]Kurdila, A. J. (editor), *Wavelet and Multiscale Methods for Partial Differential Equations*, Academic Press, San Diego, California, 1997.

[45]Ko, J., Kurdila, A. J., and Pilant, M. S., "Triangular Wavelet Based Finite Elements via Multivalued Scaling Equations," *Computer Methods in Applied Mechanics and Engineering*, Vol. 146, Nos. 1–2, July 1997, pp. 1–17.

[46]Vasilyev, O. V., and Paolucci, S., "A Dynamically Adaptive Multilevel Wavelet Collocation Method for Solving Partial Differential Equations in a Finite Domain," *Journal of Computational Physics*, Vol. 125, 1996, pp. 498–512.

[47]Teixeira de Freitas, J. A., Moitinho de Almeida, J. P., and Ribeiro Pereira, E. M. B., "Non-conventional Formulations for the Finite Element Method," *Structural Engineering and Mechanics*, Vol. 4, No. 6, 1996, pp. 655–678.

[48]Belytschko, T., Krongauz, Y., Organ, D., Fleming, M., and Krysl, P., "Meshless Methods: An Overview and Recent Developments," *Computer Methods in Applied Mechanics and Engineering*, Vol. 139, 1996, pp. 1–47.

[49]Liu, W.-K., Jun, S., Adee, J., and Belytschko, T., "Reproducing Kernel Particle Methods for Structural Dynamics," *International Journal for Numerical Methods in Engineering*, Vol. 38, No. 10, May 1995, pp. 1655–1679.

[50]Gingold, R. A., and Monaghan, J. J., "Smoothed Particle Hydrodynamics: Theory and Application to Non-Spherical Stars," *Monthly Notices of the Royal Astronomical Society*, Vol. 181, 1977, pp. 375–389.

[51]Atluri, S. N., Kim, H.-G., and Cho, J. Y., "A Critical Assessment of the Truly Meshless Local Petrov-Galerkin (MLPG) and Local Boundary Integral Equation (LBIE) Methods," *Computational Mechanics*, Vol. 24, 1998, pp. 348–372.

[52]Zienkiewicz, O. C., and Onate, E., "Finite Volumes vs Finite Elements. Is There Really a Choice?," *Nonlinear Computational Mechanics, State-of-the-Art*, P. Wriggers and W. Wagner (editors), Springer-Verlag, Berlin, 1991, pp. 240–254.

[53]Onate, E., Cervera, M., and Zienkiewicz, O. C., "A Finite Volume Format for Structural Mechanics," *International Journal for Numerical Methods in Engineering*, Vol. 37, No. 2, 1994, pp. 181–201.

[54]Bailey, C., and Cross, M., "A Finite Volume Procedure to Solve Elastic Solid Mechanics Problems in Three Dimensions on an Unstructured Mesh," *International Journal for Numerical Methods in Engineering*, Vol. 38, 1995, pp. 1757–1776.

[55]Demirdzic, I., and Muzaferija, S., "Finite Volume Methods for Stress Analysis in Complex Domains," *International Journal for Numerical Methods in Engineering*, Vol. 37, 1994, pp. 3751–3766.

[56]Demirdzic, I., and Martovic, D., "Finite Volume Method for Thermoelastic Plastic Stress Analysis," *Computer Methods in Applied Mechanics and Engineering*, Vol. 109, 1993, pp. 331–349.

[57]Dioh, N. N., Ivankovic, A., Leevers, P. S., and Williams, J. G., "Stress Wave Propagation Effects in Split Hopkinson Pressure Bar Tests," *Proceedings of the Royal Society of London*, Series A, Vol. 449, 1995, pp. 187–204.

[58]Ivankovic, A., and Venizelos, G. P., "Rapid Crack Propagation: Predicting Full-Scale Critical Pressure From S4 Test Results," *Engineering Fracture Mechanics*, Vol. 59, 1998, pp. 607–622.

[59]Ivankovic, A., "Finite Volume Modeling of Dynamic Fracture Problems," *Computer Modeling and Simulation in Engineering*, Vol. 4, No. 3, Aug. 1999, pp. 227–235.

[60]Demirdzic, I., and Muzaferija, S., "Numerical Method for Coupled Fluid Flow, Heat Transfer and Stress Analysis Using Unstructured Moving Meshes with Cells of Arbitrary

Topology," *Computer Methods in Applied Mechanics and Engineering*, Vol. 125, 1995, pp. 235–255.

[61]Song, Ch., and Wolf, J. P., "The Scaled Boundary Finite-Element Method—Alias Consistent Infinitesimal Finite Element Cell Method—for Elastodynamics," *Computer Methods in Applied Mechanics and Engineering*, Vol. 147, 1997, pp. 329–355.

[62]Demirdzic, I., Muzaferija, S., and Peric, M., "Benchmark Solutions of Some Structural Analysis Problems Using Finite Volume Method and Multigrid Acceleration," *International Journal for Numerical Methods in Engineering*, Vol. 40, 1997, pp. 1893–1908.

[63]Duarte, C. A. M., and Oden, J. T., "Hp Clouds—An *hp* Meshless Method," *Numerical Methods for Partial Differential Equations*, Vol. 12, 1996, pp. 673–705.

[64]Babuska, I., and Melenk, J. M., "The Partition of Unity Finite Element Method," *International Journal for Numerical Methods in Engineering*, Vol. 40, 1997, pp. 727–758.

[65]Strouboulis, T., Copps, K., and Babuska, I., "The Generalized Finite Element Method: An Example of Its Implementation and Illustration of Its Performance," *International Journal for Numerical Methods in Engineering*, Vol. 47, 2000, pp. 1401–1417.

[66]Chapelle, D., and Bathe, K. J., "Fundamental Considerations for the Finite Element Analysis of Shell Structures," *Computers and Structures*, Vol. 66, No. 1, 1998, pp. 19–36.

[67]MacNeal, R. H., "Perspective on Finite Elements for Shell Analysis," *Finite Elements in Analysis and Design*, Vol. 30, 1998, pp. 175–186.

[68]Noor, A. K., Belytschko, T., and Simo, J. C. (editors), *Analytical and Computational Models for Shells*, Proceedings of the symposium, ASME Winter Annual Meeting, Dec. 10–15, 1989, San Francisco, CA, CED Vol. 3, ASME, New York, 1989.

[69]Bathe, K. J., *Finite Element Procedures*, Prentice Hall, Englewood Cliffs, New Jersey, 1996.

[70]Bathe, K. J., Iosilevich, A., and Chapelle, D., "An Evaluation of the MITC Shell Elements," *Computers and Structures*, Vol. 75, No. 1, 2000, pp. 1–30.

[71]Bathe, K. J., Iosilevich, A., and Chapelle, D., "An inf-sup Test for Shell Finite Elements," *Computers and Structures*, Vol. 75, No. 5, 2000, pp. 439–456.

[72]Tvergaard, V., and Needleman, A., "Effects of Non-local Damage in Porous Plastic Solids," *International Journal of Solids and Structures*, Vol. 32, 1995, pp. 1063–1077.

[73]Owen, D. R. J., Onate, E., and Hinton, E., *Computational Plasticity: Fundamentals and Applications*, Proceedings of the Fifth International Conference, International Center for Numerical Methods in Engineering (CIMNE), Barcelona, 1997.

[74]Simo, J. C., and Hughes, T. J. R., *Computational Inelasticity*, Springer Verlag, New York, 1998.

[75]Ramaswamy, S., and Aravas, N., "Finite Element Implementation of Gradient Plasticity Models—Part I: Gradient-Dependent Yield Functions," *Computer Methods in Applied Mechanics and Engineering*, Vol. 163, Nos. 1–4, 1998, pp. 11–32.

[76]Reddy, B. D. (editor), Advances in Computational Plasticity (special issue), *Computer Methods in Applied Mechanics and Engineering*, Vol. 171, Nos. 3–4, 1999.

[77]Needleman, A., "Computational Mechanics at the Mesoscale," *Acta Materialia*, Vol. 48, 2000, pp. 105–124.

[78]Raabe, D., *Computational Materials Science*, Wiley-VCH, Weinheim, Germany, 1998.

[79]Eberhardt, J. J., Hay, P. J., and Carpenter, J. A., "Materials by Design—A Hierarchical Approach to the Design of New Materials," *Materials Research Society Symposium Proceedings*, Vol. 63, 1985, pp. 191–206.

[80]U.S. Congressional Board, *Advanced Materials by Design*, Office of Technology Assessment, Washington, D.C., 1988.

[81]Noor, A. K., and Burton, W. S., "Computational Models for High-Temperature Multilayered Composite Plates and Shells," *Applied Mechanics Reviews*, Vol. 45, No. 10, 1992, pp. 419–446.

[82]Noor, A. K., Burton, W. S., and Bert, C. W., "Computational Models for Sandwich Panels and Shells," *Applied Mechanics Reviews*, Vol. 49, No. 3, March 1996, pp. 155–199.

[83]Rolfes, R., Noor, A. K., and Sparr, H., "Evaluation of Transverse Thermal Stresses in Composite Plates Based on First-Order Shear Deformation Theory," *Computer Methods in Applied Mechanics and Engineering*, Vol. 167, Nos. 3–4, Dec. 1998, pp. 355–368.

[84]Rolfes, R., Rohwer, K., and Ballerstaedt, M., "Efficient Linear Transverse Normal Stress Analysis of Layered Composite Plates," *Computers and Structures*, Vol. 68, 1998, pp. 643–652.

[85]Noor, A. K., and Malik, M., "Accurate Determination of Transverse Normal Stresses in Sandwich Panels Subjected to Thermomechanical Loadings," *Computer Methods in Applied Mechanics and Engineering*, Vol. 178, Nos. 3–4, 1999, pp. 431–443.

[86]Rolfes, R., Noack, J., and Taeschner, M., "High Performance Three Dimensional Analysis of Thermo-mechanically Loaded Composite Structures," *Composite Structures*, Vol. 46, No. 4, 1999, pp. 367–379.

[87]Noor, A. K., "Finite Element Buckling and Postbuckling Analyses," *Buckling and Postbuckling of Composite Plates*, G. J. Turvey and I. H. Marshall (editors), Chapman & Hall, London, 1995, pp. 58–107.

[88]Noor, A. K. (editor), *Buckling and Postbuckling of Composite Structures*, Proceedings of the Symposium on Buckling and Postbuckling of Composite Structures, ASME International Mechanical Engineering Congress and Exposition, Chicago, IL, Nov. 6–11, 1994, AD-Vol. 41/PVP-Vol. 293, American Society of Mechanical Engineers, New York, 1994.

[89]Noor, A. K., and Burton, W. S., "Three Dimensional Solutions for the Thermal Buckling and Sensitivity Derivatives of Temperature-Sensitive Multilayered Angle-Ply Plates," *Journal of Applied Mechanics*, Vol. 59, No. 4, Dec. 1992, pp. 848–856.

[90]Noor, A. K., Starnes, Jr., J. H., and Peters, J. M., "Nonlinear and Postbuckling Responses of Curved Composite Panels with Cutouts," *Composite Structures*, Vol. 34, No. 2, 1996, pp. 213–240.

[91]Noor, A. K., and Kim, Y. H., "Buckling and Postbuckling of Composite Panels with Cutouts Subjected to Combined Edge Shear and Temperature Change," *Computers and Structures*, Vol. 60, No. 2, May 1996, pp. 203–222.

[92]Noor, A. K., Starnes, Jr., J. H., and Peters, J. M., "Thermomechanical Buckling and Postbuckling Responses of Composite Panels with Skewed Stiffeners," *Finite Elements in Analysis and Design*, Vol. 27, No. 2, Oct. 1997, pp. 193–214.

[93]Noor, A. K., Starnes, Jr., J. H., and Peters, J. M., "Curved Sandwich Panels Subjected to Temperature Gradient and Mechanical Loads," *Journal of Aerospace Engineering*, ASCE, Vol. 10, No. 4, Oct. 1997, pp. 143–161.

[94]Noor, A. K., and Peters, J. M., "Analysis of Curved Sandwich Panels with Cutouts Subjected to Combined Temperature Gradient and Mechanical Loads," *Journal of Sandwich Structures and Materials*, Vol. 1, Jan. 1999, pp. 42–59.

[95]Noor, A. K., and Tenek, L. H., "Steady-State Nonlinear Heat Transfer in Multilayered Composite Panels," *Journal of Engineering Mechanics*, ASCE, Vol. 118, No. 8, Aug. 1992, pp. 1661–1678.

[96]Rolfes, R., "Higher-Order Theory and Finite Element for Heat Conduction in Composites," in *Proceedings of the Seventh International Conference on Numerical Methods in Thermal Problems*, R. W. Lewis, J. H. Chin and G. M. Homsy (editors), Paris, Vol. VII, 1991, pp. 880–889.

[97]Chamis, C. C., and Hopkins, D. A., "Probabilistic Composite Design," *Composite Materials: Testing and Design*, 13th vol., ASTM STP-1242, S. J. Hooper (editor), American Society for Testing and Materials, 1997, pp. 23–42.

[98]Chamis, C. C., Murthy, P. L. N., and Minnetyan, L., "Progressive Fracture in Composite Structures," *Composite Materials: Fatigue and Fracture*, 6th vol., ASTM STP-1285, E. A. Armanios (editor), American Society for Testing and Materials, 1997, pp. 70–84.

[99]Chamis, C. C., Abumeri, G. H., and Patnaik, S. N., "Multi-scale Multi-level Multidisciplinary Analysis and Design for Engine Structures," *Proceedings of the Seventh AIAA/USAF/NASA/ISSMO Symposium on Multidisciplinary Analysis and Optimization*, St. Louis, MO, Sept. 2–4, 1998, Part 2, AIAA, 1998, pp. 1187–1197.

[100]Arnold, S. M., Bednarcyk, B. A., Trowbridge, D., and Wilt, T. E., "Micromechanics Analysis Code with Generalized Method of Cells (MAC/GMC) User Guide," Ver. 3.0, NASA TM-1999-209070, 1999.

[101]Tani, J., Takagi, T., and Qiu, J., "Intelligent Material Systems: Application of Functional Materials," *Applied Mechanics Reviews*, Vol. 51, No. 8, Aug. 1998, pp. 505–521.

[102]Brebbia, C. A., Marchetti, M., and Santini, P. (editors), *Smart Structures 98. Computational Methods for Smart Structures and Materials*, WIT Press, Southampton, U.K., Aug. 1998.

[103]Tang, Y. Y., Noor, A. K., and Xu, K.-M., "Assessment of Computational Models for Thermo-electro-elastic Multilayered Plates," *Computers and Structures*, Vol. 61, No. 5, Dec. 1996, pp. 915–933.

[104]Xu, K.-M., Noor, A. K., and Tang, Y. Y., "Three Dimensional Solutions for Free Vibrations of Initially-Stressed Thermo-electro-elastic Multilayered Plates," *Computer Methods in Applied Mechanics and Engineering*, Vol. 141, Nos. 1–2, Feb. 1997, pp. 125–139.

[105]Xu, K.-M., and Noor, A. K., "Three-Dimensional Analytical Solutions for Coupled Thermo-electro-elastic Response of Multilayered Cylindrical Shells," *AIAA Journal*, Vol. 34, No. 4, April 1996, pp. 802–812.

[106]Xu, K.-M., and Noor, A. K., "Predictor-Corrector Finite Element Approach for Electro-elastic Analysis of Hybrid Composite Plates," *Computer Methods in Applied Mechanics and Engineering*, Vol. 147, Nos. 1–2, July 1997, pp. 139–145.

[107]Mackerle, J., "Smart Materials and Structures—A Finite Element Approach: A Bibliography (1986–1997)," *Modeling Simul. Mater. Sci. Eng.*, Vol. 6, 1998, pp. 293–334.

[108]Mitchell, M. R., and Landgraf, R. W. (editors), *Advances in Fatigue Lifetime Predictive Techniques*, STP-1122, ASTM, 1992.

[109]Noor, A. K., Harris, Jr., C. E., Housner, J. M., and Hopkins, D. A. (compilers), *Computational Methods for Failure Analysis and Life Prediction*, Proceedings of the workshop, NASA Langley Research Center, Hampton, VA, Oct. 14–15, 1992, NASA CP-3230, 1993.

[110]Halford, G. R., "Cumulative Fatigue Damage Modeling—Crack Nucleation and Early Growth," *International Journal of Fatigue*, Vol. 19, Suppl. No. 1, 1997, pp. 5253–5260.

[111]Sun, J. Q., and Miles, R. N., "Acoustic Fatigue Life Prediction for Nonlinear Structure," *Journal of Sound and Vibration*, Vol. 150, No. 3, 1991, pp. 531–535.

[112]Miles, R. N., "Effect of Spectral Shape on Acoustic Fatigue Life Estimates," *Journal of Sound and Vibration*, Vol. 153, March 1992, pp. 376–386.

[113]Reifsnider, K. L., "Performance Simulation: A New Approach to the Prediction of Remaining Strength and Life of Material Systems," *Achievement in Composites in Japan and the United States*, A. Kobayashi (editor), Proceedings of the Japan-U.S. CCM-V, Tokyo, 1990, pp. 3–9.

[114]Reifsnider, K. L. (editor), *Fatigue of Composite Materials*, Elsevier Science Publishers, New York, 1990.

[115]Johnson, W. S., "Damage Development in Titanium Metal Matrix Composites Subjected to Cyclic Loading," NASA TM-107597, April 1992.

[116]Zaretsky, E. V. (editor), "MMC Life System Development (Phase I)—A NASA/Pratt & Whitney Life Prediction Cooperative Program," NASA Ref. Publication 1361, Nov. 1996.

[117]Nemeth, N. N., Janosik, L. A., and Palko, J. L., "CARES/Life Software for Characterization and Prediction of the Lifetime of Ceramic Parts," *Proceedings of the NAFEMS'99 World Congress*, April 25–28, 1999, Newport, Rhode Island, pp. 1125–1144.

[118]Shah, A. R., Mital, S. K., and Murthy, P. L. N., "PCEMCAN—Probabilistic Ceramic Matrix Composite Analyzer User's Guide," Ver. 1.0, NASA TM-206984, July 1998.

[119]Belytschko, T., Englemann, B. E., and Liu, W. K., "A Review of Recent Developments in Time Integration," *State-of-the-Art Surveys on Computational Mechanics*, A. K. Noor and J. T. Oden (editors), American Society of Mechanical Engineers, New York, 1989, pp. 185–199.

[120]Smolinski, P., "A Variable Multi-step Method for Transient Heat Conduction," *Computer Methods in Applied Mechanics and Engineering*, Vol. 86, 1991, pp. 61–71.

[121]Smolinski, P., "Stability of Variable Explicit Time Integration for Unsteady Diffusion Problems," *Computer Methods in Applied Mechanics and Engineering*, Vol. 93, 1991, pp. 247–252.

[122]Hulbert, G. M., and Hughes, T. J. R., "Space-Time Finite Element Methods for Second-Order Hyperbolic Equations," *Computer Methods in Applied Mechanics and Engineering*, Vol. 84, 1990, pp. 327–348.

[123]Hulbert, G. M., "Time Finite Element Methods for Structural Dynamics," *International Journal for Numerical Methods in Engineering*, Vol. 33, 1992, pp. 307–331.

[124]Hulbert, G. M., "Discontinuity-Capturing Operators for Elastodynamics," Computer *Methods in Applied Mechanics and Engineering*, Vol. 96, 1992, pp. 409–426.

[125]Karaoglan, L., and Noor, A. K., "Space-Time Finite Element Methods for the Sensitivity Analysis of Contact/Impact Response of Axisymmetric Composite Structures," *Computer Methods in Applied Mechanics and Engineering*, Vol. 144, Nos. 3–4, May 1997, pp. 371–389.

[126]Liu, W. K., Zhang, Y., and Ramirez, M. R., "Multiple Scale Finite Element Methods," *International Journal for Numerical Methods in Engineering*, Vol. 32, 1991, pp. 969–990.

[127]Zhong, Z.-H., and Mackerle, J., "Contact-Impact Problems: A Review with Bibliography," *Applied Mechanics Reviews*, Vol. 47, No. 2, Feb. 1994, pp. 55–76.

[128]Zhong, Z.-H., *Finite Element Procedures for Contact-Impact Problems*, Oxford University Press, Oxford, U.K., 1993.

[129]Noor, A. K., and Carden, H. D. (compilers), "Computational Methods for Crashworthiness," Proceedings of the workshop, Sept. 2–3, 1992, NASA Langley Research Center, Hampton, VA, NASA CP-3223, Aug. 1993.

[130]Wriggers, P., "Finite Element Algorithms for Contact Problems," *Archives of Computational Methods in Engineering*, Vol. 2, No. 4, 1995, pp. 1–49.

[131]Bathe, K. J., and Bouzinov, P. A., "On the Constraint Function Method for Contact Problems," *Computers and Structures*, Vol. 64, No. 5/6, 1997, pp. 1069–1085.

[132]Bathe, K. J., Guillermin, O., Walczak, J., and Chen, H.-Y., "Advances in Nonlinear Finite Element Analysis of Automobiles," *Computers and Structures*, Vol. 64, No. 5/6, 1997, pp. 881–891.

[133]Mackerle, J., "Contact Mechanics—Finite Element and Boundary Element Approaches—A Bibliography (1995–1997)," *Finite Elements in Analysis and Design*, Vol. 29, 1998, pp. 275–285.

[134]Martins, J. A. C., and Klarbring, A., "Computational Modeling of Contact and Friction" (special issue), *Computer Methods in Applied Mechanics and Engineering*, Vol. 177, Nos. 3–4, 1999.

[135]Brebbia, C. A., Jones, N., Manolis, G. D., and Talaslidis, D. G. (editors), *SUSI 98. Structures Under Shock and Impact V*, WIT Press, Southampton, U.K., June 1998.

[136]Mamalis, A. G., Manolakos, D. E., Demosthenous, G. A., and Ioannidis, M. B., *Crashworthiness of Composite Thin-Walled Structural Components*, Technomic Publishing Co., Lancaster, PA, 1998.

[137]Huston, R. L., "Multibody Dynamics—Modeling and Analysis Methods," *Applied Mechanics Reviews*, Vol. 44, No. 3, March 1991, pp. 109–117.

[138]Haug, E. J., "Dynamics of Articulated Structures," *Structural Dynamics and Aeroelasticity*, Chapter 5, Vol. 5, *Monograph on Flight-Vehicle Materials, Structures and Dynamics—Assessment and Future Directions*, A. K. Noor and S. L. Venneri (editors), American Society of Mechanical Engineers, New York, 1993, pp. 45–53.

[139]Huston, R. L., "Multibody Dynamics Since 1990," *Applied Mechanics Reviews*, Vol. 49, No. 10, Part 2, Oct. 1996, pp. S35–S40.

[140]Shabana, A. A., "Flexible Multi-body Dynamics: Review of Past and Recent Developments," *Multi-body System Dynamics*, Vol. 1, 1997, pp. 189–222.

[141]Elishakoff, I., "Essay on Uncertainties in Elastic and Viscoelastic Structures: From A. M. Freudenthal's Criticism to Modern Convex Modeling," *Computers and Structures*, Vol. 56, No. 6, 1995, pp. 871–895.

[142]Elishakoff, I., "Three Versions of the Finite Element Method Based on Concepts of Either Stochasticity, Fuzziness, or Anti-optimization," *Applied Mechanics Reviews*, Vol. 51, No. 3, March 1998, pp. 209–218.

[143]Elishakoff, I. (editor), "Whys and Hows in Uncertainty Modeling," International Centre for Mechanical Sciences, Courses and Lectures, No. 388, Springer-Verlag, Wien, 1999.

[144]Elishakoff, I., "Possible Limitations of Probabilistic Methods in Engineering," *Applied Mechanics Reviews*, Vol. 53, No. 2, Feb. 2000, pp. 19–36.

[145]"Probabilistic Structural Analysis Methods (PSAM) for Select Space Propulsion System Components," Southwest Research Institute, San Antonio, TX, NASA CR-1999-208872, January 1999.

[146]Riha, D., Thacker, B., Millwater, H., and Wu, Y.-T., "Probabilistic Engineering Analysis Using the NESSUS Software," 41st AIAA/ASME/ASCE/AHS/ASC Structures, Structural Dynamics and Materials Conference, April 3–6, 2000, Atlanta, GA, Paper 2000-1512.

[147]Araujo, J. M., and Awruch, A. M., "Stochastic Finite Elements for Structural Analysis," *Computers and Structures*, Vol. 52, No. 3, Aug. 1994, pp. 461–469.

[148]Kleiber, M., and Hien, T. D., *The Stochastic Finite Element Method*, John Wiley, New York, 1992.

[149]Schueller, G. I., "A State-of-the-Art Report on Computational Stochastic Mechanics," *Probabilistic Engineering Mechanics*, Vol. 12, No. 4, Oct. 1997, pp. 197–321.

[150]Kleiber, M. (editor), "Computational Stochastic Mechanics," (special issue), *Computer Methods in Applied Mechanics and Engineering*, Vol. 168, Nos. 1–4, 1999.

[151]Noor, A. K., and Malone, J. B. (compilers), "Computational Tools and Facilities for the Next-Generation Analysis and Design Environment," Proceedings of the workshop, Sept. 17–18, 1996, Hampton, VA, NASA CP-3346, March 1997.

[152]Wasfy, T. M., and Noor, A. K., "Application of Fuzzy Sets to Transient Analysis of Space Structures," *Finite Elements in Analysis and Design*, Vol. 29, 1998, pp. 153–171.

[153]Abdel-Tawab, K., and Noor, A. K., "A Fuzzy-Set Analysis for a Dynamic Thermo-elasto-visco-plastic Damage Response," *Computers and Structures*, Vol. 70, No. 1, Jan. 2000, pp. 91–107.

[154]Qiu, Z., and Elishakoff, I., "Anti-optimization of Structures with Large Uncertain but Non-random Parameters via Interval Analysis," *Computer Methods in Applied Mechanics and Engineering*, Vol. 152, Nos. 3–4, Jan. 1998, pp. 361–372.

[155]Kleiber, M., "Computer-Assisted Qualitative Mechanics: An Exemplary Simulation of a 'Snap-Through' Problem," *Computers and Structures*, Vol. 52, No. 6, 1994, pp. 1261–1268.

[156]Kulpa, Z., Radomski, A., Gajl, O., Kleiber, M., and Skalna, I., "Hybrid Expert System for Qualitative and Quantitative Analysis of Mechanical Structures," *Qualitative Reasoning: Eleventh International Workshop*, L. Ironi (editor), Cortona, Italy, June 3–6, 1997, pp. 287–293, Publ. No. 1036, Istituto di Analisi Numerica C.N.R., Pavia, Italy, 1997.

[157]Kulpa, Z., Pownuk, A., and Skalna, I., "Analysis of Linear Mechanical Structures with Uncertainties by Means of Interval Methods," *Computer Assisted Mechanics and Engineering Sciences*, Vol. 5, No. 4, 1998, pp. 1–38.

[158]Topping, B. H. V., and Bahreininejad, A., *Neural Computing for Structural Mechanics*, Saxe-Coburg Publications, Edinburgh, U. K., 1997.

[159]Noor, A. K., and Andersen, C. M., "Hybrid Analytical Technique for the Nonlinear Analysis of Curved Beams," *Computers and Structures*, Vol. 43, No. 5, 1992, pp. 823–830.

[160]Noor, A. K., "Recent Advances and Applications of Reduction Methods," *Applied Mechanics Reviews*, Vol. 47, No. 5, 1994, pp. 125–146.

[161]Kobayashi, A. S., "Hybrid Experimental-Numerical Stress Analysis," *Handbook of Experimental Mechanics*, second edition, Wiley-VCH, Weinheim, Germany, 1993, pp. 751–783.

[162]Szewczyk, Z. P., and Noor, A. K., "A Hybrid Neuro-computing/Numerical Strategy for Nonlinear Structural Analysis," *Computers and Structures*, Vol. 58, No. 4, Feb. 1996, pp. 661–678.

[163]Szewczyk, Z. P., and Noor, A. K., "A Hybrid Neuro-computing Strategy for Sensitivity Analysis of Nonlinear Structures," *Computers and Structures*, Vol. 65, No. 6, Dec. 1997, pp. 869–880.

[164]Ghaboussi, J., Pecknold, D. A., Zhang, M., and Haj-Ali, R. M., "Auto-progressive Training of Neural Network Constitutive Models," *International Journal for Numerical Methods in Engineering*, in press.

[165]Shephard, M. S., and Weatherill, N. P. (editors), "Adaptive Meshing" (special issue), *International Journal for Numerical Methods in Engineering*, Vol. 32, No. 4, Sept. 1991.

[166]Noor, A. K. (editor), *Adaptive, Multilevel and Hierarchical Computational Strategies*, Proceedings of the symposium, ASME Winter Annual Meeting, Nov. 8–13, 1992, Anaheim, California, AMD-157, American Society of Mechanical Engineers, New York, 1992.

[167]Demkowicz, L., Oden, J. T., and Babuska, I., "Reliability in Computational Mechanics" (special issue), *Computer Methods in Applied Mechanics and Engineering*, Vol. 101, Nos. 1–3, Dec. 1992.

[168]Lee, N.-S., and Bathe, K. J., "Error Indicators and Adaptive Remeshing in Large Deformation Finite Element Analysis," *Finite Elements in Analysis and Design*, Vol. 16, 1994, pp. 99–136.

[169]Verfurth, R., A Posteriori *Error Estimation and Adaptive Mesh-Refinement Techniques*, John Wiley & Sons, Chichester, U.K., 1996.

[170]Ladeveze, P., and Oden, J. T. (editors), *Advances in Adaptive Computational Methods*, Elsevier Science, New York, 1998.

[171]Oden, J. T., and Prudhomme, S., "Goal-Oriented Error Estimation and Adaptivity for the Finite Element Method," *Computers and Mathematics with Applications*, in press.

[172]Stewart, J. R., and Hughes, T. J. R., "A Tutorial in Elementary Finite Element Error Analysis: A Systematic Presentation of a priori and a posteriori Error Estimates," *Computer Methods in Applied Mechanics and Engineering*, Vol. 158, Nos. 1–2, May 1998, pp. 1–22.

[173]Zhu, J. Z., "A Posteriori Error Estimation—The Relationship Between Different Procedures," *Computer Methods in Applied Mechanics and Engineering*, Vol. 150, Nos. 1–4, Dec. 1997, pp. 411–422.

[174]Babuska, I., Strouboulis, T., Gangaraj, S. K., and Upadhyay, C. S., "Pollution Error in the h-version of the Finite Element Method and the Local Quality of the Recovered Derivatives," *Computer Methods in Applied Mechanics and Engineering*, Vol. 140, Nos. 1–2, Jan. 1997, pp. 1–37.

[175]Babuska, I., Strouboulis, T., Gangaraj, S. K., Copps, K., and Datta, D. K., "Practical Aspects of a-posteriori Estimation for Reliable Finite Element Analysis," *Computers and Structures*, Vol. 66, No. 5, March 1998, pp. 627–664.

[176]Krizek, M., and Neittaanmaki, P., "On Super-convergence Techniques," *Acta Applicandae Mathematicae*, Vol. 9, 1987, pp. 175–198.

[177]Babuska, I., and Miller, A., "The Postprocessing Approach in the Finite Element Method—Part 1: Calculation of Displacements, Stresses and Other Higher Derivatives of the Displacements," *International Journal for Numerical Methods in Engineering*, Vol. 20, 1984, pp. 1085–1109.

[178]Zienkiewicz, O. C., and Zhu, J. Z., "The Super-convergent Patch Recovery and a posteriori Error Estimates. Part 1: The Recovery Technique," *International Journal for Numerical Methods in Engineering*, Vol. 33, 1992, pp. 1331–1364.

[179]Zienkiewicz, O. C., Zhu, J. Z., and Wu, J., "Super-convergent Patch Recovery Techniques—Some Further Tests," *Communications in Numerical Methods in Engineering*, Vol. 9, No. 3, 1993, pp. 251–258.

[180]Tabbara, M., Blacker, T., and Belytschko, T., "Finite Element Derivative Recovery by Moving Least Square Interpolants," *Computer Methods in Applied Mechanics and Engineering*, Vol. 117, Nos. 1–2, July 1994, pp. 211–223.

[181]Zhang, Z., and Zhu, J. Z., "Analysis of the Super-convergent Patch Recovery Technique and A Posteriori Error Estimator in the Finite Element Method (I)," *Computer Methods in Applied Mechanics and Engineering*, Vol. 123, 1995, pp. 173–187.

[182]Zhang, Z., and Zhu, J. Z., "Analysis of the Super-convergent Patch Recovery Technique and A Posteriori Error Estimator in the Finite Element Method (II)," *Computer Methods in Applied Mechanics and Engineering*, Vol. 163, Nos. 1–4, 1998, pp. 159–170.

[183]Lee, T., Park, H. C., and Lee, S. W., "A Super-convergent Stress Recovery Technique with Equilibrium Constraint," *International Journal for Numerical Methods in Engineering*, Vol. 40, No. 6, March 1997, pp. 1139–1160.

[184]Tenchev, R. T., "A Study of the Accuracy of Some FEM Stress Recovery Schemes for Two Dimensional Stress Concentration Problems," *Finite Elements in Analysis and Design*, Vol. 29, 1998, pp. 105–119.

[185]Yazdani, A. A., Gakwaya, A., and Dhatt, G., "An Improved Super-convergent Patch Recovery Technique for the Axisymmetrical Problems," *Computers and Structures*, Vol. 66, No. 6, March 1998, pp. 799–821.

[186]Fish, J., and Markolefas, S., "Adaptive s-Method for Linear Elastostatics," *Computer Methods in Applied Mechanics and Engineering*, Vol. 104, 1993, pp. 363–396.

[187]Fish, J., and Markolefas, S., "Adaptive Global-Local Refinement Strategy Based on the Interior Error Estimates of the h-Method," *International Journal for Numerical Methods in Engineering*, Vol. 37, 1994, pp. 827–838.

[188]Wiberg, N.-E., Li, X. D., and Abdulwahab, F., "Adaptive Finite Element Procedures in Elasticity and Plasticity," *Engineering with Computers*, Vol. 12, 1996, pp. 120–141.

[189]Lee, C. K., and Lo, S. H., "Automatic Adaptive Three Dimensional Finite Element Refinement Using Different-Order Tetrahedral Elements," *International Journal for Numerical Methods in Engineering*, Vol. 40, 1997, pp. 2195–2226.

[190]Rashid, M. M., "The Arbitrary Local Mesh Replacement Method: An Alternative to Remeshing for Crack Propagation Analysis," *Computer Methods in Applied Mechanics and Engineering*, Vol. 154, Nos. 1–2, Feb. 1998, pp. 133–150.

[191]Dechaumphai, P., "Adaptive Unstructured Meshing for Thermal Stress Analysis of Built-up Structures," *Proceedings of the 33rd AIAA/ASME/ASCE/AHS/ASC Structures, Structural Dynamics and Materials Conference*, April 13–15, 1992, Dallas, TX, Part 2, 1992, pp. 1069–1077.

[192]Dechaumphai, P., Thornton, E. A., and Weiting, A. R., "Flow-Thermal-Structural Study of Aerodynamically Heated Leading Edges," *Journal of Spacecraft*, Vol. 26, No. 4, 1989, pp. 201–209.

[193]Wieting, A. R., Dechaumphai, P., and Bey, K. S., "Application of Integrated Fluid-Thermal-Structural Analysis Method," *Thin-Walled Structures*, Vol. 11, 1991, pp. 1–23.

[194]Koerber, S. R., "Fluid-Structure-Interaction Focused on Aspects of Aerokinetic Heating," 30th Fluid Dynamics Conference, Norfolk, VA, June 28–July 1, 1999, AIAA Paper No. 99–3797, American Institute of Aeronautics and Astronautics, Reston, VA, 1999.

[195]"Multiple Field Methods," *Spectrum Solver Theory Manual*, Ver. 2.0, Centric Engineering Systems, Sunnyvale, CA, 1997, Chapter 4.

[196]Melis, M. E., and Wang, W.-P., "Multiphysics Simulation of Active Hypersonic Cowl Lip Cooling," 30th Fluid Dynamics Conference, Norfolk, VA, June 28–July 1, 1999, AIAA Paper No. 99-3510, American Institute of Aeronautics and Astronautics, Reston, VA, 1999.

[197]Bathe, K. J., "Simulation of Structural and Fluid Flow Response in Engineering Practice," *Computer Modeling and Simulation in Engineering*, Vol. 1, 1996, pp. 47–77.

[198]Bathe, K. J., Zhang, H., and Zhang, X., "Some Advances in the Analysis of Fluid Flows," *Computers and Structures*, Vol. 64, No. 5/6, 1997, pp. 909–930.

[199]Kleiber, M., and Hisada, T. (editors), *Design-Sensitivity Analysis*, Atlanta Technology Publications, Atlanta, GA, 1993.

[200]Hinton, E., and Sienz, J., "Aspects of Adaptive Finite Element Analysis and Structural Optimization," *Advances in Structural Optimization*, B. H. V. Topping and M. Papadrakakis (editors), CIVIL-COMP, Ltd., Edinburgh, Scotland, 1994, pp. 1–25.

[201]Squire, W., and Trapp, G., "Using Complex Variables to Estimate Derivatives of Real Functions," *SIAM Review*, Vol. 40, No. 1, March 1998, pp. 110–112.

[202]Haber, R. B., Tortorelli, D. A., and Vidal, C. A., "Design Sensitivity Analysis of Nonlinear Structures—I: Large Deformation Hyperelasticity and History Dependent Material Response," *Structural Optimization: Status and Promise*, M. P. Kamat (editor), American Institute of Aeronautics and Astronautics, Washington, D.C., 1993, pp. 369–406.

[203]Kleiber, M., Antunez, H., Hien, T. D., and Kowalczyk, P., *Parameter Sensitivity in Nonlinear Mechanics: Theory and Finite Element Computations*, John Wiley and Sons, New York, 1997.

[204]Kirsch, U., "Efficient Sensitivity Analysis for Structural Optimization," *Computer Methods in Applied Mechanics and Engineering*, Vol. 117, Nos. 1–2, July 1994, pp. 143–156.

[205]Kleiber, M., Hien, T. D., and Postek, E., "Incremental Finite Element Sensitivity Analysis for Nonlinear Mechanics Applications," *International Journal for Numerical Methods in Engineering*, Vol. 37, No. 19, Oct. 1994, pp. 3291–3308.

[206]Kulkarni, M., and Noor, A. K., "Sensitivity Analysis of the Nonlinear Dynamic Viscoplastic Response of Two Dimensional Structures with Respect to Material Parameters," *International Journal for Numerical Methods in Engineering*, Vol. 38, 1995, pp. 183–198.

[207]Noor, A. K., Needleman, A., and Peters, J. M., "Sensitivity Analysis for Failure and Damage in Dynamically Loaded Tensile Bars," *Computer Methods in Applied Mechanics and Engineering*, Vol. 151, Nos. 3–4, Jan. 1998, pp. 461–478.

[208]Griewank, A., *Evaluating Derivatives: Principles and Techniques of Algorithmic Differentiation*, SIAM, April 2000, Philadelphia, PA.

[209]Chinchalker, S., "The Application of Automatic Differentiation to Problems in Engineering Analysis," *Computer Methods in Applied Mechanics and Engineering*, Vol. 118, 1994, pp. 197–207.

[210]Green, L. L., Newman, P. A., and Haigler, K. J., "Sensitivity Derivatives for Advanced CFD Algorithm and Viscous Modeling Parameters via Automatic Differentiation," *Journal of Computational Physics*, Vol. 125, 1996, pp. 313–324.

[211]Sherman, L. L., Taylor III, A. C., Green, L. L., Newman, P. A., Hou, G. W., and Korivi, V. M., "First and Second-Order Aerodynamic Sensitivity Derivatives via Automatic Differentiation with Incremental Iterative Methods," *Journal of Computational Physics*, Vol. 129, 1996, pp. 307–331.

[212]Follen, G., Evans, A. L., Naiman, C., and Lopez, I., "Numerical Propulsion System Simulation's National Cycle Program," *34th AIAA/ASME/ASCE/SAE/ASEE Joint Propulsion Conference and Exhibit*, Cleveland, Ohio, July 13–15, 1998, AIAA-98-3113.

[213]Topping, B. H. V., and Khan, A. I., *Parallel Finite Element Computations*, Saxe-Coburg Publications, Edinburgh, U.K., 1996.

[214]Hendrickson, B., and Leland, R., "An Improved Spectral Graph Partitioning Algorithm for Mapping Parallel Computations," Sandia National Labs., TR-SAND92-1460, 1992.

[215]Barnard, S. T., and Simon, H., "A Parallel Implementation of Multilevel Recursive Spectral Bisection for Application to Adaptive Unstructured Meshes," *Proceedings of the Seventh SIAM Conference on Parallel Processing for Scientific Computing*, Vol. 18, 1995, pp. 627–632.

[216]Khan, A. I., and Topping, B. H. V., "Sub-domain Generation for Parallel Finite Element Analysis," *Computing Systems in Engineering*, Vol. 4, Nos. 4–6, 1993, pp. 473–488.

[217]Watson, B. C., and Noor, A. K., "Sensitivity Analysis for Large-Deflection and Postbuckling Responses on Distributed-Memory Computers," *Computer Methods in Applied Mechanics and Engineering*, Vol. 129, 1996, pp. 393–409.

[218]Watson, B. C., and Noor, A. K., "Large Scale Contact/Impact Simulation and Sensitivity Analysis on Distributed-Memory Computers," *Computer Methods in Applied Mechanics and Engineering*, Vol. 141, Nos. 3–4, 1997, pp. 373–388.

[219]Farhat, C., and Le Tallec, P. (editors), "Vistas in Domain Decomposition and Parallel Processing in Computational Mechanics" (special issue), *Computer Methods in Applied Mechanics and Engineering*, Vol. 184, Nos. 2–4, 2000.

[220]George, P. L., *Automatic Mesh Generation. Application to Finite Element Methods*, John Wiley & Sons and Masson, 1991.

[221]Frykestig, J., "Advancing Front Mesh Generation Techniques with Application to the Finite Element Method," Chalmers University of Technology, Dept. of Structural Mechanics, Rept. 94:10, Goteborg, Sweden, 1994.

[222]Tworzydlo, W. W., and Oden, J. T., "Towards an Automated Environment in Computational Mechanics," *Computer Methods in Applied Mechanics and Engineering*, Vol. 104, 1993, pp. 87–143.

[223]Chandra, S., Woodman, N. J., and Blockley, D. I., "An Object-Oriented Structure for Transient Dynamics on Concurrent Computers," *Computers and Structures*, Vol. 51, No. 4, May 1994, pp. 437–452.

[224]Kecskemethy, A., and Hiller, M., "An Object-Oriented Approach for an Effective Formulation of Multi-body Dynamics," *Computer Methods in Applied Mechanics and Engineering*, Vol. 115, Nos. 3–4, May 1994, pp. 287–314.

[225]Besson, J., and Foerch, R., "Large Scale Object-Oriented Finite Element Code Design," *Computer Methods in Applied Mechanics and Engineering*, Vol. 142, Nos. 1–2, March 1997, pp. 165–187.

[226]Mackie, R. I., "Using Objects to Handle Complexity in Finite Element Software," *Engineering with Computers*, Vol. 13, 1997, pp. 99–111.

[227]Eyheramendy, D., and Zimmermann, Th., "Object-Oriented Finite Elements. III. Theory and Application of Automatic Programming," *Computer Methods in Applied Mechanics and Engineering*, Vol. 154, Nos. 1–2, Feb. 1998, pp. 41–68.

[228]Zimmermann, Th., Bomme, P., Eyheramendy, D., Vernier, L., and Commend, S., "Aspects of an Object-Oriented Finite Element Environment," *Computers and Structures*, Vol. 68, No. 1–3, June 1998, pp. 1–16.

[229]Dubois-Pelerin, Y., and Pegon, P., "Object-Oriented Programming in Nonlinear Finite Element Analysis," *Computers and Structures*, Vol. 68, No. 4, May 1998, pp. 225–241.

[230]Devloo, P. R. B., "Efficiency Issues in an Object-Oriented Programming Environment," *Artificial Intelligence and Object-Oriented Approaches for Structural Engineering*, B. H. V. Topping and M. Papadrakakis (editors), CIVIL-COMP, Ltd., Edinburgh, Scotland, 1994, pp. 147–151.

[231]Denning, P. J., and Metcalfe, R. M., *Beyond Calculation—The Next Fifty Years of Computing*, Springer Verlag, New York, 1997.

[232]Foster, I., and Kesselman, C. (editors), *The Grid: Blueprint for a New Computing Infrastructure*, Morgan Kaufmann Publishers, Inc., San Francisco, California, 1999.

[233]Noor, A. K., "New Computing Systems and Future High Performance Computing Environment and Their Impact on Structural Analysis and Design," *Computers and Structures*, Vol. 64, Nos. 1–4, 1997, pp. 1–30.

[234]Koike, N. (editor), *Heterogeneous Computing and Multidisciplinary Applications*, Proceedings of the Eighth NEC Research Symposium, Berlin, May 1997, SIAM, Philadelphia, PA, May 2000.

[235]Holtz, W. B., *The CAD Rating Guide*, 5th ed., PennWell Publishing, Nashua, New Hampshire, 1997.

[236]Kumar, B. (guest editor), "Information Technology in Civil and Structural Engineering Design" (special issue), *Computers and Structures*, Vol. 67, No. 5, June 1998.

[237]Generazio, E. R., and Harris, C. E., "The Role of Modeling and Simulation in the Development of Advanced Nondestructive Evaluation Systems," *The Aeronautical Journal*, July 1999, pp. 325–328.

[238]Hughes, T. J. R., and Stewart, J., "A Space Time Formulation for Multi-Scale Phenomena," *Journal of Comput. Applied Math.*, Vol. 74, 1996, pp. 217–229.

[239]Hughes, T. J. R., Feijoo, G. R., Mazzei, L., and Quincy, J.-B., "The Variational Multiscale Method—A Paradigm for Computational Mechanics," *Computer Methods in Applied Mechanics and Engineering*, Vol. 166, Nos. 1–2, 1998, pp. 3–24.

[240]Noor, A. K. (compiler), "Computational Intelligence and Its Impact on Future High Performance Engineering Systems," *Proceedings of the workshop*, June 27–28, 1995, Hampton, VA, NASA CP-3323, Jan. 1996.

[241]Schueller, G. I., "Structural Reliability—Recent Advances," *Structural Safety and Reliability*, N. Shiraishi, M. Shinozuka and Y. K. Wen (editors), Proceedings of the Seventh International Conference on Structural Safety and Reliability, Kyoto, Japan, Nov. 24–28, 1997, A. A. Balkema Publishers, Rotterdam, The Netherlands, 1998, pp. 3–35.

[242]Myers, R. H., and Montgomery, D. C., *Response Surface Methodology: Process and Product Optimization Using Designed Experiments*, John Wiley & Sons, New York, 1995.

PROGRESS IN ASTRONAUTICS AND AERONAUTICS SERIES VOLUMES

***1. Solid Propellant Rocket Research (1960)**
Martin Summerfield
Princeton University

***2. Liquid Rockets and Propellants (1960)**
Loren E. Bollinger
Ohio State University
Martin Goldsmith
The Rand Corp.
Alexis W. Lemmon Jr.
Battelle Memorial Institute

***3. Energy Conversion for Space Power (1961)**
Nathan W. Snyder
Institute for Defense Analyses

***4. Space Power Systems (1961)**
Nathan W. Snyder
Institute for Defense Analyses

***5. Electrostatic Propulsion (1961)**
David B. Langmuir
Space Technology Laboratories, Inc.
Ernst Stuhlinger
NASA George C. Marshall Space Flight Center
J. M. Sellen Jr.
Space Technology Laboratories, Inc.

***6. Detonation and Two-Phase Flow (1962)**
S. S. Penner
California Institute of Technology
F. A. Williams
Harvard University

***7. Hypersonic Flow Research (1962)**
Frederick R. Riddell
AVCO Corp.

***8. Guidance and Control (1962)**
Robert E. Roberson
Consultant
James S. Farrior
Lockheed Missiles and Space Co.

***9. Electric Propulsion Development (1963)**
Ernst Stuhlinger
NASA George C. Marshall Space Flight Center

***10. Technology of Lunar Exploration (1963)**
Clifford I. Cumming
Harold R. Lawrence
Jet Propulsion Laboratory

***11. Power Systems for Space Flight (1963)**
Morris A. Zipkin
Russell N. Edwards
General Electric Co.

***12. Ionization in High-Temperature Gases (1963)**
Kurt E. Shuler, Editor
National Bureau of Standards
John B. Fenn, Associate Editor
Princeton University

***13. Guidance and Control–II (1964)**
Robert C. Langford
General Precision Inc.
Charles J. Mundo
Institute of Naval Studies

***14. Celestial Mechanics and Astrodynamics (1964)**
Victor G. Szebehely
Yale University Observatory

***15. Heterogeneous Combustion (1964)**
Hans G. Wolfhard
Institute for Defense Analyses
Irvin Glassman
Princeton University
Leon Green Jr.
Air Force Systems Command

***16. Space Power Systems Engineering (1966)**
George C. Szego
Institute for Defense Analyses
J. Edward Taylor
TRW Inc.

***17. Methods in Astrodynamics and Celestial Mechanics (1966)**
Raynor L. Duncombe
U.S. Naval Observatory
Victor G. Szebehely
Yale University Observatory

***18. Thermophysics and Temperature Control of Spacecraft and Entry Vehicles (1966)**
Gerhard B. Heller
NASA George C. Marshall Space Flight Center

***19. Communication Satellite Systems Technology (1966)**
Richard B. Marsten
Radio Corporation of America

---
*Out of print.

***20. Thermophysics of Spacecraft and Planetary Bodies: Radiation Properties of Solids and the Electromagnetic Radiation Environment in Space (1967)**
Gerhard B. Heller
NASA George C. Marshall Space Flight Center

***21. Thermal Design Principles of Spacecraft and Entry Bodies (1969)**
Jerry T. Bevans
TRW Systems

***22. Stratospheric Circulation (1969)**
Willis L. Webb
Atmospheric Sciences Laboratory, White Sands, and University of Texas at El Paso

***23. Thermophysics: Applications to Thermal Design of Spacecraft (1970)**
Jerry T. Bevans
TRW Systems

***24. Heat Transfer and Spacecraft Thermal Control (1971)**
John W. Lucas
Jet Propulsion Laboratory

25. Communication Satellites for the 70's: Technology (1971)
Nathaniel E. Feldman
The Rand Corp.
Charles M. Kelly
The Aerospace Corp.

26. Communication Satellites for the 70's: Systems (1971)
Nathaniel E. Feldman
The Rand Corp.
Charles M. Kelly
The Aerospace Corp.

27. Thermospheric Circulation (1972)
Willis L. Webb
Atmospheric Sciences Laboratory, White Sands, and University of Texas at El Paso

28. Thermal Characteristics of the Moon (1972)
John W. Lucas
Jet Propulsion Laboratory

***29. Fundamentals of Spacecraft Thermal Design (1972)**
John W. Lucas
Jet Propulsion Laboratory

***30. Solar Activity Observations and Predictions (1972)**
Patrick S. McIntosh
Murray Dryer
Environmental Research Laboratories, National Oceanic and Atmospheric Administration

***31. Thermal Control and Radiation (1973)**
Chang-Lin Tien
University of California at Berkeley

***32. Communications Satellite Systems (1974)**
P. L. Bargellini
COMSAT Laboratories

***33. Communications Satellite Technology (1974)**
P. L. Bargellini
COMSAT Laboratories

***34. Instrumentation for Airbreathing Propulsion (1974)**
Allen E. Fuhs
Naval Postgraduate School
Marshall Kingery
Arnold Engineering Development Center

***35. Thermophysics and Spacecraft Thermal Control (1974)**
Robert G. Hering
University of Iowa

36. Thermal Pollution Analysis (1975)
Joseph A. Schetz
Virginia Polytechnic Institute
ISBN 0-915928-00-0

***37. Aeroacoustics: Jet and Combustion Noise; Duct Acoustics (1975)**
Henry T. Nagamatsu, Editor
General Electric Research and Development Center
Jack V. O'Keefe, Associate Editor
The Boeing Co.
Ira R. Schwartz, Associate Editor
NASA Ames Research Center
ISBN 0-915928-01-9

***38. Aeroacoustics: Fan, STOL, and Boundary Layer Noise; Sonic Boom; Aeroacoustics Instrumentation (1975)**
Henry T. Nagamatsu, Editor
General Electric Research and Development Center
Jack V. O'Keefe, Associate Editor
The Boeing Co.
Ira R. Schwartz, Associate Editor
NASA Ames Research Center
ISBN 0-915928-02-7

***39. Heat Transfer with Thermal Control Applications (1975)**
M. Michael Yovanovich
University of Waterloo
ISBN 0-915928-03-5

---
*Out of print.

***40. Aerodynamics of Base Combustion (1976)**
S. N. B. Murthy, Editor
J. R. Osborn,
Associate Editor
Purdue University
A. W. Barrows
J. R. Ward,
Associate Editors
Ballistics Research Laboratories
ISBN 0-915928-04-3

***41. Communications Satellite Developments: Systems (1976)**
Gilbert E. LaVean
Defense Communications Agency
William G. Schmidt
CML Satellite Corp.
ISBN 0-915928-05-1

***42. Communications Satellite Developments: Technology (1976)**
William G. Schmidt
CML Satellite Corp.
Gilbert E. LaVean
Defense Communications Agency
ISBN 0-915928-06-X

***43. Aeroacoustics: Jet Noise, Combustion and Core Engine Noise (1976)**
Ira R. Schwartz, Editor
NASA Ames Research Center
Henry T. Nagamatsu,
Associate Editor
General Electric Research and Development Center
Warren C. Strahle,
Associate Editor
Georgia Institute of Technology
ISBN 0-915928-07-8

***44. Aeroacoustics: Fan Noise and Control; Duct Acoustics; Rotor Noise (1976)**
Ira R. Schwartz, Editor
NASA Ames Research Center
Henry T. Nagamatsu,
Associate Editor
General Electric Research and Development Center
Warren C. Strahle,
Associate Editor
Georgia Institute of Technology
ISBN 0-915928-08-6

***45. Aeroacoustics: STOL Noise; Airframe and Airfoil Noise (1976)**
Ira R. Schwartz, Editor
NASA Ames Research Center
Henry T. Nagamatsu,
Associate Editor
General Electric Research and Development Center
Warren C. Strahle,
Associate Editor
Georgia Institute of Technology
ISBN 0-915928-09-4

***46. Aeroacoustics: Acoustic Wave Propagation; Aircraft Noise Prediction; Aeroacoustic Instrumentation (1976)**
Ira R. Schwartz, Editor
NASA Ames Research Center
Henry T. Nagamatsu,
Associate Editor
General Electric Research and Development Center
Warren C. Strahle,
Associate Editor
Georgia Institute of Technology
ISBN 0-915928-10-8

***47. Spacecraft Charging by Magnetospheric Plasmas (1976)**
Alan Rosen
TRW Inc.
ISBN 0-915928-11-6

***48. Scientific Investigations on the Skylab Satellite (1976)**
Marion I. Kent
Ernst Stuhlinger
NASA George C. Marshall Space Flight Center
Shi-Tsan Wu
University of Alabama
ISBN 0-915928-12-4

***49. Radiative Transfer and Thermal Control (1976)**
Allie M. Smith
ARO Inc.
ISBN 0-915928-13-2

***50. Exploration of the Outer Solar System (1976)**
Eugene W. Greenstadt
TRW Inc.
Murray Dryer
National Oceanic and Atmospheric Administration
Devrie S. Intriligator
University of Southern California
ISBN 0-915928-14-0

***51. Rarefied Gas Dynamics, Parts I and II (two volumes) (1977)**
J. Leith Potter
ARO Inc.
ISBN 0-915928-15-9

***52. Materials Sciences in Space with Application to Space Processing (1977)**
Leo Steg
General Electric Co.
ISBN 0-915928-16-7

*Out of print.

***53. Experimental Diagnostics in Gas Phase Combustion Systems (1977)**
Ben T. Zinn, Editor
Georgia Institute of Technology
Craig T. Bowman, Associate Editor
Stanford University
Daniel L. Hartley, Associate Editor
Sandia Laboratories
Edward W. Price, Associate Editor
Georgia Institute of Technology
James G. Skifstad, Associate Editor
Purdue University
ISBN 0-915928-18-3

***54. Satellite Communication: Future Systems (1977)**
David Jarett
TRW Inc.
ISBN 0-915928-18-3

***55. Satellite Communications: Advanced Technologies (1977)**
David Jarett
TRW Inc.
ISBN 0-915928-19-1

***56. Thermophysics of Spacecraft and Outer Planet Entry Probes (1977)**
Allie M. Smith
ARO Inc.
ISBN 0-915928-20-5

***57. Space-Based Manufacturing from Nonterrestrial Materials (1977)**
Gerald K. O'Neill, Editor
Brian O'Leary, Assistant Editor
Princeton University
ISBN 0-915928-21-3

***58. Turbulent Combustion (1978)**
Lawrence A. Kennedy
State University of New York at Buffalo
ISBN 0-915928-22-1

***59. Aerodynamic Heating and Thermal Protection Systems (1978)**
Leroy S. Fletcher
University of Virginia
ISBN 0-915928-23-X

***60. Heat Transfer and Thermal Control Systems (1978)**
Leroy S. Fletcher
University of Virginia
ISBN 0-915928-24-8

***61. Radiation Energy Conversion in Space (1978)**
Kenneth W. Billman
NASA Ames Research Center
ISBN 0-915928-26-4

***62. Alternative Hydrocarbon Fuels: Combustion and Chemical Kinetics (1978)**
Craig T. Bowman
Stanford University
Jorgen Birkeland
Department of Energy
ISBN 0-915928-25-6

***63. Experimental Diagnostics in Combustion of Solids (1978)**
Thomas L. Boggs
Naval Weapons Center
Ben T. Zinn
Georgia Institute of Technology
ISBN 0-915928-28-0

***64. Outer Planet Entry Heating and Thermal Protection (1979)**
Raymond Viskanta
Purdue University
ISBN 0-915928-29-9

***65. Thermophysics and Thermal Control (1979)**
Raymond Viskanta
Purdue University
ISBN 0-915928-30-2

***66. Interior Ballistics of Guns (1979)**
Herman Krier
University of Illinois at Urbana–Champaign
Martin Summerfield
New York University
ISBN 0-915928-32-9

***67. Remote Sensing of Earth from Space: Role of "Smart Sensors" (1979)**
Roger A. Breckenridge
NASA Langley Research Center
ISBN 0-915928-33-7

***68. Injection and Mixing in Turbulent Flow (1980)**
Joseph A. Schetz
Virginia Polytechnic Institute and State University
ISBN 0-915928-35-3

---
*Out of print.

***69. Entry Heating and Thermal Protection (1980)**
Walter B. Olstad
NASA Headquarters
ISBN 0-915928-38-8

***70. Heat Transfer, Thermal Control, and Heat Pipes (1980)**
Walter B. Olstad
NASA Headquarters
ISBN 0-915928-39-6

***71. Space Systems and Their Interactions with Earth's Space Environment (1980)**
Henry B. Garrett
Charles P. Pike
Hanscom Air Force Base
ISBN 0-915928-41-8

***72. Viscous Flow Drag Reduction (1980)**
Gary R. Hough
Vought Advanced Technology Center
ISBN 0-915928-44-2

***73. Combustion Experiments in a Zero-Gravity Laboratory (1981)**
Thomas H. Cochran
NASA Lewis Research Center
ISBN 0-915928-48-5

***74. Rarefied Gas Dynamics, Parts I and II (two volumes) (1981)**
Sam S. Fisher
University of Virginia
ISBN 0-915928-51-5

***75. Gasdynamics of Detonations and Explosions (1981)**
J. R. Bowen
University of Wisconsin at Madison
N. Manson
Universite de Poitiers
A. K. Oppenheim
University of California at Berkeley
R. I. Soloukhin
Institute of Heat and Mass Transfer, BSSR Academy of Sciences
ISBN 0-915928-46-9

***76. Combustion in Reactive Systems (1981)**
J. R. Bowen
University of Wisconsin at Madison
N. Manson
Universite de Poitiers
A. K. Oppenheim
University of California at Berkeley
R. I. Soloukhin
Institute of Heat and Mass Transfer, BSSR Academy of Sciences
ISBN 0-915928-47-7

***77. Aerothermodynamics and Planetary Entry (1981)**
A. L. Crosbie
University of Missouri-Rolla
ISBN 0-915928-52-3

***78. Heat Transfer and Thermal Control (1981)**
A. L. Crosbie
University of Missouri-Rolla
ISBN 0-915928-53-1

***79. Electric Propulsion and Its Applications to Space Missions (1981)**
Robert C. Finke
NASA Lewis Research Center
ISBN 0-915928-55-8

***80. Aero-Optical Phenomena (1982)**
Keith G. Gilbert
Leonard J. Otten
Air Force Weapons Laboratory
ISBN 0-915928-60-4

***81. Transonic Aerodynamics (1982)**
David Nixon
Nielsen Engineering & Research, Inc.
ISBN 0-915928-65-5

***82. Thermophysics of Atmospheric Entry (1982)**
T. E. Horton
University of Mississippi
ISBN 0-915928-66-3

***83. Spacecraft Radiative Transfer and Temperature Control (1982)**
T. E. Horton
University of Mississippi
ISBN 0-915928-67-1

***84. Liquid-Metal Flows and Magneto-hydrodynamics (1983)**
H. Branover
Ben-Gurion University of the Negev
P. S. Lykoudis
Purdue University
A. Yakhot
Ben-Gurion University of the Negev
ISBN 0-915928-70-1

---
*Out of print.

***85. Entry Vehicle Heating and Thermal Protection Systems: Space Shuttle, Solar Starprobe, Jupiter Galileo Probe (1983)**
Paul E. Bauer
McDonnell Douglas Astronautics Co.
Howard E. Collicott
The Boeing Co.
ISBN 0-915928-74-4

***86. Spacecraft Thermal Control, Design, and Operation (1983)**
Howard E. Collicott
The Boeing Co.
Paul E. Bauer
McDonnell Douglas Astronautics Co.
ISBN 0-915928-75-2

***87. Shock Waves, Explosions, and Detonations (1983)**
J. R. Bowen
University of Washington
N. Manson
Universite de Poitiers
A. K. Oppenheim
University of California at Berkeley
R. I. Soloukhin
Institute of Heat and Mass Transfer, BSSR Academy of Sciences
ISBN 0-915928-76-0

***88. Flames, Lasers, and Reactive Systems (1983)**
J. R. Bowen
University of Washington
N. Manson
Universite de Poitiers
A. K. Oppenheim
University of California at Berkeley
R. I. Soloukhin
Institute of Heat and Mass Transfer, BSSR Academy of Sciences
ISBN 0-915928-77-9

***89. Orbit-Raising and Maneuvering Propulsion: Research Status and Needs (1984)**
Leonard H. Caveny
Air Force Office of Scientific Research
ISBN 0-915928-82-5

***90. Fundamentals of Solid-Propellant Combustion (1984)**
Kenneth K. Kuo
Pennsylvania State University
Martin Summerfield
Princeton Combustion Research Laboratories, Inc.
ISBN 0-915928-84-1

91. Spacecraft Contamination: Sources and Prevention (1984)
J. A. Roux
University of Mississippi
T. D. McCay
NASA Marshall Space Flight Center
ISBN 0-915928-85-X

92. Combustion Diagnostics by Nonintrusive Methods (1984)
T. D. McCay
NASA Marshall Space Flight Center
J. A. Roux
University of Mississippi
ISBN 0-915928-86-8

93. The INTELSAT Global Satellite System (1984)
Joel Alper
COMSAT Corp.
Joseph Pelton
INTELSAT
ISBN 0-915928-90-6

94. Dynamics of Shock Waves, Explosions, and Detonations (1984)
J. R. Bowen
University of Washington
N. Manson
Universite de Poitiers
A. K. Oppenheim
University of California at Berkeley
R. I. Soloukhin
Institute of Heat and Mass Transfer, BSSR Academy of Sciences
ISBN 0-915928-91-4

95. Dynamics of Flames and Reactive Systems (1984)
J. R. Bowen
University of Washington
N. Manson
Universite de Poitiers
A. K. Oppenheim
University of California at Berkeley
R. I. Soloukhin
Institute of Heat and Mass Transfer, BSSR Academy of Sciences
ISBN 0-915928-92-2

96. Thermal Design of Aeroassisted Orbital Transfer Vehicles (1985)
H. F. Nelson
University of Missouri-Rolla
ISBN 0-915928-94-9

97. Monitoring Earth's Ocean, Land, and Atmosphere from Space—Sensors, Systems, and Applications (1985)
Abraham Schnapf
Aerospace Systems Engineering
ISBN 0-915928-98-1

---

*Out of print.

98. Thrust and Drag: Its Prediction and Verification (1985)
Eugene E. Covert
Massachusetts Institute of Technology
C. R. James
Vought Corp.
William F. Kimzey
Sverdrup Technology AEDC Group
George K. Richey
U.S. Air Force
Eugene C. Rooney
U.S. Navy Department of Defense
ISBN 0-930403-00-2

99. Space Stations and Space Platforms—Concepts, Design, Infrastructure, and Uses (1985)
Ivan Bekey
Daniel Herman
NASA Headquarters
ISBN 0-930403-01-0

100. Single- and Multi-Phase Flows in an Electromagnetic Field: Energy, Metallurgical, and Solar Applications (1985)
Herman Branover
Ben-Gurion University of the Negev
Paul S. Lykoudis
Purdue University
Michael Mond
Ben-Gurion University of the Negev
ISBN 0-930403-04-5

101. MHD Energy Conversion: Physiotechnical Problems (1986)
V. A. Kirillin
A. E. Sheyndlin
Soviet Academy of Sciences
ISBN 0-930403-05-3

102. Numerical Methods for Engine-Airframe Integration (1986)
S. N. B. Murthy
Purdue University
Gerald C. Paynter
Boeing Airplane Co.
ISBN 0-930403-09-6

103. Thermophysical Aspects of Re-Entry Flows (1986)
James N. Moss
NASA Langley Research Center
Carl D. Scott
NASA Johnson Space Center
ISBN 0-930430-10-X

***104. Tactical Missile Aerodynamics (1986)**
M. J. Hemsch
PRC Kentron, Inc.
J. N. Nielson
NASA Ames Research Center
ISBN 0-930403-13-4

105. Dynamics of Reactive Systems Part I: Flames and Configurations; Part II: Modeling and Heterogeneous Combustion (1986)
J. R. Bowen
University of Washington
J.-C. Leyer
Universite de Poitiers
R. I. Soloukhin
Institute of Heat and Mass Transfer, BSSR Academy of Sciences
ISBN 0-930403-14-2

106. Dynamics of Explosions (1986)
J. R. Bowen
University of Washington
J.-C. Leyer
Universite de Poitiers
R. I. Soloukhin
Institute of Heat and Mass Transfer, BSSR Academy of Sciences
ISBN 0-930403-15-0

***107. Spacecraft Dielectric Material Properties and Spacecraft Charging (1986)**
A. R. Frederickson
U.S. Air Force Rome Air Development Center
D. B. Cotts
SRI International
J. A. Wall
U.S. Air Force Rome Air Development Center
F. L. Bouquet
Jet Propulsion Laboratory, California Institute of Technology
ISBN 0-930403-17-7

***108. Opportunities for Academic Research in a Low-Gravity Environment (1986)**
George A. Hazelrigg
National Science Foundation
Joseph M. Reynolds
Louisiana State University
ISBN 0-930403-18-5

109. Gun Propulsion Technology (1988)
Ludwig Stiefel
U.S. Army Armament Research, Development and Engineering Center
ISBN 0-930403-20-7

---
*Out of print.

110. Commercial Opportunities in Space (1988)
F. Shahrokhi
K. E. Harwell
University of Tennessee Space Institute
C. C. Chao
National Cheng Kung University
ISBN 0-930403-39-8

111. Liquid-Metal Flows: Magnetohydrodynamics and Application (1988)
Herman Branover
Michael Mond
Yeshajahu Unger
Ben-Gurion University of the Negev
ISBN 0-930403-43-6

112. Current Trends in Turbulence Research (1988)
Herman Branover
Micheal Mond
Yeshajahu Unger
Ben-Gurion University of the Negev
ISBN 0-930403-44-4

113. Dynamics of Reactive Systems Part I: Flames; Part II: Heterogeneous Combustion and Applications (1988)
A. L. Kuhl
R&D Associates
J. R. Bowen
University of Washington
J.-C. Leyer
Universite de Poitiers
A. Borisov
USSR Academy of Sciences
ISBN 0-930403-46-0

114. Dynamics of Explosions (1988)
A. L. Kuhl
R & D Associates
J. R. Bowen
University of Washington
J.-C. Leyer
Universite de Poitiers
A. Borisov
USSR Academy of Sciences
ISBN 0-930403-47-9

115. Machine Intelligence and Autonomy for Aerospace (1988)
E. Heer
Heer Associates, Inc.
H. Lum
NASA Ames Research Center
ISBN 0-930403-48-7

116. Rarefied Gas Dynamics: Space Related Studies (1989)
E. P. Muntz
University of Southern California
D. P. Weaver
U.S. Air Force Astronautics Laboratory (AFSC)
D. H. Campbell
University of Dayton Research Institute
ISBN 0-930403-53-3

117. Rarefied Gas Dynamics: Physical Phenomena (1989)
E. P. Muntz
University of Southern California
D. P. Weaver
U.S. Air Force Astronautics Laboratory (AFSC)
D. H. Campbell
University of Dayton Research Institute
ISBN 0-930403-54-1

118. Rarefied Gas Dynamics: Theoretical and Computational Techniques (1989)
E. P. Muntz
University of Southern California
D. P. Weaver
U.S. Air Force Astronautics Laboratory (AFSC)
D. H. Campbell
University of Dayton Research Institute
ISBN 0-930403-55-X

119. Test and Evaluation of the Tactical Missile (1989)
Emil J. Eichblatt Jr.
Pacific Missile Test Center
ISBN 0-930403-56-8

120. Unsteady Transonic Aerodynamics (1989)
David Nixon
Nielsen Engineering & Research, Inc.
ISBN 0-930403-52-5

121. Orbital Debris from Upper-Stage Breakup (1989)
Joseph P. Loftus Jr.
NASA Johnson Space Center
ISBN 0-930403-58-4

122. Thermal-Hydraulics for Space Power, Propulsion and Thermal Management System Design (1990)
William J. Krotiuk
General Electric Co.
ISBN 0-930403-64-9

*Out of print.

123. Viscous Drag Reduction in Boundary Layers (1990)
Dennis M. Bushnell
Jerry N. Hefner
NASA Langley Research Center
ISBN 0-930403-66-5

***124. Tactical and Strategic Missile Guidance (1990)**
Paul Zarchan
Charles Stark Draper Laboratory, Inc.
ISBN 0-930403-68-1

125. Applied Computational Aerodynamics (1990)
P. A. Henne
Douglas Aircraft Company
ISBN 0-930403-69-X

126. Space Commercialization: Launch Vehicles and Programs (1990)
F. Shahrokhi
University of Tennessee Space Institute
J. S. Greenberg
Princeton Synergetics Inc.
T. Al-Saud
Ministry of Defense and Aviation Kingdom of Saudi Arabia
ISBN 0-930403-75-4

127. Space Commercialization: Platforms and Processing (1990)
F. Shahrokhi
University of Tennessee Space Institute
G. Hazelrigg
National Science Foundation
R. Bayuzick
Vanderbilt University
ISBN 0-930403-76-2

128. Space Commercialization: Satellite Technology (1990)
F. Shahrokhi
University of Tennessee Space Institute
N. Jasentuliyana
United Nations
N. Tarabzouni
King Abulaziz City for Science and Technology
ISBN 0-930403-77-0

***129. Mechanics and Control of Large Flexible Structures (1990)**
John L. Junkins
Texas A&M University
ISBN 0-930403-73-8

130. Low-Gravity Fluid Dynamics and Transport Phenomena (1990)
Jean N. Koster
Robert L. Sani
University of Colorado at Boulder
ISBN 0-930403-74-6

131. Dynamics of Deflagrations and Reactive Systems: Flames (1991)
A. L. Kuhl
Lawrence Livermore National Laboratory
J.-C. Leyer
Universite de Poitiers
A. A. Borisov
USSR Academy of Sciences
W. A. Sirignano
University of California
ISBN 0-930403-95-9

132. Dynamics of Deflagrations and Reactive Systems: Heterogeneous Combustion (1991)
A. L. Kuhl
Lawrence Livermore National Laboratory
J.-C. Leyer
Universite de Poitiers
A. A. Borisov
USSR Academy of Sciences
W. A. Sirignano
University of California
ISBN 0-930403-96-7

133. Dynamics of Detonations and Explosions: Detonations (1991)
A. L. Kuhl
Lawrence Livermore National Laboratory
J.-C. Leyer
Universite de Poitiers
A. A. Borisov
USSR Academy of Sciences
W. A. Sirignano
University of California
ISBN 0-930403-97-5

134. Dynamics of Detonations and Explosions: Explosion Phenomena (1991)
A. L. Kuhl
Lawrence Livermore National Laboratory
J.-C. Leyer
Universite de Poitiers
A. A. Borisov
USSR Academy of Sciences
W. A. Sirignano
University of California
ISBN 0-930403-98-3

---

*Out of print.

135. Numerical Approaches to Combustion Modeling (1991)
Elaine S. Oran
Jay P. Boris
Naval Research Laboratory
ISBN 1-56347-004-7

136. Aerospace Software Engineering (1991)
Christine Anderson
U.S. Air Force Wright Laboratory
Merlin Dorfman
Lockheed Missiles & Space Company, Inc.
ISBN 1-56347-005-0

137. High-Speed Flight Propulsion Systems (1991)
S. N. B. Murthy
Purdue University
E. T. Curran
Wright Laboratory
ISBN 1-56347-011-X

138. Propagation of Intensive Laser Radiation in Clouds (1992)
O. A. Volkovitsky
Yu. S. Sedenov
L. P. Semenov
Institute of Experimental Meteorology
ISBN 1-56347-020-9

139. Gun Muzzle Blast and Flash (1992)
Günter Klingenberg
Fraunhofer-Institut für Kurzzeitdynamik, Ernst-Mach-Institut
Joseph M. Heimerl
U.S. Army Ballistic Research Laboratory
ISBN 1-56347-012-8

140. Thermal Structures and Materials for High-Speed Flight (1992)
Earl. A. Thornton
University of Virginia
ISBN 1-56347-017-9

141. Tactical Missile Aerodynamics: General Topics (1992)
Michael J. Hemsch
Lockheed Engineering & Sciences Company
ISBN 1-56347-015-2

142. Tactical Missile Aerodynamics: Prediction Methodology (1992)
Michael R. Mendenhall
Nielsen Engineering & Research, Inc.
ISBN 1-56347-016-0

143. Nonsteady Burning and Combustion Stability of Solid Propellants (1992)
Luigi De Luca
Politecnico di Milano
Edward W. Price
Georgia Institute of Technology
Martin Summerfield
Princeton Combustion Research Laboratories, Inc.
ISBN 1-56347-014-4

144. Space Economics (1992)
Joel S. Greenberg
Princeton Synergetics, Inc.
Henry R. Hertzfeld
HRH Associates
ISBN 1-56347-042-X

145. Mars: Past, Present, and Future (1992)
E. Brian Pritchard
NASA Langley Research Center
ISBN 1-56347-043-8

146. Computational Nonlinear Mechanics in Aerospace Engineering (1992)
Satya N. Atluri
Georgia Institute of Technology
ISBN 1-56347-044-6

147. Modern Engineering for Design of Liquid-Propellant Rocket Engines (1992)
Dieter K. Huzel
David H. Huang
Rocketdyne Division of Rockwell International
ISBN 1-56347-013-6

148. Metallurgical Technologies, Energy Conversion, and Magnetohydrodynamic Flows (1993)
Herman Branover
Yeshajahu Unger
Ben-Gurion University of the Negev
ISBN 1-56347-019-5

149. Advances in Turbulence Studies (1993)
Herman Branover
Yeshajahu Unger
Ben-Gurion University of the Negev
ISBN 1-56347-018-7

150. Structural Optimization: Status and Promise (1993)
Manohar P. Kamat
Georgia Institute of Technology
ISBN 1-56347-056-X

---
*Out of print.

151. Dynamics of Gaseous Combustion (1993)
A. L. Kuhl
Lawrence Livermore National Laboratory
J.-C. Leyer
Universite de Poitiers
A. A. Borisov
USSR Academy of Sciences
W. A. Sirignano
University of California
ISBN 1-56347-060-8

152. Dynamics of Heterogeneous Gaseous Combustion and Reacting Systems (1993)
A. L. Kuhl
Lawrence Livermore National Laboratory
J.-C. Leyer
Universite de Poitiers
A. A. Borisov
USSR Academy of Sciences
W. A. Sirignano
University of California
ISBN 1-56347-058-6

153. Dynamic Aspects of Detonations (1993)
A. L. Kuhl
Lawrence Livermore National Laboratory
J.-C. Leyer
Universite de Poitiers
A. A. Borisov
USSR Academy of Sciences
W. A. Sirignano
University of California
ISBN 1-56347-057-8

154. Dynamic Aspects of Explosion Phenomena (1993)
A. L. Kuhl
Lawrence Livermore National Laboratory
J.-C. Leyer
Universite de Poitiers
A. A. Borisov
USSR Academy of Sciences
W. A. Sirignano
University of California
ISBN 1-56347-059-4

155. Tactical Missile Warheads (1993)
Joseph Carleone
Aerojet General Corporation
ISBN 1-56347-067-5

156. Toward a Science of Command, Control, and Communications (1993)
Carl R. Jones
Naval Postgraduate School
ISBN 1-56347-068-3

***157. Tactical and Strategic Missile Guidance Second Edition (1994)**
Paul Zarchan
Charles Stark Draper Laboratory, Inc.
ISBN 1-56347-077-2

158. Rarefied Gas Dynamics: Experimental Techniques and Physical Systems (1994)
Bernie D. Shizgal
University of British Columbia
David P. Weaver
Phillips Laboratory
ISBN 1-56347-079-9

159. Rarefied Gas Dynamics: Theory and Simulations (1994)
Bernie D. Shizgal
University of British Columbia
David P. Weaver
Phillips Laboratory
ISBN 1-56347-080-2

160. Rarefied Gas Dynamics: Space Sciences and Engineering (1994)
Bernie D. Shizgal
University of British Columbia
David P. Weaver
Phillips Laboratory
ISBN 1-56347-081-0

161. Teleoperation and Robotics in Space (1994)
Steven B. Skaar
University of Notre Dame
Carl F. Ruoff
Jet Propulsion Laboratory, California Institute of Technology
ISBN 1-56347-095-0

162. Progress in Turbulence Research (1994)
Herman Branover
Yeshajahu Unger
Ben-Gurion University of the Negev
ISBN 1-56347-099-3

163. Global Positioning System: Theory and Applications, Volume I (1996)
Bradford W. Parkinson
Stanford University
James J. Spilker Jr.
Stanford Telecom
Penina Axelrad,
Associate Editor
University of Colorado
Per Enge,
Associate Editor
Stanford University
ISBN 1-56347-107-8

164. Global Positioning System: Theory and Applications, Volume II (1996)
Bradford W. Parkinson
Stanford University
James J. Spilker Jr.
Stanford Telecom
Penina Axelrad,
Associate Editor
University of Colorado
Per Enge,
Associate Editor
Stanford University
ISBN 1-56347-106-X

---

*Out of print.

165. Developments in High-Speed Vehicle Propulsion Systems (1996)
S. N. B. Murthy
Purdue University
E. T. Curran
Wright Laboratory
ISBN 1-56347-176-0

166. Recent Advances in Spray Combustion: Spray Atomization and Drop Burning Phenomena, Volume I (1996)
Kenneth K. Kuo
Pennsylvania State University
ISBN 1-56347-175-2

167. Fusion Energy in Space Propulsion (1995)
Terry Kammash
University of Michigan
ISBN 1-56347-184-1

168. Aerospace Thermal Structures and Materials for a New Era (1995)
Earl A. Thornton
University of Virginia
ISBN 1-56347-182-5

169. Liquid Rocket Engine Combustion Instability (1995)
Vigor Yang
William E. Anderson
Pennsylvania State University
ISBN 1-56347-183-3

170. Tactical Missile Propulsion (1996)
G. E. Jensen
United Technologies Corporation
David W. Netzer
Naval Postgraduate School
ISBN 1-56347-118-3

171. Recent Advances in Spray Combustion: Spray Combustion Measurements and Model Simulation, Volume II (1996)
Kenneth K. Kuo
Pennsylvania State University
ISBN 1-56347-181-7

172. Future Aeronautical and Space Systems (1997)
Ahmed K. Noor
NASA Langley Research Center
Samuel L. Venneri
NASA Headquarters
ISBN 1-56347-188-4

173. Advances in Combustion Science: In Honor of Ya. B. Zel'dovich (1997)
William A. Sirignano
University of California
Alexander G. Merzhanov
Russian Academy of Sciences
Luigi De Luca
Politecnico di Milano
ISBN 1-56347-178-7

174. Fundamentals of High Accuracy Inertial Navigation (1997)
Averil B. Chatfield
ISBN 1-56347-243-0

175. Liquid Propellant Gun Technology (1997)
Günter Klingenberg
Fraunhofer-Institut für Kurzzeitdynamik, Ernst-Mach-Institut
John D. Knapton
Walter F. Morrison
Gloria P. Wren
U.S. Army Research Laboratory
ISBN 1-56347-196-5

176. Tactical and Strategic Missile Guidance Third Edition (1998)
Paul Zarchan
Charles Stark Draper Laboratory, Inc.
ISBN 1-56347-279-1

177. Orbital and Celestial Mechanics (1998)
John P. Vinti
Gim J. Der, Editor
TRW
Nino L. Bonavito, Editor
NASA Goddard Space Flight Center
ISBN 1-56347-256-2

178. Some Engineering Applications in Random Vibrations and Random Structures (1998)
Giora Maymon
RAFAEL
ISBN 1-56347-258-9

179. Conventional Warhead Systems Physics and Engineering Design (1998)
Richard M. Lloyd
Raytheon Systems Company
ISBN 1-56347-255-4

180. Advances in Missile Guidance Theory (1998)
Joseph Z. Ben-Asher
Isaac Yaesh
Israel Military Industries—Advanced Systems Division
ISBN 1-56347-275-9

181. Satellite Thermal Control for Systems Engineers (1998)
Robert D. Karam
ISBN 1-56347-276-7

---

*Out of print.

182. Progress in Fluid Flow Research: Turbulence and Applied MHD (1998)
Yeshajahu Unger
Herman Branover
Ben-Gurion University of the Negev
ISBN 1-56347-284-8

183. Aviation Weather Surveillance Systems (1999)
Pravas R. Mahapatra
Indian Institute of Science
ISBN 1-56347-340-2

184. Flight Control Systems (2000)
Rodger W. Pratt, Editor
Loughborough University
ISBN 1-56347-404-2

185. Solid Propellant Chemistry, Combustion, and Motor Interior Ballistics (2000)
Vigor Yang
Pennsylvania State University
Thomas B. Brill
University of Delaware
Wu-Zhen Ren
China Ordnance Society
ISBN 1-56347-442-5

186. Approximate Methods for Weapons Aerodynamics (2000)
Frank G. Moore
ISBN 1-56347-399-2

187. Micropropulsion for Small Spacecraft (2000)
Michael M. Micci
Pennsylvania State University
Andrew D. Ketsdever
Air Force Research Laboratory, Edwards Air Force Base
ISBN 1-56347-448-4

187. Structures Technology for Future Aerospace Systems (2000)
Ahmed K. Noor
NASA Langley Research Center
ISBN 1-56347-384-4

---
*Out of print.